DICTIONAR

DICTIONARY OF GEOLOGICAL TERMS

REVISED EDITION

Prepared under the direction of
The American Geological Institute

ANCHOR PRESS

Anchor Press/Doubleday
Garden City, New York

Library of Congress Cataloging in Publication Data

American Geological Institute.
Dictionary of geological terms.

Published in 1957 and 1960 under title: Glossary of
geology and related sciences.
1. Geology—Dictionaries.   I. Title.
QE5.A48   1975   550'.3
ISBN 0-385-08452-8
Library of Congress Catalog Card Number 73–9004

PREFACE

This newly revised edition of the AGI *Dictionary of Geological Terms* contains some 8500 of the more commonly used terms in geology and the related earth sciences. It is intended for use by students of geology; elementary and secondary school science teachers; hobbyists who are interested in rocks, minerals, and fossils; and others who have occasion to use but are unfamiliar with geological terminology.

The new dictionary contains more than 1000 new terms in addition to many newly revised terms from the 1957 edition of the *Dictionary of Geological Terms* published by Dolphin Books (Doubleday & Company, Inc.). Most of the 7500 terms in the first edition were selected from the *Glossary of Geology and Related Sciences* (2nd edition 1960), by a team of geologists from the Department of Geology, Iowa State University. Members of this editorial team were: Donald L. Biggs, Keith M. Hussey, John Lemish, aided by Mrs. Isabelle Lyons, Lyle V. Sendlein, and Leo A. Thomas.

The first edition of the AGI *Glossary* was published in 1957 as an outgrowth of a compilation project which was started early in 1953 after several years of planning. The *Glossary* Project was financed by a grant from the National Science Foundation supplemented by a loan from the Geological Society of America.

The 1972 *Glossary of Geology* was edited by Margaret Gary, Robert "Skip" McAfee, Jr., and Carol L. Wolf. The preparation of the *Glossary* was undertaken with the advice and guidance of the AGI Committee on Publications under the chairmanship of the late Lewis M. Cline (1966–68), Philip E. LaMoreaux (1968–70), and Charles W. Collinson (1971–72). The work was partially funded by the National Science Foundation. The names of those participating in the 1972 *Glossary* Project follow this preface.

The 1976 edition of the AGI *Dictionary of Geological Terms* was prepared under the guidance and editorship of William H. Matthews III (AGI Director of Education and Regent's Professor of Geology at Lamar University, Beaumont, Texas) and Robert E. Boyer, Professor of Geology and Chairman, Department of Geological Sciences, The University of Texas, Austin. Members of the revision team were: Edward C. Jonas, John Lufkin, Wulf Massell, and Keith P. Young of the Geological Sciences Department, The University of Texas, Austin; Robert E. Sheriff, representing the Society of Exploration Geophysicists; and P. B. Snyder, Departments of Geology and Secondary Education, Lamar University. The manuscript was typed by Elizabeth Powers and Cheryl Patterson.

CONTRIBUTORS TO
1972 GLOSSARY OF GEOLOGY

Although the editors, revision team, and AGI assume responsibility for content as well as for form, they respectfully thank and acknowledge the contribution of those scientists who gave their time generously and without recompense to review and write definitions relevant to their particular fields of interest:

Astrogeology: Peter Boyse, Bevan M. French, Paul D. Lowman, Jr., Brian A. Mason, and Robert L. Wildey.

Cartography, surveying, map projections, photogeology: Earle J. Fennell, Arthur D. Howard, and Douglas M. Kinney.

Crystallography, mineralogy, and gems: L. G. Berry, L. L. Copeland, Michael Fleischer, Richard T. Liddicoat, Jr., Kurt Servos, and George Switzer.

Economic geology: M. E. Hopkins, Thomas A. Howard, Joel J. Lloyd, Ross Shipman, Jack A. Simon, Robert B. Smith, Robert R. Wheeler, and Walter S. White.

Engineering geology and rock mechanics: Mary Hill, Eugene Kojan, and David J. Varnes.

Geochemistry: Z. S. Altschuler, William Back, P. L. Greenland, Mead LeRoy Jensen, and Priestly Toulmin III.

Geochronology and absolute age determination: Mead LeRoy Jensen, Richard F. Marvin, Zell E. Peterman, John N. Rosholt, Barney J. Szabo, and Robert E. Zartman.

Geomorphology: Willard Bascom, Robert F. Black, Arthur L. Bloom, Arthur D. Howard, Watson H. Monroe, J. W. Pierce, Arthur N. Strahler, and William F. Tanner.

Geophysics: Don L. Anderson, L. G. Berryman, William J. Best, John K. Costain, Jules D. Friedman, M. Charles Gilbert, Pembroke J. Hart, Douglas M. Kinney, Arthur Lachenbruch, R. J. P. Lyon, Robert Moxham, Fred Nicodemus, Otto Nuttli, Donald A. Rice, Malcolm Rigby, Frank E. Senftle, Ralph T. Shuey, John S. Summer, Ben Tsai, Stanley H. Ward, Charles A. Whitten, Frans-Erik Wickman, and George Zissis.

Glaciers, ice, snow, and glacial geology: Richard Foster Flint, C. L. McGuinness, Mark F. Meier, and Stephen C. Porter.

History and philosophy of geology: George V. Cohee and George W. White.

Hydrology: Manuel A. Benson, John Ficke, Luna Leopold, and C. L. McGuinness.

Marine geology and oceanography: Willard Bascom, A. R. Gordon, Jr., and the U. S. Naval Oceanographic Office at Suitland, Maryland, and Norman J. Hyne.

Meteorology: Charles W. Collinson and Glen Stout.

Paleoclimatology and paleoecology: Robert E. Gernant and Thor N. V. Karlstrom.

Paleobotany: Anne S. Benninghoff and William S. Benninghoff.

Paleontology of the invertebrates: Richard H. Benson, Jean Berdan, William B. N. Berry, Richard S. Boardman, Raymond C. Douglass, J. Wyatt Durham, J. Thomas Dutro, Jr., Robert M. Finks, Mackenzie Gordon, Jr., Robert R. Hessler, Dorothy Hill, Erle G. Kauffman, Robert V. Kesling, Porter M. Kier, Bernhard Kummel, Alfred R. Loeblich, Jr., Donald B. Macurda, Jr., Raymond C. Moore, Norman J. Newell, William A. Newman, W. A. Oliver, John Pojeta, Jr., A. J. Rowell, J. S. Ryland, K. Norman Sachs, Norman F. Sohl, Walter C. Sweet, Helen Tappan, Curt Teichert, Ruth M. Todd, J. Marvin Weller, John W. Wells, C. W. Wright, and Ellis L. Yochelson.

Palynology: Alfred Traverse.

Igneous and metamorphic petrology: Harold H. Banks, Jr., Paul Bateman, Phillippa M. Black, Wilfred B. Bryan, Richard V. Dietrich, William G. Melson, Anne K. Loring, Kurt E. Lowe, Lucian B. Plann, Robert I. Tilling, and Hatten S. Yoder, Jr.

Sedimentary petrology: Murray Felsher, Robert L. Folk, George deVries Klein, Roy C. Lindholm, Francis J. Pettijohn, and J. W. Pierce.

Soils: Mary Hill, Charles E. Kellog, and Arnold C. Orvedal.

Statistical methods: Peter Fenner, J. C. Gower, William C. Krumbein, Richard McCammon, Daniel F. Merriam, and G. V. Middleton.

Stratigraphy: George V. Cohee (assisted by Virginia Byers and Marjorie MacLachlan).

Structural geology and tectonics: Fred A. Donath, William C. Gussow, Marshall Kay, Philip B. King, John M. Logan, and Robert S. Young.

Volcanology and pyroclastics: Roy A. Bailey, Wilfred B. Bryan, Richard S. Fiske, Jack Green, and Alexander McBirney.

DICTIONARY OF GEOLOGICAL TERMS

A

a- (direction) *Struct. petrol: 1.* The direction of tectonic transport, similar to the direction in which cards might slide over one another. Striae in a slickensided surface are parallel to *A*. *2. Biol:* Without, as apod, without feet.

aa A Hawaiian term for basaltic lava flows typified by a rough, jagged, spinose, clinkery surface. *Cf.* PAHOEHOE

a-axis *1. Crystallog:* One of the crystallographic axes used as reference in crystal description. It is the axis that is oriented horizontally, front-to-back. *2. Struct. petrol:* The fabric axis which is the direction of maximum displacement, e.g., of tectonic transport. It is usually assumed to be perpendicular to the fold axis, or b axis.

abatement The method of reducing the degree or intensity of pollution, also the use of such a method.

abaxial Any side of an organ or organism away from the center or axis.

Abbe refractometer An instrument used for determining the refractive index of liquids, minerals, and gemstones. Its operation is based on the measurement of the critical angle.

ablation The combined processes by which a glacier wastes.

ablation cone Ice cone; ice pyramid. A debris-covered cone of ice, firn, or snow formed by differential ablation.

ablation factor The rate at which a snow or ice surface wastes away. *See* ABLATION

Abney hand level Hand level with movable bubble tube that can be used to measure vertical angles.

ab-plane *Struct. petrol:* The surface along which differential movement takes place. a is the direction of displacement—that is, the direction of tectonic transport; b lies in this surface of movement and is perpendicular to a.

abrasion pH Acidity resulting from OH⁻ or H⁺ ions being absorbed at the surfaces of finely ground minerals suspended in water.

absarokite A variety of alkalic basalt consisting of about equal amounts of olivine, augite, labradorite, and sanidine, with accessory biotite, apatite, and opaque oxides. Leucite is sometimes present in small amounts. Absarokite forms a series with shoshonite and banakite and is transitional into shoshonite with decreasing amounts of olivine and increasing amounts of plagioclase and sanidine.

absolute age *See* GEOCHRONOLOGY

absolute temperature Temperature measured in degrees Celsius from absolute zero, −273. 18° C. Absolute temperatures are given either as "degrees absolute" (e.g., 150° A.) or as "degrees Kelvin" (e.g., 150 K.).

absolute viscosity *See* VISCOSITY, ABSOLUTE

absolute zero The temperature at which all thermal motion of atoms and molecules ceases: −273. 18° C.

absorption *1.* The attraction of

molecules of gases or liquids or ions in solution to the surface of solids in contact with them. 2. In optics, the reduction of intensity of light in transmission through an absorbing substance or in reflection from a surface. In crystals, the absorption may vary with the vibration-direction of the transmitted light. *See* PLEOCHROISM. 3. The process by which energy, such as that of electromagnetic or seismic wave, is converted into other forms of energy, e.g., heat. 4. *Hydrol:* A term applied to the entrance of surface water into the lithosphere by all methods.

absorption coefficient The absorptance of a radiation as it passes through a medium divided by the length of the path through that medium.

abstraction 1. A method of shifting waterways related to the shifting of divides. 2. A stream which for any reason is able to corrade its bottom more rapidly than do its neighbors, expands its valley at their expense, and eventually "abstracts" them. And conversely, a stream which for any reason is able to corrade its bottom less rapidly than its neighbors, has its valley contracted by their encroachments, and is eventually "abstracted" by one or the other.

abtragung That part of degradation not resulting directly from stream erosion, i.e., preparation and reduction of rock debris by weathering and translocation of waste.

abutment The retaining wall, *q.v.*, that holds back unstable areas and prevents land slippage.

abysmal sea That part of the sea which occupies the ocean basins proper.

abyssal 1. Of, or pertaining to, deep within the earth; plutonic. 2. Of, or pertaining to, the oceanic deeps below 1000 fathoms (1830

m.). 3. Referring to the great depths of seas or lakes where light is absent.

abyssal hill Relatively small topographic feature of the deep ocean floor ranging to 610–915 m. high and a few miles wide.

abyssal plains Flat, nearly level areas which occupy the deepest portions of many ocean basins.

Ac Abbreviation for acmite in normative calculations of igneous rocks and for actinolite in similar calculations for metamorphic rocks.

Acadian Middle Cambrian.

Acadian orogeny A middle Paleozoic deformation, especially in the northern Appalachians. In Gaspé and adjacent areas, the climax of the orogeny is dated as early in the Late Devonian, but deformational, plutonic, and metamorphic events were prolonged over a more extended period. The Acadian had best be regarded, not as a single orogenic episode, but as an orogenic era. *Cf.* ANTLER OROGENY

acanthite A mineral, Ag_2S, monoclinic.

acceleration 1. The rate of change of velocity. 2. The appearance of modifications earlier and earlier in successive generations in the evolution of species, sometimes to the point that the steps are omitted= brachygenesis. 3. In Paleozoic corals the addition of more secondary septa in one pair of quadrants than in the other pair.

acceleration due to gravity The acceleration of a body falling freely in a vacuum due to the gravitational attraction of the earth. The International Committee on Weights and Measures has adopted as a standard or accepted value 980.665 cm./sec.^{2}, but its true value varies with latitude, altitude, and the nature of the underlying rocks.

accelerometer An instrument

used to measure acceleration; specifically, a seismograph designed to measure earth particle accelerations.

accessory elements Minor elements; trace elements, *q.v.*

accessory minerals Those mineral constituents of a rock that occur in such small amounts that they are disregarded in its classification and definition. Opposed to essential minerals.

accidental inclusions Enclosed crystals or fragments having no genetic connection with the igneous rocks in which they occur.

acclivity An ascending slope, as opposed to declivity.

accordant fold One of several similarly oriented folds.

accordant summit levels Even-crested ridges, *q.v.*

accordion fold A fold in which the limbs are straight and maintain a constant thickness but in which there is thickening and sharpening of the hinge area.

accretion *1.* The gradual addition of new land to old by the deposition of sediment carried by the water of a stream. *2.* The process by which inorganic bodies grow larger, by the addition of fresh particles to the outside. *3.* A theory of continental growth by the addition of miogeosynclines to the craton. *4.* In soils, the process of illuviation is usually one of the addition of minerals by accretion.

accretionary ridge Beach ridge located inland from the modern beach showing that the coast has been built out seaward; ridges of this kind may be accentuated by the development of dunes.

accretion hypotheses Any hypothesis of the origin of the earth which assumes that it has grown from a small nucleus by the gradual addition of solid bodies, such as meteorites, as-

teroids, or planetesimals, formerly revolving about the Sun in independent orbits, but eventually drawn by gravitation to the earth and incorporated with it.

accretion theory A theory of the origin of the solar system involving the development of the planets from vortices in a disk-shaped mass of gas.

accretion topography Topographic features built by accumulation of sediment, e.g., meander scrolls.

accretion veins Veins formed by the repeated filling of channelways and their reopening by the development of fractures in the general zone undergoing mineralization.

accumulator plant *Geobot. prospecting:* Plant or tree that acquires an abnormal content of a metal where growing in metal-bearing soil.

A C F diagram Triangular diagram showing the chemical character of a metamorphic rock in which the three components plotted are: $A = Al_2O_3 + Fe_2O_3 - (Na_2O + K_2O)$, $C = CaO$, $F = FeO + MgO + MnO$.

ac-fracture *Struct. petrol:* A tension fracture parallel with the *ac* fabric plane and normal to *b*. Where ac-fractures are well developed, *b* is usually a strong lineation coincident with flexure fold axes.

achondrite Rare stony meteorite without chondrules.

acicular Needle-shaped; slender, like a needle or bristle, as some leaves or crystals.

aciculate Needle-shaped (pertaining to shape of gastropod shells).

acid, *n.* *1.* A substance containing hydrogen which dissociates to form hydrogen ions when dissolved in water (or which reacts with water to form hydronium ions). *2.* A substance capable of donating protons to other substances. *3. Adj:* A term applied to igneous rocks having a higher percentage of silica than orthoclase, the limiting figure commonly adopted being 66%.

acid clay A clay which, when in water suspension, gives off H ions. Hydrogen clay.

acidic *1.* A descriptive term applied to those igneous rocks that contain more than 66% SiO_2 as contrasted with intermediate and basic. Sometimes loosely and incorrectly used as equivalent to felsic and to oversaturated, but these terms include rock types (e.g., nepheline, syenite, and quartz basalt, respectively) which are not generally considered acidic. *2.* Less frequently used in reference to composition of feldspars, based on their content of silica. *3.* When referring to hydrothermal, pegmatitic, or other aqueous fluids the term is used in its chemical sense of high hydrogen ion concentration (low pH); very loosely used to refer to solutions containing salts of the strong acids (chlorides, sulfates, etc.) regardless of pH. *4.* In furnace practice, a slag in which silica is present in excess of the amount required to form a "neutral" slag with the earthy bases present.

acidization The process of forcing acid into a limestone, dolomite, or sandstone in order to increase permeability and porosity by removing a part of the rock constituents. It is also used to remove mud injected during drilling. The general objective of acidization is to increase productivity.

aclinic Having no inclination or dip; situated where the compass needle does not dip, as the aclinic line, or magnetic equator.

acmite Aegirite. *See* PYROXENE

acoustical well logging Any determination of the physical proper-

ties of a borehole by acoustical means. Travel times of P-waves over a unit distance are usually measured to determine velocities of surrounding rocks.

acoustic waves The waves which contain sound energy and by the motion of which sound energy is transmitted in air, in water, or in the earth. The wave may be described in terms of change of pressure, of particle displacement, or of density.

ac-plane *Struct. petrol:* A plane at right angles to the surface of movement. The ac-plane contains *a,* the direction of tectonic transport, and *c,* the axis perpendicular to the surface of movement.

acquired character Character not inherited but acquired by an individual organism during its lifetime as a result of use or disuse according to its mode of life or the conditions under which it lived.

acre A measure of surficial land area, containing in the United States and England 43,560 square feet.

acre-foot The volume of liquid or solid required to cover 1 acre to a depth of 1 foot.

acre-inch The quantity of water, soil or other material that will cover 1 acre 1 inch deep.

acre-yield The average amount of oil or gas or water recovered from 1 acre of a reservoir.

actinolite Iron-rich amphibole.

activation energy The extra amount of energy which any particle or group of particles must have in order to go from one energy state into some other energy state. Applied to changes in phase, as in chemical reactions, and to movement of particles, as in diffusion ("activation energy migration").

active fault A fault along which there is recurrent movement, that is usually indicated by small, periodic displacements or seismic activity.

active glacier A glacier in which some of the ice is flowing.

active layer Annually thawed layer; mollisol. Layer of ground above the permafrost which thaws in the summer and freezes again in the winter.

active permafrost Permafrost which, after having been thawed due to natural or artificial causes, is able to return to permafrost under the present climate.

active processes map A derivative map that combines into map units the active geological processes, such as erosion, deposition, etc., as an affect either from or on the activities of man.

active volcano See VOLCANO

activity, chemical *1.* Tendency to react spontaneously and energetically with other substances. *2.* Effective concentration; concentration as modified by the effects of the solvent and other dissolved substances. When activities are used in place of concentrations, dissociation constants and solubility products are true constants instead of approximate ones.

activity ratio Ratio of plasticity index to percentage of clay-sized minerals in sediment.

acute bisectrix In optically biaxial minerals, the direction bisecting the acute angle between optic axes.

adamantine A type of mineral luster suggesting extreme hardness such as in diamonds

adamellite Synonymous with and perhaps a preferable name for QUARTZ MONZONITE.

adaptation Adjustment of organisms to their environments by natural selection.

adaptive norm That part of an organic population which can survive and reproduce in the environment usually occupied by the species; the remainder carries hereditary defects and diseases.

adaptive radiation The adaptation, through natural selection, of the descendants of a taxon to a multitude of environments and habits by differentiating into different species and even higher taxonomic categories.

adaptive zone A unit of environment occupied by a single kind of organism.

adhesion The attraction of the molecules in the walls of interstices for molecules of water.

adiabatic *Thermodyn:* The relationship of pressure and volume when a gas or other fluid is compressed or expanded without either giving out or receiving heat. See ISOTHERMAL

adiabatic gradient A temperature gradient in a column of material such that essentially no heat enters or leaves the system upon such processes as convection.

adit *1.* A nearly horizontal passage from the surface by which a mine is entered and unwatered. In the United States an adit is usually called a tunnel, though the latter, strictly speaking, passes entirely through a hill and is open at both ends. Frequently called drift, or adit level. *2.* As used in the Colorado statutes it may apply to a cut either open or under cover, or open in part and under cover in part, dependent on the nature of the ground.

adjusted stream Stream which flows essentially parallel to the strike of underlying beds.

adobe Unburned, sun-dried bricks. Clay and silty deposits found in the desert basins of southwestern United States and in Mexico where the material is extensively used for making sun-dried bricks. The composition is a mixture of clay and silt together with minor amounts of other materials.

adsorption *1.* Adhesion of molecules of gases, or of ions or molecules in solutions, to the surfaces of solid bodies with which they are in contact. *2.* The behavior of a multicomponent fluid system which results in a dissolved material becoming more concentrated, or less concentrated, at an interface than in the body of the solution.

adularia A variety of orthoclase.

advance *1.* **of a beach** A continuing seaward movement of the shore line; a net seaward movement of the shore line over a specified time. *Syn:* PROGRESSION. *2.* **of a glacier** The forward movement of a glacier front.

advection *Tect:* Lateral mass movements of mantle material. Such mantle movements have been proposed as the cause of the strike-slip movements along mid-oceanic ridges.

aegirite Aegirine. *See* PRYOXENE

Aeolian *Obs.* Eolian, *q.v.*

aeon A period of geologic time equal to one billion years.

aerate To expose to the action of the air. Supply or charge with air.

aeration The process of being supplied or impregnated with air. Aeration is used in waste water treatment to foster biological and chemical purification.

aeration, zone of *See* ZONE OF AERATION

aerial Relating to the air or atmosphere. "Subaerial" is applied to phenomena occurring under the atmosphere; "subaqueous" to phenomena occurring under water.

aerial magnetometer A device used to measure variations in the earth's magnetic field while being transported by an aircraft. Same as airborne magnetometer.

aerial photograph A photograph of the earth's surface taken from the air. It is usually one of a series taken from an aircraft moving in a systematic pattern at a given altitude in order to obtain a mosaic for mapping land divisions, geology, soil, vegetation, topography, etc.

aerobic *1.* Living or active only in the presence of oxygen. *2.* Pertaining to or induced by aerobic organisms.

aerolite Stony meteorite.

aeromagnetic Refers to magnetometer observations made from a moving airplane.

aff Affinity. *Paleontol:* Indicates a specimen or specimens believed to be closely related to but not exactly the same as the named species.

affine Refers to deformative movements in which, at the scale considered, individual particles move uniformly with respect to each other and originally straight lines and even planes are not distorted but a sphere is transformed into an ellipsoid.

aftershock An earthquake which follows a larger earthquake and originates at or near the focus of the larger earthquake.

Aftonian The first Pleistocene interglacial of classical glacial periods of the Pleistocene. The interglacial stage following the Nebraskan glacial.

agate A waxy variety of cryptocrystalline quartz (chalcedony) in which the colors are in bands, clouds, or distinct groups.

age *1.* Any great period of time in the history of the earth or the material universe marked by special phases of physical conditions or organic development; as the age of mammals. *2.* Formal geologic time unit corresponding to a stage; it is always capitalized. *3.* Informal geologic time unit corresponding to any stratigraphic unit (not capitalized).

age ratio The ratio of daughter

to parent isotope. Often used to indicate a ratio that is perturbed by some factor and therefore not indicative of the absolute age of the mineral.

agglomerate Contemporaneous pyroclastic rock containing a predominance of rounded or subangular fragments greater than 32 mm. in diameter.

agglutinated Refers to those foraminiferans the test of which are composed of minute pieces of substrate cemented by an animal secretion.

aggradation *1.* The process of building up a surface by deposition. *2.* The growth of a permafrost area. In both senses, the opposite of degradation.

aggrade *1.* To build up the grade or slope (of the earth) by deposition of sediment. *2.* It is suggested, in accordance with Davis' original proposal, that "graded" be used specifically for the stream in which equilibrium is maintained and that "degrading" and "aggrading" be restricted to cases of the shifting equilibrium. "Degrading" is down-cutting approximately at grade, in contradistinction to such self-explanatory terms as trench or incise. "Aggrading" is upbuilding approximately at grade.

aggregate *1.* To bring together; to collect or unite into a mass. *2.* Composed of a mixture of substances, separable by mechanical means. *3.* The mineral material, such as sand, gravel, shells, slag, or broken stone, or combinations thereof, with which cement or bituminous material is mixed to form a mortar or concrete. Fine aggregate may be considered as the material that will pass a 6.35 mm. screen, and coarse aggregate as the material that will not pass a 6.35 mm. screen.

aggregate structure A mass of

separate little crystals, scales, or grains which extinguish under the polarizing microscope at different intervals during the rotation of the stage.

aggressive or **invasive magmas** Those magmas which force their way into place.

aging cycle of a lake The life cycle of a lake, whereby it gradually fills in with sediment and plant debris (over decades or millennia, depending on the size of the lake) until it becomes a swamp and gradually terrestrial.

Agnatha Class of vertebrates; jawless fishes, e.g., lampreys. Ord.-Rec.

agonic line A line passing through points on the earth's surface at which the direction of the magnetic needle is truly north and south; a line of no magnetic declination.

agricultural pollution The liquid and solid wastes from all types of farming, including runoff from pesticides, fertilizers, and feedlots; erosion and dust from plowing, animal manure and carcasses, and crop residues and debris.

A horizon Zone of eluviation in eluviated soils. The uppermost zone in the soil profile, from which soluble salts and colloids have been leached, and in which organic matter has accumulated.

airborne magnetometer A device used to measure variations in the earth's magnetic field while being transported by an aircraft. Same as aerial magnetometer.

airborne scintillation counter Any scintillation counter especially designed to measure the ambient radioactivity from an aircraft in flight.

air damping The use of air, usually in condenser microphone detectors and in inductive geophones where the coil mass is small, to establish friction against the moving mass, causing decay of motion with time following application of impulse.

air gun A popular energy source used in marine seismic surveys. Air under high pressure is explosively released to provide the initial shock wave. Air guns have been adopted for use in borehole velocity surveys.

air shooting The act or action of applying a seismic pulse to the earth by detonating a charge or charges in the air above the surface of the earth; the process of exploration by the use of such detonations.

air wave The acoustic energy pulse transmitted through the air as a result of the detonation of a seismic shot.

Airy isostasy That hypothesis of equilibrium for the earth's solid outer crust in which the crustal density is supposed constant so that mountains are compensated by "roots" analogous to the underwater extensions of icebergs floating in the ocean.

A K F diagram Triangular diagram showing the chemical character of a metamorphic rock in which the three components plotted are: $A=Al_2O_3+Fe_2O_3-(CaO+Na_2O)$, $K=K_2O$, $F=FeO+MgO+MnO$.

alabaster Terra alba. Compact fine-grained gypsum, white or delicately shaded. Used for ornamental vessels, figures, and other carvings.

alaskite A plutonic rock consisting of orthoclase, microcline, and subordinate quartz, with few or no mafic constituents. Plagioclase may or may not be present. A leucocratic variety of granite.

alb Flat or gently inclined narrow shelf separating the nearly vertical side of an alpine glacial trough from the mountain slope above.

albedo The percentage of the incoming radiation that is reflected by a natural surface such as the ground, ice, snow, water, clouds, or particulates in the atmosphere.

Albers conical equal-area map projection An equal-area conic projection having two standard parallels, each of which lies equidistant between the central and extreme parallels of the map, and along which the scale is true. Meridians are represented by equally spaced straight lines that converge to a common point beyond the map limits, and parallels are represented by arcs of concentric circles whose center is at the common point and which are at right angles to the meridians. The meridional scale is too large between the standard parallels and too small beyond them; distances between the parallels decrease north and south of the standard parallels.

Albertan The middle of three epochs of the Cambrian of North America; also the series of strata laid down during that epoch.

albertite A dark brown to black asphaltic pyrobitumen having conchoidal fracture. It is practically insoluble in alcohol, but partially soluble in turpentine. It occurs in veins 0.3 to 4.9 meters wide in the Albert Shale of New Brunswick.

albite See PLAGIOCLASE

albite-epidote-amphibolite facies Metamorphic rocks produced under intermediate temperature and pressure conditions by regional metamorphism or in the outer contact metamorphic zone.

albitite A coarse-grained dike rock consisting almost wholly of albite. Common accessory minerals are muscovite, garnet, apatite, quartz, and opaque oxides.

Alexandrian Lower Silurian.

alexandrite A variety of the mineral chrysoberyl which may appear to be either green or red, depending on the nature of the light by which it is examined.

alfisol A soil classification order characterized by an iron-rich surface layer, a clay-rich subsurface layer.

Algae Class of thallophytes, includes single-celled plants and common seaweeds. Precamb.-Rec.

algal, *adj.* Of, pertaining to, or composed of algae.

algal bloom A proliferation of living algae on the surface of lakes, streams, or ponds. Algal blooms are stimulated by phopshate or other nutrient enrichment.

algal limestone A limestone composed largely of remains of calcium carbonate producing algae or a boundstone or biolithite in which such algae or algal deposits serve to bind together the fragments or flocules of other calcium carbonate producers.

algal structure A deposit, usually calcareous, which shows banding, irregular concentric structures, crusts, pseudopisolites, or pseudoconcretionary forms of calcium carbonate precipitated as a result of organic (usually algal) activity.

Algoman orogeny Orogeny and accompanying granitic emplacement which affected Precambrian rocks of northern Minnesota and adjacent Ontario about 2400 m.y. ago. *Syn:* KENORAN OROGENY

Algonkian Preferred usage is "Late Precambrian."

algorithm A set of step-by-step instructions designed to perform a numerical or algebraic operation as in digital computer applications.

alidade *1.* The part of surveying instrument consisting of a sighting device, index, and reading or recording devices. *2.* A straightedge ruler carrying a sighting device, such as slot sights or a telescope mounted parallel to the ruler.

alkali *1.* Sodium carbonate or potassium carbonate, or more generally any bitter-tasting salt found at or near the surface in arid and semiarid regions. *2.* A strong base, e.g., $NaOH$ or KOH. *3.* An alkali metal.

alkalic Refers to: (1) solution containing alkali metal ions; (2) igneous rock with more alkali metals than are contained in feldspars, therefore such minerals as feldspathoids are present; (3) igneous rock with more alkali metals than average for its clan; (4) igneous rock with alkali-lime index less than 51; (5) igneous rocks of Atlantic series. *Obs.*

alkali-calcic series Those igneous rock series having alkali-lime indices in the range 51–55.

alkalic igneous rocks *Petrol:* A term rather loosely used, generally meaning one of the following: *1.* More than average alkali (K_2O+Na_2O) for that clan in which they occur. *2.* Containing feldspathoids or other minerals, such as acmite, so that the molecular ratio of alkali to silica is greater than $1:6$. *3.* The term is sometimes defined also as embracing those rock series having a low alkali-lime index (51 or less).

alkali feldspar Sodium and potassium-rich feldspars. Microcline, orthoclase, albite, anorthoclase, and sanidine.

alkali flat A level lakelike plain formed in low depressions where accumulated water evaporates depositing fine sediment and dissolved minerals which form a hard surface if mechanical sediments prevail or a crumbly powdered surface if efflorescent salts are abundant.

alkali-lime index The weight percentage of silica, in a sequence of igneous rocks on a variation diagram, where the weight percentages of CaO and of (K_2O+Na_2O) are equal, i.e., the point of crossing of the curves for CaO and (K_2O+Na_2O).

alkali metal Any metal of the alkali group, as lithium, sodium, potassium, rubidium, or cesium.

alkaline, *adj. 1.* Having the qualities of a base. *Syn:* BASIC. *2.* Containing sodium and/or potassium in excess of the amount needed to form feldspar with the available silica, e.g., an alkaline rock—in this sense sometimes written alkalic. *3.* Containing ions of one or more alkali metals, e.g., an alkaline ore solution. *4.* Containing cations of the strong bases in excess of the anions of strong acids, e.g., an alkaline ore solution. (Note that geologic usage gives "alkaline solution" so many different meanings that it is ambiguous without further qualification; it is therefore recommended that alkalic be used when definitions 2 and 3 are meant.) *5.* Waters containing more than average amounts of carbonates of sodium, potassium, magnesium, and/or calcium.

alkalinity The capacity of a water to accept protons, i.e., hydrogen ions. It is usually expressed as milliequivalents per liter.

alkali rocks Igneous rocks in which the abundance of alkalies in relation to other constituents has impressed a distinctive mineralogical character; generally indicated by the presence of soda pyroxenes, soda amphiboles, and/or feldspathoids. *Cf.* CALC-ALKALIC SERIES

alkemade line A line connecting the composition points of two primary phases whose phase areas are adjacent and meet to form a boundary curve.

allanite Orthite. A mineral, a monoclinic member of the epidote group. Composition variable, formula $(Ca,Ce,La)_2(Al,Fe,Mg)_3(SiO_4)_3(OH)$. Commonly contains a little thorium and may be metamict.

Alleghenyan Lower Middle Pennsylvanian.

Allegheny orogeny An event which deformed the rocks of the Valley and Ridge province, and those of the adjacent Allegheny Plateau in the central and southern Appalachians. Most of the orogeny was probably late in the Paleozoic, but phases may have extended into the Early Triassic.

Allen's rule Warm-blooded animals generally have shorter legs, tails, and ears in cold than in warm regions.

alliaceous Applied to minerals having the odor of garlic when rubbed, scratched, or heated; e.g., arsenical minerals.

Alling scale A system of classifying size grades of sediments, for use with thin and polished sections, with subdivisions based on the fourth root of 10.

allochem Sediment formed by chemical or biochemical precipitation within a depositional basin; includes intraclasts, oolites, fossils, and pellets; *cf.* PSEUDOALLOCHEM

allochemical metamorphism Metamorphism accompanied by addition or removal of material so that bulk chemical composition of rock is changed.

allochthon Rocks that have been moved a long distance from their original place of deposition by some tectonic process, generally related to overthrusting or recumbent folding, or perhaps gravity sliding. Used in contrast to AUTOCHTHON, *q.v.*

allochthonous *1.* A term applied to rocks of which the dominant constituents have not been formed *in situ. Cf.* AUTOCHTHONOUS. *2.*

Usually refers to material resources which originated outside the division of an ecosystem under consideration.

allogene; allothigene A mineral or rock which has been transported to the site of deposition from without.

allogenic Term meaning generated elsewhere, applied to those constituents that came into existence outside of, and previously to, the rock of which they now constitute a part, e.g., the pebbles of a conglomerate. *Cf.* AUTHIGENIC

allogenic succession Succession induced by geochemical processes acting from outside the local community.

allophane An amorphous hydrated alumino-silicate developed in soils and having a variable composition.

allotriomorphic Xenomorphic.

allotropic Those substances which may exist in two or more forms, as diamond and graphite.

alluvial *1.* Pertaining to alluvium. *2.* Formerly used as a term for recent unconsolidated sediments.

alluvial dam Sedimentary deposit built by an overloaded stream which dams its channel, especially characteristic of distributaries on alluvial fans.

alluvial fan A cone-shaped deposit of alluvium made by a stream where it runs out onto a level plain or meets a slower stream. The fans generally form where streams issue from mountains upon the lowland.

alluvial plain *1.* Flood plains produced by the filling of a valley bottom are alluvial plains and consist of fine mud, sand, or gravel. *2.* A plain resulting from the deposition of alluvium by water. In the southwestern United States most alluvial plains are formed by streams having a considerable grade, and hence they are generally referred to as alluvial slopes.

alluviation The deposition of mechanical sediments by rivers anywhere along their courses.

alluvium *1.* A general term for all detrital deposits resulting from the operations of modern rivers, thus including the sediments laid down in river beds, flood plains, lakes, fans at the foot of mountain slopes, and estuaries. *2.* The rather consistent usage of the term throughout its history makes it quite clear that alluvium is intended to apply to stream deposits of comparatively recent time, that the subaqueous deposits of seas and lakes are not intended to be included, and that permanent submergence is not a criterion. Alluvium may become lithified, as has happened frequently in the past, and then may be termed ancient alluvium.

almandine, almandite *See* GARNET

alnöite A lamprophyre consisting of biotite, augite, olivine, and melilite. Apatite, perovskite, nepheline, and opaque oxides are common accessories. *See* LAMPROPHYRE

alp Topographic shoulder located high on the side of a glaciated trough.

alpha particle A helium atom lacking two electrons and therefore having a double positive charge.

alpha rays; alpha radiation Radiation consisting of alpha particles emitted during the decay of some radioactive elements.

Alpides Great east-west structural belt including Alps of Europe and Himalayas and related mountains of Asia mostly folded in Tertiary time.

alpine Of, pertaining to, or like the Alps or any lofty mountain. Resembling a great mountain range of southern Europe called the Alps. Implies high elevation, particularly above tree line, and cold climate.

alpine glacier A glacier occupying a depression within or lying on mountainous terrain. *Syn:* MOUNTAIN GLACIER

alpine orogeny Series of diastrophic movements beginning perhaps in the Late Triassic and continuing until the present. *Cf.* LARAMIDIAN

alpine range Signifies a range possessing the rugged peak-and-sierra form and the internal structures incidental to intense crumpling, metamorphism, and igneous intrusion as exemplified in the Swiss Alps.

alteration Changes in the chemical or mineralogical composition of a rock, generally produced by weathering or hydrothermal solutions.

alternation of generations *1.* The orderly succession of sexual and asexual types of reproduction in the life cycle of many species of Foraminifera, resulting in the production of different kinds of tests; also occur in other animals, such as the Coelenterata. *2.* The alternation of a spore-producing phase and gamete-producing phase in the life cycle of a plant.

altimeter An aneroid barometer used for determining elevations.

altiplanation A special phase of solifluction that, under certain conditions, expresses itself in terracelike forms and flattened summits and passes that are essentially accumulations of loose rock materials.

altithermal, *n.* Period of high temperature, particularly the postglacial thermal optimum.

altitude *1.* The vertical angle between the plane of the horizon and the line to the observed point as a star. *2.* The vertical distance between a point and a

datum surface or plane, such as mean sea level. *See* ELEVATION

alumina Aluminum oxide, Al_2O_3.

alunite A mineral $KAl_3(SO_4)_2(OH)_6$, hexagonal rhombohedral, usually in white, gray, or pink masses in hydrothermally altered feldspathic rocks.

alunitization Introduction of, or replacement by, alunite.

alunogen A mineral, $Al_2(SO_4)_3\cdot16H_2O$ triclinic, usually found as fibrous masses formed by the action of acid sulfate waters on rocks.

alveolar Having small cellular structures like a honeycomb.

amalgam *1.* A mineral, an alloy of silver and mercury. *2.* An alloy of mercury with another metal.

amazonite; Amazonstone A green variety of microcline. Used as a gem.

amber A fossil resin from coniferous trees.

amblygonite A mineral, $(Li,Na)Al(PO_4)(F,OH)$. Triclinic. An ore of lithium found in pegmatites. Used as a gemstone.

ambulacrum Area or ray in echinoderms that marks a branch of the water vascular system and generally bears numerous tube feet.

amethyst A purple or bluish-violet variety of quartz, SiO_2. Used as a gem.

ammonite One of a large extinct group of mollusks related to the living chambered nautilus. The sutures are complex and angular, whereas they are straight or simply curved in the nautilus and its relatives.

ammonoid An inclusive term for GONIATITES, CERATITES, and AMMONITES, *q.v.*

Ammonoidea Order of cephalopods, mostly coiled, whose septa meet the external shell to form folded sutures.

amorphous Without form; applied to rocks and minerals having no definite crystalline structure.

amosite An asbestos mineral with iron content higher than anthophyllite.

Amphibia Class of vertebrates, air-breathing tetrapods that develop from a water-breathing larval tadpole stage. Dev.-Rec.

amphibian A cold-blooded animal with legs, feet, and lungs, that breathes by means of gills in the early stages and by means of lungs in the later stages of life. One of the Amphibia.

amphibole A mineral group with the general formula $A_2B_5(Si,Al)_8O_{22}(OH)_2$, where A is mainly Mg, Fe, Ca, and Na; B is mainly Mg, Fe^{+2}, Al, and Fe^{+3}. Common rock-forming minerals of the inosilicate group characterized by good prismatic cleavage in two directions intersecting at angles of 56° and 124°. Most common amphibole minerals are hornblende, tremolite-actinolite, and cummingtonite-grunerite.

amphibolite A crystalloblastic rock consisting mainly of amphibole and plagioclase. Quartz is absent, or present in small amounts only. When quartz is more abundant there is a gradation to hornblende-plagioclase gneiss. See FEATHER AMPHIBOLITE; GARBENSCHIEFER; OLLENITE

amphibolite facies Rocks produced by medium- to high-grade regional metamorphism.

Amphineura Class of mollusks whose flattened body is covered by eight articulated dorsal plates, exclusively marine; chitons. Ord.-Rec.

amphoteric Having both basic and acidic properties.

amplitude The elevation of the crest of a wave or ripple above the adjacent troughs. *Hydrodyn:* One-half the wave height.

amygdale A gas cavity or vesicle in volcanic rocks, filled with secondary products such as zeolites, calcite, or silica minerals. *Adj:* AMYGDALAR *Syn:* (Rare) AMYGDULE

amygdaloid A general name for volcanic rocks (ordinarily basalts or andesites) that contain numerous gas cavities (vesicles) filled with secondary minerals such as zeolites, calcite, chalcedony, or quartz. The filled cavities are called amygdules or amygdales. *Adj:* AMYGDALOIDAL

amygdule *1.* A gas cavity or vesicle in an igneous rock which is filled with such secondary minerals as zeolite, calcite, quartz, or chalcedony. The term amygdale is preferred in British usage, in which case amygdule is applied only to small amygdales. *2.* An agate pebble.

anaerobic, *adj. 1.* Living or active in the absence of free oxygen. *2.* Pertaining to or induced by organisms that can live in the absence of free oxygen.

anaerobic sediment A highly organic sediment with no free oxygen, usually rich in hydrogen sulfide.

anaerobic zone A zone where oxygen is lacking. Substances are reduced, not oxidized.

analbite High temperature albite; inversion occurs at about 700° C.

analcime; analcite A mineral, $NaAlSi_2O_6.H_2O$, an isometric zeolite, commonly found in diabase and in alkali-rich basalts.

analog Any device which represents a range of numbers by directly measurable variable quantities such as voltages, rotations, etc., as in analog computer, or analog systems. *Cf.* DIGITAL

analytic group Rock stratigraphic unit formerly a formation but up-graded because subdivisions of the unit are now considered to be formations.

analyzer That part of a polariscope that receives the light after polarization and exhibits its properties. In a petrographic microscope, the polarizing mechanism (Nicol prism, Polaroid, etc.) which intersects the light after it has passed through the object.

anamorphic zone The zone of rock-flowage, especially characterized by silicatization involving decarbonation, dehydration, and deoxidation. *See* KATAMORPHIC ZONE

anastomosing *1.* An anastomosing stream (braided stream, *q.v.*) branching, interlacing, intercommunicating, thereby producing a netlike or braided appearance. *2.* Netted; interveined; said of leaves marked by cross veins forming a network; sometimes the vein branches meet only at the margin.

anatase Octahedrite. A mineral, TiO_2. Tetragonal, trimorphous with rutile and brookite.

anatexis *1.* A high-temperature metamorphic process by which plutonic rock in the deeper levels of the crust is dissolved and regenerated as a magma. *Cf.* SYNTEXIS. *2.* The complete melting of crustal rocks to form granitic magma, as opposed to rheomorphism or mobilization, which implies merely the development of sufficient liquid to permit movement. Some include both processes under the term anatexis.

anatexites Metamorphic rocks, formed by the process of anatexis. They show only faint schistose structure and are granitelike in composition.

anauxite A kaolinite-like mineral containing an excess of silica. Largely discredited.

anchored dune Sand dune stabilized by growth of vegetation.

anchor ice Ice that forms in the bottom of rivers when the rest of the water is not frozen. *Syn:* GROUND ICE; BOTTOM ICE

andalusite A mineral, Al_2SiO_5, trimorphous with kyanite and sillimanite. Orthorhombic. Commonly occurs in schists and gneisses.

andesine *See* PLAGIOCLASE

andesite A volcanic rock composed essentially of andesine and one or more mafic constituents. The plagioclase is usually strongly zoned and may range in composition from about An_{35} to An_{70}, but the average composition usually falls within the range of andesine. When the rock is porphyritic, the phenocrystic plagioclase is usually more calcic than the groundmass plagioclase, and in addition the groundmass may contain small amounts of microcrystalline or occult potassic feldspar and cristobalite. Pyroxene, hornblende, or biotite, or all three in various proportions may constitute the mafic constituents.

andesite line The geographic boundary between the circum-Pacific rock province (the andesite-dacite-rhyolite association of the Pacific margin) and the intro-Pacific rock province (the olivine basalt-trachyte association of the islands lying within the Pacific Basin). It is based primarily on petrographic data and runs from Alaska via Japan, the Marianas, Palau Islands, Bismarck Archipelago, and the Fiji and Tonga groups to the east of New Zealand and Chatham Islands. Along the eastern side of the Pacific the position of the line is less clearly defined, but probably it runs along the coasts of North and South America. In the South Pacific it has not yet been traced.

andosol A black or dark brown soil that is formed from volcanic material in a temperate to tropical humid climate.

andradite A mineral of the garnet group, $Ca_3Fe_2(SiO_4)_3$.

Angara Stable shield region in northern Asia.

Angiospermae Class of Spermatophyta or Pteropsida; plants with highly specialized flowers and seeds. Jur.-Rec.

angle of dip A synonym for dip.

angle of incidence In optics, the angle between the incident ray of light and the normal to the surface.

angle of repose The maximum slope or angle at which a material such as soil or loose rock remains stable. When exceeded, mass movement by slipping as well as by water erosion may be expected. *Syn:* CRITICAL SLOPE

anglesite A mineral, $PbSO_4$, orthorhombic. A common alteration product of lead sulfide ores.

Ångström unit (Often anglicized to Angstrom; abbreviated A. or Å.) A unit of length, 10^{-8} cm., commonly used in structural crystallography. *See* kX

angular, *adj.* A roundness grade showing very little or no evidence of wear, with edges and corners sharp. Secondary corners, *q.v.*, numerous (15 to 30) and sharp. Class limits 0 to 0.15.

angularity *1.* Sharpness of edges and corners of grains. A grain is angular if most of the edges or corners are sharp, and rounded if most are smooth. Not to be confused with sphericity, *q.v.* A nearly spherical particle may have sharp corners and be angular, while a flat pebble may not spherical in shape but still be well rounded as to its corners. *Cf.* SHAPE. *2. Geophys:* Stepout-moveout or moveout time, *q.v.*

angular unconformity An unconformity in which the older strata dip at a different angle (generally steeper) than the younger strata.

See DISCONFORMITY, NONCONFORMITY

anhedral A crystal showing no external faces.

anhydrite A mineral, anhydrous calcium sulfate, $CaSO_4$. Orthorhombic, commonly massive in evaporite beds.

anhydrous Completely or essentially without water, as anhydrous magma.

Animalia The animal kingdom.

anion An ion that bears a negative charge.

anisotropic Having physical properties that vary in different directions; specifically in crystal optics, showing double refraction. Characteristic of all crystalline substances except those belonging to the isometric system. *See* ISOTROPIC

anisotropy Condition of having different properties in different directions; example: the state of geologic strata of transmitting sound waves with different velocities in the vertical and in the horizontal directions.

ankerite A mineral, a ferroan variety of dolomite, $CaCO_3 \cdot (Mg,Fe,Mn)CO_3$.

annabergite Nickel bloom. A mineral, $(Ni,Co)_3(AsO_4)_2 \cdot 8H_2O$; monoclinic, usually found as green incrustations as an alteration product of nickel arsenides.

Annelida The phylum of invertebrate animals which includes the segmented worms.

annual layer *1.* Sedimentary layer deposited or presumed to have been deposited during the course of a year, e.g., glacial varve. *2.* Dark layer in stratified salt deposit containing disseminated anhydrite.

annual ring The layer of xylem (wood) formed by one year's growth of cambium.

annular drainage pattern Annular drainage, as the name implies, is ringlike in pattern. It is subsequent in origin and associated with maturely dissected dome or basin structures.

anomaly *1.* A deviation from uniformity; a local feature distinguishable in a geophysical, geochemical, or geobotanical measurement over a larger area; a feature considered capable of being associated with commercially valuable petroleum or other mineral deposits; an area or restricted portion of a geophysical survey, such as magnetic or gravitational, which is different in appearance from the survey in general; specifically, an area within which it appears that successful drilling or other search for hydrocarbons or minerals may be conducted. In seismic usage anomaly is generally synonymous with structure, but it is also used for spurious or unexplainable seismic events or for local deviations of potential functions which can be conclusively attributed to no unique cause. *2.* The departure of the local mean value of a meteorological element from the mean value for the latitude. *3.* (Gravity) In comparing any set of observational data with a computed theoretical curve, the difference of an observed value and the corresponding computed value (observed minus computed).

anorogenic A geological feature that forms during a period of tectonic quiescence between orogenic periods.

anorthite A mineral of the plagioclase feldspar group, $CaAl_2Si_2O_8$.

anorthoclase Triclinic alkali feldspars containing more sodium than $Ab_{63}Or_{37}$ (*see* PLAGIOCLASE) which invert to monoclinic symmetry when heated and reinvert to triclinic when cooled.

anorthosite A plutonic rock composed almost wholly of plagioclase.

antecedent platform A postulated submarine platform 50 meters or more below sea level from which barrier reefs and atolls grow upward to the water surface. The formation of the platform predates its colonization by corals. Hence, it is termed an antecedent platform.

antecedent stream A stream that flowed in its present course prior to the development of the existing topography; a stream that was established before local uplift began and incised its channel at the same rate the land was rising.

antecedent valley A valley which was established before orogenic movement occurred and maintained its course during orogeny.

antediluvian Formerly referred to time or deposits antedating Noah's flood.

anthophyllite A mineral of the amphibole group, $(Mg,Fe)_7Si_8O_{22}(OH)_2$.

Anthozoa Class of coelenterates represented by polyps that build solitary or colonial calcareous external skeletons; corals.

anthracite *1.* Generally a hard, black lustrous coal containing a high percentage of fixed carbon and a low percentage of volatile matter, commonly referred to as "hard coal" and mined in the United States, mostly in eastern Pennsylvania, although in small quantities in other states. *2.* Nonagglomerating anthracitic coal having 92% or more, and less than 98% of fixed carbon (dry, mineral-matter-free) and 8% or less, and more than 2% of volatile matter (dry, mineral-matter-free).

anthraxolite Refers to a highly graphitic coal—an example quoted having a percentage of 97.72 fixed carbon—such as anthracite-like asphaltic material occurring in veins in Precambrian slate of the Sudbury district.

anthraxylon [<*Gr. anthrax* coal + *xylon* wood] The vitreous-appearing components of coal, which in thin section are shown to be derived from the woody tissues of plants, such as stems, limbs, branches, twigs, roots, including both wood and cortex, changed and broken up in fragments of greatly varying sizes through biological decomposition and weathering during the peat stage, and later flattened and transformed into coal through the coalification process, but still present as definite units.

anticlinal *1. Geol:* Inclined toward each other, as, the ridge tiles of the roof of a house. *2.* Of, or pertaining to, an anticline.

anticlinal axis The plane or surface that divides an anticline as symmetrically as possible.

anticlinal fold An upwardly convex flexure in which one limb dips gently toward the apex and the other limb dips more steeply away from it. *Cf.* MONOCLINAL FLEXURE; UNICLINE; MONOCLINE

anticlinal mountain A mountain formed by an anticlinal fold.

anticlinal theory The theory that water, oil, and gas accumulate in the order named, in up-bowed strata, provided such a structure contains reservoir rocks in proper relation to source rocks and an impervious barrier.

anticlinal valleys Those which follow anticlinal axes.

anticline A fold that is convex upward or had such an attitude at some stage of development. In simple anticlines the beds are oppositely inclined, whereas in more complex types the limbs may dip in the same direction. Some anticlines are of such complicated form that no simple definition can be given. Anti-

clines may also be defined as folds with older rocks toward the center of curvature, providing the structural history has not been unusually complex.

anticlinorium A series of anticlines and synclines so arranged structurally that together they form a general arch or anticline.

antidune A transient form of ripple on the stream bed analogous to a sand dune. An antidune progressively moves upstream.

antiform An anticlinal-type structure in which the stratigraphic sequence is not known. *Cf.* ANTICLINE

antigorite A mineral of the serpentine group, $(Mg,Fe)_3Si_2O_5(OH)_4$.

antimonite The native sulfide of antimony; stibnite.

antimony A mineral, the native element, occurring in tin-white masses. Hexagonal rhombohedral.

antiperthite An intergrowth of sodic and potassic feldspar generally thought to have formed during slow cooling by unmixing of sodium and potassium ions in an originally homogeneous alkalic feldspar. In the antiperthites the potassic member (usually orthoclase) forms thin films, lamellae, strings, or irregular veinlets within the sodic member (usually albite). *See* PERTHITE

antiroot According to the Pratt isostasy hypothesis, crustal material of higher density under the oceans has isostatic compensation for its lesser mass and lower topographic elevation.

antistress mineral A mineral such as anorthite, potash feldspars, pyroxenes, forsterite, and a leucite, etc., whose formation in metamorphosed rocks is favored by conditions that are controlled, not by shearing stress, but by thermal action and by hydrostatic pressure that is probably no more than moderate.

antithetic fault Minor normal faults that are of the opposite orientation to the major fault with which they are associated. *Ant:* SYNTHETIC FAULT

Antler orogeny An orogeny which extensively deformed Paleozoic rocks of the Great Basin in Nevada during Late Devonian and Early Mississippian Time; named by R. J. Roberts (1951) for relations in the Antler Peak quadrangle near Battle Mountain, Nevada. The main expression of the orogeny is the emplacement of eugeosynclinal western rocks over miogeosynclinal eastern rocks along the Roberts Mountains thrust. Minor orogenic pulses followed the main event, extending into the Permian. It is broadly equivalent to the Acadian orogeny of eastern North America.

Ao horizon That portion of the A horizon of a soil profile which is composed of pure humus.

apatite A mineral group consisting of fluorapatite, $Ca_5(PO_4)_3(F, PH,Cl)_3$; chlorapatite, hydroxylapatite. Hexagonal. Collophane and francolite are related minerals.

aperiodic motion Any nonperiodic motion; any motion with a continuous frequency spectrum. Example: a pulse from a shot in a shot hole.

apex *1.* The tip, point, or angular summit of anything, as, the apex of a mountain. The end, edge, or crest of a vein nearest the surface. *2.* The highest point of a stratum, as, a coal seam. *3. Geol:* The top of an anticlinal fold of strata. This term, as used in United States Revised Statutes, has been the occasion of much litigation. It is supposed to mean something nearly equivalent to outcrop. *4. Paleontol:* The pointed, initial end of an elongate or conical form (as in a coral, gastropod, foraminifer,

etc.). *5.* In a conodont, the point where two limbs join. *6.* In a brachiopod, the place of initial growth.

aphanite A general term applied to dense, homogeneous rocks whose constituents are too small to be distinguished by the unaided eye. The adjective form aphanitic is currently used more frequently than the noun.

aphanitic Pertaining to a texture of rocks in which the crystalline constituents are too small to be distinguished with the unaided eye. It includes both microcrystalline and cryptocrystalline textures.

API gravity The standard American Petroleum Institute method for specifying the density of crude petroleum. The density in degrees API is equal $\frac{141.5}{P} -131.5$ where P is the specific gravity of the oil measured at 15.6° C. Note: This is one of several so-called Baumé scales for comparing lighter liquids with water.

aplite A dike rock consisting essentially of quartz and alkali feldspar, with a fine-grained, sugary texture.

apophyllite A mineral of the zeolite group, $KCa_4Si_8O_{20}(F,OH)$. $8H_2O$. Tetragonal.

apophysis *1.* A branch from a vein or dike to which it is attached; an epiphesis is the same, but not attached. *2.* A small dike or sill injected from a larger intrusive body into adjacent rocks.

Appalachian orogeny Late Paleozoic diastrophism beginning perhaps in the Late Devonian and continuing until end of Permian in the Appalachian orogenic belt.

apparent dip The dip of a rock layer as exposed in any section not at a right angle to the strike. It is a component of and hence always less than the true dip.

apparent movement of faults The apparent movement observed in any chance section across a fault is a function of several variables: (1) The attitude of the fault; (2) the attitude of the disrupted strata; (3) the attitude of the surface upon which the fault is observed; and (4) the true movement (net slip) along the fault.

apparent plunge Inclination of a normal projection of lineation in the plane of a vertical cross section.

apparent resistivity The electrical resistivity of rocks as measured by an array of current and voltage electrodes in a borehole or on the surface of the earth. It is equivalent to the actual resistivity if the earth is truly homogeneous. In practice it is a weighted average of resistivities. *See* RESISTIVITY

apparent velocity The velocity with which a fixed phase of a seismic wave, usually its front or beginning, passes an observer.

applanation All physiographic processes which tend to reduce the relief of a district and, dominantly, by adding material to the area or areas affected, cause the topography to become more and more plainlike.

apron Outwash plain, *q.v.*

aquamarine A transparent, light bluish-green variety of beryl. Used as a gem.

aqueous *1.* Pertaining to water. *2.* Pertaining to sediment deposited by water.

aqueous ripple marks Ripple marks made by waves and water currents as distinguished from ripple marks made by the wind, called aeolian ripple marks.

aquiclude A formation that will not transmit water fast enough to furnish an appreciable supply for a well or spring.

aquifer Stratum or zone below the surface of the earth capable of producing water as from a well.

aquifuge A rock which contains no interconnected openings or interstices and therefore neither absorbs nor transmits water.

arable Not to be confused with tillable. An arable soil is a soil that will satisfactorily produce cultivated crops.

Arachnida A large and varied group of specialized arthropods among which spiders, mites, ticks, scorpions, and Merostomata are living examples.

Arachnoidea Class of arthropods; includes Arachnida and Merostomata.

aragonite A mineral, orthorhombic, $CaCO_3$, trimorphous with calcite and vaterite.

arborescent Descriptive of large treelike plants. *Syn:* DENDRITIC, *q.v.*

arc Islands or mountains arranged in a great curve.

arch *1.* An anticline. *2.* In plutonic rocks, the planar or linear flow structures may form a dome that extends across the whole pluton. In an arch the flow structures are confined to the borders of the pluton.

archaeocyathid One of an extinct group of calcareous cup-shaped spongelike organisms found in the Lower and Middle Cambrian; worldwide in distribution.

Archean; Archaean The term, meaning ancient, has been generally applied to the oldest rocks of the Precambrian; Early Precambrian is preferable.

Archeocyathea *See* PLEOSPONGIA

Archeozoic The era during which, or during the latter part of which, the oldest system of rocks was laid down. Early Precambrian is preferable.

archipelago Any sea or broad sheet of water interspersed with many islands or with a group of islands; also, such a group of islands.

Arctic *1.* The region within the Arctic Circle (66°30′ N.). *2. Geog:* Lands north of the 10° C. July isotherm (or that of whichever month is warmest) provided the mean temperature for the coldest month is not higher than 0° C.

Arctic pack *1.* The drifting ice floes of the Arctic Ocean. *2.* A synonym for "polar ice."

arcuate Curved or bowed.

areal eruption Volcanic eruption resulting from collapse of the roof of a batholith; the volcanic rocks grade into parent plutonic rocks.

areal geology That branch of geology which pertains to the distribution, position, and form of the areas of the earth's surface occupied by different kinds of rock or different geologic units, and to the making of geologic maps.

areal map A geologic map showing the horizontal area or extent of rock units exposed at the surface.

arenaceous, *adj. 1.* Applied to rocks that have been derived from sand or that contain sand; not to be confused with siliceous. *2.* Applied to agglutinated Foraminifera, especially those agglutinated forms, *q.v.,* that use silt grains for construction of the test.

arenite; arenyte Consolidated rock of the texture of sand irrespective of composition. *Syn:* PSAMMITE

arête [*Fr.*] An acute and rugged crest of a mountain range, or a subsidiary ridge between two mountains or of a mountain spur such as that between two cirques. *See* MATTERHORN

argentiferous Containing silver.

argentite A mineral, Ag₂S. Isometric above 179° C., inverting to orthorhombic (acanthite) below

this temperature. Important ore of silver.

argillaceous Applied to all rocks or substances composed of clay minerals, or having a notable proportion of clay in their composition, as shale, slate, etc. Argillaceous rocks are readily distinguished by the peculiar, "earthy" odor when breathed on.

argillic Pertaining to clay or clay minerals. Said of a soil horizon characterized by an illuvial accumulation of clays.

argillite A rock whose degree of induration is somewhat higher than mudstone, claystone. Argillite is less indurated than shale.

arid, *adj.* Said of a climate characterized by dryness, variously defined as rainfall insufficient for plant life or for crops without irrigation; less than 10 inches of annual rainfall; or a higher evaporation rate than precipitation rate. *Syn:* DRY

aridisol Soils of the arid regions with carbonate, salt, and gypsum accumulations.

aridity, *n.* The state of a region in respect to its dryness or lack of moisture. The amount of rainfall is not a sure index, for the aridity of a region depends in part on temperature.

arithmetic mean particle diameter A measure of average particle size obtained by summing the products of the size grade midpoints times the frequency of particles in each class, and dividing by the total frequency.

Arkansas stone A true novaculite, *q.v.,* used as an oilstone for sharpening tools or instruments. Found in the Ozark Mountains of Arkansas.

arkose A sandstone containing 25% or more of feldspars usually derived from silicic igneous rocks.

arkosic Having wholly or in part the character of arkose.

arkosic sandstone A sandstone in which much feldspar is present. This may range from unassorted products of granular disintegration of fine- or medium-grained granite to a partly sorted river-laid or even marine arkosic sandstone. Has been used for various other kinds of rock, including graywacke.

arm *1.* An inlet of water from the sea or other body of water. *2.* An appendage of a starfish or a crinoid.

armored mud ball Rounded pebble or boulder originally composed of a mud core which became studded with small pebbles as the mass of mud rolled along.

array A configuration of similar detectors, such as electrodes or geophones, designed to enhance the detectability of certain transient events while suppressing other interfering events.

arrival (first, secondary, etc.) *Seis. explor:* Refers to the appearance of energy on a record traveling by way of some path under consideration. "First arrivals" indicates energy arriving with the earliest possible traveltime. "Secondary arrivals" refers to weaker or later energy returns by some other path. *Syn:* BREAK; KICK

arroyo *1.* The channel of an ephemeral or intermittent stream, usually with vertical banks of unconsolidated material 2 feet or more high. *Syn:* WADY. *2.* Vertical-walled, flat-floored channel of ephemeral stream of the semiarid Southwest.

arsenate A salt of arsenic acid, a compound containing the radical AsO^{-3} or AsO_3^{-1}.

arsenic A mineral, the native element, occurring in gray masses. Hexagonal rhombohedral.

arsenide A compound of arsenic with one other more positive element or radical.

arsenolite A mineral, arsenious oxide, As_2O_3. Cubic.

arsenopyrite A mineral, FeAsS, tin-white. Monoclinic, pseudo-orthorhombic. *Syn:* MISPICKEL

arterite A veined gneiss in which the vein-material was injected from a magma. Venite is a veined gneiss of similar aspect and composition, but differs from arterite in that the vein-material has been derived by secretion from the rock itself. Where it is impossible to discriminate between arterite and venite, the term phlebite is used.

artesian Refers to ground water under sufficient hydrostatic head to rise above the aquifer containing it.

artesian aquifer One that contains artesian water.

artesian basin A geologic structural feature or combination of such features in which water is confined under artesian pressure.

artesian spring One whose water issues under artesian pressure, generally through some fissure or other opening in the confining bed that overlies the aquifer.

artesian water Ground water that is under sufficient pressure to rise above the level at which it is encountered by a well, but which does not necessarily rise to or above the surface of the ground.

artesian well One in which the water level rises above the top of the aquifer, whether or not the water flows at the land surface.

arthrodiran One of a group of extinct fish that were abundant in the Devonian, with heavily armored heads which are movably jointed to similar armor covering the anterior part of the body.

Arthropoda Phylum of segmented animals encased in an external chitinous skeleton with jointed legs.

Articulata Class of brachiopods

in which shells generally have well-developed articulating teeth and sockets; shells calcareous.

articulate *1.* Jointed; provided with nodes or joints, or places where separation may naturally take place. *Paleontol: 2.* One of a subclass of the crinoids (the Articulata) in which the calyx is relatively flexible; *3.* One of a class of Brachiopoda (the Articulata) in which the shells are held together along the hinge line by means of articulating devices of various kinds.

articulation *1.* Movable joint. *2.* Manner of joining of adjacent mineral grains in a rock; contact may be smooth and plane, curved or sinuous, angularly interlocked or sutured, or one mineral may completely enclose another.

artifacts Man-made objects of prehistoric age, such as weapons or tools of flint.

ås A Swedish term for esker. *Pron:* auss, as in "Aussie" when pronounced by an Englishman. *Pl:* åsar. *See also* OS

åsar Plural of ås.

asbestos *1.* White, gray, green-gray, or blue-gray fibrous variety of amphibole, usually tremolite or actinolite, or of chrysotile. Blue asbestos is crocidolite. *2.* A mineral fiber with countless industrial uses; a hazardous air pollutant when inhaled.

ascension, infiltration by The theory of infiltration by ascension in solution from below considers that orebearing solutions come from the heated zones of the earth and that they rise through cavities and at diminished temperatures and pressures deposit their burdens. Hypogene.

aschistic Pertains to rocks of minor igneous intrusions that have not been differentiated into light and dark portions but that have essentially the same composition as the larger intrusions with which they are associated.

asexual Refers to any type of reproduction which does not involve the union of sex cells (gametes).

ash *1.* In coal, the inorganic residue after burning. Ignition generally alters both the weight and the composition of the inorganic matter. *2.* Fine pyroclastic material (under 4.0 mm. diameter; under 0.25 mm. diameter for fine ash). A term usually refers to the unconsolidated material, but is sometimes also used for its consolidated counterpart, or tuff. *Syn:* DUST; VOLCANIC ASH; VOLCANIC DUST

ash cone A volcanic cone built primarily of unconsolidated ash and generally shaped something like a saucer, with a rim in the form of a wide circle and a broad central depression often nearly at the same elevation as the surrounding country. They usually show maximum growth on the leeward side. Individual ash beds forming the cone dip both inward and outward, those in the high part of the rim approaching the angle of repose. Ash cones are believed to be the result of violent hydro-explosions caused when lava erupts under water or water-saturated rocks close to the surface. In form, ash cones bear a general resemblance to maars. Consolidated ash cones are called tuff cones or tuff rings.

ash fall *1.* A rain of airborne volcanic ash falling from an eruption cloud. Characteristic of vulcanian eruptions. *2.* A deposit of volcanic ash resulting from such a fall and lying on the ground surface.

ash flow *1.* An avalanche of volcanic ash, generally a highly heated mixture of volcanic gases and ash, traveling down the flanks of a volcano or along the surface of the ground and produced by the explosive disintegration of viscous lava in a volcanic crater or by the explosive emission of gas-charged ash from a fissure or group of fissures. Ash flows of the type described at Mount Pelée are considered to represent the feeblest type of the nuée ardente. The solid materials contained in a typical ash flow are generally unsorted and ordinarily include volcanic dust, pumice, scoria, and blocks in addition to ash. 2. A deposit of volcanic ash and other debris resulting from such a flow and lying on the ground surface. *Syn:* IGNIMBRITE

ash shower A rain of airborne volcanic ash falling from an eruption cloud, generally of short duration. *See* ASH FALL

ash, volcanic Uncemented pyroclastic material consisting of fragments mostly under four millimeters in diameter.

asphalt A brown to black solid or semisolid bituminous substance occurring in nature, but also obtained as a residue from the refining of certain petroleum Asphalt melts between 65° and 95° C. and is soluble in carbon disulfide.

asphalt-base petroleum Crude oils which, upon processing, yield relatively large amounts of asphaltic residues.

asphaltic sand Natural mixtures of asphalts with varying proportions of loose sand grains. The quantity of bituminous cementing material extracted from the sand may run as high as 10% and this bitumen is composed of a soft asphalt which rarely has a penetration as low as 15.6° C.

asphaltite *1.* A dark-colored, solid, difficulty fusible, naturally occurring hydrocarbon complex, insoluble in water, but more or less completely soluble in carbon disulfide, benzol, etc. 2. One of the harder of the solid hydrocarbons with melting points

between 121° and 316° C. Examples are gilsonite and grahamite.

assay 1. To test ores or minerals by chemical or blowpipe examination. To determine the proportion of metals in ores by smelting in the way appropriate to each. Gold and silver require an additional process called cupelling, for the purpose of separating them from the base metals. *See* FIRE ASSAY. 2. An examination of a mineral, an ore, or alloy differing from a complete analysis in that it determines only certain ingredients in the substance examined, whereas an analysis determines everything it contains.

assay foot The assay value multiplied by the number of feet across which the sample is taken.

assay inch The assay value multiplied by the number of inches over which the sample was taken.

assay limit The limits of an ore body as determined by assay, rather than by structural, stratigraphic, or other geologic controls.

assay ton A weight of 29.166+ grams used in assaying, for convenience. Since it bears the same relation to the milligram that a ton of 2000 pounds does to the troy ounce the weight in milligrams of precious metal obtained from the assay of an ore gives directly the number of ounces to the ton.

assay value The amount of the gold or silver, in ounces per ton of ore, as shown by assay of any given sample. Average assay value. The weighted result obtained from a number of samples, by multiplying the assay value of each sample by the width or thickness of the ore face over which it is taken, and then dividing the sum of these products by the total width of cross section sampled. The result obtained would represent an average face sample.

assemblage zone Biostratigraphic unit defined and identified by a group of associated fossils rather than by a single index fossil. *Syn:* CENOZONE. *Cf.* RANGE ZONE, FAUNIZONE

assimilation *1.* The incorporation into a magma, of material originally present in the wall rock. The term does not specify the exact mechanism or results; the "assimilated" material may be present as crystals from the original wall rocks, newly formed crystals including wall rock elements, or as a true solution in the liquid phase of the magma. The resulting rock is called hybrid. Also termed magmatic assimilation. *2.* The uptake of food material for production of new biomass.

association *1.* A climax community that is the largest subdivision of a climax, biome, or formation. *2.* Loosely, any stable community.

associations of igneous rocks Kindreds. Groups of rocks having chemical and petrographic characteristics in common, and usually occurring together. *See* PETROGRAPHIC PROVINCE

asterism *Mineral:* A starlike effect seen either by transmitted or reflected light.

Asteroidea A subclass of invertebrate animals, belonging to the phylum Echinodermata; the starfish.

asthenolith *1.* Body of magma locally melted anywhere at any time within any solid portion of the earth. *2.* Local radiogenic magma pocket. *3.* Accumulation of sialic magma of low viscosity and very small residual strength at the upper surface of the salsima layer.

asthenosphere A zone within the earth's mantle where plastic movements take place to permit isostatic adjustments. The asthenosphere begins some 50–100 km. below the surface and extends to a depth of perhaps 500 km. Presumably, plates of the lithosphere move over the asthenosphere.

astrobleme An ancient erosional scar on the earth's surface, produced by the impact of a cosmic body, and usually characterized by a circular outline and highly disturbed rocks showing evidence of intense shock. An eroded remnant of a meteoritic or cometary impact crater.

astrolabe Instrument for measuring altitudes of celestial objects. Three general types used in surveying: pendulum, planispheric, prismatic.

asymmetrical *1.* Without proper proportion of parts; unsymmetrical. *2. Crystallog:* Having no center, plane, or axis of symmetry.

asymmetrical ripple mark The normal form of ripple mark, with short downstream slopes and comparatively long gentle upstream slopes. *See* WATER CURRENT RIPPLE MARK

asymmetric fold A fold in which one limb dips more steeply than the other. If one limb becomes overturned, the term overfold or overturned fold is used.

atacamite A mineral, $Cu_4Cl_2(OH)_6$, orthorhombic, blackish green.

atectonic An adjective to describe an event that occurs when orogeny is not taking place. *Syn:* NONTECTONIC

Athabasca tar sand A tar sand, *q.v.*, of tremendous proportions, along the Peace and Athabasca rivers in Alberta, Canada.

Atlantic series, province, or suite One of two great groups of igneous rocks (along with the Pacific group), based on their tectonic setting. The Atlantic series are found in nonorogenic areas, often associated with block sinking and great crustal instability, and erupted along faults and fissures or through explosion vents. The Atlantic series was originally described as occurring in the coastal districts of the Atlantic basin. Later it became evident that there was no intrinsic connection with the Atlantic Ocean, the Hawaiian lavas, for example, being of "Atlantic" type, and hence the name intra-Pacific province is synonymous with Atlantic province.

Atlantic-type coastline A discordant coastline, especially one as developed in many areas around the Atlantic Ocean; e.g., the S.W. coastline of Ireland and the N.W. coastlines of France and Spain. *Ant:* PACIFIC-TYPE COASTLINE

Atlantis A "lost" continent once thought to exist in the Atlantic Ocean.

atmometer An instrument for measuring the rate of evaporation; an atmidometer or evaporimeter. Four main classes of atmometers may be distinguished: (1) large evaporation tanks sunk in the ground or floating on protected waters; (2) small open evaporation pans; (3) porous porcelain bodies; (4) atmometers with wet paper surfaces, represented by the Piché evaporimeter.

atmophile elements *1.* The most typical elements of the atmosphere (H, C, N, O, I, Hg, and inert gases). *2.* Elements which occur either in the uncombined state, or which, as volatile compounds, will concentrate in the gaseous primordial atmosphere.

atmosphere *1.* The gaseous envelope surrounding the earth. The atmosphere is odorless, colorless, tasteless; very mobile, flowing readily under even a slight pressure gradient; elastic, compressible, capable of unlim-

ited expansion, a poor conductor of heat, but able to transmit vibrations with considerable velocity. Its weight has been calculated as 5.9×10^{15} tons. One-half the mass of the atmosphere lies below 3.46 miles. The ordinary term for the mixture of gases comprising the atmosphere is air, which also includes water vapor and solid and liquid particles. *2.* A unit of pressure: A normal atmosphere is equal to the pressure exerted by a vertical column of mercury 760 mm. in height, at 0° C., and with gravity taken at 980. 665 cm./sec.², equal to about 14.7 pounds per square inch.

atmospheric pressure The force per unit area exerted by the atmosphere in any part of the atmospheric envelope. Some of the expressions for the normal value of the atmospheric pressure at sea level are: 76.0 centimeters of mercury; 29.92 inches of mercury; 1033.3 centimeters of water; 33.9 feet of water; 1033.3 grams per square centimeter; 1,013,250.0 dynes per square centimeter; 14.66 pounds per square inch; 1.01325 bars (1 bar=1,000,000 dynes/cm.²); 1013.25 millibars.

atmospheric radiation The radiation emitted by the atmosphere in two directions, upward to space and downward to the earth, and consisting mainly of the long-wave terrestrial radiation plus the small amount of short-wave solar radiation absorbed in the atmosphere. Figuring on the basis of a year and using a heat unit of 10^{22} calories it has been calculated that of the 201 heat units absorbed in the atmosphere 134 are returned to the earth as the so-called back radiation, and 67 are lost to space. In summer this back radiation equals or exceeds one-half of the incoming solar radiation in all northern latitudes; in winter, it exceeds the total incoming solar radiation at all latitudes above 15° N.

atmospheric water Water which exists in the atmosphere in gaseous, liquid, or solid state.

Atokan Lower Pennsylvanian, above Morrowan.

atoll A ringlike island or islands encircling or nearly encircling a lagoon. Generally composed of coral and/or calcareous algae (*Lithothamnion*).

atoll texture A ring of one mineral enclosing, or enclosed by another mineral.

atomic bond Attraction exerted between atoms and ions. Four types are: metallic, ionic or polar, homopolar or co-ordinate, residual or van der Waals. Bonding may be intermediate between these types.

atomic mass Variously but not commonly used as a synonym for atomic weight, mass number, or the mass of an individual atom.

atomic number The number of positive charges on the nucleus of an atom; the number of protons in the nucleus.

atomic proportions or ratios The ratios or proportions in which the various atomic species occur in a substance, obtained by dividing weight per cent of each substance by the atomic weight of the substance. When recalculated to atoms per 100 atoms total, the values are atom per cent.

atomic radius The radius of an atom (average distance from the center to the outermost electron of the neutral atom), commonly expressed in Angstrom units (10^{-8} cm.).

atomic weight Average relative weight of the atoms of an element referred to an arbitrary

standard of 16.0000 for the atomic weight of oxygen. The atomic weight scale used by chemists takes 16.0000 as the average atomic weight of oxygen atoms as they occur in nature; the scale used by physicists takes 16.00435 as the atomic weight of the most abundant oxygen isotope. Division by a factor of 1.000272 converts an atomic weight on the physicists' scale to the weight on the chemists' scale.

atom per cent *See* ATOMIC PROPORTIONS

attapulgite *See* PALYGORSKITE

attenuation constant A term used to describe a mathematical parameter in a material where a physical quantity of value x_0 is changed to a value x_1 by virtue of traveling a unit distance through a medium, or by virtue of the elapse of a unit time, and x_0 and x_1 are related by the equation $x_1 = x_0 e^{-p}$ where $p = \alpha + j\beta$; $\alpha : \beta$ are real and $j = \sqrt{-1}$; α is the attenuation constant; β is the phase shift or unit phase angle. A more specific geophysical definition is given in Dobrin. The relation between the initial amplitude I_0 of a seismic disturbance and the amplitude I at a distance r is given by $I = \dfrac{I_0 e^{-qr}}{r}$ where q is the attenuation constant.

Atterberg limits A collective term including liquid limit, plastic limit, and plasticity index, *q.v.*

Atterberg scale A proposed grade scale for the classification of sediments based on a decimal system beginning with 2 mm. The limits of the subclass are found by taking the square root of the product of the larger grade limits. The subdivision thus made follows the logarithmic rule. This has become the accepted European standard for classification of particle size.

attitude A general term to describe the relations of some directional feature in a rock to a horizontal plane. The attitude of planar features (bedding, foliation, joints, etc.) is described by giving the strike and dip. The attitude of a linear feature (fold axis, lineation, etc.) is described by giving the strike of the horizontal projection of the linear feature and its plunge.

attraction, gravitational A reciprocal attractive force existing between two point masses or particles of matter. The gravitational force between two bodies is directly proportional to the product of their masses and inversely proportional to the square of the distance between their centers of gravity (true of spheres of special type). Einstein's theory modifies this simple relationship expressed by Newton.

attrition 1. Wearing away by friction. 2. The wear and tear that rock particles in transit undergo through mutual rubbing, grinding, knocking, scraping, and bumping with resulting comminution in size.

aufeis [*Ger.*] A sheet of ice formed on a river flood plain in winter when shoals in the river freeze solid or are otherwise dammed so that water spreads over the flood plain and freezes.

augen [*Ger.* eyes] *Petrol:* Applied to large, lenticular-shaped minerals which in cross section have the shape of an eye. Typically include alkali feldspars in metamorphic rocks, especially gneisses.

augen gneiss A general term for gneissose rocks containing phacoidal crystals or aggregates. The augen ("eyes") may represent uncrushed fragments, or porphyroblasts.

augen schist A mylonitic rock characterized by the presence of recrystallized minerals in schistose streaks and lenticles.

augen structure A structure found in some gneisses and granites in which certain of the constituents are squeezed into elliptical or lens-shaped forms and, especially if surrounded by parallel flakes of mica, resemble eyes.

auger Any drilling device in which the cuttings are mechanically continuously removed from the bottom of the bore during the drilling operation without the use of fluids. A rotary drilling device used to drill shot holes or geophone holes in which the cuttings are removed by the device itself without the use of fluids.

augite A mineral of the pyroxene group, $(Ca,Na)(Mg,Fe,Al)(Si,-Al)_2O_6$.

aureole *Geol:* A zone surrounding an igneous intrusion in which contact metamorphism of the country rock has taken place. *Syn:* CONTACT AUREOLE; CONTACT ZONE

auri-argentiferous Containing both gold and silver, applied to minerals.

auric Of, pertaining to, or containing gold, especially when combined in its highest or triad valency, as auric chloride. $AuCl_3$.

auriferous Containing gold.

austral Southern.

autecology The study of the individual organism or species rather than the community. Life history and behavior, rather than adaptation to environment, are usually emphasized.

authigenic 1. Generated on the spot. A term applied to growth in place of occurrence. It includes secondary enlargement. 2. Pertaining to minerals formed on the spot where they are now found, before burial and consolidation of the sediment. They are the products of chemical and biochemical action.

autochthon 1. *Alpine geol:* A succession of beds that have been moved comparatively little from their original site of formation, although they may be intensely folded and faulted. 2. A fossil now occurring where the organism once lived; not transported.

autochthonous, *adj.* A term applied to rocks of which the dominant constituents have been formed *in situ,* e.g., rock salt. *Cf.* ALLOCHTHONOUS

autoclastic A term applied to rocks that have been brecciated in place by mechanical processes, e.g., brush breccias. *Syn:* PROTOCLASTIC. *See* CRUSH CONGLOMERATE

autocorrelation A special case of correlation in which a function is correlated with itself. *See* CORRELATION

autogenic succession Succession induced by biotic processes acting from within the system.

autogeosyncline A parageosyncline that subsides as an elliptical basin or trough but without associated highlands.

autointrusion A process wherein the residual liquid of a differentiating magma is injected into rifts formed in the crystallized fraction at a late stage by deformation of unspecified origin.

autolith Cognate xenolith. An inclusion or fragment of older igneous rock that is genetically related to the rock including it.

autometamorphism; automorphism 1. A type of metamorphism caused by decrease in temperature in newly congealed igneous rock in which residual hydrothermal solutions are able to react with the igneous minerals, e.g., the albitization of basalt to form spilite. 2. The alteration of an igneous rock by its own residual liquors.

autometasomatism Process of alteration of newly crystallized igneous rock by its own last, water-rich, liquid fraction which is trapped within the rock generally by an impervious chilled border.

automorphic A term applied to those minerals of igneous rocks that are bounded by their own crystal faces. Rocks that consist predominantly of an automorphic mineral assemblage are said to have an automorphic-granular or pandiomorphic-granular texture. *Syn:* IDIOMORPHIC; EUHEDRAL. *Cf.* ALLOTRIOMORPHIC; XENOMORPHIC; ANHEDRAL

autotype 1. Hypotype illustrated later by the author of a species. 2. Genotype species by original designation.

autunite A mineral, $Ca(UO_2)_2(PO_4)_2.8-12H_2O$, a common secondary mineral occurring in yellow plates. Tetragonal. Fluorescent.

auxiliary fault Branch fault; minor fault ending against a major one.

auxiliary minerals In the Johannsen classification of igneous rocks, those light-colored, relatively rare or unimportant minerals such as apatite, muscovite, corundum, fluorite, and topaz.

available moisture Moisture in soil that is available for use by plants.

available relief The vertical distance between the altitude of the original surface, after uplift, and the level at which grade is first attained.

avalanche A large mass of snow or ice, sometimes accompanied by other material, moving rapidly down a mountain slope. Avalanches are usually classified by the type of snow involved as climax, combination, damp snow, delayed action, direct action, dry snow, hangfire, and windslab avalanche. Sometimes avalanche is used to describe those landslides in which the landslide material caught a pocket of air beneath, reducing friction and resulting in incredibly rapid movement as in snow or ice avalanches.

aven A vertical shaft leading upward from a cave passage, at times connecting with passages above.

aventurine A translucent or transparent variety of quartz or feldspar containing shiny green inclusions.

average igneous rock A theoretical rock whose chemical composition is believed to be similar to the average composition of the outermost shell of the earth extending to a depth of about ten miles. This composition is calculated in different ways and there is not complete agreement as to how an average should be reached or its significance.

average velocity *Seismol:* The ratio of the distance traversed along a ray by a seismic pulse to the time required for that traverse. The average velocity is usually measured or expressed for a ray perpendicular to the reference datum plane.

Aves Class of vertebrates; birds. Jur.-Rec.

Axes, Beta *Struct. petrol:* An axis defined on a Schmidt net by the intersection of a group of great circles representing foliation surfaces. It may or may not coincide with a lineation.

axes, fabric *Struct. petrol:* Three mutually perpendicular directions in tectonites, usually denoted *a*, *b*, and *c*, which refer to the movement pattern.

axes, tectonic A general term for the a, b, c, fabric coordinates used by structural geologists and petrologists.

axial angle The acute angle between the two optic axes of a biax-

ial crystal (symbol 2V). The angle measured in air after leaving the crystal is larger than 2V and bears the symbol 2E.

axial compression *See* COMPRESSION, AXIAL

axial elements In crystallography, the ratio of unit distances along crystallographic axes and the angles between these axes.

axial plane *1.* A crystallographic plane that includes two of the crystallographic axes. *2.* As applied to folds, it is a plane that intersects the crest or trough in such a manner that the limbs or sides of the fold are more or less symmetrically arranged with reference to it. *3.* The plane of the optic axes of an optically biaxial crystal.

axial-plane cleavage Rock cleavage essentially parallel to the axial plane of a fold.

axial-plane folding Large-scale secondary folding of preexisting folds in response to movements which varied considerably from those which caused the original folding. Thus the axial planes have been folded.

axial-plane foliation Foliation developed in rocks parallel to the axial plane of a fold and perpendicular to the chief deformational pressure.

axial-plane schistosity Schistosity developed parallel to the axial planes of folds.

axial-plane separation Distance between axial planes of adjacent anticline and syncline.

axial ratio The ratio obtained by comparing the length of a crystallographic axis with one of the lateral axes taken as unity.

axial symmetry Spheroidal symmetry. *Struct. petrol:* Refers to symmetry of fabric or symmetry of movement. Spheroidal symmetry of fabric is characterized by an axis of symmetry, like an oblate or prolate spheroid. Axial symmetry of movement is typified by the settling of sediments in a body of stagnant water or the deformation of a sphere into an oblate spheroid.

axial trace The intersection of the axial plane of a fold with the surface of the earth or any other specified surface. Sometimes such a line is loosely and incorrectly called the axis.

axial trough Distortion of a fold axis downward into a form similar to a syncline.

axinite A mineral, $(Ca,Mn,Fe)_3$-$Al_2BSi_4O_{15}(OH)$, in brown, violet, or green triclinic crystals.

axis *1.* A straight line, real or imaginary, passing through a body, on which it revolves or may be supposed to revolve; a line passing through a body or system around which the parts are symmetrically arranged. *2. Crystallog:* One of the imaginary lines in a crystal that is used for reference in determining the positions and symbols of the crystal plane.

axis of a fold The line following the apex of an anticline or the lowest part of a syncline.

axis of rotation The imaginary line about which all the parts of a rotating body turn.

axis of symmetry An imaginary line in a crystal, about which it may be rotated so as to occupy the same position in space 2, 3, 4, or 6 times in a complete 360° revolution.

azeotropic mixtures A special case of gas-liquid equilibria, in binary or higher systems in which certain mixtures, upon boiling or condensation, have gas and liquid phases of identical compositions.

azimuth *1. of a body:* That arc of the horizon that is included between the meridian circle at the given place and a vertical plane passing through the body. It is measured, in surveying, from due north around to the right. *2.* The horizontal direction reckoned clockwise from the meridian plane. In this country the basic control surveys measure azimuths from the south. This is not true for all countries or all surveys. *3. of a line:* The angle which a line forms with the true meridian as measured on an imaginary horizontal circle.

azimuthal or **zenithal map projection** A map projection on which the azimuths or directions of all lines radiating from a central point or pole are the same as the azimuths or directions of the corresponding lines on the sphere.

azimuth compass A magnetic compass supplied with sights, for measuring the angle that a line on the earth's surface, or the vertical circle through a heavenly body, makes with the magnetic meridian.

azonal soils Soils which are so young that development is not complete, or soils which are developed on such barren bedrock (such as sand dunes) that development is incomplete.

azurite A mineral, $Cu_2(CO_3)$-$(OH)_2$. Monoclinic. Azure blue. A common secondary mineral found in the oxidized zone of copper deposits.

B

b- (direction) *Struct. petrol:* That direction in the plane of movement at right angles to the direction of tectonic transport. In a slickensided surface, b lies in this surface but is at right angles to the striae.

back *Min:* 1. The top or roof of an underground passage; 2. That part of a lode which is nearest the surface in relation to any portion of the workings of the mine; thus the back of the level or stope is that part of the unstoped lode which is above. 3. *Meteor.* (*verb*): To change direction counterclockwise; applied to the wind when it so changes, as for example from the north to the northwest, east to northeast, etc., in the Northern Hemisphere.

backdeep An oceanic depression on the concave side of an island arc.

backfill The material used to refill a ditch or other excavation, or the process of doing so.

background 1. The normal slight radioactivity shown by a counter, not due to abnormal amounts of radioactive elements in adjacent rocks, soils, or waters. The background count is contributed from three sources: cosmic rays, radioactive impurities in the counter, and the usual trace amounts of radioactivity in the vicinity of the counter. 2. In geochemical prospecting it refers to the range in values representing the normal concentration of a given element in a given material under investigation such as rock, soil, plants, and water.

background level With respect to air pollution, amounts of pollutants present in the ambient air due to natural sources.

backlimb More gently dipping side of an asymmetrical anticline produced by lateral thrusting.

back reef 1. Area behind a reef, between it and the land. Used with different meanings by different authors for: the reef flat, the lagoonal deposits, or for areas of deposits of sediments of land origin connecting the land with the reef. 2. The low energy side of a reef, whether landward or not, with all of the other implications.

backset eddy The ocean circulation is made up of great eddies, which in turn set up smaller eddies between the main current and the coastal border. These smaller currents revolve in the reverse direction to that of the great circulation. Eddies of this kind are appropriately called backset eddies.

backshore 1. Upper shore zone beyond the reach of ordinary waves and tides. 2. One or more nearly horizontal surfaces called berms formed landward from the beach crest; may slope inland.

backsight 1. Backsight method (plane table). A method in plane table traversing wherein orientation of the table is effected by aligning the alidade on an established map line and then rotating the table until the line of sight is coincident with the corresponding ground line. 2. A sight on a previously established survey point.

back slope *Geol:* The less-sloping side of a ridge. Contrasted with escarpment, the steeper slope. *Syn:* STRUCTURAL PLAIN

back thrusting Thrusting toward the interior of an orogenic belt. In the Appalachian Valley and Ridge Province the relative direction of thrusting has generally been northwest. Thrusts in which the relative direction of thrusting has been southeast are back thrusts.

backwash Return flow of water on a beach after the advance of a wave.

backwater *1.* A creek or series of connected lagoons parallel to a coast, separated narrowly from the sea, and communicating with it by barred outlets. *2.* A currentless body of water of the same trend as a river and fed from it at the lower end by a back flow; usually in the plural.

bacteria Single-celled microorganisms that lack chlorophyll. Some bacteria are capable of causing human, animal, or plant diseases, others are essential in pollution control because they break down organic matter in the air and in the water.

baddeleyite A mineral, ZrO_2. Monoclinic.

badlands A region nearly devoid of vegetation where erosion, instead of carving hills and valleys of the ordinary type, has cut the land into an intricate maze of narrow ravines and sharp crests and pinnacles.

baffle Any deflector device used to change the direction of flow or the velocity of water, sewage, or products of combustion such as fly ash or coarse particulate matter. Also used in deadening sound.

bahamite Consolidated limestone composed of sediment similar to that now accumulating in the Bahamas; high purity, generally fine-grained, massively bedded, widely extensive, without abundant fossils.

bail *1.* To dip or throw out; as, to bail water. *2.* To clear of water by dipping or throwing it out; as, to bail a boat. *3.* The handle of a bucket used for hoisting ore, rock, water, etc., from a mine.

bailer *1.* A long cylindrical container with a valve at the bottom, used in cable tool drilling for removing water, cuttings, mud, and oil from a well. *2.* A person who removes water from a mine by dipping it up with a bucket. *3.* A metal tank, or skip, with a valve in the bottom, used for unwatering a mine.

bajada *1.* The nearly flat surface of a continuous apron consisting of confluent alluvial fans which, together with the pediment, make up the piedmont slope in a basin. Anglicized spelling is bahada. *2.* A series of confluent alluvial fans along the base of a mountain range. It is underlain entirely by gravelly detritus that is ill sorted and poorly stratified. The convexities of the component fans impart to the bajada an undulating surface. *Syn:* COMPOUND ALLUVIAL FAN; ALLUVIAL SLOPE, *q.v.*

balanced forces A system of forces in which all forces are balanced so that no acceleration occurs.

balas A rose-red variety of spinel. Corruption of Badakhshan, a locality in Afghanistan, where it is found.

bald A high rounded knob or mountain top, bare of forest. Local in southern states.

ballas A hard globular variety of diamond.

ballast Broken stone, gravel, sand, etc., used for keeping railroad ties in place.

ball clay; pipe clay A plastic white-burning clay used as a bond in chinaware.

banco An oxbow lake or meander cut off from a river by an alteration in its course. Local in Texas.

band *1.* A stratum or lamina conspicuous because it differs in color from adjacent layers. A group of layers displaying color differences is described as being banded. *2.* A range of frequencies between prescribed limits.

banded The texture of rocks having thin and nearly parallel bands of different textures, colors, or minerals. Eutaxitic.

banded iron formation A finely banded, siliceous, hematitic (ironoxide bearing) rock of Precambrian age.

banded ore Banded texture. Ore composed of bands as layers that may be composed of the same minerals differing in color or textures or proportions, or they may be composed of different minerals.

banded structure A term applied to veins having distinct layers or bands. This may be due to successive periods of deposition, or replacement of some earlier rock.

banded textures Banded ores, *q.v.*

banded vein A vein made up of layers of different minerals parallel with the walls. Also called ribbon vein.

bank *1.* The rising ground bordering a lake, river, or sea; on a river, designated as right or left as it would appear facing downstream. *2.* An elevation of the sea floor of large area, surrounded by deeper water, but safe for surface navigation; a submerged plateau or shelf, a shoal, or shallow.

bank deposits Shoal water, local mounds, ridges and terraces of sediments, rising above the surrounding sea bottom and of more limited extent than the blanket deposits of shelf areas.

bankfull or **bankfull stage** The water surface elevation attained by the stream when flowing at capacity, i.e., stage above which banks are overflowed.

bar *1.* A mass of sand, gravel, or alluvium deposited on the bed of a stream, sea, or lake, or at the mouth of a stream forming an obstruction to water navigation. *2.* A term used in a generic sense to include various types of submerged or emergent embankments of sand and gravel built on the sea floor by waves and currents. *3.* An offshore ridge or mound of sand, gravel, or other unconsolidated material submerged at least at high tide, especially at the mouth of a river or estuary, or lying a short distance from and usually parallel to, the beach. *4. Meteor:* A unit of pressure equal to 10^6 dynes/cm.^{2}; equivalent to a mercurial barometer reading of 750.076 mm. at 0° C. (or 29.5306 inches at 32° F.), gravity being equal to 980.616 cm./sec.^{2}. It is equal to the mean atmospheric pressure at about 100 meters above mean sea level. The standard atmospheric pressure of 760 mm. or 29.921 inches is equal to 1,013,250.1444 dynes/cm.^{2} or 1,013.3 milibar. *5. Paleontol:* Any conodont with one usually large denticle at one end above main part of escutcheon, with discrete denticles and generally with an anticusp, anterior process or lateral process. That portion of the conodont which holds the denticles.

Barbados earth A deposit consisting of fossil radiolarians. *See* TRIPOLI

barchan *1.* A dune having crescentic ground plan, with the convex side facing the wind; the profile is asymmetric, with the gentler slope on the convex side, and the steeper slope on the concave or leeward side. 2. The crescent or barchan type is most characteristic of the inland desert regions. It presents a gently convex surface to the wind, while the lee side is steep and abrupt.

bar finger A long narrow sand body of lenticular cross section underlying a distributary channel in a bird foot delta. The sand body, which is several times wider than the distributary channel, is produced by the seaward advance of the lunate bar at the distributary mouth.

barite; baryte A mineral, $BaSO_4$. Orthorhombic. Sp. gr. 4.5. The principal ore of barium, also used in paints and drilling muds.

barite dollar Rounded disk-shaped masses of barite formed in a sandstone or sandy shale. *Cf.* BARITE ROSE; PETRIFIED ROSE

barite rose; barite rosette Petrified rose, *q.v.*

barnacle Member of the Cirripedia; sessile, shelled or burrowing in shells of other animals, an arthropod.

barograph A barometer which makes a continuous record of barometric changes. Barographs may be of the mercurial or aneroid variety, but are generally of the latter type.

barometer An instrument for measuring atmospheric pressure.

barometric elevation In surveying, an elevation above mean sea level established by the use of instruments which involve measuring the difference in air pressure between the point in question and some reference base of known value, whose elevation is based on a more precise type of data.

barometric pressure Atmospheric pressure as indicated or measured by a barometer.

barred basin *See* SILLED BASIN

barrel, oil A volumetric unit of measurement equivalent to 42 U.S. gallons (158.76 liters).

barrens An area relatively barren of vegetation in comparison with adjacent areas because of adverse soil or climatic conditions, or wind, or other adverse environmental factors—for example, sand barrens or rock barrens.

barrier beach Offshore bar. This term refers to a single elongate sand ridge rising slightly above the high-tide level and extending generally parallel with the coast, but separated from it by a lagoon. The term should apply to islands and spits. *Cf.* BARRIER ISLAND

barrier flat The relatively flat area, often occupied by pools of water, separating the exposed or seaward edge of a barrier and the lagoon behind the barrier.

barrier ice Shelf ice.

barrier iceberg Tabular iceberg broken off from an ice shelf or piedmont ice afloat.

barrier island Preferred by some to offshore bar. 1. Similar to a barrier beach but consisting of multiple instead of single ridges and commonly having dunes, vegetated zones, and swampy terraces extending lagoonward from the beach. 2. A detached portion of a barrier beach between two inlets.

barrier reef A coral reef that is separated from the coast by a lagoon that is too deep for coral growth. Generally, barrier reefs follow the coasts for long distances, often with short interruptions, termed passes.

barriers 1. A continuous offshore ridge built by the shore drift. The barrier follows the line of breakers instead of the shoreline. *Cf.* BARRIER BEACH, *q.v.* 2. Ice shelf. *3.* Ice shelf in some particular locality, e.g., Ross Barrier. Barrier is no longer used in official British publications and maps, being replaced by ice shelf and ice front. *4.* In an ecological sense, those physical, chemical, climatic, or biologic factors which are foreign to an organism's native habitat and exceed the tolerance of that organism.

Barrovian metamorphism Regional metamorphism that can be zoned into metamorphic facies.

bar screen In waste water treatment, a screen that removes large floating and suspended solids.

bar theory A theory advanced by Ochsenius in 1877 to account for thick deposits of salt, gypsum, and other evaporites. The theory assumes a lagoon separated from the ocean proper by a bar. As water is lost by evaporation and evaporites are formed in the lagoon, additional water of normal salinity flows from the ocean. Because some water in the lagoon is evaporating, the salinity there constantly increases, and finally reaches a point where gypsum, salt, and other evaporites are deposited.

barysphere Centrosphere, *q.v.*

basal cleavage A type of isometric crystal cleavage in which the cleavage occurs parallel to the basal pinacoid.

basal conglomerate A coarse, usually well-sorted and lithologically homogeneous sedimentary deposit which is found just above an erosional break.

basal pinacoid A crystal form consisting of only 2 parallel facies, so oriented as to cut the vertical axis $\underline{c}$ and to be parallel with planes of the lateral axes $\underline{a}$ and $\underline{b}$.

basal plane Basal pinacoid, *q.v.*

basalt *1.* An extrusive rock composed primarily of calcic plagioclase and pryoxene, with or without olivine. The plagioclase is normally zoned and usually ranges in composition from bytownite to labradorite, but less calcic varieties are known. Augite, pigeonite, and hypersthene or bronzite are the common pyroxenes. Apatite and magnetite are almost always present as accessories. Basalts rich in olivine and calcic augite are generally classified as olivine basalts; those poor in olivine and containing orthopyroxene and/or pigeonite are generally classified as tholeiitic basalts or tholeiites. The groundmass of tholeiitic basalts is commonly glassy, or if crystallized, usually contains quartz and alkalic feldspar. *2.* More generally, any fine-grained, dark-colored igneous rock.

basaltic crystal layer or shell An inner layer of worldwide extent, composed of basalt, underlying the oceans, and the granitic continents.

basal till Till carried at or deposited from the under surface of a glacier.

basanite An extrusive rock composed of calcic plagioclase, augite, olivine, and a feldspathoid (nepheline, leucite, or analcime). Essentially a feldspathoidal olivine basalt.

base *1.* A substance whose water solution has a bitter taste and a soapy feel and changes the color of certain organic dyes. *2.* A substance containing the OH radical which dissociates to form OH^- ions when dissolved in water. *3.* A substance capable of accepting protons from a donor. *4.* A cation (e.g., base exchange). *5.* A base metal or a base metal oxide (e.g., a rock with a high content of bases). *6.* A substance with basic properties. *7.* A substance capable of combining with silica in a rock, such as lime, potash, etc.

base correction In exploration,

particularly in magnetics or gravity or barometric surveys, where a base station is used, the base correction is the adjustment required to reduce measurements made in the field so that they can be expressed with reference to the base station values.

base exchange A reaction in which cations absorbed on the surface of clay or zeolite crystals are replaced from cations in the surrounding solution.

base level The theoretical lowest level of erosion of a portion of the earth's land surface. In general, the ultimate base level of erosion of the land surface is sea level, but this limit is rarely attained. Temporary base levels may occur along a stream course by establishment of lakes.

base line A very accurately surveyed line on the earth's surface, the exact length and position of which have been precisely determined. This survey line is used as a reference for accurately computing the distances and positions of remote points and objects.

base map A map on which information may be placed for purposes of comparison or geographical correlation.

basement *1.* Complex, generally of igneous and metamorphic rocks, overlain unconformably by sedimentary strata. *2.* Crustal layer beneath a sedimentary layer and above the Mohorovičić discontinuity.

basement complex A series of rocks generally with complex structure beneath the dominantly sedimentary rocks. In many places they are igneous and metamorphic rocks of either Early or Late Precambrian, but in some places may be much younger, as Paleozoic, Mesozoic, or even Cenozoic.

base metal A metal commonly used in industry by itself rather than alloyed with other metals. Generally considered to be one of the following: copper, lead, zinc, tin, or mercury.

base net (triangulation) The triangle formed by sighting a third point from the two ends of a base line; or two adjacent triangles with the base line being a common side to each. The base net is the initial figure in a triangulation system.

base of weathering In seismic interpretation, the boundary between the low-velocity surface layer and an underlying comparatively high-velocity layer. This may correspond to the geologic base of weathering, but not necessarily. The boundary may vary with time. The boundary is considered in deriving time corrections for seismic records.

base station In exploration, particularly magnetic or gravity or barometric surveys, a reference station where quantities under investigation have known values or may be under repeated or continuous measurement in order to establish additional stations in relation to it.

basic *1. Chem:* Performing the office of a base in a salt; having the base in excess. *2.* Having more than one equivalent of the base for each equivalent of acid. *3. Geol:* A general descriptive term for those igneous rocks that are comparatively low in silica. About 55% or 50% is the superior limit. *Cf.* ACIDIC. *4.* In furnace practice, a slag in which the earthy bases are in excess of the amount required to form a "neutral" slag with the silica present. *5.* In water solutions, having a pH greater than 7 (not necessarily identical with *2* above).

basic borders in igneous rocks

Refers to the occurrence of more basic rocks at the margins of igneous intrusions. Variously interpreted as chilled zones, basic fronts, etc.

basic front In granitization, an advancing zone enriched in calcium, magnesium and iron, which is said to represent those elements in the sediments being granitized, over and above those necessary to form granite.

basic rock A term rather loosely used in lithology to mean generally one of the following: *1.* An igneous rock containing 45% to 52% of silica, free or combined. *2.* An igneous rock in which minerals comparatively low in silica and rich in the metallic bases, such as the amphiboles, the pyroxenes, biotite, and olivine are dominant. *3.* Very loosely, an igneous rock composed dominantly of dark-colored minerals. In all three senses contrasted with acid.

The term is misleading and undesirable and is going out of use. As used in the first sense above it is being replaced by subsilicic and as used in the second sense it should be replaced by mafic or by some term denoting the dominant mineral or minerals. *See* BASIC. *2.* As used in the third sense it should be replaced by melanocratic.

basification The development of a more basic rock, commonly richer in hornblende, biotite, and oligoclase, presumably by the contamination of a granitic magma by assimilation of country rock. This phenomenon occurs chiefly at the margins of the granite mass.

basin *1.* An amphitheater, cirque, or corrie. Local in Rocky Mountains. *2.* An extensive depressed area into which the adjacent land drains, and having no surface outlet. Use confined almost wholly to the arid West. *3.* The drainage or catchment area of a stream or lake. *4. Struct geol:* A syncline that is circular or elliptical in plan; i.e., the outcrop of each formation is essentially circular or elliptical, and the beds dip inward.

basin, starved A depositional basin which received a thinner section of deposits than adjoining areas because the rate of subsidence was materially greater than the rate of deposition.

basin, structural A synclinal tract or area in which the rocks dip generally toward a central point, and in which folding occurred subsequent to deposition.

basin and range landscape Landscape consisting of fault-block mountains and intervening basins.

basin and range structure Regional structure dominated by fault-block mountains separated by sediment-filled basins.

basin folds Anticlinal and synclinal folds occurring in structural basins and regarded by some as due to differential settling.

basining *Geol:* A settlement of the ground in the form of basins, in many cases, at least, due to the solution and transportation of underground deposits of salt and gypsum. Such basining produces numerous depressions, from those of a few square yards to those 129.5 km.^{2} in area, in the high-plains region east of the Rocky Mountains.

basin order *Geomorph:* A first order basin contains all of the drainage area of a first order stream; a second order basin contains all of the drainage area of a second order stream, etc. *See* STREAM ORDER

basin range A kind of mountain range characteristic of the Great

Basin province and formed by a faulted and tilted block of strata.

basin sedimentary A segment of the earth's crust which has been downwarped, usually for a considerable time, but with intermittent risings and sinkings. The sediments in such basins increase in thickness toward the center of the basin.

batholith *1.* Originally defined in 1895 as a stock-shaped or shield-shaped mass of igneous rock intruded as the fusion of older formations. On removal of its rock cover and on continued denudation, this mass holds its diameter or grows broader to unknown depths. *2.* A body of intrusive rock at least 40 sq. mi. in area.

bathometer An instrument for measuring depths of water.

bathyal *1.* Pertaining to the benthonic environment on the continental slope, ranging in depth from 200 to 2000 meters. *2.* Pertaining to the bottom and overlying waters between 100 and 1000 fathoms (600 and 6000 feet). *3.* Of, or pertaining to, the deeper parts of the ocean; deep sea.

bathymetric Relating to measurement of depths; usually applied to the ocean.

bathymetric chart A topographic map of the bed of the ocean.

bathypelagic Referring to that portion of the deep waters of the ocean which lie between depths of 200 and 2000 meters.

bathyplankton The plankton of the greater depths, especially the abyssal zone.

bathythermograph A torpedo-shaped instrument towed behind a ship that continuously registers the temperature of the sea water.

battery ore Manganese oxide ore suitable for use in dry cells.

batture Elevated river bed, as where a river is confined by natural levees above flood-plain level.

Baumé gravity *See* GRAVITY, BAUMÉ

bauxite *1.* A rock composed of one or more aluminum hydroxide minerals (boehmite, gibbsite, or diaspore) and impurities in the form of silica, clay, silt, and iron hydroxides. Generally formed in tropical and subtropical latitudes under conditions of good surface drainage. The principal ore of aluminum. *2.* Used collectively for lateritic aluminous ores.

bauxitic latosols Red and reddish soils leached of soluble minerals and of iron and silica, but retaining oxides of aluminum.

bauxitization Development of bauxite from either primary aluminum silicates or secondary clay minerals.

Baveno law *See* TWIN LAW

b-axis *Struct. petrol:* A fabric axis normal to the direction of movement or a-axis. It lies in the foliation surface (ab), if such is developed, and in many instances coincides with a lineation, as along fold-axes or intersections of s-planes. In a slickensided surface, b is at right angles to the striae (lineation).

bay *1.* A recess in the shore or an inlet of a sea or lake between two capes or headlands, not as large as a gulf but larger than a cove. *Cf.* BRIGHT; EMBAYMENT. *2.* A swampy area, usually oval-shaped and covered with brush; local on South Atlantic Coast. *3. Min:* An open space for waste between two packs in a longwall working.

bay bar *See* BAYMOUTH BAR

bay head Southern United States. A swamp at the head of a bay.

bayhead bar A bar built a short distance out from the shore at the head of a bay.

baymouth bar A bar extending partially or entirely across the mouth of a bay.

bayou A lake, or small sluggish secondary stream, often in an abandoned channel or a river delta. Local on Gulf Coast. One of the half-closed channels of a river delta. Local on Mississippi Delta.

bc-fracture *Struct. petrol:* A tension fracture parallel with the bc fabric plane and normal to a. The orientation of these fractures subnormal to fabric axis a affords a criterion for direction sense of shear.

bc-plane *Struct. petrol:* A plane that is perpendicular to the plane of movement and parallel to the b-direction in that plane, i.e., it is perpendicular to a, the direction of tectonic transport.

Bé° Abbreviation for Baumé degree.

beach The gently sloping shore of a body of water which is washed by waves or tides, especially the parts covered by sand or pebbles.

beach berm Nearly horizontal bench or narrow terrace formed by wave action in unconsolidated material on the backshore of a beach with surface rising behind it and sloping off in front. Some beaches have no berm, others have more than one.

beach cusp Cuspate deposits of beach material built by wave action along the foreshore. Sand, gravel, or coarse cobblestones are heaped together in rather uniformly spaced ridges which trend at right angles to the sea margin, tapering out to a point near the water's edge.

beach face The section of the beach normally exposed to the action of the wave uprush. The foreshore zone of a beach. *See* SHORE FACE

beach placers Placer deposits either on a present or ancient sea beach. There are a series of these at Nome, Alaska, known as first, second, or third beach, etc., due to change of shore line.

beach plain An irregular surface consisting of successive embankments added to a growing compound spit by longshore currents. The embankments may be closely spaced or widespread with lagoons between them.

beach profile of equilibrium A profile normal to the length of a beach and concave upward. The slope is steep above normal high water and more gentle seaward.

beach ridge An essentially continuous mound of beach material behind the beach that has been heaped up by wave or other action. Ridges may occur singly or as a series of approximately parallel deposits. In England they are called fulls.

beach scarp An almost vertical slope along the beach caused by erosion by wave action. It may vary in height from a few inches to several feet, depending on wave action and the nature and composition of the beach.

bead *1.* The globule of precious metal obtained by the cupellation process. *2.* A glassy drop of flux, as borax, used as a solvent for a color test for various elements before the blowpipe.

beaded drainage Pattern of short minor streams connecting small pools, characteristic of an area underlain by permafrost.

beak *Paleontol:* 1. The generally pointed extremity of a brachiopod or pelecypod shell which marks the beginning of shell growth; 2. The prolongation of certain univalve shells containing the canal (as in a gastropod). (Not in general usage in this sense by paleontologists.) 3. *Bot:* A long, prominent, and substantial point; applied particularly to prolongations of fruits and pistils.

Beaman stadia arc; Beaman arc An auxiliary attachment on an alidade consisting of a stadia arc, mounted on the outer side of the ordinary vertical arc, and enabling the observer to determine differences in elevation of the instrument and the stadia rod without use of vertical angles.

bean ore A name of limonite, when found in lenticular aggregations. Called also pea ore, when found in small, rounded masses. A coarse-grained pisolitic iron ore.

bearing The direction of a line with reference to the cardinal points of the compass. True bearing: The horizontal angle between a ground line and a geography meridian. A bearing may be referred to either the south or north point. (N. 30° E., or S. 30° W.) Magnetic bearing: The horizontal angle between a ground line and the magnetic meridian. A magnetic bearing differs from a true bearing by the exact angle of magnetic declination of the locality.

beat A periodic pulsation caused by the simultaneous occurrence of two waves, currents, or sounds of slightly different frequency; to cause two waves of slightly different frequency to be opposed; "beat frequency," that frequency which is the difference of two different frequencies.

Beaufort wind scale A system of estimating wind velocities, originally based (1806) by its inventor, Admiral Sir Francis Beaufort of the British Navy, on the effects of various wind speeds on the amount of canvas which a full-rigged frigate of the early 19th century could carry; since modified and widely used in international meteorology.

Becke line A bright line, visible under the microscope, that separates substances of different refractive indices.

Becke test *Opt. mineral:* A test used under the microscope for comparing indices of refraction. The so-called Becke line appears to move toward the material (i.e., mineral or immersion liquid) of higher refractivity as the tube of the microscope is raised.

bed *1.* The smallest division of a stratified series, and marked by a more or less well-defined divisional plane from its neighbors above and below. *2.* A seam or deposit of mineral, later in origin than the rock below and older than the rock above, i.e., a regular member of the series of formations, and not an intrusion. A deposit, as of ore (or coal), parallel to the stratification. *3.* That portion of an outcrop or face of a quarry which occurs between two bedding planes. *4.* The level surface of rock upon which a curb or crib is laid. *5.* *Geophys:* A rock mass usually of large horizontal extent compared to vertical or near-vertical thickness, bounded, especially on its upper side, by material with different physical properties. *6.* The floor or bottom on which any body of water rests.

bedded Applied to rocks resulting from consolidated sediments, and accordingly exhibiting planes of separation designated bedding planes.

bedded deposit *1.* Any stratified deposit. *2.* *Econ. geol:* Blanket deposit.

bedding Collective term signifying existence of beds or laminae. Planes dividing sedimentary rocks of the same or different lithology. Structure occurring in granite and similar rocks evident in a tendency to split more or less horizontally or parallel to the land surface.

bedding cleavage Cleavage that is parallel to the bedding.

bedding fault A fault that is parallel to the bedding.

bedding fissility A term generally restricted to primary foliation parallel to the bedding of sedimentary rocks, i.e., it forms while the sediment is being deposited and compacted. It is the result of the parallelism of the platy minerals to the bedding plane, partly because they were deposited that way and partly because they were rotated into this position during compaction.

bedding joint Joint parallel to bedding.

bedding plane In sedimentary or stratified rocks, the division planes which separate the individual layers, beds, or strata.

bedding-plane slip The slipping of sedimentary strata along bedding planes during folding.

bedding schistosity Schistosity that is parallel to the bedding.

Bedford limestone Mississippian limestone quarried extensively near Bedford, Indiana, for building purposes.

bed load 1. Soil, rock particles, or other debris rolled along the bottom of a stream by the moving water, as contrasted with the "silt load" carried in suspension. 2. That part of the total sediment load of a stream composed of all particles greater than a limiting size whether moving on the bed or in suspension; includes all bed material in movement.

bed material The material of which the bed is composed, and may be the result of either suspended or bed-load movement, or both, or, in some cases, may be even residual.

bedrock 1. The solid rock underlying auriferous gravel, sand, clay, etc., and upon which the alluvial gold rests. 2. Any solid rock exposed at the surface of the earth or overlain by unconsolidated material.

beekite 1. A concretionary form of calcite, occurring commonly in small rings on the surface of a fossil shell (coral, sponge, etc.), which has weathered out of its matrix. 2. Chalcedony occurring in the form of subspherical discoid, rosettelike or doughnut-shaped accretions, generally intervoluted as bands or layers and commonly found on silicified fossils and on joint planes.

behead *Geol:* To cut off and capture by erosion the upper portion of a watercourse. Said of the encroachment of a stronger stream upon a weaker one.

beheaded stream In stream piracy the stream from which water has been diverted.

beidellite An aluminum-rich member of the montmorillonite (smectite) group.

belemnite An exinct type of cephalopod known from cigar-shaped fossils.

belt A zone or band of a particular kind of rock strata exposed on the surface. *Cf.* ZONE. An elongated area of mineralization.

belted plain A coastal plain feature found in an area of essentially horizontal or slightly dipping strata where differential erosion causes the durable strata to form belts of hilly land a few feet higher than the lower surface. Such a land surface, found both on recent coastal plains, as in eastern United States, and on older plains, as in interior New York, and in the Paris Basin, France, is known as a belted plain.

bench 1. A strip of relatively level earth or rock, raised and narrow. A small terrace or comparatively level platform breaking the continuity of a declivity. 2. A level or gently sloping erosion plane inclined seaward. 3. A nearly horizontal area at about the level of maximum high water on the sea side of a dike. 4.

One of two or more divisions of a coal seam separated by slate, etc., or simply separated by the process of cutting the coal, one bench or layer being cut before the adjacent one. 5. A level layer worked separately in a mine. 6. An elongated area of mineralization usually marked by a characteristic mineralogy or structure.

bench mark A relatively permanent material object, natural or artificial, bearing a marked point whose elevation above or below an adopted datum (such as sea level) is known. The usual designation is B.M. or P.B.M. (permanent bench mark). A temporary or supplemental bench mark (T.B.M.) is of a less permanent nature and the elevation may be less precise.

bench placers Placers in ancient stream deposits from 50 to 92 m. above present streams.

bend 1. A curve in a river channel whose lateral changes involve a decrease in radius. Bends generally grow into meanders. 2. In Cornwall, indurated clay; a term applied by the miner to any hardened argillaceous substance.

beneficiate To improve the grade or ore by milling, sintering, etc.

Benioff zone Subduction zone; a dipping zone containing earthquake foci.

benthic Benthonic. Includes all of the bottom terrain from the shore line to the greatest deeps.

benthonic 1. Refers to the bottom of a body of standing water. *Cf.* PELAGIC. 2. Of or pertaining to sea-floor types of life or marine bottom-dwelling forms of life. Pertaining to the benthos, *q.v.*

benthos 1. The life dwelling on the bottom of the sea. Also applied to deepest part of a sea or ocean. 2. Bottom-dwelling forms of marine life, either in fixed position or in attachment to the substratum, or capable of crawling, burrowing, or swimming on, in, or above the substratum. 3. The bottom of the sea, especially of the deep oceans.

bentonite A sedimentary rock formed from the alteration in place of volcanic ash. Largely composed of the clay mineral montmorillonite. The rock commonly has great ability to absorb water and swell. Used commercially in drilling fluids, catalysts, paint, and plastic fillers, etc.

berg 1. A hill or mountain. Local in Hudson River Valley. 2. An iceberg.

bergschrund The crevasse occurring at the head of a mountain glacier, which separates the moving snow and ice of the glacier from the relatively immobile snow and ice adhering to the headwall of the valley. It commonly penetrates to the rock face of the headwall.

berg till A glacial deposit having the resemblance of both till and lacustrine clays which formed from materials rafted into ice border lakes by icebergs.

berm 1. Terraces which originate from the interruption of an erosion cycle with rejuvenation of a stream in the mature stage of its development and renewed dissection, leaving remnants of the earlier valley floor above flood level. 2. A nearly horizontal portion of the beach or backshore formed by the deposit of material by wave action. Some beaches have no berms, others have one or several.

berm crest The seaward limit and generally the highest point of a marine berm. *Syn:* BERM EDGE

Bertrand lens A removable lens in the tube of a petrographic microscope that is used in conjunction with convergent light to form interference figures.

beryl A mineral, $Be_3Al_2Si_6O_{18}$,

but commonly containing 6% or less of total Na_2O, Li_2O, and Cs_2O, and about 2% H_2O. Hexagonal. The main ore of beryllium, occurring in granitic pegmatites. Emerald and aquamarine are gem varieties.

beryllium A metal that when airborne has adverse effects on human health; it has been declared a hazardous air pollutant.

beta particle An electron.

beta quartz Quartz formed at a temperature between 573° C. and 870° C. The commonest examples are the bipyramidal quartz crystals found as phenocrysts in quartz porphyries.

beta radiation The emission of either an electron or a positron by an atomic nucleus. If the particle emitted is an electron, the emission is said to be $\beta-$ emission; if a positron, it is said to be $\beta+$ emission. That portion of radiation from a radioactive source which could be strongly deflected by a perpendicular magnetic field.

bev Abbreviation for billion electric volts.

beveling The planing by erosion of the outcropping edges of strata. When the observer travels in the direction of dip, he crosses successively younger beds.

B-girdle Circular pattern in petrofabric diagrams indicating a b-axis.

B horizon Illuvial horizon in soils that are illuviated. The lower soil zone which is enriched by the deposition or precipitation (accretion) of material from the overlying zone or A horizon.

biaxial Crystals having two optic axes and 3 indices of refraction. A property possessed by all those crystals in the orthorhombic, monoclinic, and triclinic systems.

biaxial indicatrix Ellipsoid whose three axes at right angles to each other are proportional in length to the indices of refraction of a biaxial crystal.

bicarbonate A salt containing a cation and the radical HCO_3, e.g., $NaHCO_3$.

bichromate Dichromate.

bifurcate, *adj.* Forked as some Y-shaped hairs, stigmas, or styles.

"big bang" hypothesis The hypothesis that the presently observed expansion of the Universe may be extrapolated back to a primeval cosmic fireball. Depending on the ration of the initial expansion velocity to the mass of the Universe, which is relatable to presently observable parameters (the deceleration parameter), the Universe may or may not reach a maximum distension and collapse in on itself again. *Syn:* FIREBALL HYPOTHESIS; PRIMEVAL-FIREBALL HYPOTHESIS

bight A bend or curve, as in a river or mountain chain; a bend in a coast forming an open bay; also the bay itself.

bilateral symmetry With the individual parts arranged symmetrically along the two sides of an elongate axis, as in the earthworm, cat, etc.

billow A wave, especially a great wave or surge of water.

binary, *adj.* Composed of two elements, of an element and a radical that acts like an element, or of two such radicals. Thus $NaCl, Na_2O, Na_2SO_4$ and $(NH_4)_2$-SO_4 are all binary compounds.

binary granite *1.* A granite consisting of quartz and feldspar only. *2.* A granite containing both biotite and muscovite mica.

binary system A system consisting of two components, e.g., the system $MgO-SiO_2$.

binomen Name consisting of two words such as the name of a species, first a generic name and second a specific name.

binomial Irrespective of the nature of the concept by which species are circumscribed, the unit must fit into the binomial system of nomenclature. It must have a Latin name and that Latin name, composed of two words —the generic name and the specific name—is the binomial.

binomial system System by which organisms are known by first a generic name and second a specific or trivial name; subspecies or varieties receive a third name.

biochemical Refers to chemical processes or substances related to or produced by the activity of living organisms.

biochemical deposit A precipitated deposit resulting directly or indirectly from vital activities of an organism, such as bacterial iron ores and limestones.

Biochemical Oxygen Demand (BOD) The amount of oxygen required by the biological population of a water sample to oxidize the organic matter in that water. It is usually determined over a 5-day period under standardized laboratory conditions and hence may not represent actual field conditions.

biochron Geologic time unit corresponding to biostratigraphic range zone.

biochronology 1. Geologic time scale based on fossils. 2. Study and relations between geologic time and organic evolution. 3. Dating of geologic events by biostratigraphic evidence.

bioclastic Refers to rocks consisting of fragmental organic remains.

biocoenose An assemblage of organisms that live together as an interrelated community. A natural ecological unit.

biodegradable The process of decomposing quickly as a result of the action of microorganisms.

bioecology A collective term that includes plant and animal ecology as one discipline.

biofacies 1. Lateral variations in the biologic aspect of a stratigraphic unit. 2. Assemblages of animals or plants formed at the same time under different conditions.

biogenesis 1. Formation by the action of organisms. 2. The doctrine that all life has been derived from previously living organisms.

biogenetic law Ontogeny. The so-called "law" of recapitulation: Ontogeny recapitulates phylogeny.

biogenic Pertaining to a deposit resulting from the physiological activities of organisms. The rock thus formed is designated a biolith.

biogeochemical cycling The cycling of chemical constituents through a biological system.

biogeography The study of geographical distribution of plants and animals and the reasons for their distribution.

bioherm 1. A moundlike or circumscribed mass built exclusively or mainly by sedentary organisms such as corals, stromatoporoids, algae, etc., and enclosed in normal rock of different lithological character. 2. An organic reef or mound built by corals, stromatoporoids, gastropods, echinoderms, Foraminifera, mollusks, and other organisms.

biolithite A limestone bound together by the frames of organisms; a boundstone; reef rock that is bound by an organic framework.

biological magnification The concentration of certain substances up a food chain. A very important mechanism in concentrating pesticides and heavy metals in organisms such as fish.

biologic facies 1. Particular association of organisms. 2. Paleontologic nature of a stratigraphic unit. 3. Rocks or sediments characterized by their biologic content.

biologic species Species whose recognition is based on biologic relations, particularly the willingness or ability of individuals to interbreed.

biology The study of all organisms; includes neontology and paleontology.

biomass Total mass of living organisms per unit area or unit volume, preferably the latter, per unit time.

biome A major climax community composed of plants and animals. Equivalent to climax or formation. A major ecologic zone or region corresponding to a climatic zone or region. A major community of plants and animals associated with a stable environmental life zone or region (e.g., Northern Coniferous Forest or Great Plains).

biomechanical deposit A deposit due to the detrital accumulation of organic material, as in the cases of limestones and coal.

biometrics The application of measurement and statistics to biologic studies.

biomicrite A limestone consisting of a variable proportion of fossil skeletal debris and carbonate mud. When using the term, the major organism should be specified; e.g., "crinoid biomicrite." *See* MICRITE

biophages Those organisms which obtain nourishment from other organisms (e.g., predators, parasites, and pathogens).

biophile An element which is required by or is found in the bodies of living organisms. The list of such elements includes C, H, O, N, P, S, Cl, I, Br, Ca, Mg, K, Na, V, Fe, Mn, and Cu. All may belong also to the chalcophile or lithophile groups.

biosome A body of sediment or three-dimensional rock mass consisting of uniform paleontologic content and deposited under uniform biological conditions; the biostratigraphic equivalent of lithosome.

biosparite A sparite, *q.v.*, with allochems derived from fossils.

biospecies *1.* A species living at the present time all of whose characters, relations, reactions, and activities can be observed. *2.* Group of gamodemes capable of interbreeding.

biosphere *1.* Zone at and adjacent to the earth's surface where all life exists. *2.* All living organisms of the earth.

biostratigraphic Referring to stratigraphic phenomena defined on fossil content.

biostratigraphic unit Rock stratigraphic unit defined and identified by contained fossils without regard to lithologic or other physical features or relations; *cf.* LITHOSTRATIGRAPHIC UNIT

biostratigraphic zone A set of strata characterized by a particular fossil or group of fossils.

biostratigraphy The study of layered rocks by utilizing fossils.

biostrome A bedded structure composed of shell beds, crinoid beds, coral beds, etc., which was built by sedentary organisms grown and preserved in place. In contrast to bioherms they lack moundlike or lenslike form. *Cf.* LUMACHELLE

biota The animal and plant life of a region; flora and fauna collectively.

biotic assemblages map A derivative map for planning that combines for any one area the reaction of soil and substrate and biota to the activities of man.

biotic community *1.* All of the plant and animal populations occupying a given area, usually named after the dominant (in size or numbers) plant or animal in area. *2.* An integrated, mutually adjusted assemblage of organisms inhabiting a natural area. The assemblage may or may not be self-sufficient and is considered to be in a state of dynamic equilibrium.

biotic factors Factors of a biological nature such as availability of food, competition between species, predator-prey relationship, etc., which, besides the purely physical and chemical factors, also affect the distribution and abundance of species.

biotic potential *1.* The maximum reproduction power or ability. The inherent property of an organism to reproduce and survive in greater numbers. *2.* The ability of an organism to reproduce in an optimum, unrestricted, noncompetitive environment.

biotic succession The natural replacement of one or more groups of organisms occupying a specific habitat by new groups. The preceding groups in some ways prepare or favorably modify the habitat for succeeding groups.

biotite A mineral $K(Mg,Fe^{+2})_3$- $(Al,Fe^{+3})Si_3O_{10}(OH)_2$, member of the mica group. A common rock-forming mineral. Monoclinic, perfect basal (001) cleavage. Black in hand specimen, brown to green in thin section.

biotope *1.* A term used by ecologists and biologists to designate an area of uniform ecology and organic adaptation; the living representative of a biofacies. *2.* An ecological term matching "biofacies" and which signifies organic environment. *3.* An area inhabited by a uniform community adapted to its environment.

bioturbation The churning and stirring of a sediment by organisms.

biozone *1.* Biostratigraphic unit including all strata deposited during the existence of a particular kind of fossil; *cf.* TEILZONE. *2.* Biostratigraphic unit identified by the actual occurrence of a

particular kind of fossil (not recommended). *Syn:* RANGE ZONE. *3.* Originally proposed as a geologic time unit corresponding to *1* above (obsolete). *Syn:* BIO-CHRON

Birch discontinuity Seismic discontinuity within the earth's mantle at a depth of about 900 km. caused perhaps by phase or chemical change or both.

bird *1.* Those parts of the airborne magnetometer, including the case but excluding the cables, which are streamed behind an aircraft. Also, the bomb used in air seismic shooting. *2.* Member of the vertebrate class Aves.

birdfoot delta A delta formed by the outgrowth of fingers or pairs of natural levees at the mouths of river distributaries making the digitate or "birdfoot" form typified by the Mississippi River delta.

birdseye limestone Very fine grained limestone containing spots or tubes of crystalline calcite.

birefringence The property of splitting a beam of ordinary light into two beams which traverse the crystal at different speeds. Possessed by crystals belonging to any crystallographic system other than the isometric.

biscuit-board topography Topography characterized by a rolling upland out of which cirques have been cut like big bites, and which represents an early or partial stage in glaciation.

bisector A plane or line of symmetry.

bisectrix A line bisecting the angle between the two optic axes of a biaxial crystal, designated acute or obtuse, depending on which of the supplementary angles in being bisected.

biserial Consisting of a double series as with the plates of cystoid brachioles and some crinoid arms.

bisexual Describing an organism that produces both eggs and sperm or a flower that bears both stamens and pistils.

bismuthinite; bismuth glance A mineral, Bi_2S_3. Orthorhombic.

bisphenoid A crystal form similar to a tetrahedron which has been elongated parallel to one of the crystal axes.

bit, drilling The cutting device at the lower end of cable drilling tools or rotary drill pipe, the function of which is to accomplish the actual boring or cutting.

bitter lakes Lakes rich in sulfates and alkaline carbonates, as distinct from salt lakes.

bittern 1. The bitter mother liquor that remains in saltworks after the salt has crystallized out. 2. Natural solutions in evaporite basins which resemble saltwork liquors, especially in their high magnesium content.

bitumen 1. A general name for various solid and semisolid hydrocarbons which are soluble in carbon bisulfide, whether gases, easily mobile liquids, viscous liquids, or solids. See ASPHALT. 2. Originally, native mineral pitch, tar, or asphalt. The term is generally applied to any of the flammable, viscid, liquid, or solid hydrocarbon mixtures soluble in carbon disulfide; often used interchangeably with hydrocarbons.

bituminous 1. Yielding bitumen, or holding bitumen in composition. This term is also commonly used for certain varieties of coal which burn freely with flame, although they really contain no bitumen. 2. Containing much organic or at least carbonaceous matter, mostly in the form of the tarry hydrocarbons which are usually described as bitumen. 3. Having the odor of bitumen. Often applied to minerals. 4. Yielding volatile bituminous matter on heating (e.g., bituminous coal). See HUMIC

bivalve Common term for the pelecypods, q.v.

bivariant equilibrium Said of a system having two degrees of freedom.

black body An ideal body, the surface of which absorbs all the radiation that falls upon it; i.e., it neither reflects nor transmits any of the incident radiation. The nearest approach to such a body among natural substances is soot, though the sun is often considered as a black body in meteorological studies of its radiation.

black damp A poisonous gas containing carbon monoxide that escapes from coal in coal mines.

black diamond 1. Carbonado; a black gem diamond; dense black hematite that takes a polish like metal. 2. A synonym of coal.

blackjack 1. A dark-colored, ironrich variety of sphalerite, $(Zn,Fe)S$. Isometric. It has a resinous luster and a light-colored streak or powder. Cf. BLENDE; SPHALERITE. 2. Crude black oil used to lubricate mine-car wheels. 3. Soft black, carbonaceous clay or earth associated with coal. 4. In Derbyshire, a kind of cannel coal. 5. In Illinois, a thin stratum of coal interbedded with layers of slate. A poor, bony coal.

black mud A mud formed in lagoons, sounds, or bays, in which there is poor circulation or weak tides. The color is generally due to black sulfides of iron, and to organic matter.

black sands Local deposits of heavy minerals concentrated by wave and current action on beaches. The heavy minerals consist largely of magnetite, ilmenite, and hematite associated with other minerals such as garnet, rutile, zircon, chromite, amphiboles, and pyroxenes.

Blancan The lowermost age of the continental Pleistocene sequence in North America; also the stage of strata deposited during that age.

blanket deposit *1.* A flat deposit of ore of which the length and breadth are relatively great as compared with the thickness. The term is current among miners, but it has no very exact scientific meaning. More or less synonymous terms are flat sheets, bedded veins, beds, or flat masses. *2.* Sedimentary deposit of great lateral extent and relatively uniform thickness, particularly a sandstone.

blanket sand A body of sand or sandstone that covers a considerable two-dimensional area. Often called a sheet sand.

blastetrix In an anisotropic medium, any surface perpendicular to which is a direction of greatest ease of growth.

blasting Abrasion effected by the movement of fine particles against a stationary fragment. In a sandblast or dry blasting the carrying agent is air; in wet blasting it is a current of water.

Blastoidea Class of stemmed budlike echinoderms with body enclosed by 13 plates regularly arranged in 3 circlets, and ambulacral areas borne on the body surface. Ord.-Perm.

blastophitic A metamorphosed rock which originally contained lath-shaped crystals partly or wholly enclosed in augite and in which part of the original texture remains.

blastoporphyritic A term applied to the textures of metamorphic rocks derived from porphyritic rocks, and in which the porphyritic character still remains as a relict feature, veiled but not obliterated by subsequent recrystallization.

blastopsammite A relict fragment of sandstone contained in a metamorphosed conglomerate.

bleaching clay Any clay in its natural state or after chemical activation that has the capacity for absorbing coloring matter from oil. Generally montmorillonite-rich clays.

bleach spot Deoxidation sphere. A greenish or yellowish area in red-colored rocks developed by the reduction of ferric oxide around an organic particle.

bleb A small, usually rounded inclusion of one material in another, as blebs of olivine, poikilitically enclosed in pyroxene.

bleed In England, to give off water or gas, as from coal or other stratum.

bleeding core In oil-field usage, one which shows little or no evidence of petroleum when first removed from the core barrel, but after a short time turns brown and exudes a film or drops of oil.

blende Sphalerite, *q.v.*

blind valley *1.* A feature in karst areas where a stream flows or disappears into a tunnel at the closed end of a valley. *2.* A type of valley in which a spring emerges from an underground channel to form a surface stream whose valley is enclosed at the head by steep and possibly precipitous walls.

blister cone A domelike cone on a lava flow which formed when the cooling crust buckled over caves resulting from the flow of molten lava beneath the hardening crust.

block An angular fragment over 256 mm. in diameter showing little or no modification in form due to transportation; similar in size to a boulder.

block caving A method of mining ore from the top down in successive layers of much greater thickness than characteristic of top slicing. Each block is undercut over the greater part of its bottom area and the supporting pillars blasted out. As the block caves and settles, the cover follows.

block diagram Three-dimensional perspective representation of geologic or topographic features showing a surface area and generally two vertical cross sections.

block folding Folding in an uplifted block bounded by steep faults that results from lateral spreading over lower blocks.

block mountains Mountains carved by erosion from large uplifted earth blocks bounded on one side or both by fault scarps. *See* FAULT BLOCK MOUNTAINS

blocks, volcanic Essential, accessory, or accidental volcanic ejecta, usually angular and larger than 32 mm. in diameter, erupted in a solid state.

blockstripes Forms transitional to stone stripes but containing material coarser and of less uniform size than in stone stripes.

bloom *1.* An earthy mineral that is frequently found as an efflorescence, as cobalt bloom. Also called blossom. *2.* The fluorescence of petroleum. *3. V:* To form an efflorescence; as salts with which alkali soils are impregnated bloom out on the surface of the earth in dry weather, after a rain or irrigation. *See* PLANKTON BLOOM, ALGAL BLOOM

blowhole *1.* A hole in a sea cliff from which columns of spray often accompanied with noise are forced upward. Blowholes are formed by wave erosion which extends sea caves along joints or other cracks to the surface. *2.* Blowholes are the minute craters formed on the surfaces of thick lava flows. They are often visible on driblet cones.

blown or **eolian sands** Those produced by the action of windborne particles of rocks; chemical composition of blown sands depends to a large extent on the

original rocks from which they have been derived. The sorting is not good.

blowout *1.* A general term for various saucer-, cup-, or trough-shaped hollows formed by wind erosion on a preexisting dune or other sand deposit; the adjoining accumulation of sand derived from the depression, where readily recognizable, is commonly included. *See* BLOWOUT DUNE. *2.* A term much used by prospectors and miners for any surface exposure of strongly altered, discolored rock associated, or thought to be associated, with a mineral deposit. *3.* In drilling a well by the rotary method an unexpected volume of gas under pressure sometimes "blows" the mud-laden drilling fluid from the hole, thus putting an end to drilling until controlled. The term is also used in standard-tool drilling when the flow of gas is sufficient to interfere with drilling operation.

blowout dune A term sometimes applied to the accumulations of sand derived from blowout troughs or basins, particularly where the accumulation is of large size and rises to considerable height above the source area.

blowpipe A tube through which air is blown into a flame to direct it and increase its temperature.

blue band *1.* A layer of dense, bubble-free ice in a glacier. *2.* The dark ribbon effect produced on the surface of the glacier by the exposure of these layers. *3.* A thin but persistent bed of bluish clay that is found near base of No. 6 coal throughout the Illinois-Indiana coal basin.

blue ground A term applied to the slaty-blue or blue-green kimberlite breccia of the diamond pipes of South Africa, occurring beneath a superficial oxidized covering known as yellow ground.

blue mud *1.* An ocean-bottom deposit containing up to 75% of terrigenous materials, of dimensions below 0.03 mm. The depth-range of occurrence is about 229 to 5124 meters. Colors range from reddish to brownish at the surface, but beneath the surface the colors of the wet muds are gray to blue. *2.* A common variety of deep-sea mud having a bluish-gray color due to presence of organic matter and finely divided iron sulfides. Calcium carbonate is present in variable amounts up to 35%.

blue vitriol Chalcanthite.

bluff *1.* Any high headland, or bank presenting a precipitous front. *2.* In America, the name given to the high vertical banks of certain rivers. *3.* A high steep bank or cliff. *4.* Altered country rock filling a lode. Analogous to mullock of Australia.

board coal In England, coal having a fibrous or woody appearance.

B.O.D. *See* BIOCHEMICAL OXYGEN DEMAND

bodily tides Tides or tilts of the surface of the earth caused by the gravitational field of the sun and moon. Earth tides.

body waves Either transverse or longitudinal seismic waves transmitted in the interior of an elastic solid or fluid and not related to a boundary surface.

boehmite A mineral, AlO(OH), dimorph of diaspore. Orthorhombic. A major constituent of some bauxites.

bog Morass; swamp. *1.* Common name in Scotland and Ireland for a wet spongy morass, chiefly composed of decaying vegetable matter or peat. *2.* A swamp or tract of wet land, covered in many cases with peat.

bog burst Refers to a bog built up into a low dome higher than the surrounding land and dammed by the organic material growing around the margin of the bogs.

boghead coal *1.* A variety of bituminous or subbituminous coal resembling cannel coal in appearance and in behavior during combustion. It is characterized by a high percentage of algal remains and volatile matter. Upon distillation it gives exceptionally high yields of tar and oil. *Cf.* KEROSENE SHALE. *2.* A non-banded coal with the translucent attritus consisting predominately of algae, and having less than 5% of anthraxylon.

bog iron A spongy variety of hydrated oxide of iron or limonite. Found in layers and lumps on level sandy soils which have been covered with swamp or bog. *Cf.* BROWN IRON ORE

bog manganese Wad.

bogue Bayou.

boiling point *1.* The temperature at which a liquid begins to boil, or to be converted into vapor by bubbles forming within its mass. It varies with the pressure. *2.* The temperature at which crude oil on being heated begins to give forth its different distillates. The boiling point of crude oils and the amounts of distillates obtained at specified temperatures differ considerably.

boiling spring *1.* A spring or fountain which gives out water at the boiling point or at a high temperature. *2.* A spring rising from the bottom of residual clay basins at the head of interior valleys. (Jamaica)

bojite *1.* Hornblende gabbro with primary hornblende. *2.* Hornblende diorite.

bolson (both syllables accented equally) In southwest United States and northern Mexico a basin, depression, or valley having no outlet and which, geologically, as a closed basin has received great thicknesses of sediments (e.g., in the sense that it was geologically an undrained basin).

bombs, volcanic Pyroclastic ejecta consisting of fragments of lava that were liquid, or plastic at the time of ejection, and having forms, surface markings, or internal structures acquired during flight through the air or at the time of landing after flight. They range from a few millimeters to several decimeters in length. Bombs less than 4 mm. in diameter are better classed as volcanic ash or dust.

bonanza [Sp. fair weather] In miners' phrase, good luck, or a body of rich ore. A mine is in bonanza when it is profitably producing ore. A body of rich ore.

bond, van der Waals See VAN DER WAALS FORCES

bone bed In England, a term applied to several thin strata or layers, from their containing innumerable fragments of fossil bones, scales, teeth, coprolites, and also organic remains.

bone phosphate The calcium phosphate obtained from bones; also in commerce, applied to calcium phosphate obtained from phosphatic rocks, e.g., those of North Carolina.

hook structure 1. A peculiar rock structure resulting from numerous parallel sheets of slate alternating with quartz. 2. Alternation of parallel slabs or slivers of rock with quartz or other gangue mineral in a vein.

boomer 1. A marine seismic energy source utilizing a high voltage discharge through a transducer which is towed under water. 2. A very strong, usually low frequency reflection event on a seismic reflection section.

boracite A mineral, $Mg_6Cl_2B_{14}$-O_{26}, found in evaporites. Orthorhombic, pseudocubic.

borate A salt of boric acid; a compound containing the radical BO_3^{-3}.

borax A mineral, $Na_2B_4O_7.10H_2O$, on ore of boron. Monoclinic.

borax bead In blowpipe analysis, a drop of borax, which, fused with a small quantity of a metallic oxide, will show the characteristic color of the element; as, a blue borax bead indicates the presence of cobalt.

border facies of igneous rocks See BASIC BORDERS IN IGNEOUS ROCKS

borderland Long relatively narrow land mass adjacent to a continental border that contributed sediment to a geosyncline or confined an epicontinental sea. See CONTINENTAL SHELF, CONTINENTAL BORDERLAND

bore Egre; Eager. 1. A violent rush of tidal water; the advancing edge or front of the tidal wave as it ascends a river or estuary. 2. A tidal flood with a high, abrupt front (such as occurs in the Amazon in South America, the Hugli in India, and the Bay of Fundy). 3. Submarine sand ridge in very shallow water that may rise to intertidal level. 4. A borehole or boring.

boreal Northern.

borehole A hole drilled into the earth, often to a great depth, as a prospective oil well or for exploratory purposes.

bornhardt An inselberg, q.v., of large size.

bornite A mineral, Cu_5FeS_4, isometric. Reddish-brown, readily tarnishes to iridescent blue or purple ("peacock ore"). An ore of copper.

bort 1. A trade name for diamonds too badly flawed or too off-color to be used in jewelry. 2. Carbonado or black diamond. Useful for abrasion.

boss *1.* A mass of intrusive rock which forms at the surface rounded, craggy, or variously shaped eminences, having a circular, elliptical, or irregular ground plan, and descending into the earth with vertical or steeply inclined sides. *Syn:* STOCK. *2.* In Foraminifera, a round and raised or knoblike structure. *3.* Coarse, short nodules occurring on the spire of a gastropod. *4.* The cone which supports the spheroidal summit of a tubercle on an echinoid plate.

botryoidal The natural habit of crystal growth resembling the form of a bunch of grapes. Commonly dispayed by goethite and smithsonite.

bottom *1.* Bottom is a word frequently heard in the Mississippi Valley and farther west, and used to designate the alluvial tracts along the river courses, which are sometimes called bottom lands, and sometimes simply bottoms. *2.* The bed of a body of still or running water. *3.* The floor of an underground passage; also sometimes referred to as the sill of a level. *4.* Bottom flow= underflow. A density current denser than any part of the surrounding fluid and which flows along the bottom of the body of water.

bottom-hole pressure The hydraulic pressure existing at the bottom or test position in a borehole as determined by one of several geophysical devices.

bottom land Lowland formed by alluvial deposit along a stream or in a lake basin; a flood plain.

bottom load Material rolled, pushed, or bounced along the bottom of a stream. *Cf.* BED LOAD; TRACTIONAL LOAD

bottom-set beds The layers of finer material carried out and deposited on the bottom of the sea or a lake in front of a delta. As the delta grows forward they are covered by the fore-set beds. *See* FORE-SET BEDS; TOP-SET BEDS

boudin One of a series of sausage-shaped segments occurring in a boudinage structure.

boudinage A structure common in strongly deformed sedimentary and metamorphic rocks, in which an original continuous competent layer or bed between less competent layers has been stretched, thinned, and broken at regular intervals into bodies resembling boudins or sausages, elongated parallel to the fold axes. *Syn:* SAUSAGE STRUCTURE

Bouguer anomaly The gravity value existing after the Bouguer corrections to a level datum have been applied.

Bouguer reduction Bouguer correction. A correction made in gravity survey data to take account of the elevation of the station and the rock between the station and some level datum, usually sea level.

boulangerite A mineral, $Pb_5Sb_4S_{11}$. Monoclinic.

boulder; bowlder *1.* A term for large rounded blocks of stone lying on the surface of the ground, or sometimes embedded in loose soil, different in composition from the rocks in the vicinity, and which have been therefore transported from a distance. *2.* A fragment of rock brought by natural means from a distance (though this notion of transportation from a distance is not always, in later usage, involved) and usually large and rounded in shape. Cobblestones taken from river beds are, in some American localities, called boulders. About 256 mm. (Wentworth scale).

boulder barricade A belt of boulders along a shore line visible between low and half tides.

boulder clay Till; drift clay; drift. Boulder clay is an unstratified or little stratified and unassorted deposit of silty and clayey materials in which are embedded particles in the size range from sands to boulders.

boulder pavement *1.* Surface of boulder-rich till abraded to flatness by glacier movement. *2.* Boulders in till, when grouped in an approximately horizontal plane and striated on their upper surfaces in a common direction constitute a boulder pavement according to the usage of the Scottish geologists.

boulder rampart A narrow ridge of boulders thrown up along part of the edge of a reef flat, especially on the side from which the prevailing winds blow. The rampart, which seldom exceeds 1 or 2 m. in height, occurs close behind the lithothamnion ridge where it is present.

boulder train Boulder trains take their origin from knobs or prominences of rock which lay in the path of glacial advance and gave off boulders readily and abundantly to the overriding ice. Such trains lie in the line of glacial movement, but the boulders are not carried forward in strictly parallel lines. They may therefore appropriately be called boulder fans. The boulders are of a single kind, or at least of the few kinds represented by the parent knob. They usually grow smaller and more worn as traced away from it. They mingle with the underlying drift, and in this respect differ from the boulder belts.

boundary *1.* A line between areas of the earth's surface occupied by rocks or formations of different type and age; especially used in connection with geologic mapping, hence, also, a line between two formations or cartographic units on a geologic map. *2.* That which indicates or fixes a limit or extent or marks a bound, as of territory. *3.* A plane separating two formations or other rock units.

boundary monument A material object placed on or near a boundary line to preserve and identify the location of the boundary line on the ground.

boundary tension A general term used to designate all surface and interfacial tensions at boundary surfaces, such as liquid-gas, liquid-liquid, and liquid-solid.

boundary waves Seismic waves which are propagated along free surfaces or acoustic interfaces, and which depend on layering for their existence.

bournonite Cog-wheel ore. A mineral, $PbCuSbS_3$. Orthorhombic.

Bowen's reaction series A series of minerals, wherein any early formed mineral phase tends to react with the melt, later in the differentiation, to yield a new mineral further down in the series. Thus early formed crystals of olivine react with later liquids to form pyroxene crystals; these in turn may further react with still later liquids to form amphiboles. There are two different series, a continuous reaction series, *q.v.*, and a discontinuous reaction series, *q.v.*

box canyon A canyon having steep rock sides and a zigzag course, presenting a view from its bottom of four almost vertical walls.

box folds A fold in which the broad flat top of an anticline (or the broad flat bottom of a syncline) is bordered on either side by steeply dipping limbs.

box work Limonite and other minerals which originally formed as blades or plates along cleavage or fracture planes and then the intervening material dissolved leaving the intersecting blades or plates as a network. Usually found on the ceilings of caves.

Brachiopoda A phylum of marine, shelled animals with two unequal shells or valves each of which normally is bilaterally symmetrical. Also called lamp shells.

brachyanticline A term used in the U.S.S.R. to designate a long, narrow anticline.

brachy axis The shorter lateral axis in the crystals of the orthorhombic and triclinic systems. *Obs.*

brachydome Side dome. *Crystallog:* A dome parallel to the shorter lateral axis. *Obs.*

brachygenesis The phenomenon in evolution in which part of a presumed recapitulated sequence has evolved out and no longer appears.

brachyhaline Polyhaline. *See* BRACKISH

brachypinacoid Side pinacoid. A pinacoid parallel to the vertical and the shorter lateral axis.

brachysyncline Short broad syncline.

brackish *1.* Slightly salty. *2.* Term applied to waters whose saline content is intermediate between that of streams and sea water; neither fresh nor salty, but in between. *3.* Salty, generally less so than sea water, variously defined as less than 30 to 15 parts per thousand salinity.

Bradygenesis Retardation of development of a group of organisms which may progressively diminish the rate so that certain individuals in some or all of their characters may fall more and more behind the normal rate of progress.

Bragg's law The numerical relationship among wavelength, crystal spacing, d, and diffraction angle, θ. $N\lambda = 2d \sin \theta$.

braid To branch and rejoin producing a netlike pattern, as with some streams.

braided stream *1.* A braided stream is one flowing in several dividing and reuniting channels resembling the strands of a braid, the cause of division being the obstruction by sediment deposited by the stream. *2.* Where more sediment is being brought to any part of a stream than it can remove, the building of bars becomes excessive, and the stream develops an intricate network of interlacing channels, and is said to be braided. *See* ANASTOMOSING

branch *1.* A creek or brook, as used locally in southern states. Also used to designate one of the bifurcations of a stream, as, a fork. *2.* Subdivision of an igneous rock series.

branch fault Auxiliary fault.

braunite A mineral, $3Mn_2O_3$.-$MnSiO_3$. Tetragonal, commonly in brownish-black masses.

Bravais lattice Fourteen uniquely symmetrical ways in which points can be arranged in a three-dimensional array.

breached anticline An anticline that has been more deeply eroded in the center. Consequently, erosional scarps face inward toward the center of the anticline.

breached cone A cinder cone in which lava has broken through the sides and carried away the broken materials.

breached crater Volcanic craters have generally a complete rim, but in some cases the heavy lava rises in the crater until its weight is so great that it bursts through one wall of the crater and flows out through the breach. Such craters are known as breached craters.

breadcrust bomb A volcanic bomb with a checkered and cracked exterior resulting from the shrinkage of the skin upon congealing. *See* BOMBS, VOLCANIC

break Arrival; event; kick. *1.* On a seismogram trace, a sudden displacement indicating the arrival of energy over some path of interest, used especially for near surface paths, as "first breaks" of energy refracted along the "weathered layer" base in exploration seismic operations. *2.* Abrupt change in surface slope.

breaker A wave breaking on the shore, over a reef, etc. Breakers may be (roughly) classified into three kinds although there is much overlapping. Spilling breakers break gradually over quite a distance; plunging breakers tend to curl over and break with a crash; and surging breakers peak up, but then instead of spilling or plunging they surge up the beach face.

breaker depth Breaking depth. The still-water depth at the point where the wave breaks.

breaker height Average height of breaking waves from trough to crest.

breaker terrace In glacial or artificial lakes where water rises against a bank of kame gravels, a characteristic terrace develops, composed of the heaviest stones handled by the waves. The front has a steep slope and has no admixture of fine gravels. Similar forms are built of lighter gravels by moderate waves, but these are destroyed in heavy storms. Those built by the heaviest cobbles which can be moved and are persistent forms and are very characteristic of lakes in the glacial drift, especially of those in the kame areas.

breaks *1.* The broken land at the border of an upland that is dissected by ravines. *2.* An area in rolling land eroded by small ravines and gullies; also used to indicate any sudden change in topography, as from a plain to hilly country.

break thrust A thrust fault that cuts across one limb of a fold.

breccia [*It.*] *1.* Fragmental rock whose components are angular and therefore, as distinguished from conglomerates, are not waterworn. There are friction or fault breccias, talus breccias, and eruptive breccias. *2.* A rock made up of highly angular coarse fragments. May be sedimentary or formed by crushing or grinding along faults.

breccia, volcanic *See* VOLCANIC BRECCIA

breeder reactor A nuclear fuel generator that burns uranium-235 and, in the process, captures neutrons which convert thorium or the more abundant uranium-238 into fissionable materials. In this way, additional fuel is generated and original fuel consumption is greatly reduced.

Brevard zone A continuous belt of cataclastic rocks, approximately 1.6 to 6 km. wide, which roughly parallels about 600 km. of the southeastern side of the Crystalline Blue Ridge province in the southern Appalachians.

bridal-veil fall A cataract of great height and such small volume that the falling water is dissipated in spray before reaching the lower stream bed.

bridge *1.* A solutional remnant forming a rock span across a cave inclined less than 45° from the horizontal. *2.* Collapsed walls or large fragments of rock falling and lodging in a drill hole in such a way as to obstruct passage of drilling tools. Also a mechanical obstruction placed intentionally to obstruct a drill hole. *3.* A circuit composed of four or more elements connected in a loop, usually having a source of electrical power connected across at least two elements of the loop, and used for measuring electrical impedance. Also, a jumper or wire connector used to short-circuit or to offer a path of negligible impedance around an electrical circuit or circuit element. Also, a plug or obstruction in a shot hole above the bottom of the hole, usually formed by caving or an exploding charge whether accidentally or intentionally. *V:* To form a bridge or to be bridged.

bright coal *1.* Anthraxylon. The constituent of banded coal which is of a jet-black pitchy appearance, more compact than dull coal, and breaking with a conchoidal fracture when viewed macroscopically, and which in thin section always shows preserved cell structure of woody plant tissue, either of stem, branch, or root. *2.* A coal composed of anthraxylon and attritus, in which the translucent cell-wall degradation matter or translucent humic matter predominates. *3.* A banded coal containing less than 20% opaque attritus and more than 5% anthraxylon. *4.* A type of banded coal containing from 100% to 81% pure bright ingredients (vitrain, clarain, and fusain), the remainder consisting of claro-durain and durain.

brine A highly saline solution. A solution containing appreciable amounts of NaCl and other salts.

British thermal unit Abbreviated B.t.u. A unit of heat which is 1/180 part of that required to raise the temperature of one pound of water from 32° F. to 212° F. at sea level. Usually considered as that amount of heat required to raise the temperature of one pound of water from 63° F. to 64° F.

brittleness Property of material that ruptures easily with little or no plastic flow.

brochantite A mineral, $Cu_4(SO_4)$-$(OH)_6$, common in the oxidation zone of copper sulfide deposits. Monoclinic, emerald to dark green.

brodel *1.* A bulbous mass of silt, without horizontal continuity and completely enclosed by clay except through "necks" by which they are connected with overlying silt beds. *2.* Highly irregular, aimlessly contorted interpenetrating structures in Pleistocene sediments. *Cf.* INVOLUTION; CRYOTURBATION

bromide A compound of bromine with one or more other more positive elements than bromine.

bromoform $CHBr_3$, tribromomethane, m.p. 5° C., sp.gr. 2.9; a colorless liquid, of narcotic odor. Used especially in mineral separation.

bronzite A variety of the enstatite-hypersthene series. *See* PYROXENE

brookite A mineral, TiO_2. Orthorhombic, trimorphous with rutile and anatase.

brow *1.* The edge of the top of a hill or mountain; the point at which a gentle slope changes to an abrupt one; the top of a bluff or cliff. *2.* The frontal portion of a nappe, in which the strata commonly are overrolled.

brown coal Lignite. A low-rank coal which is brown, brownish black, but rarely black. It commonly retains the structures of the original wood. It is high in moisture, low in heat value, and decomposes upon dehydration.

brown iron ore Its approximate formula is $2Fe_2O_3.3H_2O$, equivalent to about 59.8% iron. Probably a mixture of hydrous iron oxides. Residual and replacement limonite, mainly at erosion surfaces. *Syn:* BROWN HEMATITE; LIMONITE

brown soils A zonal group of soils having a brown surface horizon which grades below into lighter colored soil, and finally into a layer of carbonate accumulation, developed under short grasses, bunch grasses, and shrubs in a temperate to cool, semiarid climate.

brownstone Ferruginous sandstone in which the grains are generally coated with iron oxide. Applied almost exclusively to a dark brown sandstone derived from the Triassic of the Connecticut River Valley.

brucite A mineral, $Mg(OH)_2$. Commonly in foliated masses. Hexagonal rhombohedral.

Brunton A small pocket compass with sights and a reflector attached, used in sketching mine workings, as in mine examinations, or in preliminary surveys.

Bryophyta Division of non-vascular plants that may have differentiated stems and leaves but no true roots; includes liverworts and mosses.

Bryozoa Phylum of small colonial animals, equipped with a lophophore, and that build calcareous structures of many kinds, mostly marine. The phylum of invertebrate animals which are popularly called "moss animals." They are called Polyzoa by some zoologists.

B-tectonite *1.* Tectonite whose fabric indicates an axial direction rather than a slip surface. *2. Struct. petrol:* A tectonite is a deformed rock the fabric of which is the result of the systematic movement of the individual units under a common external force. In a B-tectonite a lineation is prominent and s-planes may be absent. The point diagram commonly shows a girdle about the b-axis.

B.t.u. An abbreviation for British thermal unit.

bubble point A state of fluids characterized by the coexistence of a liquid phase with an infinitesimal quantity of gas phase in equilibrium.

bubble train (in lava) A string or strings of vesicles, marking the path followed by the rising gas escaping from a lava flow.

bubble trend Planar or linear bubble zone as in glacial ice.

bucking plate; bucking iron An iron plate on which ore is ground by hand by means of a muller. Extensively used for the final reduction of ore samples for assaying.

buffalo wallow These are the minor, very shallow depressions on the upland surface. They may be from a few meters to several meters in diameter and from a fraction of a meter to several meters in depth.

buffer solution A solution to which large amounts of acid or base may be added with only a very small resultant change in the hydrogen ion concentration.

buhrstone; burrstone; burstone A silicified fossiliferous limestone, with abundant cavities which were formerly occupied by fossil shells. Its cellular character and toughness occasioned its extensive use as a millstone in former years.

building stone Any stone used in masonry construction, generally stone of superior quality that is quarried and trimmed or cut to form regular blocks.

bulb glacier Where valley glaciers descend to the foot of a mountain and out upon an open slope, as in a broad valley or upon a plain, the glacier ends spread into an ice fan, or bulb glacier, or piedmont bulb.

bulk modulus Volume elasticity; incompressibility modulus. Under increasing force per unit area a body will decrease in size but increase in density.

bulk sample A large sample consisting of tons or hundreds of tons which is then milled and the grade computed from the results.

bullion *1.* Concretion found in some types of coal; composed of carbonate or silica stained by brown humic derivatives; often well-preserved plant structures form the nuclei. *2.* Nodules of clay, ironstone, pyrite, shale, etc., which generally enclose a fossil.

bull quartz A miner's or prospector's term for white, coarse-grained, barren quartz.

Bunter Lower Triassic.

buoyancy The resultant of upward forces, exerted by the water on a submerged or floating body, equal to the weight of the water displaced by this body.

buried hill Hills consisting of resistant older rock over which later sediments were deposited.

buried placers Old placer deposits which have become buried beneath lava flows or other strata.

burrow A cylindrical or near-cylindrical tube, often filled with clay or sand, which may lie along the bedding plane or penetrate the rock, made by an animal that lived in the mud.

butane A gaseous hydrocarbon of the paraffin series, formula C_4H_{10}.

butte [*Fr.*] Mesa. A conspicuous isolated hill or small mountain, especially one with very steep or precipitous sides, or a turretlike formation such as those found in the badlands.

button The globule of metal remaining on an assay-cupel or in a crucible, at the end of the fusion.

buttress *1.* A ridge on the inner surface of a valve of a pelecypod that serves as a support for part of the hinge. *2.* Aboral extension of apical denticle, generally on inner and outer sides of a conodont.

b.y. Billion years=1 aeon.

bysmalith A more or less vertical cylindrical body of igneous rock that crosscuts the adjacent sediments and has been injected by pushing up the overlying strata along steep faults. The term was introduced by J. P. Iddings.

bytownite *See* PLAGIOCLASE

C

cable tools The equipment used in the standard, percussion, or cable tool method of drilling. It consists essentially of a steel bit fastened below drill stem and jars, all suspended on a wire line. The whole can be lowered or raised by controlling machinery near the mouth of the hole. In drilling, the tools are alternately lifted and dropped, the rock being cut by the repeated blows of the bit.

cabochon A cut and polished unfaceted gemstone having a domed or convex form.

cacholong An opaque or feebly translucent variety of common opal containing a minor amount of alumina. Its color may be bluish-white, pale yellow, or red.

cadastral map A map showing the boundaries of subdivisions of land, usually with the bearings and lengths thereof and the areas of individual tracts, for purposes of describing and recording ownership.

cadastral survey A survey relating to land boundaries and subdivisions, made to create units suitable for transfer or to define limitations of title, for example, the surveys of city or town lots.

cadmium *See* HEAVY METALS

cafemic A mnemonic term used collectively for the calcium, ferrous-ferric, and magnesium constituents of rocks or magmas; an extension of the CIPW (1902) term "femic."

Cainozoic Cenozoic.

calamine Hemimorphite. Sometimes used in Europe for smithsonite.

calaverite A mineral, $AuTe_2$, which commonly contains some Ag. Monoclinic.

calc- [*Lat.*] Prefix meaning limy, i.e., containing calcium carbonate.

calc-alkalic series Those igneous rock series having an alkali-lime index of 55 to 61.

Calcarea Calcispongiae. A group of sponges (Porifera) in which the spicules are composed of calcium carbonate.

calcarenite A deposit composed of cemeted sand-size grains of calcium carbonate; usually a biosparite or a grainstone.

calcareous Containing calcium carbonate.

calcareous algae A seaweed that removes carbon dioxide from the shallow water in which it lives and as a consequence secretes or deposits a more or less solid calcareous structure.

calcareous ooze *See* OOZE, CALCAREOUS

calcareous tufa *See* TUFA

calcic *1.* Containing calcium, as, calcic plagioclase, calcic pyroxene. Also said of igneous rocks containing such minerals. *2.* Refers to igneous rocks having an alkali-lime index of more than 61.

calcic series Those igneous rock series having an alkali-lime index of 61 or greater.

calcification Replacement of the original hard parts of an animal or plant by calcium carbonate.

calcify To make or become hard or stony by the deposit of calcium salts.

calcilutite *1.* A name suggested

by A. W. Grabau for a limestone or dolomite made up of calcareous rock flour, the composition of which is typically nonsiliceous, though many calcilutites have an intermixture of clayey material. 2. A calcium carbonate rock made up of grains or crystals with an average diameter less than 0.0625 mm.

calcirudite A name for a "limestone or dolomite composed of broken or worn fragments of coral or shells or of limestone fragments, the interstices filled with calcite, sand, or mud, and with a calcareous cement." The word is derived from the Latin for lime and rubble.

calcisiltite Limestone composed of calcareous sediment of silt size.

Calcispongia Class of sponges with skeletons composed of separate or fused calcareous spicules. Dev.-Rec.

calcite A mineral $CaCO_3$, hexagonal, rhombohedral, trimorphous with aragonite and vaterite; one of the most common minerals; the principal constituent of limestone.

calcitic dolomite A carbonate rock in which the percentage of calcite is between 10 and 50, and the percentage of dolomite between 50 and 90.

calcium carbonate A solid, $CaCO_3$, occurring in nature, as calcite, etc.

calc-schist A marble with a more or less distinct schistosity due to the parallelism of platy crystals of calcite. See MARBLE; SCHIST

calc-silicate hornfels A fine-grained metamorphic rock rich in calc-silicate minerals.

calc-silicate marble A marble with conspicuous calcium and/or magnesium silicate minerals.

calc-sinter Stalactitic or stalagmitic carbonate of lime. This is so called from the German *Kalk*

(lime) and *sintern* (to drop). It is deposited from thermal springs holding carbonate of lime in solution.

calcspar Calcite.

caldera A large basin-shaped volcanic depression, more or less circular or cirquelike in form, the diameter of which is many times greater than that of the included volcanic vent or vents, no matter what the steepness of the walls or form of the floor. Three major types: explosion, collapse, erosion.

caldera complex The diverse rock assemblage underlying a caldera, comprising dikes, sills, stocks, and vents breccias; craterfills of lava; talus beds of tuff, cinder, and agglomerate; fault gouge and fault breccias; talus fans along fault escarpments; cinder cones; and other products laid down in a caldera.

Caledonian orogeny A name commonly used for the early Paleozoic deformation in Europe which created an orogenic belt, the Caledonides, extending from Ireland and Scotland northwestward through Scandinavia.

Caledonides The orogenic belt extending from Ireland and Scotland northwestward through Scandinavia, formed by the early Paleozoic Caledonian orogeny.

calf A piece of floating ice which has broken away from a larger piece of sea ice or land ice.

caliche 1. In Chile and Peru, impure native nitrate of soda. 2. A desert soil formed by the near surface crystallization of calcite and/or other soluble minerals by upward-moving solutions. 3. In Chile, whitish clay in the selvage of veins. In Mexico: 4. Feldspar, a white clay; 5. A compact transition limestone. In Colombia: 6. A mineral vein recently discovered; 7. In placer mining, a bank composed of clay, sand, and gravel. *8.* In Mexico and southwest U.S., gravel, sand, or desert debris cemented by porous calcium carbonate; also the calcium carbonate itself. *See* DURICRUST

californite A compact, massive vesuvianite. Used as an ornamental stone.

caliper log A graphic record which shows, to scale, the diameter of a drilled hole.

calomel Horn quicksilver; mercurial horn ore. Mercurous chloride, Hg_2Cl_2.

calorie A unit of heat; the amount of heat needed to raise the temperature of one gram of water from 3.5° to 4.5° centigrade. Various prefixes are used if the degree of temperature change is at different magnitudes.

calving Breaking off and floating away as icebergs of large masses of a glacier that reaches the sea.

calyx *1.* Upper part of a corallite in which a coral polyp sits. *2.* Plated structure of a crinoid body; excludes stem and arms.

camber *1.* A convex terminal shoulder of the continental shelf. *2.* A structure that forms in areas of flat-lying rocks. A plastic clay beneath a more competent bed flows toward a valley so that the competent bed sags downward and appears to be draped over the sides of the valley.

Cambrian The oldest of the periods of the Paleozoic Era; also the system of strata deposited during that period.

camouflage Substitution in a crystal lattice of a trace element for a common element of the same valence, e.g., Ga^{+3} for Al^{+3}. The trace element is said to be camouflaged by the common element.

Campanian The next to youngest of nine ages of the Cretaceous Period; also the stage of strata deposited during that age.

Campbell's law Where two streams that head opposite to each other are affected by an even lengthwise tilting movement, that one whose declivity is increased cuts down vigorously and grows in length headward at the expense of the other. If the tilting that affects them is part of a general warping, the divide migrates toward an axis of upwarping.

camptonite A lamprophyre having pyroxene, sodic hornblende and olivine as dark constituents and labradorite as its light constituent. Sodic orthoclase may also be present.

Canada balsam A yellowish, transparent balsam yielded by a North American species of silver fir; used for mounting microscopic preparations and as an adhesive for glass in optical instruments. When exposed to the air, it becomes brittle and discolored, and its refractive index gradually increases. The average values for the index of refraction are 1.524 (uncooked), 1.538 (slightly undercooked), and 1.543 (overcooked). For normally cooked balsam the refractive index is between 1.534 and 1.540.

Canadian The oldest of three epochs of the Ordovician of North America; also the series of strata deposited during that epoch.

canal *1.* An artificial watercourse cut through a land area for navigation, irrigation. *2.* A long narrow arm of the sea extending far inland. *3.* On the Atlantic Coast, a sluggish coastal stream. *4.* A long, fairly straight natural channel with steeply sloping sides, generally a mile or more in width. *5.* A cave passage partly filled with water. *6.* A hollow, gutterlike extension of the lower (anterior) end of a gastropod shell which carries within it the siphon. 7. The tubes which run lengthwise of the walls of the test as in the foraminiferal family the Camerinidae.

cancrinite A mineral of the feldspathoid group, $(Na,K,Ca)_{6-8}(Al,Si)_{12}O_{24}(SO_4CO_3,Cl).nH_2O)$. Hexagonal. Yellow.

candle ice Elongate prismatic crystals of ice arranged perpendicular to the surface.

cannel coal A variety of bituminous coal of uniform and compact fine-grained texture with a general absence of banded structure. It is dark gray to black in color, has a greasy luster, and is noticeably of conchoidal or shell-like fracture. It is noncaking, yields a high percentage of volatile matter, ignites easily, and burns with a luminous smoky flame.

cannel shale A black shale formed by the accumulation of sapropels accompanied by a considerable quantity of inorganic material, chiefly silt and clay.

canyon 1. A steep-walled chasm, gorge, or ravine; a channel cut by running water in the surface of the earth, the sides of which are composed of cliffs or series of cliffs rising from its bed. Sometimes spelled cañon. 2. Oceanog: A deep submarine depression of valley form with relatively steep sides.

capacitance Electrical capacity. That property of a pair of electrical conductors whereby an electrical charge is stored when a difference of potential is applied to the conductors.

capacity The ability of a water or wind current to transport detritus, as measured by the quantity it can carry past a given point in a unit of time.

cape 1. A point of land extending into the sea or a lake; a headland. 2. A relatively extensive land area jutting seaward from a continent or large island which prominently marks a change in, or interrupts notably, the coastal trend; a prominent feature.

capillarity The attractive force between two unlike molecules, illustrated by the rising of water in capillary tubes of hairlike diameters or the drawing-up of water in small interstices, as those between the grains of a rock.

capillary Resembling a hair; fine, minute; having a very small bore.

capillary conductivity (soil water) 1. Physical property related to the readiness with which unsaturated soil transmits water. 2. Ratio of water flow velocity to driving force in unsaturated soil.

capillary fringe A zone, in which the pressure is less than atmospheric, overlying the zone of saturation and containing capillary interstices some or all of which are filled with water that is continuous with the water in the zone of saturation but is held above that zone by capillarity acting against gravity.

capillary interstice 1. An opening small enough for water to be held in it by capillarity at a considerable height above the level at which it is held by hydrostatic pressure. 2. An opening or void small enough to produce appreciable capillary rise.

capillary migration Movement of liquid water produced by the molecular attraction of the rock material for the water.

capillary pressure 1. The difference in pressure existing between two phases, air-fluid, gas-fluid or between fluids, measured at points of the interface and occurring in the interconnected phases in a rock. 2. The difference in pressure across the interface between two immiscible fluid phases jointly occupying the pores of a rock.

capillary water (soils) The portion of soil water which is held by cohesion as a continuous film around the particles and in the capillary spaces.

cap rock *1.* A disklike plate over all or part of the top of most salt domes in the Gulf Coast states, and some in Germany, composed of anhydrite, gypsum, limestone, and occasionally sulfur. *2.* A comparatively impervious stratum immediately overlying an oil- or gas-bearing rock. *3. See* CALICHE

capture *1.* Piracy, *q.v. 2.* Substitution in a crystal lattice of a trace element for a common element of lower valence, e.g., Pb^{++} for K^+.

Caradocian The next to youngest of six epochs of the Ordovician Period; also the series of strata deposited during that epoch.

carapace *1. Paleontol:* A bony or chitinous case covering the dorsal part of an animal. *2. Structure:* Upper normal limb of a recumbent fold.

carat *(Fr.) 1.* A unit employed in weighing diamonds, and equal to 3 1/6 troy grains (205 mg.). A carat-grain is one-fourth of a carat. The international metric carat (C.M.) of 200 mg. has been made the standard in Great Britain, France, Germany, Holland, and the United States. *2.* A term employed to distinguish the fineness, or purity, of a gold alloy, and meaning one-twenty-fourth. Fine gold is 24-carat gold. Goldsmiths' standard is 22 carats fine, i.e., contains 22 parts gold, 1 copper, and 1 silver. *Symbol:* k.

carbide A compound of carbon with one other more positive element or radical.

carboid General name for a group of pyrobitumens that are insoluble in carbon disulfide.

carbonaceous *1.* Coaly. *2.* Pertaining to, or composed largely of, carbon. *3.* The carbonaceous sediments include original organic tissues and subsequently produced derivatives of which the composition is chemically organic.

carbonate A salt of carbonic acid; a compound containing the radical CO_3^{+2}.

carbonation *1.* Process of introducing carbon dioxide into water. *2.* A process of chemical weathering by which minerals are replaced by carbonates.

carbonatite The term is not synonymous with limestone. Carbonatites are intrusive carbonate rocks which are associated with alkaline igneous intrusive activity in many localities. They have been regarded as: *1.* Surely intrusive and just as surely not igneous. *2.* Rocks formed as a phase of a magma that was rich both in soda and lime, as well as being in a highly carbonated condition. *3.* Mobilized or hydrothermally redistributed sedimentary limestones. *4.* Almost pure carbonate rocks, which appear to have been formed by reaction of basic magma with walls of limestone and dolomite.

carbon cycle One of the mineral cycles, *q.v.*

carbon dioxide (CO₂) A colorless, odorless, nonpoisonous gas that is a normal part of the ambient air. CO_2 is a product of fossil fuel combustion, and some researchers have theorized that excess CO_2 raises atmospheric temperatures.

carbon 14 A radioactive isotope of carbon with atomic weight 14, produced by collisions between neutrons and atmospheric nitrogen. Useful in determining the age of carbonaceous material younger than 30,000 years. Half-life 5700 years $+30$.

Carboniferous The fifth of six periods of the Paleozoic of areas other than North America; also the system of rocks deposited during that period.

carbonization *1.* In the coalification process, carbonization characterizes the progressive changes undergone by the preserved organic matter and biochemical decomposition products between the death of the plant or animal and the stage of essentially complete reduction to residual carbon, *in situ*. 2. The slow decay under water of organic material, plant or animal, resulting in a concentration of carbon as a film of carbon showing more or less distinctly the form and structure of the original tissue. 3. The process of converting to carbon, by removing other ingredients, a substance containing carbon, as in the charring of wood.

carbon monoxide (CO) A colorless, odorless, highly toxic gas that is a major air pollutant and the normal byproduct of incomplete fossil fuel combustion.

carbon ratio 1. The ratio of the fixed carbon in any coal to the fixed carbon plus the volatile hydrocarbons; expressed in percentage. 2. The ratio of the most common carbon isotope (C^{12}) to either of the less common isotopes (C^{13} or C^{14}), or the reciprocal of one of these ratios. If unspecified, the term generally refers to the ratio (C^{12}/C^{13}).

carbon-ratio theory The theory that in any area the gravity of the oil varies inversely as the carbon ratio of the coal.

carborundum A crystalline compound, SiC, consisting of silicon and carbon. It is produced in an electric furnace and used as an abrasive. Silicon carbide.

cardinal 1. Pertains to the hinge in brachiopods and pelecypods. 2. In corals, pertains to the generally longer of the two septa which develop from the single axial septum of the early growth stages (located opposite the counter septum).

cardinal points The four principal points of the compass: North, South, East, and West.

carina(-nae) 1. A raised ridge or keel. 2. In some corals, one of the vertical strengthening plates which extend a short distance from the septa; these appear in cross section as septal spines. 3. In a barnacle, one of the two unpaired plates of the fixed tubular portion of the shell which adjoins the terga. 4. A keel or flange as found around the edge of some foraminiferal tests. 5. In conodonts, the central denticulated, nodose, or smooth ridge extending down the middle of the platform or blade.

Carlsbad twin A type of crystal twinning common in feldspars. *See* TWIN LAW

carnallite A mineral, $KMgCl_3 \cdot 6H_2O$. Orthorhombic. An ore of potassium.

carnotite A mineral, $K(UO_2)(VO_4)_2 \cdot 1-3H_2O$. Monoclinic. Usually occurs as a fine-grained canary-yellow incrustation. An ore of U and V.

Carolina bay Ovate depression, generally marshy, of a type occurring abundantly on the coastal plain from New Jersey to Florida; origin attributed to fallen meteorites, upwelling springs, eddy currents, etc.

Carrara marble A general name given to all the marbles quarried near Carrara, Italy. The prevailing colors are white to bluish, or white with blue veins; a fine grade of statuary marble is here included.

carrier bed Deep, porous, and permeable beds, such as coarse sheet sands, through which oil can migrate for long distances.

cartography 1. The science and art of expressing graphically, by means of maps and charts, the visible physical features of the earth's surface, both natural and man-made. *2.* The science and art of map construction.

cascade *1.* A waterfall, usually a small waterfall; especially one of a series of small falls, formed by water in its descent over rocks. *2.* A gravity-collapse structure. A bed that buckles into a series of recumbent folds as it slides down the flanks of an anticline. *3.* Series of small closely spaced waterfalls or very steep rapids.

cascalho Alluvial material including gravel and ferruginous sand in which Brazilian diamonds are found.

case-hardening The process by which the surface of a porous rock, especially sandstone or tuff is coated by a cement, or desert varnish, *q.v.*, formed by evaporation of mineral-bearing solutions.

casing-head gas Unprocessed natural gas produced from a reservoir containing oil. Such gas contains gasoline vapors and is so called because it is usually produced under low pressure through the casing head of an oil well.

Caspian Masses of salt water included in the dry land; so called from the Caspian Sea, the largest of them.

cassiterite Tin-stone. A mineral, SnO_2. Tetragonal. The most important ore of tin.

cast *1.* The mineral or other substance that fills a hole in a rock which has been formed by the solution of the orignal hard material of which the shell or skeleton consisted. *2.* A natural mold which has been filled naturally with some mineral substance. (This is usage as opposed to artificial cast.) *3.* Restricted to casts in which the filling is of an uncrystallized substance, as opposed to pseudomorphs in which the replacing matter is a crystallized mineral. *4.* A second type of plant

fossil that is closely related to the compression is the cast. A cast results from the filling of a cavity formed by the decay of some or all of the tissues of a plant part. *5.* Deposit of fecal material with attendant undigested sediment. Frequently found as pellets or strands.

castings Fecal pellets; coprolite, *q.v.*

cataclasis Rock deformation accomplished by fracture and rotation of mineral grains or aggregates; granulation.

cataclasite A rock that has been formed by shattering (cataclasis) which has been less extreme than in the case of a mylonite. *See* AUGEN GNEISS; AUGEN SCHIST; CRUSH BRECCIA; CRUSH CONGLOMERATE; FLASER GABBRO; MYLONITE; PROTOCLASTIC

cataclastic *1.* Pertaining to a texture found in metamorphic rocks in which brittle minerals have been broken and flattened in a direction at a right angle to the pressure stress. *2.* Refers to coarse fragmentation of rock in transit, e.g., glacial action.

cataclysm *1.* Any overwhelming flood of water; especially, the Noachian deluge. *2.* Any violent and extensive subversion of the ordinary phenomena of nature; an extensive stratigraphic catastrophe. *3.* Any violent flood or inundation that overspreads or sweeps over a country. *Syn:* DELUGE; DEBACLE. The term is now obsolete but is found in many early geological works.

catagenesis Evolutionary process resulting in decadence and decreased vigor. *Cf.* ANAGENESIS

cataract A waterfall, usually of great volume; a cascade in which the vertical fall has been concentrated in one sheer drop or overflow.

catastrophe *1. Geol:* A sudden, violent change in the physical conditions of the earth's surface; a cataclysm. *2. Min:* A disaster in which many lives are lost or much property damaged, as by a mine fire, explosion, inrush of water, etc.

catastrophism *1.* The doctrine that explained the differences between fossils in successive stratigraphic horizons by assuming a general catastrophe followed by creation of the different organisms found in the next younger beds. *2.* The doctrine that sudden, violent, short-lived more or less worldwide events outside our present experience of knowledge of nature have greatly modified the earth's crust.

catazone The deepest zone of rock metamorphism where very high pressures and temperatures both prevail.

catchment area As applied to an aquifer the term includes its intake area and all areas that contribute surface water to the intake area.

catchment basin Drainage basin.

cathode Electronegative pole.

cathode rays Rays that, when a Crookes's tube is excited by an alternating high potential current of electricity, or by a series of spark discharges, pass in straight lines from the cathode to the opposite wall of the tube, producing a fluorescent area.

cation An ion that bears a positive charge.

cation exchange Base exchange.

catoctin Monadnock; residual inselberg. Erosional knobs and ridges found on the piedmont plain.

catogene A general term for sedimentary rocks, since they were formed by deposition from above, as of suspended material. *Cf.* HYPOGENE

cat's-eye A greenish, chatoyant variety of chrysoberyl.

catstep Narrow, generally backward tilted terrace on steep hillside produced by slumping.

cauldron subsidence *1.* A structure resulting from the lowering along a steep ring fracture of a more or less cylindrical block, usually 1.6 to 16 km. in diameter, into a magma chamber. Usually associated with ring-dikes. In surface cauldron subsidence the ring fracture penetrates the surface of the earth, as a result of which volcanic rocks are lowered. In underground cauldron subsidence, the ring fracture does not penetrate the surface. *2.* The process of forming a cauldron subsidence.

caustic metamorphism Metamorphism in contact rock produced by the heat of lava flows and small dikes.

cave *1.* A natural cavity, recess, chamber, or series of chambers and galleries beneath the surface of the earth, within a mountain, a ledge of rocks, etc.; sometimes a similar cavity artificially excavated. *2.* Any hollow cavity. *3.* A cellar or underground room. *4.* The ash pit in a glass furnace. *5.* The partial or complete falling in of a mine; called also cave-in. *6.* Underground opening generally produced by solution of limestone large enough to be entered by a man.

cave breccia Angular fragments of limestone forming a fill.

cave coral A rough, knobby growth of calcite resembling coral in shape. *Syn:* CORAL FORMATION

cave earth *1.* Deposits of clay, silt, sand, or gravel flooring or filling a cave passage. In a more restricted sense cave earth includes only the finer fractions, i.e., clay, silt, and fine sand deposits. *Syn:* FILL

cave ice Ice formed in a cave by natural processes.

cave marble Cryptocrystalline banded deposit of calcite or aragonite capable of taking a high polish. *Syn:* CAVE ONYX

cave pearl Smooth, rounded concretion of calcite or aragonite formed by concentric precipitation around a nucleus. Usually found in caves. *Syn:* PISOLITE

cavern A subterranean hollow; an underground cavity; a cave. Often used, as distinguished from cave, with the implication of largeness or indefinite extent.

cavernous Containing cavities or caverns, sometimes quite large. Most frequent in limestones and dolomites.

cave system The underground network of passages, chambers, etc., or of caves in a given area whether continuous or discontinuous from a single opening.

caving system A method of mining in which the ore, the support of a great block being removed, is allowed to cave or fall, and in falling is broken sufficiently to be handled; the overlying strata subside as the ore is withdrawn. There are several varieties of the system. *See* BLOCK CAVING

cavitation *1.* Corrasive and corrosive effect of collapsing of bubbles produced by decrease of pressure due to increase of velocity (Bernoulli effect) at point where pressure is increased due to decrease of velocity. *2. Geophys:* The formation of a bubble by a charge detonated in water.

cavity Solutional concavity in limestone caves, the outline of which is determined by a joint or joints. Also applied to small hollows in cavernous lava.

cay *1.* A flat mound of sand built up on a reef flat slightly above high tide level. In some such mounds there is a large admixture of coral fragments, and the surfaces of the mounds may show a number of concentric ridges formed by successive additions along the peripheries of the mounds. *2.* A key; a comparatively small and low coastal island of sand or coral. Pro-

nounced "key." The spelling "kay" is common in the West Indies. *3.* A low insular bank of sand, coral, etc., awash or drying at low water.

Cayugan The youngest of three epochs in the Silurian of North America; also the series of strata deposited during that epoch.

CDP Abbreviation for common depth point; also referred to as CRP, or, common reflection point.

CDP stack Abbreviation for common depth point stack.

celadonite A mineral of the mica group, $K(Mg,Fe^{+2})(F^{+3},Al)Si_4O_{10}$-$(OH_2)$. Green, often occurring in basaltic rocks.

celestite A mineral, $SrSO_4$. Orthorhombic. The principal ore of strontium.

cellular *1.* Characterized by small openings or cells which may or may not be connected. *Cf.* VESICULAR; POROUS. *2. Petrol:* Applied to igneous rocks, especially lavas, containing numerous gas cavities. *Syn:* VESICULAR; SCORIACEOUS

Celsius scale A thermometric scale, proposed in 1742 by Anders Celsius with 0° as the boiling point of water and 100° as the melting point of ice, just the reverse of the centigrade scale, *q.v.*

cement *1.* Chemically precipitated material occurring in the interstices between allogenic particles of clastic rocks. Silica, carbonates, iron oxides and hydroxides, gypsum, and barite are the most common. Clay minerals and other fine clastic particles should not be considered cement. *2.* The word is also used in gold-mining regions to describe various consolidated, fragmental aggregates, such as breccia, conglomerate, and the like, that are auriferous. *3.* A finely divided metal obtained by precipitation. *Cf.* PORTLAND CEMENT

cementation The process of precipitation of a binding material around grains or minerals in rocks. Quartz, calcite, dolomite, siderite, and iron oxide are common cementing materials.

cement deposits The Cambrian conglomerates occupying supposed old beaches or channels. Gold bearing in the Black Hills.

cement rock An argillaceous limestone used in the manufacture of natural hydraulic cement. Contains lime, silica, and alumina in varying proportions, and usually more or less magnesia.

cement texture A result of the replacement of cementing matter of sandstones by ore minerals.

Cenomanian The fourth of nine ages of the Cretaceous Period; also the stage of strata deposited during that age.

cenote A type of sink developed in limestone areas by the collapse of caverns which cuts off natural channels of circulation and allows water to fill the depression.

Cenozoic The latest of the four eras into which geologic time, as recorded by the stratified rocks of the earth's crust, is divided; it extends from the close of the Mesozoic Era to and including the present. Also the whole group of stratified rocks deposited during the Cenozoic Era. The Cenozoic Era includes the periods called Tertiary and Quaternary in the nomenclature of the U. S. Geological Survey; some European authorities divide it, on a different basis, into the Paleogene and Neogene periods, and still others extend the Tertiary Period to include the whole.

center counter *Struct. petrol:* A circular hole in a piece of cardboard or plastic, used to count the number or percentage of points on a point diagram. The area of the hole is ordinarily 1% of the area of the larger projection and is usually 1 cm. in radius.

center line In U.S. public land surveys, the line connecting opposite quarter-section or sixteenth-section corners.

center of gravity That point in a body or system of bodies through which the resultant attraction of gravity acts when the body or system of bodies is in any position; that point from which the body can be suspended or poised in equilibrium in any position.

center of instrument The point of the vertical axis of rotation at the same elevation as the axis of collimation when that axis is in a horizontal position. A point at, or near, the intersection of the horizontal and vertical axes of a transit.

center of symmetry A point within an object through which any straight line drawn extends to similar point on the object at equal distances on opposite ends of the line.

centigrade scale A modification of the thermometric scale introduced by Anders Celsius. It has its zero at the melting point of ice, while 100° represents the boiling point of pure water at a pressure of 760 mm. of mercury. A centigrade degree is 9/5 of a Fahrenheit degree.

centipoise Unit for measuring viscosity, which is the tendency of a fluid to resist change of form. The centipoise is 1/100 poise. *See* POISE; VISCOSITY, ABSOLUTE.

centrifugal force The upward component of the kinetic reaction mv^2/ρ, where m is the mass, v the velocity, and ρ is the radius of curvature of the path of a moving particle, from Newton's second law of motion. This force is equal to a balancing component, F sin θ, called the centripetal force, or acceleration, where the vector F is the resultant of applied forces and θ is the angular motion in polar coordinates.

centrifugal replacement Replacement of a mineral which begins in the central part of the host mineral and works outward.

centripetal force The component F sin θ of an applied force which balances the centrifugal force acting on a moving particle changing direction. The vector F is the resultant of applied forces and θ is the angular motion in polar coordinates.

centripetal replacement Replacement of a mineral from the periphery inward.

centrosphere Barysphere. The central core of the earth, composed of heavy material and making up most of its mass.

cephalon Dorsal head shield of some arthropods consisting of several fused segments.

cephalopod One of the Cephalopoda. A marine invertebrate characterized by a head surrounded by tentacles and, in most fossil forms, by the presence of a straight or spirally coiled, calcareous shell divided into numerous interior chambers.

Cephalopoda Most highly developed class of mollusks that swam by ejecting a net of water from the mantle cavity through a muscular funnel. Most of those preserved as fossils had straight to symmetrically coiled shells divided into chambers by transverse septa.

ceramic Pertaining to pottery, including porcelain and terra cotta or its manufacture. Often taken to include all products made by heating natural materials—pottery,

chinaware, glass, bricks, cement, etc.

cerargyrite Horn silver. A mineral, AgCl. Isometric. A secondary mineral; an ore of silver.

ceratite A type of ammonoid with sutures in which the lobes are subdivided into subordinate crenulations although the saddles remain smoothly rounded, and undivided.

cerussite A mineral, $PbCO_3$, a member of the aragonite group. Orthorhombic. An ore of lead, commonly formed by the oxidation of lead sulfide.

cf [*Lat.* conferre] To compare. Used in paleontology to indicate that a specimen or specimens are closely comparable to but not the same as a named species.

cfr [*Lat.* conformalis] Used in paleontology by many authors to indicate that a specimen is similar, but too poorly preserved, to identify accurately.

cfs Cubic feet per second, a measure of the amount of water passing a given point.

chabazite A mineral of the zeolite group, $CaAl_2Si_4O_{12}.6H_2O$.

chain *1.* The legal unit of length for the survey of public lands of the United States. The chain is the equivalent of 20.13 m. The name is derived from Edmund Gunter's chain, which was a series of links connected by rings. Advantage in measuring in chains is that 10 sq. chains=1 acre. *2.* Any series of related, interconnected, or similar natural features, e.g., chain of mountains, islands, lakes.

chain coral A colonial coral found commonly in the Silurian; the name refers to the chainlike appearance of the upper surface of the colony.

chaining The operation of measuring a distance on the earth by means of a tape, commonly called a chain.

chain structure See INOSILICATES

chalcanthite Blue vitriol. A mineral, $CuSO_4.5H_2O$. Triclinic.

chalcedony Cryptocrystalline quartz and much chert, commonly microscopically fibrous. The material of agate.

chalcocite A mineral, Cu_2S. Orthorhombic. An important ore of copper, common in the zone of secondary enrichment.

chalcophile elements Elements which show a strong affinity for sulfur, and which are readily soluble in molten iron monosulfide; elements commonly found in sulfide ores.

chalcopyrite Copper pyrites. A mineral, $CuFeS_2$. Tetragonal. An important ore of copper.

chalk A very soft, white to light gray, unindurated limestone composed of the tests of floating microorganisms and some bottom-dwelling forms (ammonoids and pelecypods) in a matrix of finely crystalline calcite; some chalk may be almost devoid of organic remains.

chalybite Siderite.

chamber 1. An enlargement of a cave passage forming a cavity of relatively large size. Syn: HALL. 2. In Foraminifera, the unit of which all foraminiferal tests are composed, consisting of a cavity and the wall surrounding it. 3. The internal divisions of a cephalopod shell formed by septa which partition off the inner part of the cephalopod shell.

chamosite A mineral of the kaolinite group, $(Fe^{+2},Mg,Al,Fe^{+3})_6$ $(AlSi_3)O_{10}(OH)_8$. Monoclinic, an important constituent of many oolitic iron ores.

Champlainian The middle of three Ordovician epochs; also the series of strata deposited during that epoch.

chance packing A random combination of systematically packed grain colonies surrounded by, or alternating with, colonies packed haphazardly. The average porosity of chance-packed aggregates of uniform spheres is slightly less than 40%.

Chandler wobble Movement of the earth's axis that completes a cycle in about 428 days.

channel 1. The deepest portion of a stream, bay, or strait through which the main volume or current of water flows. 2. The part of a body of water deep enough to be used for navigation through an area otherwise too shallow for navigation. 3. A large strait, as the English Channel. 4. Metal: A sow or runner. 5. A cut along the line where rock or stone is to be split. 6. Paleontol: A groove, such as the groove that winds down the columella near its base in some shells and terminates in the siphonal notch or in the canal.

channel capacity The maximum flow which a given channel is capable of transmitting without overtopping its banks. See BANK-FULL STAGE

channeled upland Grooved upland.

channel-fill deposit Deposit which has accumulated in a stream channel where the transporting capacity of the stream has been insufficient to remove the sand and detritus as rapidly as it has been delivered.

channel flow Movement of surface runoff in long narrow depressions or troughs bounded by banks or valley walls that slope toward the channel.

channelization The straightening and deepening of streams to permit water to move faster, to reduce flooding or to drain marshy acreage for farming. However, channelization may reduce the organic waste assimilation capacity of the stream and may disturb fish breeding and destroy the stream's natural beauty.

channel-lag deposit Relatively

coarse materials that have been sorted out and left as a residual accumulation in the normal processes of a stream.

channel-mouth bar Bar built where a stream enters a body of standing water, resulting from decrease in velocity.

channel recording A system, chain, or cascade of interconnected devices through which geophysical data may flow from source to recorder, e.g., geophone, amplifier (with filters and gain control), galvanometer, and optical system.

channel sample A sample taken at a given spot but covering a relatively long distance and narrow width. It yields an average value and masks details. A composite collection, which is usually taken across the face of a formation or vein to give an average sample.

channel sands Sandstone deposited in a stream bed or other channel eroded into the underlying bed. Frequently contain oil, gold, or other valuable minerals.

channel storage The volume of water in definite stream channels above a given measuring point or "outlet" at a given time during the progress of runoff.

channel wave Any elastic wave which is propagated in a sound channel due to a low-velocity layer in the solid earth, the ocean, or the atmosphere.

channel width *Geomorph:* Width of a channel or stream near bankfull stage. *Symbol:* w.

chaos An exceedingly coarse breccia in which many blocks are 61 m. long, some are 402 m. long, and a few more than 804 km. long. Found in the Death Valley area of California and in adjacent areas, it has been interpreted by some as a gigantic fault breccia.

chaos structure Imbricated series of relatively minor lenticular

thrust blocks occurring beneath a major thrust block.

char The solid, carbonaceous residue that results from incomplete combustion or pyrolysis of organic material. It can be burned for its energy content or, if free from large amounts of impurities, processed further for production of activated carbon for use as a filtering medium. Char produced from coal is generally called coke, while that produced from wood or bone is called charcoal.

character In seismic work, a recognizable appearance of an event or reflection on records which serves to identify it or permit its correlation—often indefinable and without basis.

characteristic radiation A spectrum of definite wave lengths of electromagnetic radiation characteristic of the atomic number of the emitting element.

charge In seismic work, the explosive combination employed for a shot defined by the quantity and type of explosive used.

charnockite *1.* A quartzo-feldspathic gneiss or granulite with hypersthene. Regarded by some petrologists as igneous. *2.* A granitic rock with hypersthene as its chief mafic constituent. Originally applied to the granitic member of a series of hypersthene-bearing rocks ranging in composition from granite through norite to hypersthene pyroxenite. Some petrographers regard charnockites as the product of deep-seated metamorphism. *3.* Granulites characterized by the mineral assemblage quartz-orthoclase-hypersthene with or without garnet and plagioclase feldspar.

chart datum The plane or level to which soundings on a chart are referred, usually taken to correspond to a low water stage of the tide. *Cf.* DATUM PLANE

chasm *1.* A deep breach in the earth's surface; an abyss; a gorge;

a deep canyon. *2.* A deep recess extending below the floor of a cave.

chatoyant Having a luster resembling the changing luster of the eye of a cat. Chatoyancy is generally a property of translucent materials containing parallel fibrous structures capable of scattering light.

chattermark A scar made by vibratory chipping of a bedrock surface by drift carried in the base of a glacier.

Chazyan The oldest of three epochs of the Ordovician Period; also the series of strata deposited during that epoch.

chelation Decomposition or disintegration of rocks or minerals resulting from the action of organisms or organic substances.

chemical activity *See* ACTIVITY, CHEMICAL

chemical equilibrium A state of balance between two opposing chemical reactions.

chemical limestone A rock composed predominantly of calcite, formed by direct chemical precipitation.

chemical oxygen demand (COD) A measure of the amount of oxygen required to oxidize organic and oxidizable inorganic compounds in water. The COD test, like the BOD test, is used to determine the degree of pollution in an effluent.

chemical potential *1.* The chemical potential of a component i in a system is equal to the change of the Gibbs free energy of the system with the change in the number of moles m_i, the temperature, pressure, and number of moles of all other components being kept constant. The chemical potential is an intensive quantity, and the chemical potential of a component is defined at each point of the system. *See* GIBBS FREE ENERGY; INTENSIVE VARIABLE. *2.* Partial molal free energy.

chemical precipitate A sediment formed of material precipitated from solution or colloidal suspension, as distinguished from material transported and deposited as detrital particles, which is clastic.

chemosphere Atmospheric zone about 25 to 100 km. above the earth's surface containing a concentration of ozone.

chenier A long, low (3–6 m. high), narrow, wooded beach ridge or sandy hummock, forming roughly parallel to a prograding shoreline seaward of marsh and mud-flat deposits (such as along the coast of southern Louisiana), enclosed on the seaward side by fine-grained sediments, and resting on peat or clay. It is well drained and fertile, often supporting large evergreen oaks or pines on higher areas; its width varies from 45 to 450 m. and its length may be several tens of kilometers.

chernozem; tchornozem; tschernosem (*Russ.* black mold) Very black soil rich in humus and carbonates, which forms under cool to temperate, semiarid conditions; first distinguished in Russia where it covers most of the Aralo-Caspian plain and much of European Russia. An accreted soil with a good humic A horizon and a B horizon accreted with calcium and other mineral nutrients.

chert *1. Mineral:* A cryptocrystalline variety of quartz. Composed of interlocking grains generally not discernible under the microscope. *2. Rock:* A compact siliceous rock of varying color composed of microorganisms or precipitated silica grains. Occurs as nodules, lenses, or layers in limestone and shales.

chertification Essentially silicification, especially by fine-grained quartz or chalcedony, used mainly in the descriptions of the Mississippi Valley lead-zinc deposits.

Chesterian Youngest of four epochs of the Mississippian of North America; also the series of strata deposited during that epoch.

chestnut soil An old soil classification term, denoting a zonal soil having a dark brown surface horizon below which is a lighter colored horizon and an accumulation of lime. It is developed under temperate to cool, subhumid to semiarid climatic conditions.

chevron fold An accordion fold, the limbs of which are of equal length. *Cf.* ZIGZAG FOLD

chiastolite A variety of andalusite, in which carbonaceous impurities are arranged in a regular manner along the longer axis of the crystal, in some varieties like the letter X (Greek "chi"), whence the name.

Chile saltpeter Sodium nitrate.

chilled contact That part of an igneous rock that is finer grained nearer the contact than the rest of the igneous rock.

chilling effect The lowering of the earth's temperature due to the increase of atmospheric particulates that inhibit penetration of the Sun's energy.

chimney *1.* An ore shoot. *Cf.* CHUTE. *2.* A natural vent or opening in the earth as a volcano. *3.* A vertical shaft in the roof of a cave passage, smaller than an aven.

chimney rock *1.* An erosional feature formed where waves wear away materials on all sides, particularly along joint planes leaving angular steep-sided remnants. *Cf.* CHIMNEY

china clay Clay derived from decomposition of feldspar and suitable for the manufacture of chinaware or porcelain. *See* KAOLIN

chip A flat fragment with maximum dimension between 4 and 64 mm.

chip sample A series of chips of ore or rock taken either in a continuous line across an exposure or at uniformly distributed intervals.

chi-square test Statistical test that measures the probability of randomness in a distribution.

chitin A nitrogenous polysaccharide forming the basic part of the arthropod exoskeleton.

chiton Popularly called "coat of mail" shells. Invertebrate marine molluscan animals the shells of which consist of eight overlapping calcareous valves or plates.

chloanthite A mineral, $(Ni,Co)-As_{2-3}$. Isometric. An ore of nickel.

chloride A compound of chlorine and a positive radical of one or more elements.

chlorinated hydrocarbons A class of generally long-lasting, broad-spectrum pesticides of which the best known is DDT, first used for insect control during World War II. Other similar compounds include aldrin, dieldrin, heptachlor, chlordane, lindane, endrin, mirex, benzene hexachloride (BHC), and toxaphene.

chlorinity Originally defined as the total amount of chlorine, bromine, and iodine in grams contained in one kilogram of sea water, assuming that the bromine and iodine have been replaced by chlorine. Now defined as identical with the mass in grams of "atomic weight silver" just necessary to precipitate the halogens in 0.3285233 kg. of sea water.

chloritization The replacement by, conversion into, or introduction of chlorite.

chloritoid A dark-green, brittle mica found in metamorphic rocks. $(Fe'',Mg)_2Al_4Si_2O_{10}(OH)_4$.

chlorophaeite A mineral closely related to chlorite in composition and found in the groundmass of tholeiitic basalts where it occupies spaces between feldspar laths, forms pseudomorphs after olivine, or occurs in veinlets and amygdules.

chlorophyll The green pigment used in photosynthesis, found primarily in plants.

Chondrichthyes Class of vertebrates consisting of fish with skeletons of cartilage rather than bone; sharks.

chondrite Stony meteorite containing chondrules embedded in a fine-grained matrix of pyroxene, olivine, and nickel-iron with or without glass.

chondrodite A mineral, $Mg_5(SiO_4)_2(OH,F)_2$. Monoclinic. Commonly occurs in contact-metamorphosed dolomites.

chondrule A small rounded body of various materials, though chiefly olivine or enstatite, found embedded in a usually fragmental base in certain of the stony meteorites.

chonolith An intrusive body of igneous rock, so irregular in form, and in its relation to the invaded formations so obscure, that it cannot be properly designated as a dike, sill, or laccolith.

Chordata Phylum of animals with a notochord which in most is replaced by a bony spinal column. Vertebrata. May or may not be considered to include the Protochordata which do not develop a spinal column.

C horizon The layer of weathered bedrock at the base of a soil. It has undergone little alteration by organisms and is presumed to be similar in chemical, physical, and mineralogical composition to the material from which at least a portion of the overlying solum has developed.

chromate A salt of chromic acid; a compound containing the radical $(CrO_4)^{-2}$.

chromatography Method of qualitative chemical analysis in which a solution is tested by applying it to treated porous paper and identifications made on the basis of the nature and location of resulting colored spots.

chromite A mineral of the spinel group, formula $(Mg,Fe'')-(Cr,Al,Fe'')_2O_4$. Isometric. The principal ore of chromium.

chron A term originally introduced designating an indefinite division of geologic time. More recently proposed as the time unit equivalent to the stratigraphic unit, "subseries," and geologic name, "Mohawkian."

chronofauna Geographically restricted natural assemblage of interacting animal populations that maintained its basic structure over a geologically significant period of time.

chronolithologic unit Time-rock unit.

chronostratigraphic unit Geologic time unit; in order of decreasing magnitude; era, period, epoch, age, stage.

chronotaxis Similarity in age. *Cf.* HOMOTAXIS.

chronozone A time-stratigraphic unit or body of strata representing the rocks formed during any minor interval of geologic time.

chrysoberyl A mineral, $BeAl_2O_4$. Orthorhombic. Used as a gem, alexandrite, cat's-eye.

chrysocolla A mineral $CuSiO_3 \cdot 2H_2O$. Usually present as green to blue-green masses in the oxidized zone of copper sulfide deposits.

chrysolite Olivine.

chrysoprase An apple-green chalcedony, used as a gem.

chrysotile A mineral of the serpentine group, $Mg_3Si_2O_5(OH)_4$. An important constituent of asbestos.

chrystocrene A mass of ice

formed in the interstices of a talus by the freezing of the waters of a subjacent stream.

churn drill *See* CABLE TOOLS

chute *1.* Chutes are the narrow passages of water between an island and the mainland. *2.* Of water, a fall; a quick descent, as in a river, or a steep channel, or narrow sloping passage by which water falls to a lower level; a rapid, a shoot. *3.* A channel or shaft underground, or an inclined trough above ground, through which ore falls or is "shot" by gravity from a higher to a lower level. *4.* In Pennsylvania, a crosscut connecting a gangway with a heading. *5.* An inclined watercourse, natural or artificial, especially one through which boats or timber are carried, as in a dam. *6.* A narrow channel with a free current, especially on the lower Mississippi River. *7.* A body of ore, usually of elongated form, extending downward within a vein.

chute cutoff Chute, *q.v.*

cienaga *1.* An area where the water table is at or near the surface of the ground. Standing water occurs in depressions in the area, and it is covered with grass or sometimes with heavy vegetation. The term is usually applied to areas ranging in size from several hundred square feet to several hundred or more acres. Sometimes springs or small streams originate in the cienaga and flow from it for short distances. *2.* An elevated or hillside marsh containing springs. Local in Southwest.

Cincinnatian The youngest of three epochs of the Ordovician Period; also the series of strata deposited during that epoch.

cinder Scoriaceous lava from a volcano; volcanic scoria.

cinder coal *1.* In England, coal altered by heat from an intrusion of lava. *2.* In Australia, a very

inferior natural coke, little better than ash.

cinder cone A conical elevation formed by the accumulation of volcanic ash or clinkerlike material around a vent.

cinders, volcanic Primarily uncemented, essential, glassy, and vesicular volcanic ejecta ranging chiefly from 4 to 3 mm. in diameter.

cinnabar A mineral, HgS. Hexagonal rhombohedral. Color vermilion. The principal ore of mercury.

C.I.P.W. classification Norm system, *q.v.* From the initial letters of the names of the men who originated it: Cross, Iddings, Pirsson, and Washington.

circular sections Although most sections through ellipsoids are ellipses, two sections that contain the intermediate axis and are symmetrical, disposed relative to the long and short axes, are circles. The circular sections are often referred to in discussing rock deformation in terms of the strain ellipsoid.

circulation In rotary drilling, the process of pumping mud-laden or other fluid down drill pipe, through the drilling bit, and upward to the surface through the annulus between drill-hole walls and drill pipe.

circumferential wave Seismic wave that travels parallel to the earth's surface.

Circum-Pacific province *See* PACIFIC SERIES

cirque *1.* A hollow, shaped like a Roman armchair. It is bowl-like, open in front, and its back and sides are formed by arêtes which rise like arms. In the bowl of the cirque there is commonly a small round lake or tarn. *2.* A deep, steep-walled recess in a mountain, caused by glacial erosion. *Syn:* CWM (Wales); CORRIE (Scotland); BOTN (Sweden); KAR (Germany)

cirque glacier A small glacier occupying a cirque or resting on the headwall of a cirque.

cirque lake Small body of water occupying a cirque depression, dammed by a rock lip, small moraine, or both. *See* TARN

citrine A yellow variety of crystalline quartz.

cladogenesis *1.* Phyletic splitting, or branching speciation—speciation by division of the phyletic line. *2.* Progressive evolutionary specialization.

claim The portion of mining ground held under the Federal and local laws by one claimant or association, by virtue of one location and record. Lode claims, maximum size 183 by 457.5 m. Placer claims 201.3 by 402.6 m. A claim is sometimes called a "location." *See* MINING CLAIMS

clam A pelecypod, generally one that is not attached and that has two similar valves.

clan *Petrog:* A clan of rocks is one bound by resemblances in composition. *Paleon:* A hierarchy of classification used by some zoologists; ranks below the subfamily and above the genus.

Clapeyron's equation As usually used, this is an expression relating the pressure and temperature of a phase transition in a closed system (such as melting) when the two modifications of the same substance coexist in equilibrium. It includes the heat of the transformation and the volume change of the transformation. It is sometimes also referred to as the Clausius-Clapeyron equation or the Clapeyron-Clausius equation.

clarain That ingredient of banded coal which appears megascopically as thin or very thick bands, intrinsically stratified parallel to the bedding plane; most often has a silky luster and scattered or diffuse reflection markedly less intense than the specular reflection of vitrain under the same illumination. It has no conchoidal fracture, but splits in sheets or irregular directions. Less friable than vitrain.

clarification In waste water treatment, the removal of turbidity and suspended solids by settling, often aided by centrifugal action and chemically induced coagulation.

clarifier In waste water treatment, a settling tank which mechanically removes settleable solids from wastes.

clarke The average percentage of an element in the earth's crust.

clarke of concentration Measure of the amount of an element present in a particular deposit or mineral.

class A biologic unit; a subdivision of a phylum.

classification *Biol.* and *Paleontol:* The formal arrangement of organisms in the groups of a hierarchy of taxonomic categories.

clast An individual constituent of detrital sediment or sedimentary rock produced by the physical disintegration of a larger mass.

clastation The act or method or means of disrupting rocks to form clastic sediments.

clastic Consisting of fragments of rocks or of organic structures that have been moved individually from their places of origin. *Cf.* DETRITAL, FRAGMENTAL

clastic dike A tabular body of clastic material transecting the bedding of a sedimentary formation, representing extraneous material that has invaded the containing formation along a crack either from below or from above. Sandstone dike, *q.v.*

clasticity index Measure of the maximum apparent grain size of a sediment.

clastic ratio The statistical relationship of the percentage of clastic rocks in a given geologic section compared with the percentage of nonclastic rocks in the same section.

clastic rock A consolidated sedimentary rock composed of the cemented fragments broken from preexisting rocks of any origin by chemical or mechanical weathering; e.g., conglomerate, sandstone, shale.

clay 1. A size term denoting particles, regardless of mineral composition, with diameter less than 1/256 mm. (4 microns). 2. A group of hydrous alumino-silicate minerals related to the micas (clay minerals). 3. A sediment of soft plastic consistency composed primarily of fine-grained minerals. 4. In engineering, any surficial material that is unconsolidated.

clay gall A dry, usually curled "clay-shaving" resulting from the drying and cracking of mud, which is later embedded in a sand stratum. Named because it resembles the gall of the mud wasp.

clay gouge A thin seam of clay separating ore, or ore and rock.

clay mineral A finely crystalline hydrous silicate of aluminum, iron manganese, magnesium, and other metals belonging to the phyllosilicate group. The principal clay mineral groups are kaolinite, smectite (montmorillonite), illite, and vermiculite.

claypan A soil term used to describe a dense, heavy, relatively impervious subsoil layer that owes its character to a relatively high clay content. Formed by the concentration of clay carried in downward percolating waters. See HARD PAN

clay plug Sediment with much organic muck deposited in a cutoff river meander.

clay shale Shale composed wholly or chiefly of argillaceous material, which again becomes clay on weathering.

clay slate 1. Slate derived from

shale. 2. Very hard consolidated shale.

claystone An indurated clay having the composition of shale, but lacking the fine lamination or fissility.

clay vein A body of clay, usually roughly tabular in form, like an ore vein, which fills a crevice in a coal seam.

cleavage 1. Crystallog: The splitting, or tendency to split, along planes determined by the crystal structure. Cleavage is always parallel to a possible crystal face, i.e., to a rational lattice plane of the crystal, and is generally designated by the name of the face or form, as basal, pinacoidal, cubic, etc. 2. Petrol: A tendency to cleave or split along definite, parallel, closely spaced planes, which may be highly inclined to the bedding planes. It is a secondary structure, commonly confined to bedded rocks, is developed by pressure, and ordinarily is accompanied by at least some recrystallization of the rocks.

cleavelandite A white, lamellar variety of albite common in pegmatites.

clerici solution A solution having a specific gravity of 4.25, used for separating heavy and light minerals. The double formate-malonate.

cliachite Amorphous generally brownish material that constitutes most so called bauxite.

cliff A high, steep face of rock; a precipice. Cf. SEA CLIFF

climate The sum total of the meteorological elements that characterize the average and extreme condition of the atmosphere over a long period of time at any one place or region of the earth's surface.

climate, continental The type of climate characteristic of land areas separated from the moderating influence of oceans by distance or mountain barriers. It is marked by relatively large daily and seasonal changes in temperature.

climate, oceanic The type of climate characteristic of land areas near oceans which contribute to the humidity and at the same time have a moderating influence on temperature and range of temperature variation. *Syn:* MARINE CLIMATE

climate-stratigraphic unit Non-uniform time unit (geochron) corresponding to an important climatic interval in an alternating or changing series.

climatic classification Classification of the climates of the different regions of the earth's surface, based on one or more of the climatic elements such as (1) temperature, (2) rainfall, (3) humidity, (4) wind, (5) temperature and rainfall, (6) nearness to land and sea, and many others. Classifications may also be based on the distribution of vegetation, on physiological effects, or may be on any basis suitable for the particular purpose or investigation.

climatic cycles Actual or supposed recurrences of such weather phenomena as wet and dry years, hot and cold years, at more or less regular intervals in response to long-range terrestrial and solar influences, such as volcanic dust and sunspots. The best known of these cycles, which have been discovered in great numbers, are the Brückner Cycle, and the glacial-interglacial periods.

climatic factors Certain physical conditions—other than the climatic elements—which currently control climate, or may by their changes over long periods cause climatic changes.

The current factors which exercise immediate control are latitude, altitude, distribution of land and sea, and topography. In addition to the above, some climatologists include many other factors, such as ocean currents, the semipermanent high- and low-pressure areas, prevailing wind, etc.

Some of the long-range factors are: obliquity of the ecliptic, extent and composition of the atmosphere, land elevation, land and water distribution, ocean circulation, etc.

climatic optimum Period of relatively high temperature since the retreat of the last Pleistocene glacier, a climatic warming between 5000 and 7000 years ago. *Cf.* ALTITHERMAL

climatic province An area of the earth which has a definite climatological character, according to a certain climatic classification, *q.v.*

climatic zone One of many latitudinally oriented areas typified by similar climate, as tropical, temperate, etc.

climax The terminal community of a sere which is in dynamic equilibrium with the prevailing climate. The major world climaxes are equivalent to formations and biomes. The term is also used in connection with any subdivision, such as climax association.

climbing bog In regions characterized by a short summer and a considerable amount of rainfall this plant sphagnum frequently extends upward from the original level of the swamp, carrying the marsh conditions to higher land.

climbing dune An active dune capable of moving over obstructions.

cline Group of organisms varying from place to place or from time to time that show no breaks in morphology or ability to interbreed.

clinker *1.* Burnt-looking, vitrified or slaggy material thrown out by a volcano. *2.* Rough, jagged lava, generally basic, typically occurring at the surface of lava flows. *3.* Slaggy or vitreous masses of coal ash.

clinker, volcanic Rough, jagged, scoriaceous, spinose fragments of lava, usually of basic composition and typically found on the surface of lava flows.

clinkstone An extrusive rock which is sonorous when struck with a hammer. *See* PHONOLITE

clino A term proposed for the environment of the sloping part of the floor of the sea which extends from wave base down to the more or less level deeper parts.

clinoaxis The inclined lateral axis in the monoclinic system, designated a.

clinochlore A mineral, a member of the chlorite group, composition approximately $(Mg,Fe'')_4Al_2$-$(Al_2Si_2)O_{10}(OH)_8$. Monoclinic.

clinodome Monoclinic crystal form whose faces are parallel to the inclined a-axis and intersect the other two.

clinoenstatite A mineral of the pyroxene group similar to enstatite but containing some calcium. Monoclinic.

clinoferrosilite A mineral of the pryoxene group similar to ferrisilite but containing calcium. Monoclinic.

clinoform The subaqueous land form analogous to the well-known "continental slope" of the oceans or to the fore-set beds of a delta.

clinometer A simple apparatus for measuring vertical angles, particularly dips, by means of a pendulum or spirit level and circular scale.

clinopinacoid In crystallography, the pair of facies, both of which are parallel to the a and b crystallographic axes.

clinopyroxene Any of a group of pyroxenes crystallizing in the monoclinic system and sometimes containing considerable calcium with or without aluminum and the alkalies.

clinothem The deposit that is laid down in the clino environment and the surface of which produces a clinoform; e.g., deposited on the continental slope, or its counterpart slope environment on deltas, in lakes, in cratonic basins, etc.

clinozoisite A mineral, a member of the epidote group, Ca_2Al_3-$(SiO_4)_3(OH)$. Monoclinic.

Clinton ore A red, fossiliferous, iron ore of the Clinton formation of the United States, with lenticular grains. *Syn:* OYESTONE ORE; FOSSIL ORE; FLAXSEED ORE

clod A term applied by miners to loosely consolidated shale commonly found in close conjunction with a coal bed.

closed basin A district draining to some depression or lake within its area, from which water escapes only by evaporation.

closed fold A fold, the limbs of which have been so compressed that they are parallel. A fold in which the beds cannot be bent closer without introducing flowage. A fold can be described on a relative scale from closed to open. *Ant:* OPEN FOLD. *Syn:* TIGHT FOLD

closed form *Crystallog:* A crystal form that encloses a finite volume of space.

closed system A system is closed if during the process under consideration no transfer of matter either into or out of the system takes place.

close-grained Having fine and closely arranged fibers, crystals, or texture. Usually said of rocks.

close-jointed A term applied to joints that are very near together.

close packing A three-dimensional stacking of equal spheres in which each sphere has contact with twelve others.

closure *Struct. geol:* In a fold, dome, or other structural trap, the vertical distance between the structure's highest point and its lowest contour that encloses itself. It is used in the estimation of petroleum reserves. *Syn:* STRUCTURAL CLOSURE. *Surv:* A cumulative measure of the various individual errors in survey measurements.

coal A readily combustible rock containing more than 50% by weight and more than 70% by volume of carbonaceous material including inherent moisture, formed from compaction and induration of variously altered plant remains similar to those in peat. Differences in the kinds of plant materials (type), in degree of metamorphism (rank), and in the range of impurity (grade) are characteristic of coal and are used in classification. *Syn:* BLACK DIAMOND

coal ball Concretions of mineralized plant debris, occurring in coal seams or in adjacent rocks.

coal basin Depressions in the older rock formations, in which coal-bearing strata have been deposited.

coal bed A coal seam. *Also spelled* COALBED

coalescing pediment The result of the union of individual pediments which results in a continuous pediment surrounding a mountain range.

coal field A region in which deposits of coal occur. Also called coal basin when of basinlike structure.

coal gas The fuel gas produced from a high-volatile bituminous coal. The average composition of coal gas by volume is: 50% hydrogen, 30% methane, 8% carbon monoxide, 4% other hydrocarbons, and 8% carbon dioxide, nitrogen, and oxygen.

coalification Those processes involved in the genetic and metamorphic history of coal beds.

coal land Land of the public domain which contains coal beds.

coal measures *1.* Strata containing coal beds, particularly those of the Carboniferous. *2.* Capitalized and used as a proper name for a stratigraphic unit more or less equivalent to the Pennsylvanian or Upper Carboniferous.

coal plant A fossil plant found in association with, or contributing by its substance to the formation of, beds of coal, esp. in the coal measures.

coal seam A stratum or bed of coal, usually of such quality and thickness (at least 30 cm.) as to be mined with profit. *Syn:* COAL BED

coal types Those differences due to variations in the kind of plant material of which the coal is composed, whereby such varieties as common banded coal, cannel coal, algal coal, and splint coal are produced.

coarse sand Sand with a diameter between 0.5 and 1 mm.

coarse topography An incompletely dissected surface, or one in which the erosional features are on a large scale.

coast A strip of land of indefinite width (may be several miles) that extends from the seashore inland to the first major change in terrain features.

coastal comb Sea ice thrown on the coast by high tide, surf, or compression.

coastal current One of the offshore currents flowing generally parallel to the shoreline with a relatively uniform velocity (as compared to the littoral currents).

coastal plains Any plain which

has its margin on the shore of a large body of water, particularly the sea, and generally represents a strip of recently emerged sea bottom.

coastal zone Coastal waters and adjacent lands that exert a measurable influence on the uses of the sea and its ecology.

coastline *1.* Technically, the line that forms the boundary between the coast and the shore. *2.* Commonly, the line that forms the boundary between the land and the water.

coast of emergence Shoreline of emergence. Made by an elevation of the sea bottom, which is added to the land, causing the sea to withdraw correspondingly; the new shoreline is determined by the amount of upheaval and the slope of the sea bottom.

coast of submergence Shoreline of submergence. Produced by a depression of the land and invasion by the sea, which fills the lower valleys.

Coast Range orogeny Major deformation, metamorphism, and plutonism during Jurassic and Early Cretaceous Time in the Coast Mountains of the Cordillera of British Columbia.

coast shelf Submerged coastal plain.

cobaltite A mineral, CoAsS, the principal ore of cobalt. Isometric.

cobble Boulderet; cobblestone. A rock fragment between 64 and 256 mm. in diameter, thus larger than a pebble and smaller than a boulder, rounded or otherwise abraded in the course of aqueous, eolian or glacial transport.

cobblestone *1.* Cobble. *2.* A rounded stone suitable for paving a street or road.

coccolith Very tiny calcareous plates, generally oval and perforated, borne on the surfaces of some marine flagellate organisms.

coccolithophore Flagellate organism that produces coccoliths.

cockade structure *1.* Concentric mineral growths on inclusions or breccia fragments. *2.* A term applied to successive crusts or unlike minerals deposited on breccia fragments in a vein.

COD *See* CHEMICAL OXYGEN DEMAND

codeclination The complement of the declination (astronomic), 90° minus the declination. Same as polar distance.

coefficient of haze (COH) A measurement of visibility interference in the atmosphere.

coelacanth One of the ancient (Devonian to Cretaceous as fossils, and one recent species) group of true or bony fishes, belonging to the Crossopterygii, allies of the lungfishes and near the ancestry of the amphibians.

Coelenterata *1.* Cnidaria. *2.* Phylum of solitary or colonial animals whose bodies consist of ectodermal and endodermal layers but lack a mesoderm; two forms, polyps and medusae, occur and may characterize alternate generations.

coelome A body cavity separate from the enteron or the intestinal tract; it occurs in multicelled animals other than Metazoa, sponges, and coelenterates. It is usually not an open cavity, but that cavity which contains the organs of the body.

coesite A mineral, SiO_2, a polymorphic form of quartz generally considered to form at very high pressures. Found in impact craters and associated structures. Monoclinic.

coffin A thick-walled container (usually lead) used for transporting radioactive materials.

cognate Term applied to a block of solidified lava which had been broken by later eruptions.

cognate inclusion Cognate xenolith; autolith, *q.v.* A term applied to a xenocryst or xenolith occurring in an igneous rock to which it is genetically related.

COH *See* COEFFICIENT OF HAZE

cohesion *1.* The resistance of a material, rock, or sediment against shear along a surface which is under no pressure. *2.* The capacity of sticking or adhering together. In effect the cohesion of soil or rock is that part of its shear strength which does not depend upon interparticle friction.

coke Bituminous coal from which the volatile constituents have been driven off by heat, so that the fixed carbon and the ash are fused together. Commonly artificial, but natural coke is also known.

coking coal Coal which can be converted into useful coke that must be strong enough to withstand handling. There is no direct relation between the elementary composition of coal and coking quality but generally coals with 80 to 90% carbon on a dry, ash-free basis are most satisfactory.

col [*Fr.*] A saddle or gap across a ridge or between two peaks; also, in a valley in which streams flow both ways from a divide, that part of the valley at the divide, especially if the valley slopes rather steeply away from the divide. *See* PASS

colatitude The complement of the latitude, or 90° minus the latitude.

colemanite A mineral, $Ca_2B_6O_{11}$.-$5H_2O$. Monoclinic.

Coleoidea Subclass of Cephalopoda having two gills, among other distinguishing features. Dibranchiata.

coliform index An index of the purity of water based on a count of its coliform bacteria.

coliform organism Any of a number of organisms common to the intestinal tract of man and animals, and therefore not harmful themselves, but whose presence in waste water is an indicator of pollution and of potentially dangerous bacterial contamination.

collapse breccia Founder breccia, *q.v.*

collapse caldera A caldera resulting primarily from collapse occasioned by the withdrawal of magmatic support of a depth or, more rarely, by the internal solution of a volcanic cone. *See* CALDERA

collapse sink Caverns may become so enlarged by solution and erosion that they may locally collapse, thus giving rise to another class of sinkholes which may be called collapse sinks.

collapse structures Gravity-collapse structures. Structures resulting from the downhill sliding of rocks under the influence of gravity to produce small klippe or folds.

collective diagram *Struct. petrol:* A point or contour diagram prepared by collecting onto one diagram the data from two or more other diagrams.

collimate *1.* To bring into line, as, the axes of two lenses or of two telescopes. Also to make parallel, as, refracted or reflected rays. *2.* To determine or correct the direction of the line of sight of a telescope by use of a collimator, or by vertical reflection from the surface of a basin of mercury.

collimation axis The straight line passing through the optical center of the object glass of a transit and the horizontal rotation axis perpendicular to the latter.

collimation error The angle between the line of collimation (line of sight) of a telescope and its collimation axis.

collimation line The line through the nodal point of the objective lens of a telescope and the center of the reticle (intersection of the crosshairs).

collimation plane The plane described by the collimation axis during the revolution of a transit.

collimator A fixed telescope with spider lines in its focus, used to adjust a second telescope by looking through it in a reverse direction with the latter, so that images of the spider lines are formed in the focus of the second telescope, as if they originated in a distant point.

colloform A textural term applied to finely crystalline, concentric mineral layering. Individual layers commonly feature radial crystal growth, as in wood-tin, chalcedony, psilomelane. *Syn:* BOTRYOIDAL; RENIFORM; GLOBULAR

colloid A substance in a state of fine subdivision having peculiar properties because of its extremely high surface area. A common colloid in nature is clay having unusual properties such as plasticity, thixotropy, and swelling. A fine-grained material that is held in suspension.

colloidal dispersion A suspension of particles of colloidal size in a medium, usually liquid. A sol or colloidal suspension, also aerosol when suspended in a gas instead of liquid.

collophane A white powdery amorphous substance having the composition of apatite.

collophanite Collophane. A dull, colorless or snow-white hydrous, calcium phosphate. $Ca_3P_2O_8 + H_2O$.

colluvial Consisting of alluvium in part and also containing angular fragments of the original rocks. Contrasted with alluvial and diluvial. Also, talus and cliff debris; material of avalanches.

colluvium A general term applied to loose and incoherent deposits, usually at the foot of a slope or cliff and brought there chiefly by gravity. Talus and cliff debris are included in such deposits.

colonial coral *See* CORAL, COLONIAL

colonization A natural phenomenon where a species invades an area previously unoccupied by that species and becomes established.

colony *1.* A group of similar organisms living together in close association; more specifically, a group of associated unicellular organisms among which there are no marked structural differences and little or no division of labor. 2. A group of individuals of a given species of animal or plant which is organized together so that the independence of individual members is partly or wholly lost.

color index *Petrol:* The sum of the dark or colored minerals in a rock expressed in percentages. It is especially applied in the classification of igneous rocks. According to this index, rocks may be divided into leucocratic (color index, 0–30), mesotype or mesocratic (color index, 30–60), and melanocratic (color index, 60–100). Shand recognizes a fourth subdivision, namely hypermelanic (color index, 90–100).

color ratio Color index, *q.v.*

columbite A mineral, the Nb-rich end member of the columbite-tantalite series, $(Fe,Mn)(Nb,Ta)_2O_6$, orthorhombic. The principal ore of niobium found in certain pegmatites.

columella A dome-shaped structure in the sporangia of bread mold and related fungi; also, a central mass of sterile tissue in sporophytes of mosses and liverworts.

columnar jointing That variety of jointing that breaks the rock into columns. Usually the joints form a more or less clearly defined hexagonal pattern. Most characteristic of basaltic rocks. Generally considered to be shrinkage cracks because of cooling.

columnar section A graphic expression of the sequence and stratigraphic relations of rock units in a region. In a vertical column, lithology is shown by standard symbols and thicknesses of rock units are drawn to scale.

columnar structure *1.* A mineralogical structure in which the unit is made up of slender columns, as in some amphiboles. *2.* Columnar jointing, *q.v.*

comagmatic A term applied to igneous rocks (or to the region in which they occur) having a common set of chemical, mineralogical, and textural features, and hence regarded as having been derived from a common parent magma. Essentially synonymous with consanguineous.

Comanchean Lower Cretaceous.

comb; combed vein (comb structure) The place, in a fissure which has been filled by successive depositions of mineral on the walls, where the two sets of layers thus deposited approach most nearly or meet, closing the fissure and exhibiting either a drusy central cavity, or an interlocking of crystals.

comber *1.* A deep-water wave whose crest is pushed forward by a strong wind, much larger than a whitecap. *2.* A longperiod spilling breaker.

combined sewers A sewerage system that carries both sanitary sewage and storm water runoff.

combustible shale Tasmanite.

combustion Burning. Technically, a rapid oxidation accompanied by the release of energy in the form of heat and light. It is one of the

three basic contributing factors causing air pollution; the others are attrition and vaporization.

commensal Said of an organism which lives with another as a tenant or as a coinhabitant, but not as a parasite.

commensalism The growth together of different species of organisms in a manner helpful to one or both without being detrimental to either.

comminution *1.* The reduction of a substance to a fine powder; pulverization; trituration. *2.* Mechanical shredding or pulverizing of waste, a process that converts it into a homogeneous and more manageable material. Used in solid waste management and in the primary stage of waste water treatment.

common depth point A subsurface reflecting point from which reflections are recorded that are common to several different shot point to geophone distances. A CDP seismic reflection survey is designed so that successive shot points are properly positioned relative to a fixed arrangement of recording units to provide a multiplicity of reflections from the same subsurface point.

common depth point stack A sum of traces whose reflection events correspond to the same common depth points. The summing is done after appropriate statics and normal moveout corrections have been applied to each trace. The objective is to attenuate random and multiple seismic signals by successive summing of reflection events.

common lead Lead having four isotopes (mass numbers 204, 206, 207, 208) in the proportions generally obtained by analyzing lead from rocks and lead minerals which are associated with little or no radioactive material; commonly considered to be the lead present at the time of the earth's formation, as distinguished from lead produced later by radioactive decay.

community An organized group of plants or animals, or both. The term is employed when it is not necessary or desirable to use a more specific designation such as association, associates, etc.

comorphism Occurrence in more than one state of coordination in a crystal, e.g., aluminum in four- or six-fold coordination in silicates.

compactability Property of sedimentary material permitting decrease in volume or thickness under load accomplished by closer crowding of constituent particles and accompanied by decrease in porosity and increase in density.

compaction 1. Decrease in volume of sediments, as a result of compressive stress, usually resulting from continued deposition above them, but also from drying and other causes. 2. Reducing the bulk of solid waste by rolling and tamping.

compass 1. An instrument for determining directions, usually by the pointing of a magnetic needle free to turn in a horizontal plane, as, the ordinary surveyor's compass though sometimes having a clinometer attached. Also, a dip compass, for tracing magnetic iron ore, having a needle hung to move in a vertical plane. 2. An instrument for describing circles, transferring measurements, etc.

compensation, isostatic A theory of equilibrium of the earth's crust assuming that columns of rock and water standing on bases of equal area have equal weights irrespective of the elevation and configuration of their upper surfaces.

compensation point Point at

which the color of a mineral in thin section between crossed Nicols is compensated (becomes dark gray) by the introduction of a quartz wedge.

competency; competence *Hydraul:* Refers to the maximum size of particles of given specific gravity, which, at a given velocity, the stream will move. Thus, a small, rapid stream can move a relatively large particle, and while its competence is great the amount of material transported is small. Conversely a large, slow-moving stream may carry in suspension a great quantity of small particles, its competence being small, but its capacity great.

competent 1. *Hydraul:* Pertaining to the competence of a stream or current of air. 2. *Struc. geol:* Said of a bed or stratum that is able to withstand the pressure of folding without flowage or change in original thickness. 3. Said of a fold in which the strata have not flowed or changed their original thickness.

complementary Refers to (1) rock types differentiated from a common magma; (2) associated dikes or other minor intrusions regarded as leucocratic and melanocratic differentiates of a common magma; (3) small differentiated igneous bodies whose compositions if mixed would approximate composition of presumed source magma.

complete glacier A complete glacier consists of two parts, the reservoir, where there is accumulation, and the dissipator, where there is wastage; these regions are separated by the névé line, and there is continual flow from the first to the second.

complex 1. *Mineral:* Containing many ingredients; compound or composite. 2. An assemblage of rocks of any age or origin that has been folded together, intricately mixed, involved, or otherwise complicated. *3.* Stratigraphic equivalent of the geologic time-unit eon.

complex ion Any ion consisting of several atoms, other than the common radicals like SO_4^- or NO_3^-. Examples are $Cu(NH_3)_4^{++}$, HgS_2^{--}, $AgCl_2^-$, $Au(CN)_2^-$.

complex ripple mark An interference ripple pattern of any kind.

complex spit A large spit with minor or secondary spits developed on the ends or points of the large spit.

component One of the independent substances present in each phase of a heterogeneous equilibrium. The number of components in a system is the minimum number of chemical constituents which must be specified in order to describe the composition of each phase present.

componental movement Mechanical deformation that is composed of many movements of component parts in such a way that the continuity of the rock is not impaired.

components The smallest number of independently variable chemical individuals (gaseous, liquid, or solid elements or compounds) by means of which the composition of each phase present may be quantitatively expressed.

composite coast An initial coast resulting from upwarping and subsidence of coastal blocks along lines transverse to the coast. Upwarping produces coastal salients, whereas downwarping produces embayments.

composite cone A volcanic cone, usually of large dimension, built of alternating layers of lava and pyroclastic material. *Syn:* STRATIFIED CONE; STRATOVOLCANO. *See* VOLCANIC CONE

composite dike A dike composed of two or more intrusions of different chemical and mineralogical compositions.

composite fault scarp A scarp whose height is due partly to differential erosion and partly to fault movement.

composite fold A fold with small folds on its limbs, regardless of dimensions.

composite gneiss A layered rock resulting from intimate penetration of magma, usually granitic, into country rocks. *See* INJECTION GNEISS; MIGMATITE

composite mobile belt Mobile belt including two or more parallel geosynclinal troughs.

composite sample A sample present a type such that more than a single set of characteristics inherent in the original material may be combined.

composite sill A sill composed of two or more intrusions of different chemical and mineralogical compositions.

composite stream A river may be composite when drainage areas of different structure are included in the basin of a single stream.

composite topography The combination of topographic features from an earlier incomplete cycle of denudation with those of the current cycle of denudation.

composite vein A large fracture zone, up to many tens of feet in width, consisting of several parallel ore-filled fissures and converging diagonals, whose walls and intervening country rock have undergone some replacement.

composition *1.* An aggregate, mixture, mass, or body formed by combining two or more substances; a composite substance. *2.* The chemical constitution of a rock or mineral. *3.* The mineralogical constitution of a rock.

composition plane Composition face; composition surface. A plane by which the two individuals of a contact twin are united. *See* TWIN

composition point In any plot of phase equilibria, that point whose coordinates represent the chemical composition of a phase or mixture.

composition triangle In ternary systems, used in several connotations: *1.* The triangle formed by connecting the composition points of any three phases that may be in equilibrium with each other. *2.* The triangle formed by connecting the composition points of any three primary phases whose liquidus surfaces meet at a point. *3.* The triangle formed by connecting the composition points of any three phases in equilibrium at the liquidus, e.g., two solid solutions and a liquid solution.

compound alluvial fan Bajada; piedmont alluvial plain, *q.v.* The fans made by neighboring streams often grow laterally until they merge. The union of several such fans makes a compound alluvial fan, or a piedmont alluvial plain.

compound coral *See* CORAL, COMPOUND

compound ripple mark Complex ripple mark consisting of one set of ripples modified by another differently oriented set.

compound sample A mixture of a number of spot samples to form an aggregate single sample. Spot samples usually are taken according to some plan and some fixed number. They may be close together or scattered. The result is an average sample or composite sample.

compound twins *See* TWIN

compound valley A valley in which part of its course belongs to one class or stage of development and in the other part to another class.

compound valley glacier One composed of two or more individual ice streams coming from different tributary valleys.

compound vein *1.* A vein or lode consisting of a number of parallel fissures united by cross fissures, usually diagonally. *2.* A vein composed of several minerals.

compound volcano A volcano that consists of a complex of two or more cones, or a volcano that has an associated volcanic dome, either in its crater or on its flanks. Examples are Vesuvius and Mount Pelée.

compressibility The change of specific volume and density under hydrostatic pressure; reciprocal of bulk modulus, *q.v.*

compression A system of forces or stresses that tends to decrease the volume or shorten a substance, or the change of volume produced by such a system of forces.

compression, axial In experimental work with cylinders, a compression applied parallel with the cylinder axis; in interpretation of deformed rocks to be used in an appropriate sense only.

compressional wave Longitudinal wave; dilatational wave; P-wave; pressure wave; irrotational wave.

compressive strength The load per unit of area under which a block fails by shear or splitting.

compressive stress A stress that tends to push together the material on opposite sides of a real or imaginary plane.

concentrate *1.* To increase the strength by diminishing the bulk as of a liquid or an ore; to intensify or purify by getting rid of useless material. *2.* To separate metal or ore from the gangue or associated rock.

concentric faults Faults that are concentrically arranged in plan.

concentric fold *See* PARALLEL FOLD

concentric fractures A system of fractures more or less concentrically disposed about a center.

concentric weathering Spheroidal weathering, q.v.

conchiolin A nitrogenous substance constituting the organic basis of most molluscan shells.

conchoidal A type of rock or mineral fracture giving smoothly curved surfaces. Characteristic of quartz and obsidian.

concordant A term used to describe those intrusive igneous bodies in which the contacts are parallel to the bedding (or foliation) of the country rock.

concordant injection Concordant injected body.

concordant pluton An intrusive igneous body, the contacts of which are parallel to the bedding (or foliation) of the country rock.

concretion A nodular or irregular concentration of certain authigenic constituents of sedimentary rocks and tuffs; developed by the localized deposition of material from solution, generally about a central nucleus. Harder than enclosing rock.

concretionary Tending to grow together. Particles of like chemical composition, when free to move, come together and form nodules of various sizes and shapes which are called concretions. Clay and ironstone nodules, balls of iron pyrite, turtle stones, etc., are good examples.

condensate Known sometimes as distillate. A heavier hydrocarbon occurring usually in gas reservoirs of great depth and high pressure. It is normally in the vapor phase but condenses as reservoir pressure is reduced by production of gas. Cf. RETRO-GRADE CONDENSATION

condensation 1. Hydrol: The process by which water changes from the gaseous state into the liquid or solid state. It is the reverse of evaporation. 2. Stratig: The phenomenon represented by a very thin section which contains all of the known faunal elements, often without succession, but jumbled. Also the process by which thinning of a stratigraphic unit occurs without loss of known zonal elements.

condensed system 1. A system in which the vapor pressure is negligibly small and hence can be ignored. 2. A system in which the pressure maintained on the system is greater than the vapor pressure of any portion.

conduction Transmission through or by means of a conductor. Distinguished, in the case of heat, from convection and radiation.

conductivity A property of an electrical conductor defined as the electrical current per unit area divided by the voltage drop per unit length.

conduit 1. A passage, generally small, that is filled with water under hydrostatic head. 2. The vertical, cylindrical passageway through which magma moves upward in a volcano.

cone 1. Steep-sided pile of sand, gravel, and perhaps boulders with fanlike outwash base deposited against the melting front of a glacier. 2. Volcanic cone. 3. Bot: Specialized branch bearing an aggregate of sporophylls.

cone-in-cone structure A concretionary structure occurring in marls, ironstones, coals, etc., characterized by the development of a succession of cones one within another.

Conemaughian The third of four series of the Pennsylvanian of the Appalachian region.

cone of depression The depression, roughly conical in shape, produced in a water table or piezometric surface by pumping or artesian flow.

cone sheet Funnel-shaped zone of fissures or dikes generally surrounding an igneous intrusion. *Cf.* RING DIKE

configuration The relative position of all electrodes as they are placed in electrical prospecting. In the Wenner configuration, four electrodes are placed in a straight line, the outer two being power electrodes and the inner potential electrodes, with equal spacing between electrodes. The Lee configuration is the same as the Wenner, with the addition of a center electrode.

confined ground water Artesian water, *q.v.*

confining bed One which, because of its position and its impermeability or low permeability relative to that of the aquifer, gives the water in the aquifer artesian head.

confining pressure An equal, all-sided pressure. In the crust of the earth the confining pressure is lithostatic pressure resulting from the load of overlying rocks. In experimental work, the confining pressure is a hydrostatic pressure, generally produced by liquids.

confluence The point where two streams meet.

confluence plain Plain formed by the merging of the valley floors of two or more streams.

confluence step The floors of main glaciated valleys often have giant steps which are evidently due to differential glacial erosion. Where two glaciated valleys of about the same size come together there may be a step up to the mouth of each one. This is called a confluence step.

conformability; conformity The mutual relation of conformable beds.

conformable *1.* Strata or groups of strata lying one above another in parallel order are said to be conformable. *2.* When beds or strata lie upon one another in unbroken and parallel order, and this arrangement shows that no disturbance or denudation has taken place at the locality while their deposition was going on, they are said to be conformable.

conformal map projection A map projection on which the shape of any small area of the surface mapped is preserved unchanged.

conformity *1. Stratig:* The relations of adjacent beds not separated by a sedimentary discontinuity. *2.* In dikes, relations of flow structure parallel to walls.

congelifluction Progressive and literal earth flow occurring under conditions of perennially frozen ground.

congelifraction Frost splitting.

congeliturbation Frost action in soil including heaving, solifluction, sludging, etc.

congeneric Belonging to the same genus.

conglomerate Puddingstone. *1.* Rounded waterworn fragments of rock or pebbles, cemented together by another mineral substance. *2.* A cemented clastic rock containing rounded fragments corresponding in their grade sizes to gravel or pebbles. Monogenetic and polygenetic types are recognized, according to the uniformity or variability of the composition and source of the pebbles.

conglomerate, volcanic *See* VOLCANIC CONGLOMERATE

congruent melting point The temperature at a specified pressure at which a solid phase changes to a liquid phase of the same composition as the solid phase.

congruous drag folds Drag folds

that bear a distinctive relationship to the major folds: (1) the axis of the drag fold is approximately parallel to the axis of the major fold; (2) the axial plane of the drag fold is approximately parallel to the axial plane of the major fold; and (3) the relative movement shown by the drag folds indicates that the beds nearer the major synclinal axis were moving upward relative to the beds nearer the anticlinal axis. Contrasts with incongruous drag folds, *q.v.*

conic map projection A map projection produced by projecting the geographic meridians and parallels onto a cone which is tangent to, or intersects the surface of a sphere, and then developing the cone into a plane.

coniferous Having cones as a reproductive structure.

conjugated fractures or veins *1.* Two sets of veins or joints occur which have the same strike, but dip in opposite directions. *2.* Any two sets of perpendicular sets of joints.

conjugate joint system A system of joints consisting of two sets that are symmetrically disposed about some other structural feature or about an inferred stress axis.

conjugate liquids Any two liquids which are immiscible in each other but are in equilibrium.

conjugation line A special case of the line connecting the composition points of two immiscible liquids which are in equilibrium with each other.

connate *1.* Born, produced or originated together; connascent. *2.* United or joined; in particular, said of like or similar structures joined as one body or organ.

connate water *1.* Water entrapped in the interstices of a sedimentary rock at the time the rock was deposited. *2.* Interstitial

water, *q.v.* *3.* Water adsorbed on mineral grains of reservoir rock and not produced with oil or gas. *4.* Water that has got into a rock formation by being entrapped in the interstices of the rock material (either sedimentary or extrusive igneous) at the time the material was deposited. It may be derived from either ocean water or land water.

conodont Tiny tooth- or jawlike fossil composed of calcium phosphate of uncertain zoological affinity.

Conodontophorida Group of small jawlike fossils of uncertain zoological affinities.

conoscope A polarizing microscope using convergent light with the Bertrand lens inserted and used to test the interference figures of crystals.

Conrad discontinuity A sharp increase in P-wave velocity observed in some areas of the earth's crust, commonly at a depth of 17–20 km.

consanguinity *Petrol:* Denotes a genetic relationship between igneous rocks which are presumably derived from a common parent magma. Such rocks are closely associated in space and time and ordinarily they show a likeness with respect to chemical, mineralogical, and textural features and geological occurrence. Rock series which show consanguineous characteristics are also called kindreds, series, suites, tribes, or clans. *See* COMAGMATIC

consecutive calderas Nested calderas.

consequent *1.* Pertaining to or characterizing the earth movements which result from the external transfer of material in the process of gradation. *Cf.* ANTECEDENT. *2.* Having a course or direction dependent on, or controlled by, the geologic structure or by the form and slope of the surface. Said chiefly of streams and drainage.

consequent fault scarp The face of an initial scarp is rapidly changed into a young consequent scarp by the loss of waste, which slips, creeps, and washes down to the scarp base, there accumulating in landslides, talus slopes, and alluvial fans, which bury the fault line.

consequent stream One which follows a course that is a direct consequence of the original slope of the surface on which it is developed.

consolidation *1. Geol:* Any or all of the processes whereby loose, soft, or liquid earth materials become firm and coherent. *2. Soil Mech:* The adjustment of a saturated soil in response to increased load. Involves the squeezing of water from the pores and decrease in void ratio.

consortium A group of individuals of different species, generally belonging to different phyla, which live together in close association.

conspecific Belonging to the same species.

constructional *Geol:* Owing its form, position, direction, or general character to building-up processes, such as accumulation by deposition or by volcanic extrusion.

contact *1.* The place or surface where two different kinds of rocks come together. Although used for sedimentary rocks, as the contact between a limestone and sandstone, it is yet more especially employed as between igneous intrusions and their walls. The word is of wide use in western mining regions on account of the frequent occurrence of ore bodies along contacts. *2.* In South Africa, a lode of great length and be-

tween two kinds of rocks, one of which is generally an igneous intrusive. *3.* The surface between two fluids in a reservoir, as gas-oil contact; water-oil contact. *4.* The bedding plane bounding a formation, often called "top" in petroleum geology. *5.* As a prefix, used to qualify an igneous rock name by implying that the texture is poikiloblastic.

contact deposit A mineral deposit found between two unlike rocks, usually applied to an ore body at the contact between a sedimentary rock and an igneous rock. A contact lode or vein.

contact metamorphism Metamorphism genetically related to the intrusion (or extrusion) of magmas and taking place in rocks at or near their contact with a body of igneous rock. *See* THERMAL METAMORPHISM

contact metasomatism A mass change in the composition of the rock other than the elimination of gases involved in simple metamorphism.

contact minerals Minerals formed by contact metamorphism.

contact twin *See* TWIN

contact vein A variety of fissure vein, between different kinds of rock occupying a typical fracture from faulting, or it may be a replacement vein formed by mineralized solutions percolating along the surface of the contact where the rock is usually more permeable and there replacing one or both of the walls by metasomatic process. Contact deposit.

contact zone *See* AURÉOLE

contamination *Petrol:* Usually applied to magmas and denoting the addition of foreign rock material, as by assimilation of wall rock.

contemporaneous deformation Deformation that takes place in sediment during or immediately following their deposition. Includes many varieties of soft-sediment deformation such as small-scale slumps, crumpling and brecciation. *Syn:* PENECONTEMPORANEOUS DEFORMATION

contemporaneous fault A growth fault; a fault in sedimentary rock that forms contemporaneously and continually as deposition occurs.

continent Large landmass rising more or less abruptly above the deep ocean floor; includes marginal areas that are shallowly submerged. At present continents constitute about one-third of the earth's surface.

continental apron The gentle incline at the base of the continental slope leading to the deep oceanic basins.

continental basin A region in the interior of a continent comprising one or several closed basins.

continental borderland *1.* Zone bordering a continent, below sea level, which is highly irregular and includes depths well in excess of those typical of a continental shelf. *2.* Terraced area or submerged plateau adjacent to a continental shelf but at greater depth. *3.* Borderland.

continental crust The type of crustal rocks underlying the continents and continental shelves; equivalent to the sial, having a thickness of 25–60 km. and a density of approximately 2.7 g./cm.³.

continental deposits Sedimentary deposits laid down within a general land area and deposited in lakes or streams or by the wind, as contrasted with marine deposits, laid down in the sea.

continental displacement Wegener hypothesis now referred to as "continental drift theory." This theory holds that an original single supercontinent, called Pangaea, split into several major

pieces which "drifted" apart to form the present-day continental land areas.

continental divide A drainage divide which separates streams flowing toward opposite sides of a continent.

continental drift The concept that the continents can drift on the surface of the earth because of the weakness of the suboceanic crust, much as ice can drift through water. As proposed by the German meteorologist Alfred Wegener in 1912, the theory that the continents were once joined into a landmass which broke apart into several landmasses which then drifted, their shapes changing somewhat, and eventually arriving at their present positions.

continental glacier An ice sheet covering a large part of a continent, e.g., the Antarctic ice sheet.

continental island *1.* An island which is near and geologically related to a continent, as are the British Isles. *2.* Continental islands are merely detached fragments of the continent near which they stand and from which they are separated, in almost all cases, by shoal water. The limit between deep and shallow water is drawn at the 100-fathom line (183 m.), and nearly all continental islands rest upon a submarine platforms which are under water less than 100 fathoms (183 m.) deep and run into the submerged continental shelf.

continental margin Zone separating the emergent continents from the deep sea bottom; generally consists of continental shelf, continental slope and continental rise.

continental nuclei Craton, *q.v.*

continental platform Platform-like mass of a continent that stands above the surrounding oceanic basins. The continental shelf is part of the continental platform.

continental rise Submarine surface beyond base of continental slope, generally with gradient less than 1 to 1000, occurring at depths from about 1373 to 5185 m. and leading down to abyssal plains.

continental rock A rock unit laid down on land as opposed to one laid down in marine water. It may be lacustrine, palustrine, eolian, fluvial, or volcanic.

continental shelf Gently sloping, shallowly submerged marginal zone of the continents extending from the shore to an abrupt increase in bottom inclination; greatest average depth less than 183 meters, slope generally less than 1 to 1000, local relief less than 18.3 meters, width ranging from very narrow to more than 321.8 kilometers.

continental slope Continuously sloping portion of the continental margin with gradient of more than 1 to 40, beginning at the outer edge of the continental shelf and bounded on the outside by a rather abrupt decrease in slope where the continental rise begins at depths ranging from about 1373 meters to 3050 meters formerly considered to extend to abyssal plains.

continuous deformation Deformation accomplished by flowage rather than by rupture.

continuous permafrost zone Regional zone predominantly underlain by permafrost, with no permafrost at widely scattered sites.

continuous profiling A seismic method of shooting in which seismometer stations are placed uniformly along the length of a line and shot from holes also spaced along the line so that each hole records seismic ray paths identical geometrically with those from immediately adjacent holes, so that events may be carried continuously by equal time comparisons.

continuous reaction series That branch of Bowen's reaction series, *q.v.*, comprising the plagioclase group of minerals, in which reaction of early formed crystals with later liquids takes place without abrupt phase changes, i.e., continuously.

contorted Bent or twisted together. Used where strata are very much folded or crumpled on a considerable scale. If on a small scale they are said to be corrugated.

contour *1.* Outline of an object. *2.* Line connecting points of equal value on a map or diagram, most commonly points of equal elevation on a map.

contour, structural An imaginary line of equal elevation on a selected stratigraphic horizon, called the structural datum.

contour, topographic An imaginary line on the ground, all points of which are at the same elevation above (or below) a specified datum surface.

contour diagram A type of petrofabric diagram prepared by contouring of a point diagram; its purpose is to obtain easier visualization of the results of the petrofabric study.

contour interval The difference in value between two adjacent contour lines. Generally, it refers to the difference in elevation between two adjacent contour lines.

contour line *1.* A line on a map representing a contour. Present usage makes contour and contour line synonymous. *2.* A line connecting points of equal value (generally elevation) above or below some reference value such as a datum plane. Contour lines are commonly used to depict topographic or structural shapes.

The quantified properties of sediments or other phenomena can be recorded by contour lines.

contour map A map showing by contour lines, *q.v.*, topographic or structural or thickness or facies differences in the area mapped.

contraction Shrinking. Rocks in passing from a vitreous to a crystalline texture shrink considerably, which may account for the subsidence of certain areas.

contraction hypothesis Theory that compression causing folding and thrusting is a result of a shrinking of the earth. The crust must decrease in size to accommodate itself to the shrinking interior of the earth.

contrails Long narrow clouds caused by the disturbance of the atmosphere during passage of high-flying jets.

controlled mosaic A mosaic fitted to a control plot by rephotographing the component vertical photographs to compensate for scale variations resulting from tilt and for variations in flight altitude.

control station A point on the ground whose position (horizontal, vertical) is used as a base for a dependent survey. *Geol:* Any surveyed point (side shot, rod shot, etc.) used for vertical or horizontal control for geologic features.

Conularida Group of four-sided more or less chitinous fossils of elongated pyramidal form with flaplike folds closing the large terminal opening; of uncertain zoologic affinities.

conularlid A group of extinct marine animals that had chitinous, pyramidal, or flattened conical shells with marked quadrilateral symmetry. Generally considered a subordinate division of the Coelenterata.

convection *1.* A process of mass movement of portions of any fluid medium (liquid or gas) in a gravitational field as a consequence of different temperatures in the medium and hence different densities. The process thus moves both the medium and the heat, and the term convection is used to signify either or both. *2.* A supposed mass movement of subcrustal or mantle material, either laterally, or in upward- or downward-directed convection cells, mainly as a result of heat variations. Theories have been proposed utilizing convection currents to explain deep-sea trenches, island arcs, geosynclines, plate motions, subduction zones, and the like.

convection cell In tectonics, a pattern of mass movement of mantle material in which the central area is uprising and the outer area is downflowing, due to heat variations.

convection currents Transfer of material due to differences in density, generally brought about by heating. Characteristic of the atmosphere and bodies of water.

convergence *1.* A term applied to the diminishing interval between geologic horizons, in some cases due to unconformable relationship and in others to variable rates of deposition, the latter applying especially to basin deposits. *See* ISOPACH. *2.* The line of demarcation between turbid river water and clear lake water, which denotes a downstream movement of water on the lake bottom and an upstream movement of water at the surface. *3.* In refraction phenomena, the decreasing of the distance between orthogonals in the direction of wave travel. This denotes an area of increasing wave height and energy concentration. *4.* In wind-setup phenomena, the increase in setup observed over that which would occur in an equivalent rectangular basin of uniform depth, caused by changes in planform or depth; also the decrease in basin width or depth causing such increase in setup. 5. *Paleontol:* Resemblance which is not due to direct relationship or genetic affinity. *See* CONVERGENT EVOLUTION

convergence map *See* ISOCHORE MAP

convergent evolution *1.* The process whereby phylogenetic stocks, which are not closely related genetically, produce similar appearing forms. *2.* The evolution toward a common adaptation (form is related to function) when it occurs in forms that have independently developed similar adaptive features even though they are far removed from each other genetically.

converted wave Seismic energy which has been converted from a P-wave to an S-wave or vice versa by reflection at an angle.

convolution A linear mathematical operation that results in a change in the shape of a wave form. Used extensively in digital seismic data processing. Often denoted by $w(t)*f(t)=g(t)$, where $w(t)$ is the original wave form, $f(t)$ is the convolution operator or linear filter, and $g(t)$ is the new wave form.

coolant A substance, usually liquid or gas, used for cooling any part of a reactor in which heat is generated, including the core, the reflector, shield, and other elements that may be heated by absorption of radiation.

coordinates Linear or angular quantities (usually two-dimensional) which designate the position which a point occupies in a given reference plane or system.

coordination number The number of atoms, ions, groups, or molecules that can be directly attached to a central atom. *Geol:*

Refers most commonly to the number of oxygen atoms that can surround a central cation.

copper A mineral, native copper. Isometric. A minor but formerly important ore of the metal.

copperas Melanterite, *q.v.*

copper glance Chalcocite. *Obs.*

coprecipitation The carrying down by a precipitate of substances normally soluble under the conditions of precipitation.

coprolite Fecal pellets or castings. The petrified, undigestible residue that has been eaten and passed through the alimentary tract of some animal. Petrified excrement.

copropel Dark-brown or gray coprogenic ooze, containing chitinous exoskeletons of benthonic arthropods in addition to reworked organic matter. *Cf.* GYTTJA

coquina Limestone composed of broken shells, corals, and other organic debris.

coral A bottom-dwelling, sessile, marine coelenterate; some are solitary individuals, but the majority grow in colonies; they secrete external skeletons of calcium carbonate. Calcareous skeleton of a coral or group of corals.

coral, colonial A coral in which the individuals are attached together as a unit and do not exist as separate animals.

coral, compound Skeleton of a colonial coral.

coral formation A formation, generally developed on cave walls, nodular in form with a rough or granular surface resembling coral. *Syn:* CAVE CORAL

coralgal Refers to carbonate sediment derived from corals and algae.

coralline *1.* Pertaining to, composed of, or having the structure of corals, as, coralline limestone. *2.* Pertaining to the lime-secreting types of red algae. *Syn:*

NULLIPORES. *3.* Applied to Bryozoa before their true nature was discovered. *Syn:* ZOOPHYTES; BRYOZOA; POLYZOA

corallite *1.* Skeleton of an individual coral animal. *2.* Skeleton of an individual coral in a colony.

corallum The calcareous exoskeleton of an individual or compound coral.

coral mud and sand Deposits formed around coral islands and coasts bordered by coral reefs, containing abundant fragments of corals.

coral reef *1.* A reef formed by the action of reef-building coral polyps, which separate carbonate of lime from the waters of the sea to form their internal skeleton. *2.* Coral reef signifies the usual complex of skeletal and shell growths, of which the framework is true coral *in situ,* though a large part may be composed of nullipore or other algal material, molluscan or other debris of littoral species, shells of the plankton, and chemically precipitated carbonates of calcium and magnesium. The corals themselves may make up less than one-half of a reef, yet its existence and increase depends on the successful growth of these animals in spite of a constant battle with the surf. *3.* A coral reef is a ridge or mount of limestone, the upper surface of which lies, or lay at the time of its formation, near the level of the sea, and is predominantly composed of calcium carbonate secreted by organisms, of which the most important are corals. *4.* The modern definition includes the idea that the deposit must be firm enough to resist erosion by waves, that the skeleton of the corals form a wave-resistant framework.

coral rock Reef limestone differs from the majority of fragmental and organic deposits in the fact that it is largely built up in a solid coherent form from the first, and therefore constitutes a rock mass in the strict sense, without any process of cementation.

coral zone The depth of the sea at which corals abound.

cordierite A mineral, $(Mg,Fe)_2$-$Al_4Si_5O_{18}$. Orthorhombic. A common mineral of metamorphic rocks.

cordillera [*Sp.*] A group of mountain ranges including valleys, plains, rivers, lakes, etc.; its component ranges may have various trends but the cordillera will have one general direction. A mountain range or system in some cases the main mountain axis of a continent; specifically the great mountain region of western North America between the Central Lowland and the Pacific Ocean.

core *1.* Central part of the earth beginning at a depth of about 2900 km., probably consisting of iron-nickel alloy; divisible into an outer core that may be liquid and an inner core about 1300 km. in radius that may be solid. *2.* Sample of rock obtained in core drilling. *3.* The central part of something, especially the filling of a hollow object. *4.* The heart of a nuclear reactor where energy is released.

core barrel A hollow cylinder attached to a specially designed bit and used to obtain and preserve a continuous section or core of the rocks penetrated in drilling.

core drilling Drilling with a hollow bit and core barrel in order to obtain a rock core.

core test A hole drilled with a core drill, usually for the purpose of securing geological information, and sometimes with the purpose of investigating geological structure.

Coriolis force The apparent force

caused by the earth's rotation which serves to deflect a moving body on the surface of the earth to the right in the northern hemisphere, but to the left in the southern hemisphere.

corner A point on a land boundary at which two or more boundary lines meet. Not the same as monument, which refers to the physical evidence of the corner's location on the ground.

corona A zone of minerals, usually radial around another mineral, or at the contact between two minerals. The term has been applied to reaction rims, corrosion rims, and originally crystallized minerals. *Syn:* KELYPHITE

corrading stream When the debris supplied to a stream is less than its capacity for carrying load the stream abrades its bed and is said to be a corrading, downcutting, or degrading stream.

corrasion Mechanical erosion performed by moving agents such as wear by glacial ice, wind, running water, etc.

correlation *1.* The determination of the equivalence in geologic age and stratigraphic position of two formations or other stratigraphic units in separated areas; or, more broadly, the determination of the contemporaneity of events in the geologic histories of two areas. Correlations may be based on paleontologic or physical evidence. *2.* In seismic interpretation, the picking of corresponding phases, obtained at two or more separated seismometer spreads, of those seismic events which appear to originate at the same geologic formation boundary. *3.* A linear mathematical operation that results in a measure of similarity between two functions, such as seismic traces, well logs, and maps.

corrosion Most commonly used for chemical erosion, whether accomplished by motionless or moving agents.

corrosion rim or border A modification of the outlines of phenocrysts due to the corrosive action of a magma upon minerals which under different conditions were previously stable. A special case of the reaction rim, *q.v.* Cf. CORONA

corry Cirque in Scotland.

cortlandtite A peridotite composed of large hornblende crystals with poikilitically included olivine crystals.

corundolite A rock consisting of corundum and iron oxides. *See* EMERY ROCK

corundum A mineral, Al_2O_3. Hexagonal. Sapphire and ruby are gem varieties.

cosmic dust Fine particles of cosmic or meteoric origin, or the remains of small meteorites which have been decomposed on passing through the earth's atmosphere.

cosmic radiation Very high-energy subatomic particles which originate in outer space and continuously bombard the earth's atmosphere. Primary cosmic rays are atomic nuclei which are almost completely absorbed in the upper atmosphere. Secondary atomic rays are those which reach the earth's surface, but have less energy than the primary type. Cosmic rays are a part of the natural background radiation.

cosmochemistry Study of the distribution of elements in the universe.

cosmogony Speculation regarding origin of the universe including origin of the earth.

cosmology Science of the universe.

cosmopolitan species An organism distributed widely throughout the world and occurring in various geographic provinces.

cospecific Conspecific.

coteau [*Fr.*] A small hill or hillock. In the northern part of the United States it was generally applied by early French travelers to a range of hills or to an escarpment forming the edge of a plateau.

cotectic surface A curved surface in a quaternary system, representing the intersection of two primary phase volumes, one or both of which are solid solution series. It is the bivariant equivalent of the univariant cotectic line in ternary systems.

cotype *1.* Used sometimes as equivalent to syntype, *q.v.* *2.* Used sometimes as equivalent to paratype, *q.v.* Not recommended for usage in either sense. *3.* Cotypes are plants that were in a different collection from that which contained the type specimens but were also used in writing the description. The term is not a proper synonym for isotypes.

coulee [*Fr.*] *1.* A short, blocky, steep-sided lava flow, generally of glassy rhyolite or obsidian, issuing from the flank of a volcanic dome or from the summit crater of a volcano. *2.* The term coulee is generally applied throughout the northern tier of states to any steep-sided gulch or water channel and at times even to a stream valley of considerable length. A solidified stream or sheet of lava.

coulee lake A lake formed when a lava flow acts as a dam across some stream valley.

couloir A deep gorge; a gully on a mountainside, especially in the Swiss Alps.

Coulomb attraction Attraction between ions of opposite electric charges.

country rock A general term applied to the rock surrounding and penetrated by mineral veins; in a wider sense applied to the rocks invaded by and surrounding an igneous intrusion.

coupled wave C-wave. A surface seismic wave of complex motion in an elastic medium, described only by mathematical explanation.

cove *1.* A small bay or open harbor; it is also popularly applied to small areas of plain or valley that extend into mountains or plateaus. In New Mexico and Arizona the re-entrants in the borders of mesas and plateaus are also called rincons [*Sp.* an inner corner]. *2.* Precipitously walled, cirquelike opening at the head of a small steep valley produced by erosion of shale below a thick massive sandstone.

covellite A mineral, CuS, of indigo-blue color. Hexagonal. A common secondary mineral; an ore of copper.

cover material Soil that is used to cover compacted solid waste in a sanitary landfill.

cracking Process of breaking down large complex molecules, particularly hydrocarbons, to produce smaller simpler ones.

crag *1.* A fossiliferous sandy marl of marine origin; generally used, capitalized, as part of the names of several formations of Pliocene age in eastern England. *2.* A steep, rugged rock; a rough broken cliff or projecting point of rock. *3.* A detached fragment of rock.

crag-and-tail *1.* A streamline hill or ridge, resulting from glaciation and consisting of a knob of resistant bedrock (the "crag"), with an elongate body (the "tail") of more erodible bedrock, or till, or both, on its lee side.

crater *1.* Bowl-shaped topographic depression with steep slopes, generally of considerable size. *2.* Volcanic orifice.

crater, volcanic A steep-walled depression at the top of a volcanic cone or on the flanks of a volcano, directly above a pipe or vent that feeds the volcano, and out of which volcanic materials are ejected. In its simplest form, usually a flat-bottomed or pointed, inverted cone more or less circular in plan. The diameter of the floor is seldom over 305m.; the depth may be as much as several hundred feet. Primarily the result of explosions or collapse at the top of a volcanic conduit. *See* EXPLOSION CRATER

crater lake A lake, generally of fresh water, formed by the accumulation of rain and ground water in a volcanic crater or caldera with relatively impermeable floor and walls. An example is Crater Lake, Oregon.

craterlet A small crater.

cratogenic *1.* Formed in or in relation to a craton. *2.* Of or pertaining to a craton.

craton A relatively immobile part of the earth, generally of large size.

cratonic shelf Zone lying between more positive and negative areas of a craton.

creek *Topog: 1.* A stream of less volume than a river. *2.* A small tidal channel through a coastal marsh. *3.* In Maryland and Virginia, a wide arm of a river or bay. *4.* A long shallow stream of intermittent flow or an arroyo in southwestern United States.

creekology An ironical term for unscientific methods of choosing drilling sites or prospective oil or gas acreage and particularly applied to selection based on the general appearance of outcrops, topography, drainage, etc.

creep *1.* An imperceptibly slow, more or less continuous downward and outward movement of slope-forming soil or rock. The movement is essentially viscous, under gravity-produced shear stresses sufficient to produce permanent deformation but too small to produce shear failure, as in a landslide. *2.* Slow deformation that results from long application of a stress. By many it is limited to stresses below the elastic limit. Part of the creep is a permanent deformation. Part of the deformation is elastic; and from this part the specimen recovers.

creep recovery The gradual recovery of elastic strain when stress is released. *Syn:* ELASTIC AFTEREFFECT; ELASTIC AFTERWORKING

crenulations Wrinkles. Small folds, with a wavelength of a few millimeters, chiefly in metamorphic rocks.

crescent beaches Crescent-shaped beaches concave toward the sea which form at the heads of bays and at the mouths of streams entering these bays along hilly and mountainous coasts.

crest *1. Topog:* The summit land of any eminence; the highest natural projection which crowns a hill or mountain, from which the surface dips downward in opposite directions. *2.* The highest point on an anticline. *See* CREST LINE. *3.* The line connecting the highest points on the same bed in an infinite number of cross sections across a fold.

crestal plane The plane formed by joining the crests of all beds in an anticline.

crest length The length of a wave along its crest.

crest line In an anticline the line connecting the highest points on the same bed in an infinite number of cross sections. Not necessarily the same as the axis of a fold.

crest of wave *1.* The highest part of a wave. *2.* That part of the wave above still-water level.

Cretaceous *1.* Of the nature of chalk; relating to chalk. *2.* The third and latest of the periods included in the Mesozoic Era; also the system of strata deposited in the Cretaceous Period.

crevasse *1.* A fissure in the ice formed under the influence of various strains. *2.* A nearly vertical fissure in a glacier. *3.* *Topog:* A break in a levee or other stream embankment.

crevasse hoar Depth hoar, *q.v.,* found in crevasses or other open spaces below the surface of a snow field or glacier.

crevice *1.* A shallow fissure in the bedrock under a gold placer, in which small but highly concentrated deposits of gold are found. *2.* The fissure containing a vein. As employed in the Colorado statute relative to a discovery shaft, a crevice is a mineral-bearing vein. *3.* Narrow, deep opening in a cavern floor. A narrow, high passageway. *4.* An enlarged joint whether mineralized or not. *5.* Corruption of crevasse.

crinoid One of the Crinoidea.

crinoidal limestone A rock composed in great part of crystalline joints of encrinites, with Foraminifera, corals, and mollusks.

Crinoidea A type of echinoderm consisting of a cup or "head" containing the vital organs, numerous radiating arms, an elongate, jointed stem, and roots by which it is attached to the sea bottom while the body, stem, and arms float.

cristobalite A mineral, SiO_2, tetragonal, polymorphous with quartz, tridymite, coesite, stishovite. The high temperature form is isometric.

critical angle Angle of incidence, measured from the perpendicular to the discontinuity, at which the refracted ray makes an angle of 90°, thus just grazing the surface of contact between the two media. Applies to optic, acoustic, and electromagnetic waves.

critical density The density of a substance at its critical temperature and under its critical pressure.

critical mass The minimum amount of a fissionable material, such as uranium-235 or plutonium-239, that is required to sustain fission in a nuclear reactor. For ^{239}Pu, the critical mass is about 5 kilograms.

critical materials Those materials vital to the national defense, the main source of which is within the continental limits of the United States, which may not be produced in quality and quantity sufficient to meet requirements.

critical point *1.* The point at which the properties of a liquid and its vapor become indistinguishable. *2.* That point on the surface of the earth where the travel times of two waves having different apparent surface velocities are the same.

critical pressure The pressure required to condense a gas at the critical temperature, above which, regardless of pressure, the gas cannot be liquefied.

critical slope Angle of repose.

critical temperature That temperature above which a substance can exist only in the gaseous state, no matter what pressure is exerted.

critical velocity That rate of flow, in a pipe, at which laminar or streamline changes to turbulent flow. In the latter, particles move in sinuous and erratic courses. *See* REYNOLDS NUMBER

crocidolite A sodium-containing variety of amphibole. Blue asbestos.

crocoite A mineral, $PbCrO_4$, orange. Monoclinic.

Croixan The youngest of three epochs of the Cambrian Period of North America; also the series of strata deposited during that epoch.

crop out To be exposed at the surface; referring to strata. *See* OUTCROP

cross-bedding 1. The arrangement of laminations of strata transverse or oblique to the main planes of stratification of the strata concerned; inclined, often lenticular, beds between the main bedding planes; found only in granular sediments. 2. Term should be applied to inclined bedding found only in profiles at right angles to the current direction.

cross-bedding, torrential Fine, horizontally laminated strata alternating with uniformly cross-bedded strata composed of coarser materials. Believed to originate under desert conditions of concentrated rainfall, abundant wind action, and playa lake deposition.

cross-correlation A comparison measure of two different mathematical functions or ordered data points. *See* CORRELATION

crosscut 1. A small passageway driven at right angles to the main entry to connect it with a parallel entry or air course. Also used in Arkansas instead of "breaking through." 2. A level, driven across the course of a vein or in general across the direction of the main workings or across the "grain of coal."

crossed Nicols Two Nicol prisms placed so that their vibration planes are mutually at right angles. *Opt. mineral:* An anisotropic crystal is interposed between the Nicol prisms in order to observe optical interference effects. The petrographic microscope is normally used with nicol prisms (or equivalent polarizing devices) in the crossed position.

crossed twinning Repeated twin-ning after two laws. Shown in microcline.

cross fault A fault that strikes diagonally or perpendicularly to the strike of a strata.

cross-lamination Cross-bedding; false bedding. The structure commonly present in granular sedimentary rocks, which consists of tabular, irregularly lenticular, or wedge-shaped bodies lying essentially parallel to the general stratification and which themselves show a pronounced laminated structure in which the laminae are steeply inclined to the general bedding.

Crossopterygii A large group of true fishes, including among others the coelacanths, the early fishes ancestral to amphibians and related to some Devonian lung-fishes.

cross section A profile portraying an interpretation of a vertical section of the earth explored by geophysical and/or geological methods.

cross-stratification The arrangement of layers at one or more angles to the dip of the formation. A cross-stratified unit is one with layers deposited at an angle to the original dip of the formation. Many authors have used "cross-bedding" and "cross-lamination" as synonymous with cross-stratification, but it is proposed to restrict the terms cross-bedding and cross-lamination to a quantitative meaning depending on the thickness of the individual layers, or cross-strata.

crude oil Petroleum liquids as they come from the ground. Also called simply "crude."

crush breccia A breccia formed essentially *in situ* by cataclasis. *See* CATACLASITE; CRUSH CONGLOMERATE

crush conglomerate 1. These beds are very much faulted, crushed, and folded and every stage can be traced between a continuous limestone-band and one which has been broken up into small rounded fragments, so as to present precisely the appearance of conglomerate. 2. Similar to a fault breccia except that the fragments are more rounded in a crush conglomerate. 3. Tectonic conglomerate, q.v.

crushing strength The compressive stress necessary to cause a solid to fail by fracture.

crust 1. A hard layer at the surface of softer material. 2. Outer layer of the earth, originally considered to overlie a molten interior, now defined in various ways: lithosphere; sial; material above Mohorovičić discontinuity (favored); tectonosphere; etc. Commonly used in a figurative and imprecise sense. 3. A layer of hard snow lying upon a soft layer.

Crustacea Subphylum of arthropods with two pairs of antennae and generally some biramous appendages. Camb.-Rec.

crustal cycle That cycle of an ecosystem that restores minerals to the soils; it involves erosions of rocks, their deposition, lithification, metamorphism, melting, mountain building, and return to erosion. The length of the cycle ranges from 150 to 600 million years.

crustal plate One of the six major blocks into which the lithosphere is divided, according to the scheme of global tectonics. It is about 100 km. thick.

crustification Those deposits of minerals and ores that are in layers or crusts and that, therefore, have been distinctively deposited from solution.

cryogenic lake Lake in a region of permanently frozen ground produced by local thawing.

cryogenic period Informal designation for a time interval in geologic history during which large bodies of ice occurred at or near the poles and climate was generally suitable for growth of continental glaciers.

cryolite A mineral, Na_3AlF_6, used in the reduction of aluminum ore. Monoclinic.

cryology 1. In the United States, the study of refrigeration. 2. In Europe, a synonym for glaciology. 3. The study of ice and snow. 4. The study of sea ice.

cryoluminescence Low temperature increase of weak luminescence or its production in normally nonfluorescent material.

cryopedology The science of intensive frost action and permanently frozen ground including studies of the processes and their occurrence and also the engineering devices which may be invented to avoid or overcome difficulties induced by them.

cryoplanation Reduction of land surfaces by processes associated with frost action.

cryosphere All of the earth's surface that is permanently frozen.

cryoturbation Frost action including frost heaving.

cryptobatholithic The first of six stages in the erosion of a batholith; it is not exposed, but its presence is indicated by dikes, sills, and mineral veins in the roof, or by areas of alteration in the overlying rock.

cryptocrystalline The texture of a rock or mineral consisting of crystals that are too small to be recognized and easily distinguished under the light microscope.

cryptoexplosion structure A nongenetic descriptive term to designate a roughly circular structure formed by the sudden, explosive release of energy and exhibiting intense, often localized rock deformation with no obvious relation to volcanic or tectonic activity. Many cryptoexplosion structures are believed to be the results of hypervelocity impact of crater-forming meteorites of asteroidal dimensions; others may have been produced by obscure volcanic activity. The term largely replaces the earlier term, cryptovolcanic structure.

cryptomelane A mineral, K-$(Mn,^{+4}Mn^{+2})_8O_{16}$. Monoclinic, usually massive.

cryptoperthite An extremely fine-grained intergrowth of potassic and sodic feldspar detectable only by means of X rays; not visible either with the naked eye or with the aid of the microscope. *See* PERTHITE

cryptovolcanic (structure) A small, nearly circular area of highly disturbed strata in which there is no trace of volcanic materials to confirm a volcanic origin. Hence the name "crypto," meaning hidden. Many of these structures are now believed to have been formed by meteorite impact, and the nongenetic term crypto-explosion structure is preferred.

Cryptozoic Time of hidden life (an obsolete term).

cryptozoon Presumed problematical fossils. Most of them have now been determined to be either concretions, *q.v.*, or some form of stromatolite, *q.v.*

crystal The regular polyhedral form, bounded by plane surfaces, which is the outward expression of a periodic or regularly repeating internal arrangement of atoms.

crystal axis A reference axis used for the description of the vectorial properties of a crystal. There are generally three noncoplanar axes, chosen parallel to the edges of the unit cell of the crystal structure so as to be parallel to symmetry directions if possible.

crystal chemistry The study of the relations among chemical compo-

sitions, internal structure, and the physical properties of crystalline matter.

crystal class One of the 32 possible combinations of crystal symmetry operations, all of which have one point in common. Often referred to as "point group symmetry."

crystal flotation *Petrol:* The act or process of floating of light crystals in a body of magma. Contrasted with crystal settling.

crystal fractionation Magmatic differentiation resulting from the settling out of crystals from a melt.

crystal gliding Translation gliding.

crystal lattice The regular and repeated three-dimensional arrangement of atoms or ions in a crystal.

crystalline Of or pertaining to the nature of a crystal, having regular molecular structure. Contrasted with amorphous.

crystalline rock *1.* Rock consisting of minerals in an obviously crystalline state. *2.* An inexact general term for igneous and metamorphic rocks as opposed to sedimentary.

crystallinity *Petrol:* The degree of crystallization exhibited by an igneous rock; expressed by terms such as holocrystalline, hypocrystalline, holohyaline.

crystallization The process through which crystalline phases separate from a fluid, viscous, or dispersed state (gas, liquid solution, or rigid solution).

crystallization, heat of *See* HEAT OF CRYSTALLIZATION

crystallization interval *1.* The interval of temperature (or less frequently, pressure), between the formation of the first crystal and the disappearance of the last drop of liquid from a magma upon cooling, usually excluding late stage aqueous fluids. *2.* More specifically, when referring to a given mineral, the range or ranges of temperatures over which that particular phase is in equilibrium with liquid. In the case of equilibria along reaction lines or reaction surfaces, crystallization intervals as thus defined include temperature ranges where certain solid phases are actually decreasing in amount with decrease in temperature.

crystallization magnetization Chemical magnetization.

crystallization nuclei Small particles of any kind around which ice crystals begin to form when a substance freezes.

crystallization schistosity Fissility resulting from the preferred orientation of crystals that grew in the easiest direction.

crystallizing force *1.* Potentiality of minerals to develop crystal form within a solid medium. *2.* Expansive force of minerals growing in a solid medium; force is different in different crystallographic directions. *3.* Tendency of minerals to grow more rapidly in one crystallographic direction than in another.

crystalloblastesis Deformation accomplished by metamorphic recrystallization.

crystalloblastic A crystalline texture due to metamorphic recrystallization. A characteristic of this texture is that the essential constituents are simultaneous crystallizations and are not formed in sequence, so that each may be found as inclusions in all the others.

crystallography The science of the interatomic arrangement of solid matter, its causes, its nature, and its consequences.

crystalloluminescence Emission of light from a substance during crystallization.

crystal mush Partially crystallized magma.

crystal optics The science which treats of the transmission of light in crystals.

crystal sedimentation Settling of crystals in a liquid magma.

crystal seeding *See* NUCLEATION

crystal settling Gravitative sinking of crystals from the liquid in which they formed, by virtue of their greater density.

crystal structure The periodic or repeating arrangement of atoms in a crystal.

crystal system One of the six subdivisions of the 32 possible crystal classes. Each system is characterized by the relative lengths and inclinations of the crystallographic axes. The six systems are isometric, tetragonal, hexagonal, orthorhombic, monoclinic, and triclinic.

crystal tuff. *1.* Tuff that consists dominantly of ejected volcanic crystals and single crystal fragments. The fragments show typical volcanic ash structures. *2.* The crystals usually are broken euhedra of the common phenocrysts of lava, and may be sheathed in an envelope of glass. *3.* An indurated deposit of volcanic ash dominantly composed of intratelluric crystals blown out during a volcanic eruption. The term should properly be restricted to tuffs containing more than 75% by volume of crystals.

cubanite A mineral, $CuFe_2S_3$. Orthorhombic.

cube *Crystallog:* A form in the isometric system enclosed by six symmetrically equivalent faces at right angles to one another.

cubic cleavage A type of isometric crystal cleavage having cleavage planes parallel to the faces of a cube; e.g., galena.

cubic packing The three-dimensional arrangement of equal spheres in which the centers of adjacent spheres occupy the corners of a cube. This arrangement is more accurately referred to as simple cubic, whereas face-centered and body-centered cubic consists of additional spheres whose centers occupy the geometric centers of each cube face, or the cube center.

cubic system Isometric system, q.v.

cuesta [Sp.] 1. Used in the southwestern United States for a sloping plain which is terminated on one side by a steep slope.

cul-de-sac Partially filled, abandoned swallow hole. A cavern passage that connects with another passage at only one point and ends abruptly in a rock wall or is blocked by cave fill or debris.

culm A vernacular term variously applied, according to the locality, to carbonaceous shale, or to fissile varieties of anthracite coal.

culmination 1. Term applied to the highest point on the crown of a nappe. Syn: AXIS CULMINATION. 2. Portion of a fold system, generally more or less at right angles to the folds, away from which the folds plunge. 3. The position of a heavenly body which is continually above the horizon, the position of lowest apparent altitude. 4. The highest or lowest altitude attained by a heavenly body as it crosses the meridian.

cumberlandite An ultramafic rock composed largely of magnetite, ilmenite, and olivine with minor plagioclase.

cummingtonite A mineral of the amphibole group, $(Mg,Fe,Mn)_7$-$Si_8O_{22}(OH)_2$.

cumulo dome Protrusion of viscous lava from a volcanic vent with little lateral spreading.

cumulo-volcano Cumulo dome.

cup coral A solitary coral, as opposed to a colonial coral.

Syn: FOSSIL COW'S HORN; FOUR-PART CORAL; HORN CORAL; SIMPLE CORAL; SOLITARY CORAL

cupellation Process of assaying for precious metals with a cupel.

cupola 1. Roof protrusion, or vertical extension of intrusive igneous rock. 2. An isolated body of plutonic rock that lies near a larger body. Both bodies are presumed to enlarge downward and join at depth. 3. A large vaulted dome in the Radiolaria.

cupriferous Copper-bearing.

cuprite A mineral, Cu_2O. Isometric. Red.

cuprous Of, pertaining to, or containing copper. Especially compounds in which copper is monovalent.

curie A measure of radioactivity.

Curie point or temperature Temperature below which a substance ceases to be paramagnetic.

Curie's law The law that magnetic susceptibility is inversely proportional to the absolute temperature.

current 1. The flowing of water, or other liquid or gas. 2. The vertical component of air motion; an air current is thus distinguished from the wind, which is the horizontal component. 3. A large-scale horizontal motion of air. 4. A large stream of ocean water moving continuously in about the same path, and distinguished from the water through which it flows mainly by temperature and salinity differences.

current, coastal See COASTAL CURRENT

current base Maximum depth in standing water below which currents are ineffective in moving sediment.

current bedding Crossbedding, q.v.

current cross-ripples Ripples which result from the interference of a current with a pre-existing set of ripples, if the action of the current is sufficiently weak and only of very short duration. As there is no oscillation of the current, there is no reason for a transformation into the hexagonal pattern, and the two sets of ripples may be found intersecting at any angle.

current land-use map A map that shows the current land use of an area.

current mark An irregular structure produced by erosion on tidal flats where the falling tidal waters erode numerous channels and leave uneroded areas as flat "plateaus" between channels.

current ripple A ripple mark produced by the action of a current flowing steadily in one direction over a bed of sand. These "current ripples" have a long, gentle slope toward the direction from which the current comes, and a shorter steeper slope on the lee side.

curvature correction, of earth An adjustment applied to a long line of sight in the computation of difference in elevation. Atmospheric refraction partially compensates for earth curvature; hence, correction tables take both curvature and refraction into account.

cusp *1.* One of a series of naturally formed low mounds of beach material separated by crescent-shaped troughs spaced at more or less regular intervals along the beach face. *Also* BEACH CUSP. *2.* In conodonts, the large main denticle which is ordinarily above the aboral attachment scar or escutcheon, sometimes called the superior fang or main cusp.

cuspidate, *adj.* With an apex somewhat abruptly and sharply concavely constricted into an elongated, sharply pointed tip.

cutan A concentration of a particular soil constituent along natural parting surfaces within the soil.

cut-and-fill *1.* During the development of the meanders the lateral planation on the one side is accompanied by deposition on the other. *2.* A structure resulting from the removal of a small portion of a bed or lamina prior to deposition of an overlying bed or lamina.

cutbank The concave wall of a meandering stream that is maintained as a steep or even overhanging cliff by the impinging of water at its base.

cutoff A new and relatively short channel formed when a stream cuts through the neck of an oxbow or horseshoe bend.

cutoff grade The lowest grade of mineralized material that qualifies as ore in a given deposit, i.e., material of the lowest assay that is included in an ore estimate.

cutoff spur When along the course of an entrenched river lateral erosion becomes predominant, a spur projecting into a loop may ultimately be worn away at the root, so that the river breaks through at this point and, abandoning its curved course, leaves an elevation isolated from the high ground on either bank which is known as a cutoff spur.

cut terrace Shelves carved in the shores of lakes by the action of waves and currents; they are bounded on both their shoreward and lakeward margins by steeper slopes; the former inclines upward and forms a sea cliff, the latter slopes downward and forms a terrace scarp. Their upper limit is a horizontal line marking the level of the water at the time they were formed, their surface slopes gently lakeward.

cuttings The fragmental rock samples broken or torn from the penetrated rock during the course of drilling.

Cuvier's principle Some very different characters of complex organisms are commonly associated, e.g., kinds of feet and teeth among the vertebrates.

C-wave Coupled wave, *q.v.*

cyanite *See* KYANITE

cycle This is the period in which a continent or any part of it would be reduced from its initial form of uplift to base level.

cycle of denudation The alternate upheaval and wearing down, together constitute a cycle of denudation, from base level back to base level. *Cf.* CYCLE OF EROSION

cycle of development The whole region presents the most emphatic expression not only of its structure, but also of the more recent cycles of development through which it has passed.

cycle of erosion The time involved in the reduction of a recently uplifted land area to a base level. Geographical cycle, *q.v.*

cycle of sedimentation *1.* A sequence of related processes and conditions repeated in the same order that is recorded in a sedimentary deposit. *2.* The cycle of sediment formation, transportation, and deposition.

cycle of shore development The progressive changes from the time when the water first assumes its level and rests against the new shore to the time when it has brought its boundaries into harmony with its movements so that no more work can be done.

cycle of topographical development Each cycle begins with the uplift of an area approximately at base level, the processes of denudation working with minimum efficiency and extreme slowness. The movement of upheaval revivifies the destructive agencies, and the work of carving out a surface of relief begins afresh, only to terminate, unless interrupted by renewed elevation, in once more base-leveling the region. A complete cycle is thus from base level back to base level, though, as it is a cycle, a beginning may be selected in any part of it.

cycle of underground drainage It begins with surface drainage and in its youth develops subterranean drainage near the points of easy escape for the water. In its maturity there is the maximum of subterranean drainage and the lower parts of the caverns have begun to retreat by collapse while in the uppermost reaches of the stream the transformation from surface to subsurface drainage may still be in progress. Old age is shown by the more general condition of collapse and the return to surface drainage. Briefly, it may be stated that the cycle is: surface drainage, partial subterranean drainage, and a return to surface drainage. The final state is peneplanation or base leveling.

cycles, igneous Refers to the usual sequence of events in which there are flows first, then large intrusions, and finally dikes.

cyclic evolution Evolution, supposed by some to have occurred in many lineages, involving successively (1) initial rapid and vigorous expansion, (2) long stable or slowly changing phase, and (3) final short episode in which overspecialized, degenerate, or inadaptive forms led to extinction.

cyclic twinning Repeated twinning of three or more individuals according to the same twin law but with the twinning axes not parallel. Often simulates high symmetry as fourfold, fivefold, sixfold, or eightfold axes, some of which are impossible in an untwinned crystal.

cyclosilicates *See* SILICATES, CLASSIFICATION

cyclothem A depositional cycle

containing a coal bed and in order from base to top rocks representing a series of environments starting with a fluvial sandstone and conformably passing through fluvial, brackish, and marine before starting the next cycle.

cylindrical map projection A map projection produced by projecting the geographic meridians and parallels onto a cylinder which is tangent to (or intersects) the surface of a sphere, and then developing the cylinder into a plane.

cymoid structure Vein-shaped like a reverse curve. Cymoid loop: splitting of a vein on its dip or strike into two branches, both of which curve away from the general trend and then unite to resume a direction parallel to but not in line with the original trend.

Cystoidea Mostly stemmed echinoderms with irregularly to regularly arranged body plates pierced by pores and biserial brachioles that appear to be outgrowths of structures originating at the mouth. Blastoids are included by some in this class.

d The spacing between successive identical planes in a crystal structure. The list of d's obtained by X-ray methods is characteristic of each substance and is widely used for mineral identification.

dacite The extrusive equivalent of quartz diorite (tonalite). The principal minerals are plagioclase (andesine and oligoclase), quartz, pyroxene or hornblende or both with minor biotite and sanidine.

dahllite A carbonate-apatite mineral.

dale A vale or small valley.

dalles [Fr.] In the Northwest, the nearly vertical walls of a canyon or gorge, usually containing a rapid. *Also* DELLS

darcy A standard unit of permeability. One darcy is equivalent to the passage of one cubic centimeter of fluid of one centipoise viscosity flowing in one second under a pressure differential of one atmosphere through a porous medium having an area of cross section of one square centimeter and a length of one centimeter. A millidarcy is one one-thousandth.

Darcy's law A derived formula for the flow of fluids on the assumption that the flow is laminar and that inertia can be neglected. The numerical formulation of this law is used generally in studies of gas, oil, and water production from underground formations.

dark mineral Any one of a group of rock-forming minerals that are dark-colored in thin section; biotite, hornblende, augite.

dark ruby silver *See* PYRARGYRITE

Darwinism Doctrine that organic evolution resulted from variation and the selection of favored individuals through natural selection.

datolite A mineral, $CaBSiO_4$-(OH). Monoclinic. Commonly occurs in cavities in diabase or basalt.

datum Any numerical or geometrical quantity or set of such quantities which may serve as a reference or base for other quantities.

datum elevation A level reference elevation used in mapping, usually sea level. In seismic mapping, the time datum used is near the surface.

datum level The level (usually sea level or mean level of nearest considerable body of water) from which altitudes are measured in surveys.

datum plane 1. The horizontal plane to which soundings, ground elevations, or water surface elevations are referred. *Also* REFERENCE PLANE. 2. An arbitrary reference surface, used in seismic mapping to minimize or eliminate local topographic effects, to which seismic times and velocity determinations are referred.

datum water level The level at which water is first struck in a shaft.

daughter element An element formed from another by radioactive decay; e.g., radon is the daughter element of radium.

D-coal Coal material that predominates in durain; occurs as microscopic particles in the lungs of miners.

DDT The first of the modern chlorinated hydrocarbon insecticides whose chemical name is 1,1,1-tricholoro-2,2-bis (p-chloriphenyl)-ethane. It has a half-life of 15 years, and its residues can become concentrated in the fatty tissues of certain organisms, especially fish. Because of its persistence in the environment and its ability to accumulate and magnify in the food chain, EPA has banned the registration and interstate sale of DDT for nearly all uses in the United States effective December 31, 1972.

dead cave A cave wherein the formations are dry; a cave in which deposition and excavation have ceased.

dead coral reef A coral reef or part of a coral reef with no living corals.

dead ground Rock in a mine, which, although producing no ore, requires to be removed in order to get at productive ground.

dead line Base of barren core of a metalliferous batholith, exposed during epibatholithic, *q.v.*, stage of erosion.

dead reckoning Determining a position by knowing the speed, direction, and lapsed time since being at a known position.

debouchure *1.* The mouth of a river or channel; the point from which a spring bursts. *2.* Point at which tubular passages connect with larger passages or chambers. Point of issuance of an underground stream.

debris *1.* [*Fr.*] Wreck, ruins, remains. *Geol:* Applied to a collection of the larger fragments of rocks and strata, to distinguish them from detritus, or those which are pulverized. *2.* The material resulting from the decay and disintegration of rocks.

It may occur in the place where it was produced, or it may be transported by streams of water or ice and deposited in other localities. Rock waste. *3. Glaciol:* Rocks, earth, and other material lying on the surface of a glacier.

debris avalanche The sudden movement downslope of the soil mantle on steep slopes caused by its complete saturation through protracted heavy rains.

debris cone A fan-shaped deposit of soil, sand, gravel, and boulders up at the point where a mountain stream meets a valley, or otherwise where its velocity is reduced sufficiently to cause such deposits. Similar to alluvial fan, but consists of coarser material lying on steeper slopes.

debris fall The relatively free falling of predominantly unconsolidated earth or debris from a vertical or overhanging cliff, cave, or arch.

debris flow A general designation for all types of rapid flowage involving debris of various kinds and conditions.

debris line Landward limit of debris washed upon a beach by storm waves.

debris slide The rapid downward movement of predominantly unconsolidated and incoherent earth and debris in which the mass does not show backward rotation but slides or rolls forward, forming an irregular hummocky deposit which may resemble morainal topography.

decay constant The fraction of a large number of atoms of a radioactive element which decays per unit time; generally denoted by the symbol λ.

Deccan basalt; Deccan trap Fine-grained, nonporphyritic, tholeiitic basaltic lava, covering an area of about 518,000 km.^{2} in the Deccan region of southeast India and consisting essentially of labradorite, clino-pyroxene, and iron ore. Olivine is generally absent, or is present in minor amount, usually near the bottom of flows. Corresponds to the plateau basalt of the Pacific Northwest and the Thulean province of western Scotland, northeast Ireland, and Iceland.

decke [*Ger.*] Nappe [*Fr.*], *q.v.*

decken structure Nappe structure. A series of large recumbent folds or overthrust sheets, or both, lying one above the other.

declination The angle, variable with geographic position, between the direction in which the magnetic needle points and the true meridian.

declination arc A graduated arc attached to the alidade of a surveyor's compass or transit, on which the magnetic declination is set off.

decline curve A graphic representation of the decline in production of an oil or gas well or group of wells. Production rates (ordinates) are plotted against time (abscissae) and the curve drawn approximately through the points. The projected curve is integrated to predict ultimate future production.

declinometer An instrument, often self-registering, for measuring or recording the declination of the magnetic needle.

declivity A descending slope, as opposed to acclivity.

declivity, law of In homogeneous material and with equal quantities of water, the rate of erosion of slopes is dependent on their declivities, the steeper being degraded the faster.

décollement Detachment structure of strata due to deformation, resulting in independent styles of deformation in the rocks above and below. It is associated with folding and with overthrusting, but is merely a descriptive term.

decomposers Usually microconsumer organisms that break down organic matter and thus aid in recycling nutrients.

decomposition Reduction of the net energy level and change in chemical composition of organic matter because of the actions of aerobic or anaerobic microorganisms.

deconvolution Any number of mathematical processes designed to restore a wave shape to the form it had before it underwent a convolution. A popular data processing technique applied to seismic reflection data for the purpose of improving the visibility and resolution of reflected events.

decrepitation The breaking-up, usually violent and with a crackling noise, of mineral substances when exposed to heat, as when common salt is thrown upon the fire.

decussate structure A microstructure in which the axes of contiguous crystals lie in diverse directions. This crisscross structure is most noticeable in rocks composed dominantly of minerals with a columnar habit.

dedolomitization The destruction of dolomite to form calcite and periclase, the latter usually hydrating to form brucite as in brucite marble or predazzite. Presumably this takes place by contact metamorphism at rather low pressures. Dedolomite also occurs in unmetamorphosed carbonate terrains where it is usually represented by calcitic limestones of special types.

deep Oceanic areas of exceptional depth, representing depressions in the ocean floor, often troughlike or synclinal in form. The term is generally understood to be restricted to depths greater than 18,000 feet (3000 fathoms). Secondary and smaller bounded areas within the great ocean basins with depths exceeding 5000 to 6000 meters.

deep inland seas Adjacent seas in restricted communication with the open ocean. Depths are greater than 200 meters; the floors are commonly 2000 to 5000 meters below sea level. The Mediterranean, East and West Indian basins, the Red Sea, and the Black Sea are examples. *Also* MEDITERRANEAN SEAS; DEEP-SEA BASINS

deep marginal seas Adjacent seas with depths exceeding 200 meters. Deep marginal seas are widely open to the oceans at the water surface but are restricted at depth by submarine ridges with or without islands. Examples are the eastern Asiatic marginal seas.

Deep Sea Drilling Project A program sponsored by the U. S. Government, with the combined effort of major oceanographic institutions to sample the ocean bottom at numerous places throughout the world using the research vessel *Glomar Challenger.*

deep-sea plain A broad and nearly level area forming by far the greater part of the ocean floor. Its depth varies from about 2000 to 3000 fathoms, but its slopes are always very gentle.

deep-sea terrace A benchlike feature bordering an elevation of the deep-sea floor at depths greater than 300 fathoms.

defect lattice A crystal lattice in which the expected systemic repetition is interrupted by an omission, an inclusion of an extra item, or the substitution of an unexpected item.

deflation The removal of material from a beach or other land surface by wind action.

deflation basin Topographic basin resulting from wind erosion.

deflection *1.* One type of pattern shown by maps of mountain ranges; it is a sudden change in the trend of one or more branches of the mountain range. *2. Geophys:* Applied to the vertical direction, the angle between the geoid and the spheroid, or the angle between the vertical and the normal to the spheroid; more properly, the deviation of the vertical.

deflocculate To break up from a flocculated state; to convert into very fine particles. *Syn:* PEPTIZE

deflocculating agent A chemical additive which produces deflocculation, e.g., ammonium hydroxide added to a water suspension of clay.

deformation Any change in the original form or volume of rock masses produced by tectonic forces; folding, faulting, and solid flow are common modes of deformation.

deformation ellipsoid *See* STRAIN ELLIPSOID

deformation fabric *1.* The orientation in space of the elements of a rock produced by external stress on the rock. It results from a rotation or movement of the constituent elements under stress, or from growth of new elements in a common orientation controlled by the stress conditions. *2. Struct. petrol:* A deformation fabric results where the components of the rock owe their orientation to the operation of stress which has produced a penetrative differential movement of the components.

deformation plane Plane normal to flow surface and parallel to direction of movement; the a-c plane of structural petrology.

deformation twinning Twinning produced by deformation and gliding within a preexisting crystal.

degassing The process whereby water and various gases were released from earth's interior during an early stage of its formation.

deglaciation The uncovering of an area from beneath glacier ice as a result of shrinkage of a glacier.

degradation The general lowering of the surface of the land by erosive processes, especially by the removal of material through erosion and transportation by flowing water.

degree day A departure of one degree per day in the mean daily temperature, from an adopted standard reference temperature, usually 18.3° C. When the mean temperature on a given day is 18.3° C. or higher, that day is not considered in making the monthly total. 18.3° C. was chosen the standard because, according to heating engineers, the minimum temperature of bodily comfort in the home is reached when the mean daily temperature falls below that value.

degree of freedom Capability of variation of a system. The number of degrees of freedom in a system is the number of independent variables, temperature, pressure, and concentration in the different phases, which must be specified in order to define the system completely. Alternatively, the number of degrees of freedom is the number of variables which may be changed independently without causing the appearance or disappearance of a phase. See VARIANCE OF A SYSTEM

degrees of frost A phrase used in England; it means the number of degrees that the temperature falls below the freezing point: thus, a day with a temperature of −2.8° C. may be designated as a day of 2.8° C. of frost.

dehydration water Chemically or physically combined water lost during mineral reactions or transformations.

delayed runoff Most of that rain water which sinks into the ground eventually returns to the surface by slow seepage and from springs; it is the delayed runoff.

delay time Intercept time. Additional time for any segment of a ray path over the time which would be required to traverse the horizontal component of that segment at highest velocity encountered on the trajectory.

deliquescent Capable of becoming liquid by the absorption of water from the air.

dellenite A variety of rhyolite transitional into quartz latite.

delta *1.* An alluvial deposit, usually triangular, at the mouth of a river. *2.* A tidal delta is a similar deposit at the mouth of a tidal inlet, put there by tidal currents. *3.* A wave delta is a deposit made by large waves which run over the top of a spit or bar beach and down the landward side. *4.* A deposit of sediment formed at the mouth of a river either in the ocean or a lake which results in progradation of the shore line. The name delta [<*Gr.* capital letter "delta," which is triangular] has been applied because of the triangular shape assumed by the saliants formed by deltas built out from straight coast.

deltageosyncline See EXOGEOSYNCLINE

delta lake Rivers are partly or wholly responsible for a class of lakes which may be called delta lakes. Lake Pontchartrain in Louisiana is an example.

delta plain Plains formed by the accumulation of silt at the mouths of streams, or by overflow along their lower courses.

deltoid Triangular; deltalike.

deluge The term applied by

early writers to the Noachian flood.

demersal Refers to fish and other nektonic animals that live on or adjacent to the seabottom and feed on benthonic organisms.

demonstrated reserves A collective term for the sum of measured and indicated reserves or resources.

dendrite *1.* A branching figure resembling a shrub or tree, produced on or in a mineral or rock by the crystallization of a foreign mineral, usually an oxide of manganese, as in the moss agate; also the mineral or rock so marked. *2.* A crystallized arborescent form, as of gold or silver; an arborization.

dendritic drainage pattern The dendritic drainage pattern is characterized by irregular branching in all directions with the tributaries joining the main stream at all angles.

dendrochronology Study and matching of tree rings with the object of dating events in the recent past.

dense, *adj. 1.* Having its parts massed or crowded together; close; compact. *2. Petrol:* A textural term applied to fine-grained, aphanitic rocks in which the grain size generally averages less than 0.05 to 0.1 mm.

densilog Record of the varying density of rocks penetrated in drilling.

density *1.* The mass or quantity of a substance in grams per cubic centimeter. *2.* The quality of being dense, close, or compact. *3.* The quantity of electricity per unit of volume at a point in space, or the quantity of electricity per unit of area at a point on a surface. *4.* In an ecological sense density is usually subdivided into two categories: (a) crude density—the number of biomass per unit of total area and (b) specific density—the

number or biomass per unit of habitable (colonizable) area.

density current Turbidity current. A highly turbid and relatively dense current which moves along the bottom slope of a body of standing water.

dentate, *adj.* With sharp, spreading, rather coarse indentations or teeth that are perpendicular to the margin.

denudation *1.* A laying bare; the process of washing away of the covering of strata. *2.* That process which, if continued far enough, would reduce all surface inequalities of the globe to a uniform base level.

denude To wear away or remove overlying matter from underlying rocks, exposing them to view.

deoxidation spheres Bleach spots, *q.v.*

departure (plane surveying) The orthographic projection of a line on an east-west axis of reference. The departure (dep.) of a line is the difference of the meridian distances or longitudes of the ends of the line. It is east or positive, and sometimes called the easting, for a line whose azimuth or bearing is in the northeast or southeast quadrant; it is west or negative, and sometimes called the westing, for a line whose azimuth or bearing is in the southwest or northwest quadrant.

depauperate fauna A fauna that is usually characterized by only small forms—micromorphs. Parts of depauperate faunas may consist of dwarf forms of normal species, but most seem to be small species, or the preserved inner parts of large species. *Cf.* DWARF FAUNA

depergelation The act or process of thawing permanently frozen ground.

deplanation A term that includes all physiographic processes which tend to reduce the relief of a district, and so cause the topography eventually to become more and more plainlike in contour, dominantly by subtracting material from the area or areas affected.

depletion *1.* The act of emptying, reducing, or exhausting, as the depletion of natural resources. *Min:* Specifically said of ore reserves. *2.* The loss sustained by its owner through the progressive exhaustion of a mineral deposit.

depletion allowance A proportion of income derived from mining or oil production that is considered to be a return of capital not subject to income tax.

depocenter An area or site of maximum deposition.

deposit *1.* Anything laid down. Formerly applied to suspended matter left by the agency of water, but now made to include also mineral matter in any form, and precipitated by chemical or other agencies, as the ores, etc., in veins. *2.* The term mineral deposit or ore deposit is arbitrarily used to designate a natural occurrence of a useful mineral or ore in sufficient extent and degree of concentration to invite exploitation.

deposition *1.* The laying down of potential rock-forming material; sedimentation. *2.* The precipitation of mineral matter from solution as the deposition of agate, vein quartz, etc.

depositional magnetization Remanent magnetization of sedimentary rock resulting from the depositional alinement of previously magnetized grains.

depreciation The waste of assets due to exhaustion, wear and tear, and obsolescence of property. Not to be confused with depletion, *q.v.*

depression A low place of any

size on a plain surface, with drainage underground or by evaporation; a hollow completely surrounded by higher ground and having no natural outlet for surface drainage.

depth *1.* The vertical distance from the still-water level (or datum as specified) to the bottom. *2.* A term which may be used for a few of the deepest soundings.

depth ice *1.* Bottom ice. *2.* Small particles of ice formed below the surface of the sea when it is churned by wave action.

depth of compensation, isostatic The depth at which major density differences of the earth's crust are compensated, isostatically. This depth is variously calculated to lie between 100 and 113.7 km.

depth section An interpreted seismic time section plotted to represent a geologic cross section. *See* MIGRATION

depth zone within the earth Zones giving rise to different metamorphic assemblages. *See* EPIZONE; MESOZONE; KATAZONE

dermal gliding Extensive horizontal shearing and displacement within the earth's crust.

de-roofing Foundering of part of the roof of a batholith resulting in the extrusion of magma onto the earth's surface.

derrick *1.* The framework or tower over a deep drill hole, such as that of an oil well, for supporting the tackle for boring, hoisting, or lowering. *2.* Any of various hoisting apparatus employing a tackle rigged at the end of a spar or beam.

desalination Salt removal from sea or brackish water.

descriptive mineralogy That branch of mineralogy devoted to the description of the physical and chemical properties and occurrences and uses of minerals.

desert A region so devoid of vegetation as to be incapable of supporting any considerable population. Four kinds of deserts, may be distinguished: (1) the polar ice and snow deserts, marked by perpetual snow cover and intense cold; (2) the middle latitude deserts, in the basinlike interiors of the continents, such as the Gobi, characterized by scant rainfall and high summer temperatures; (3) the trade wind deserts, notably the Sahara, the distinguishing features of which are negligible precipitation and large daily temperature range; and (4) coastal deserts where there is a cold current on the western coast of a large land mass such as occurs in Peru.

desert crust See DESERT PAVEMENT

desert dome Convex-upward rock surface representing penultimate stage in desert erosion of a granitic mountain mass.

desert pavement Desert crust. When loose material containing pebbles or larger stones is exposed to wind action the finer dust and sand are blown away and the pebbles gradually accumulate on the surface, forming a sort of mosaic which protects the finer material underneath from attack. This is the desert pavement. See PEBBLE MOSAICS

desert polish A smooth and shining surface imparted to rocks or other hard substances by the wind-blown sand and dust of desert regions.

desert soil Reddish desert soil and sierozem.

desert topography In arid regions the rain and frost do relatively little work, and rock destruction is almost entirely mechanical and carried on by the wind, and by the heating of the rocks by a desert sun followed by a swift chill that comes after sunset.

desert varnish A surface stain or crust of manganese or iron oxide, of brown or black color and usually with a glistening luster, which characterizes many exposed rock surfaces in the desert. It coats not only ledges of rock in place, but also boulders and pebbles that are scattered over the surface of the ground.

desiccate To dry up; to deprive or exhaust of moisture; to preserve by drying.

desiccation breccia 1. Dried, mud-cracked polygons which have broken into fragments and then have been deposited with other sediments. 2. A type of intraformational breccia, *q.v.* See EDGEWISE CONGLOMERATE

desiccation conglomerate A deposit of coarse rounded fragments formed by the fragmentation, erosion, and transportation of the plates of a mud-cracked (desiccation-cracked) layer of sediment.

desiccation crack Crack formed by shrinkage of clay or clayey beds in the course of drying under the influence of the sun's heat.

desiccation polygons Nonsorted polygons produced by drying.

desilication 1. The removal of silica from a rock; freeing of silica by the breakdown of silicates is particularly important. 2. Removal of silica from a magma by reaction with wall rock, as with limestone, to form solid lime silicates.

Desmoinesian Lower Middle Pennsylvanian.

Desmospongia Class of sponges with skeletons composed of siliceous or horny spicules of many forms other than hexactinellid. Camb.-Rec.

desorption Removal of adsorbed material.

destructional Pertaining to destruction or shaped by destructive forces; as in geology, a plain which has been formed by erosion.

detached core In some tight folds the beds in the center of a fold are so squeezed that only detached remnants of the bed remain in the center of the fold.

detail log An electric log of a well bore with a scale expanded beyond the conventional 1 inch per 100 feet of depth, made in order to portray more clearly minor variations in the formations penetrated by the hole.

detector See SEISMOMETER

detector spread In seismic work, the layout of detectors or seismometers to obtain information by shooting.

determinative mineralogy That branch of mineralogy which comprises the determination of the nature, composition, and classification of minerals, by all or any means available, as physical tests, blowpipe or wet analyses, and the crystallographic and the optical properties.

detrital Clastic; allogenic, *q.v.* Said of minerals occurring in sedimentary rocks, which were derived from preexisting igneous, sedimentary, or metamorphic rocks.

detrital ratio Clastic ratio.

detritus *1.* Material produced by the disintegration and weathering of rocks that has been moved from its site of origin. *2.* A deposit of such material. *3.* Any fine particulate debris, usually of organic origin, but sometimes defined as organic and inorganic debris.

deuteric A term applied to alterations in an igneous rock produced during the later stages of, and as a direct consequence of, the consolidation of the magma or lava. The term discriminates such alterations from the more strictly secondary changes due to a later period of alteration. *Syn:* PAULOPOST

deuteric effects Metasomatic effects which have taken place in direct continuation of the magma of the rock itself.

deuterium *1.* An isotope of hydrogen, H², containing one neutron and one proton in its nucleus. The hydrogen in "heavy water." *2.* A proposed fuel for fusion.

development *1. Geol:* Applied to those progressive changes in fossil genera and species that have followed one another during the deposition of the strata of the earth. *2.* Work done in a mine to open up ore bodies, as sinking shafts and driving levels, etc.

development well A well drilled within the known or proved productive area of an oil field, with the expectation of obtaining oil or gas from the producing formation or formations in that field. *Cf.* EXPLORATORY WELL

development work Work undertaken to open up ore bodies as distinguished from the work of actual ore extraction. Sometimes development work is distinguished from exploratory work on the one hand and from stope preparation on the other.

deviation Applied to the vertical direction, the angle between the geoid and the spheroid of the earth, or the angle between the vertical and the normal to the spheroid. *Syn:* DEFLECTION

devitrification The process by which glassy rocks break up into definite minerals. The solid state transformation from a glass to a crystalline mineral.

Devonian In the ordinarily accepted classification, the fourth in order of age of the periods comprising the Paleozoic Era, following the Silurian and preceding the Mississippian. Also the system of strata deposited at that time. Sometimes called the Age of Fishes.

dextral drag fold A drag fold such that the trace of a given bed on the surface is displaced to the right by the drag folds.

dextral fault Right-lateral fault.

dextral fold Asymmetric fold in which the long limb is apparently offset to the right by the short limb as one looks along the long limb (i.e., offset in the same manner as in a right-hand fault, q.v.).

diabase A rock of basaltic composition, consisting essentially of labradorite and pyroxene, and characterized by ophitic texture. Rocks containing significant amounts of olivine are olivine diabases. In Great Britain basaltic rocks with ophitic texture are called dolerites, and the term diabase is restricted to altered dolerites.

diabase amphibolite Amphibolite formed by dynamic metamorphism of diabase.

diabasic A textural term applied to igneous rocks in which discrete crystals or grains of pyroxene (usually augite) fill the interstices between lath-shaped feldspar (usually plagioclase) crystals. Characteristic of diabases and some gabbros. *Syn:* OPHITIC

diadochy Replacement or replaceability of one atom or ion in a crystal lattice by another.

diagenesis Process involving physical and chemical changes in sediment after deposition that converts it to consolidated rock; includes compaction, cementation, recrystallization, and perhaps replacement as in the development of dolomite.

diagnostic minerals Symptomatic minerals. Those minerals, such as olivine, quartz, etc., which indicate a rock to be under- or over-saturated.

diagonal fault Oblique fault. A fault that strikes diagonally to the strike of the adjacent strata.

diagonal joint Joint in igneous rock crossing flow lines or layers at angle of about 45°, corresponding to shear planes.

diagonal-slip fault Oblique-slip fault. A fault in which the net slip is diagonal down or up the fault plane; that is, a fault which is neither a strike-slip nor a dip-slip fault.

diagram, orientation *Struct. petrol:* A diagram, usually on a Schmidt net, which shows the degree of concentration or dispersion of selected features of rock fabric such as lineation, crystal axes, cleavage or twinning planes, etc.

diagram, scatter *Struct. petrol:* An orientation diagram which has not been contoured; lineations, axes, or poles of planes are represented by points.

diallage The variety of monoclinic pyroxene characterized by conspicuous parting parallel to the front pinacoid a {100}.

dialysis Separation of a colloid from ions and molecules in true solution, e.g., permitting ions to diffuse through a membrane which is impervious to the colloidal particles.

diamagnetic Pertaining to substances having a permeability less than that of a vacuum, i.e., less than 1. Such materials are repelled by a magnetic field, e.g., metallic bismuth. *Ant:* PARAMAGNETIC

diamond A mineral, the element carbon, the hardest substance known. Isometric. Used as a gem and industrially in cutting tools.

Bort and carbonado (black diamond) are black or dark-colored diamond aggregates.

diamond bit A rotary drilling bit studded with borts-type diamonds.

diaphanous Permitting the light to shine through.

diapir A dome or anticlinal fold, the overlying rocks of which have been ruptured by the squeezing out of the plastic core material. Diapirs in sedimentary strata usually contain cores of salt or shale. Igneous intrusions may also show diapiric structure.

diapir fold Piercing fold. Anticlines in which a mobile core, such as salt, has injected the more brittle overlying rock.

diaspore A mineral, AlO(OH). Orthorhombic. A constituent of some bauxites.

diastem A depositional break of minor extent presumed to represent a hiatus of minor duration. There should be no floral or faunal change across a diastem.

diastrophism The process or processes by which the crust of the earth is deformed, producing continents and ocean basins, plateaus and mountains, flexures and folds of strata, and faults. Also the results of these processes.

diathermic Allowing a free passage of heat.

diatom Microscopic, single-celled plant growing in marine or fresh water. Diatoms secrete siliceous frustules in a great variety of forms that may accumulate in sediments in enormous numbers.

diatomaceous earth A friable earthy deposit composed of nearly pure silica and consisting essentially of the frustules of the microscopic plants called diatoms. Diatomite, *q.v.* Sometimes wrongly called infusorial earth, *q.v.*

diatomite The silica of diatoms dried to a fine powder and used in the manufacture of dynamite, pottery glaze, etc.

diatom ooze A soft siliceous deposit found on the bottom of the deep sea made largely or partly of the shells of diatoms; similar deposits are formed from the shells of Radiolaria.

diatreme A general term for a volcanic vent or pipe emplaced in rocks (usually flat-lying sedimentary rocks) by the explosive energy of gas-charged magmas. The diamond-bearing kimberlite pipes of South Africa are diatremes.

Dibranchiata Subclass of cephalopods with or without internal shells; includes belemnites, squids, and octopods.

dichroism *See* PLEOCHROISM.

dichromate A salt of dichromic acid; a compound containing the radical $(Cr_2O_7)^{-2}$.

dichroscope An instrument for observing pleochroism in minerals, especially gems.

dickite A clay mineral of the kaolin group. It has the same composition, $Al_2Si_2O_5(OH)_4$, as kaolinite and nacrite, but is structurally distinct. Usually occurs in hydrothermal veins.

differential compaction The relative change in thickness of mud and sand (or limestone) after burial due to reduction in pore space.

differential erosion The more rapid erosion of one portion of the earth's surface as compared with another.

differential forces Forces that are not equal on all sides of a body and hence produce distortion. In contrast to hydrostatic pressure which is equal on all sides of the body and causes dilation.

differential melting A partial melting process, in which a portion of the material remains as solid crystals.

differential pressure The difference in pressure between the two sides of an orifice; the difference between reservoir and sand-face pressure; between pressure at the bottom of a well and at the wellhead; between flowing pressure at the wellhead and that in the gathering line. Any difference in pressure between that upstream and downstream where a restriction to flow exists.

differential thermal analysis A method of analysis by determining the temperature at which thermal reactions take place in a material when it is heated continuously to an elevated temperature, and also the intensity and general character of such reactions.

differential weathering When rocks are not uniform in character but are softer or more soluble in some places than in others, an uneven surface may be developed; in deserts by the action of the wind and in moist regions by solution. Columns of rock which have been isolated in any way show the effect of differential weathering.

differentiated dike A dike that consists of more than one kind of rock because of magmatic differentiation of an originally homogeneous magma into two or more fractions.

differentiated sill A sill that consists of more than one kind of rock because of magmatic differentiation of an originally homogeneous magma into two or more fractions.

differentiation, gravitational Differentiation accomplished through the existence of a gravitational field, as by sinking of a heavy phase (liquid or crystals), or the rising of a light phase (liquid, crystals, or gases) through the magma.

diffraction *1.* Term applied to the bending of rays not in accord with Snell's law, as, for example, the bending of rays around obstacles. *2.* The interference of light or other forms of radiant energy, such as X rays, electrons, and neutrons, produced by scattering due to edges or points of material objects. The essential feature in each case is that the scattering centers be separated by distances that are comparable to the wave length of the radiation. *3. Seismol:* Scattered seismic energy emanating from abrupt discontinuities of rock types in the subsurface. Particularity common where faults cut reflecting interfaces. A reflection can be thought of as the interference result of diffractions from the points defining the reflecting horizon.

diffraction of water waves The phenomenon by which energy is transmitted laterally along a wave crest.

diffraction spacing *See* D-SPACING

diffusion The spreading out of molecules, atoms, or ions into a vacuum, a fluid, or a porous medium, in a direction tending to equalize concentrations in all parts of a system.

diffusion coefficient A parameter in diffusion calculations, having the dimensions distance-squared divided by time. It varies with the nature of the particles diffusing, the nature of the diffusion medium, and temperature. It is expressed by the formula: $[l^2t^{-1}]$ where l=length, t=time.

diffusivity, thermal Coefficient of thermal diffusion. A thermal property of matter, with the dimensions of area per unit time.

digestion The biochemical decomposition of organic matter. Digestion of sewage sludge takes place in tanks where the sludge decomposes, resulting in partial gasification, liquefaction, and mineralization of pollutants.

diggings Applicable to all mineral deposits and mining camps, but in usage in the United States applied to placer mining only.

digital Representation of measured quantities in discrete or quantized units. A digital system is one in which the information is stored and manipulated as a series of discrete numbers, as opposed to an analog system.

digitation A subsidiary recumbent anticline emanating from a much larger recumbent anticline. Where several such smaller folds are associated, they resemble the fingers of a hand.

dike *1.* A tabular body of igneous rock that cuts across the structure of adjacent rocks or cuts massive rocks. Although most dikes result from the intrusion of magma, some are the result of metasomatic replacement. *2.* A wall or mound built around a low-lying area to prevent flooding.

dikelet A small offshoot or apophysis from a dike.

dike set A group of parallel dikes.

dike swarm A set of parallel dikes, but more numerous than in a dike set.

dike wall Sometimes crevices in the rocks are filled with lavas from below; then the lavas cool into rocks of a firmer texture than those of the adjacent formations; afterward, by degradation, the softer rocks on either side are carried away, and the lava rocks stand in walls. Lavas intruded in this manner are called dikes, and dike walls are common in volcanic regions.

dilatational wave P-wave, *q.v.*

dilation *1.* The expansion of ice from the freezing of water in fissures. *2.* Dilatation. Deformation that is change in volume, but not in shape. *3. Volcanol:* The process of widening of an initial fissure concomitant with the injection of magma.

dilation dike Dike resulting from the intrusion of magma into a fracture whose walls moved away from each other.

dilation veins Fat lenses in schists thought to be caused by bulging of schistose rocks due to pressure transmitted by mineralizing solutions.

dilution ratio The ratio of the volume of water of a stream to the volume of incoming waste. The capacity of a stream to assimilate waste is partially dependent upon the dilution ratio.

diluvial *1.* Pertaining to floods. *2.* Related to or consisting of diluvium.

diluvium Name given to all coarse superficial accumulations which were formerly supposed to have resulted from a general deluge; now employed as a general term for all the glacial and fluvioglacial deposits of the Ice Age.

dimensional orientation *Struct. petrol:* A preferred orientation that is shown by the shape of the individual grains.

dimension stone Stone that is quarried or cut in accordance with required dimensions.

dimorph One of the two forms of a dimorphic crystalline chemical compound or organism.

dimorphism *1. Paleontol:* The characteristic of having two distinct forms in the same species, as male and female, megaspheric and microspheric stages. *2. Mineral:* The crystallization in two crystal forms of the same chemical compound, e.g., pyrite and marcasite.

dioctahedral Phyllosilicate minerals in which only two-thirds of the possible octahedral structural positions are occupied by ions.

diopside A mineral of the pyroxene group, $CaMgSi_2O_6$.

diorite A plutonic rock composed essentially of sodic plagioclase (usually andesine) and hornblende, biotite, or pyroxene. Small amounts of quartz and orthoclase may be present.

dioxide An oxide containing two atoms of oxygen per molecule, e.g., MnO_2, ZrO_2.

dip, *n.* The angle at which a stratum or any planar feature is inclined from the horizontal. The dip is at a right angle to the strike.

dip-and-fault structure A structure in which an inclined series of beds, dipping in one direction, is cut by gravity (normal) faults dipping in the opposite direction.

dip calculation Any of a number of methods of converting observed seismic arrival time values to the dip of a reflector; most commonly the conversion of delta T values to dip values by a conversion factor based upon the geometry of the seismic array and approximate seismic propagational velocity.

dip compass Dipping compass; dip needle, *q.v.*

dip fault A fault that strikes parallel with the dip of the strata involved.

dip joint A joint that strikes approximately perpendicularly to the strike of the bedding or cleavage.

dip logging Any of several methods, electrical and caliper, of measuring the dip of formations traversed by boreholes.

diploid *1.* Isometric crystal form composed of 24 faces each meeting the axes at unequal distances. *2.* Refers to cells furnished with two sets of chromosomes.

dip needle In volcanic and earth-

quake prediction, an instrument that is used to determine minute changes in the slope of the ground surface.

dipole Any object that has different properties at two points. Commonly refers to a molecule that has concentrations of negative and positive charges at widely spaced points.

dip separation The distance of apparent relative displacement of formerly adjacent beds on either side of a fault surface, measured along the dip of the fault.

dip shift Dip slip, *q.v.*

dip shooting Any system of seismic surveying where the primary concern, both instrumental and computational, is the registration and computation of reflections for dip values, with minor emphasis on correlation of records from shot point to shot point.

dip slip The component of the slip parallel with the fault dip, or its projection on a line in the fault surface perpendicular to the fault strike.

dip-slip fault A fault in which the net slip is practically in the line of the fault dip.

dip slope A slope of the land surface which conforms approximately to the dip of the underlying rocks.

dip-strike symbol The symbol used on geological maps to show the strike and dip of some planar feature, such as bedding, foliation, joints, etc.

dip throw The component of the slip measured parallel with the dip of the strata.

dipyramid Crystal form consisting of two pyramids meeting at a plane of symmetry.

directional drilling The art of drilling a borehole wherein the course of the hole is planned before drilling. Such holes are usually drilled with rotary equipment, and are useful in drilling divergent tests from one location, tests which otherwise might be inaccessible, as, controls for fire and wild wells, etc.

directional structure All those sedimentary structures which have directional significance, and including cross-bedding, flow markings, and ripple marks.

direct runoff That part of the runoff which consists of water that has not passed beneath the surface since it was last precipitated out of the atmosphere.

dirt cone Debris cone, q.v. 1. Protecting as it does the ice below, a local ice hillock rises upon its dirt patch site as the surrounding surface is lowered, and as this grows in height its declivities increase and a portion of the dirt slides down the side. The first product of this shaping is an almost perfectly conical ice hill encased in dirt and known as a debris, sand, or dirt cone. 2. A cone of ice or snow formed through the protection from ablation afforded by a thin veneer of dirt or debris.

disappearing stream A surface stream that disappears underground in a sink.

discharge Rate of flow at a given instant in terms of volume per unit of time.

disconformity 1. Unconformity between parallel strata, e.g., with strata below not dipping at an angle to those above. 2. Local contact plane in dike where flow structures are discordant.

discontinuities in earth structure Sudden or rapid changes with depth in one or more of the physical properties of the materials constituting the earth, as evidenced by seismic data.

discontinuity Seismol: A surface at which seismic-wave velocities abruptly change; a boundary between seismic layers of the earth.

Syn: INTERFACE, SEISMIC DISCONTINUITY, VELOCITY DISCONTINUITY

discontinuous deformation Deformation of rocks accomplished by rupture rather than by flowage.

discontinuous permafrost zone Regional zone, intermediate between continuous permafrost zone and sporadic permafrost zone, underlain by permafrost in some areas and free of permafrost in other areas.

discontinuous reaction series That branch of Bowen's reaction series, q.v., including the minerals olivine, pyroxene, amphibole, and biotite, each change in the series representing an abrupt phase change.

discordance Geol: A lack of parallelism between contiguous planar bodies.

discordant 1. A term used to describe an igneous contact that cuts across the bedding or foliation of adjacent rock. 2. The lack of parallelism between sets of planar bodies.

discordant basin A shallow negative area in an island arc region which cuts across the other structural trends.

discovery claim The first claim in which a mineral deposit is found, and when this is within a gulch or on a stream the claims are simply marked or numbered from the discovery claim either by letters or figures up or down the gulch or stream.

discovery well A well discovering oil or gas in a pool hitherto unknown and unproductive.

disequilibrium assemblage Associated minerals not in thermodynamic equilibrium.

disharmonic fold 1. A fold in which abrupt changes in geometric relations occur in passing from one bed to another, especially where alternations of plastic and rigid beds occur. 2. A fold that changes in form with depth.

disharmonic relations Deviation from parallelism in beds or planes as inclinations vary.

disintegration 1. A term often applied to the natural mechanical breaking down of a rock on weathering. 2. That stage in the decomposition of vegetable and animal substances which takes place in the presence of oxygen and moisture and which may be regarded as a slow combustion of organic substance, leaving no solid carbon compounds and producing only volatile substances, namely carbon dioxide and water. *Cf.* MOLDERING; PEAT FORMATION; PUTREFACTION. 3. The breaking down of striae in a brachiopod.

disjunctive fold A fold in which the more brittle strata have fractured and separated and the more plastic beds have flowed under the forces of deformation.

dislocation 1. Displacement of rocks on opposite sides of a fracture relative to each other. Sometimes used synonymously with fault. 2. *Crystallog.* and *Metal:* A structural defect on a crystal.

dismembered river system A river system consisting of a trunk river and tributaries, the lower part of which has been flooded by the sea. As a result of flooding, the streams which were formerly tributaries of the trunk river enter the sea by separate mouths.

disorder In crystals, a statistical randomness in the filling of certain available sites by given atoms, or in the choice of between two or more atomic varieties for a given site.

dispersal Movement of individuals or their disseminants or propagules (seed, spores, larvae) out of a population or area.

dispersal pattern *Geochem. pros-*

pecting: Pattern of distribution of metal content of soil, rock, water, or vegetation.

dispersed phase Solid material in the form of a colloid suspended in a fluid referred to as the dispersion medium.

dispersion 1. *Opt. mineral:* The optical constants for different parts of the spectrum. The indices of refraction of a substance vary with the wavelength of the transmitted light in crystals of low symmetry (monoclinic and triclinic). The principal optical directions within the crystal vary with the wave length of light. 2. *Seismol:* Distortion in the shape of a seismic wave train because of the variation of velocity with frequency. Most prominent in surface waves in the presence of near surface velocity layering. *See* PHASE VELOCITY. 3. *Biol:* The process whereby organisms occupy wider and wider geographic areas through time.

dispersion halo A zone or region surrounding a mineral deposit in which the metal concentration is intermediate between that of the mineral deposit and that of the unmineralized, country rock.

disphotic zone The dimly lit zone in bodies of water where there is insufficient light for photosynthesis.

displacement 1. The word "displacement" should receive no technical meaning, but is reserved for general use; it may be applied to a relative movement of the two sides of the fault, measured in any direction, when that direction is specified; for instance, the displacement of a stratum along a drift in a mine would be the distance between the two sections of the stratum measured along the drift. The word "dislocation" will also be most useful in a general sense. 2. *Chem:* A reaction in which an elementary substance displaces and sets free a constituent element from a compound.

displacement plane The a-b fabric plane of structural petrology.

displacement pressure The minimum pressure required to force the entry of a nonwetting fluid into a porous medium saturated with a wetting liquid.

disposal well A well drilled or used for disposal of brines or other fluids in order to prevent contamination of the surface by such wastes.

disrupted gouge Similar to chattermarks, *q.v.*, but of more striking character. They mark the action of a large mass of rock carried forcibly over a rock surface whose nature favors a coarse, rough breakage rather than a smooth gouge.

dissected *1.* Cut by erosion into hills and valleys or into flat upland areas separated by valleys. Applicable especially to plains or peneplains in process of erosion after an uplift. *2. Paleobot:* Divided into many slender segments.

dissection *Geol:* The work of erosion in destroying the continuity of a relatively even surface by cutting ravines or valleys into it.

disseminated ore A scattered distribution of generally fine-grained ore minerals throughout a rock body, or vein.

dissociation The breakdown of a substance into several others, as $CaCO_3 \rightarrow CaO + CO_2$ by heat, or $NaCl \rightarrow Na^+ + Cl^-$ by solution in water.

dissociation constant The equilibrium constant for a dissociation reaction; the product of activities (or less accurately, of concentrations) of the products of dissociation divided by the activity (or concentration) of the original substance.

dissociation point The temperature at which a compound breaks up reversibly to form two or more other substances, e.g., $CaCO_3 \rightarrow CaO + CO_2$.

dissociation temperature A presumed fixed temperature "point" at which a given dissociation occurs; actually, it is usually a range due to the compositional or pressure variance, and may refer merely to the temperature at which the rate of a given dissociation becomes appreciable, under stated conditions.

dissolution The process of dissolving or, more rarely, of melting.

dissolved load That part of the total stream load that is carried in solution.

dissolved oxygen (DO) The amount of oxygen dissolved in water or sewage, expressed in parts per million (ppm) by weight, or milligrams per liter (mg/l).

dissolved solids The total amount of dissolved material, organic and inorganic, contained in water or wastes.

disthene *Syn:* KYANITE

distillation *1.* The process of decomposition whereby the original chitinous material of certain fossils has lost its nitrogen, oxygen, and hydrogen, and is now represented by a film of carbonaceous material. *2.* The removal of impurities from liquids by boiling. The steam, condensed back into liquid, is almost pure water; the pollutants remain in the concentrated residue. *3.* The most commonly used method to desalt water whereby the saline water is vaporized and then condensed. Distillation is generally done in several stages so that the heat released during condensation of the steam can be utilized to vaporize more sea water.

distortion waves Equivolumnar

waves; secondary waves; shear waves; transverse waves; S-waves, q.v.

distributary 1. An outflowing branch of a river, such as occurs characteristically on a delta. 2. A river branch flowing away from the main stream and not rejoining it. Contrasted with tributary.

distribution scatter Graphic representation of relations of specimens with respect to chosen variable characters. Cf. SCATTER DIAGRAM

distributive province The environment embracing all rocks that contribute to the formation of a contemporaneous sedimentary deposit and the agents responsible for their distribution.

disturbance 1. The bending or faulting of rock or stratum from its original position. 2. Folding and/or faulting that affects a large area but is not extensive enough to be called a revolution. The distinction between a disturbance and a revolution is very arbitrary.

divergence 1. In refraction phenomena the spreading of orthogonals in the direction of wave travel. This denotes an area of decreasing wave height and energy concentration. 2. In wind-setup phenomena, the decrease in setup observed under that which would occur in an equivalent rectangular basin of uniform depth, caused by changes in platform depth. Also the increase in basin width or depth causing such decrease in setup. 3. That process of natural selection whereby lineages diverge genetically and physically; at its most diverse divergence implies adaptive radiation, q.v.

divergent plate boundary A boundary between two plates which are moving apart, with new oceanic-type lithosphere being

created at the seam. Syn: ACCRETING PLATE BOUNDARY

diversity Refers to the number of different kinds of species in an area.

diverted stream In stream piracy the stream that was diverted from the beheaded stream to the pirate stream. See PIRACY

divide The line of separation between drainage systems; the summit of an interfluve. The highest summit of a pass or gap.

diviner One who purportedly divines the location of oil, gas, water, or ore deposits in the earth; a dowser.

divining rod Dowsing rod; wiggle stick. Forked twig considered by some to have magic properties when held in the hands of a dowser. Twig is supposed to dip when held over ore, oil, or water deposits, depending upon the specialty of the dowser.

D layer Atmospheric zone within the chemosphere at height of about 50 km. that reflects low frequency radio waves.

DNA Deoxyribonucleic acid concentrated mainly in the nuclear structures of organisms.

do- *Petrol:* A prefix, indicating that one factor dominates over another within the ratios 7/1 and 5/3 (7 and 1.67); e.g., docrystalline, docalcic, etc.

doab [*Hind.*] 1. The name given in India to the tongue of land that lies between the confluence of two or more rivers, as the doabs of the Punjab, or plains that lie between the rivers of that region. 2. A dark sandy clay found in the vicinity of many Irish bogs.

dodecahedron 1. The isometric crystal form composed of twelve rhombic faces, each parallel to one crystallographic axis and intersecting the other two at equal distances; specifically called the rhombic dodecahedron. 2. Any isometric form consisting of twelve congruent faces, as the pentagonal dodecahedron or pyritohedron, the tetrahedral pentagonal dodecahedron or tetartoid, etc.

Dogger The middle of three epochs of the Jurassic Period; also the series of strata deposited during that epoch.

dogtooth spar A variety of calcite with sharp-pointed crystals.

dolarenite Dolomite rock consisting of sand sized grains.

dolerite Applied in Great Britain to fresh basaltic rocks with ophitic texture. Used in the same sense as diabase is used in the U.S.A.

doleritic See OPHITIC; DIABASIC.

dolimorphic A term applied to rocks in which released minerals are prominent. An example is a lamprophyre from Madagascar consisting essentially of biotite and quartz with a little hornblende.

dolinen; doline; dolina 1. The dolinen (called by English writers swallow holes, sinkholes, or cockpits) are rounded hollows varying from 9.15 to 915 m. in diameter, and from 1.83 to 100.65 m. in depth. They may be either dish-, funnel-, or well-shaped. 2. [It.] Name given to the funnel-shaped cavities which communicate with the underground drainage-system in limestone regions. Similar cavities are known in this country as sinks and swallow holes.

dolomite 1. A mineral, CaMg-(CO₃)₂, commonly with some Fe replacing Mg (ankerite). Hexagonal rhombohedral. A common rock-forming mineral. 2. A term applied to those rocks that approximate the mineral dolomite in composition. Syn: MAGNESIAN LIMESTONE. It occurs in a great many crystalline and noncrystalline forms the same as pure limestone, and among rocks of all

geological ages. When the carbonate of magnesia is not present in the above proportion the rock may still be called a magnesian limestone, but not a dolomite, strictly speaking.

dolomite limestone Term applied to a carbonate rock composed predominantly of dolomite. (Not recommended.)

dolomitization The process whereby limestone becomes dolomite by the substitution of magnesium carbonate for a portion of the original calcium carbonate.

dolomold (*Insol. residue*) Originally called dolocast. *Obs.* Rhombohedral cavities of any size left in chert, pyrite, shale, limonite, glauconite, etc., by the solution of dolomite (or calcite) crystals.

dolomorphic In an insoluble residue a condition in which calcite or dolomite has been replaced by an insoluble mineral which fills the rhombohedral dolomoldic cavity in chert or other matrix.

dolostone A term proposed for a sedimentary rock composed of fragmental, concretionary, or precipitated dolomite of organic or inorganic origin.

domain By domain is meant the areal extent of a given lithology or environment. Thus the domain of a facies of sedimentation refers to the area wherein a given set of physical controls combined to produce a distinctive facies.

dome 1. A roughly symmetrical upfold, the beds dipping in all directions, more or less equally, from a point. 2. A smoothly rounded, rock-capped mountain summit, roughly resembling the dome or cupola of a building. 3. An open crystal form consisting of two faces astride a symmetry plane. 4. *Oceanog:* The dome [*Ger.* Kuppe; *Fr.* dome], an elevation of small area, but rising

with a steep angle to a depth more than 200 meters from the surface.
5. Any structural deformation characterized by local uplift approximately circular in outline, e.g., the salt domes of Louisiana and Texas.

dome mountain Mountain formed from pressure below rather than from lateral compression.

dominants A species or group of species which largely control the energy flow and strongly affect the environment within a community or association.

dominant vitrain A field term to denote, in accordance with an arbitrary scale established for use in describing banded coal, a frequency of occurrence of vitrain bands comprising more than 60% of the total coal layer.

doodlebug Any one of a large number of unscientific devices with which it is claimed minerals and oil deposits can be located.

dornick; dornock In the United States, a small rock or boulder; specifically a boulder of iron ore found in limonite mines.

dorsal, *adj.* Back; relating to the back or outer surface of a part or organ; as, the lower side of a leaf; the opposite of ventral.

dot chart 1. Graphical aid used in correction of station gravity for terrain effect, or computing gravity effects of irregular masses; can be used also in magnetic interpretation. A graticule. 2. A graphical, transparent chart used in the calculation of the gravity effects of various structures; dots on the chart represent unit areas.

double layer Ionic atmosphere containing enough ions of one charge to equal the opposite charge of a surrounded particle.

double plunging fold A fold that plunges in opposite directions from a central point. In a doubly plunging anticline, the plunge is away from this point; in a doubly plunging syncline, the plunge is toward this point.

double refraction Generally defined as the numerical difference between the greatest and the least indices of refraction.

douse; douce; dowse 1. To beat out or to extinguish an ignited jet of fire damp. To search for deposits of ore, for lodes, or water, by aid of the dowsing or divining rod.

downbuckle Abrupt downward bending of the earth's crust.

downcutting stream See CORRADING STREAM

downdip block Block on the downdip side of a strike fault.

downdrift The direction of predominant movement of littoral materials.

downs 1. A term usually applied to hillocks of sand thrown up by the sea, or the wind along the seacoast. It is also a general name for any undulating tract of upland too light for cultivation and covered with short grass. 2. A hill; especially a bank or hillock of sand thrown up by the wind in or near the shore; a flattish-topped hill. A tract of open upland, often undulating and covered with a fine turf which serves chiefly for the grazing of sheep.

downthrow The wall of a fault that has moved relatively downward. "Downthrown" is preferred by USGS.

down-to-basin fault A normal fault whose downthrown side is toward the basin. A "down-to-coast" fault is one whose downthrown side is toward the coast.

down-valley migration (meanders) As soon as the initial bends are developed in a young river into somewhat systematic curves or meanders, it not only deepens its valley, but also widens the meander belt, and pushes the whole system of curves down valley.

downward enrichment A term which is synonymous with "secondary enrichment" as the latter has applied to enrichment of ore bodies by the downward percolation of waters.

downwarp Term meaning opposite of upwarp, *q.v.*

downwasting The diminishing of glacier ice in thickness during ablation.

dowser One who operates a divining rod, *q.v.*

dowsing Searching for water, oil, or minerals with a dowsing rod.

dowsing rod Divining rod, *q.v.* Wiggle stick.

drag *1.* Minor folding or strata along the walls of a fault in in which the "drag" of displacement has produced flexures in the beds on either side. *2.* Fragments of ore torn from a lode by faulting and remaining in the fault zone. *3.* A name applied to occurrences of roof shale penetrating downward into a coal seam.

drag folds In the narrow sense, minor folds that form in an incompetent bed when the competent beds on either side of it move in such a way as to subject it to a couple. The axes of the minor folds are perpendicular to the direction in which the beds slip; the acute angle between the main bedding and the axial planes of the drag folds indicate the direction of the shear. In a broad sense, used for any fold that is a subsidiary part of a larger fold.

drag mark *1.* Long even mark commonly with longitudinal striations produced by current drag of an object across a sedimentary surface. *2.* Impression or cast of such a mark on the under surface of an overlying bed.

drainage *1.* The processes of discharge of water from an area by stream or sheet flow and removal of excess water from soil by downward flow. *2.* The means of effecting the removal of water.

drainage area oil: A term applied to that area of a reservoir contributing oil or gas to a well. It is a poor descriptive term because it suggests gravity rather than pressure as the agent of movement.

drainage basin *Hydrol:* A part of the surface of the lithosphere that is occupied by a drainage system or contributes surface water to that system.

drainage characteristics The hydrology and related features of streams, lakes, swamps, marshes, and canals within an area. Includes consideration, in their seasonal aspects, of depths, widths, banks, bottom conditions, velocities, gradients, turbidity, sedimentation, temperatures, ice conditions, and other pertinent related items.

drainage density *Geomorph:* Symbol D, D_d. Ratio of total length of all channels within a drainage basin to the area of that basin; a measure of topographic scale or texture.

drainage divide A drainage divide is the rim of a drainage basin. It is the boundary between adjacent drainage basins. The term watershed has been used to mean both drainage basin and drainage divide, and the uncertainty of meaning entailed by this double usage makes the term undesirable.

drainage pattern Arrangement of natural drainage lines within an area; patterns are related to local geology and geologic history.

drainage system A stream and its tributaries constitute a drainage system, and the area drained by a river system through a valley system is a drainage basin.

drapery Curtainlike forms of travertine, usually formed through the union of a row of stalactites.

draw *1.* In the United States, a ravine, usually dry, but forming a watercourse in a freshet, furrowed vertically by torrents. *2.* A valley or basin readily convertible into an irrigation reservoir by constructing a dam across its outlet. *3.* A natural depression or swale; a small natural drainageway.

drawdown The lowering of the water table or piezometric surface caused by pumping (or artesian flow).

draw works In rotary drilling, that part of the equipment functioning as a hoist to raise or lower drill pipe, and in some types to transmit power to the rotary table.

dredging A method of obtaining resources or for deepening streams, swamps, or coastal waters by scraping and removing solids from the bottom.

dreikanter Pebble shaped by eolian sandblasting with plane faces bounded by three sharp edges or angles.

drewite Drewite is a title proposed for the calcareous ooze consisting of impalpable calcareous material, most of which is probably precipitated through the agency of denitrifying bacteria. Formed of minute aragonite needles.

drift *1.* A horizontal passage underground. A drift follows the vein, as distinguished from a crosscut, which intersects it, or a level or gallery, which may do either. *2. Coal min:* A gangway or entry above water level, driven from the surface in the seam. *3.* Any rock material, such as boulders, till, gravel, sand, or clay, transported by a glacier and deposited by or from the ice or by or in water derived from the melting of the ice. Generally used of the glacial deposits of the Pleistocene Epoch. Detrital deposits. *4.* Detrital material washed into a cave from a sink or vertical entrance. *5.* Floating material deposited on a beach. Driftwood. *6.* The motion, of sea, ice or of vessels, resulting from ocean currents and wind stress. *7.* Wind-driven snow in motion along the surface, sometimes rising to heights of 30.5 m. or more. *8.* Snow lodged in the lee of surface irregularities under the influence of the wind. *9. Geophys:* A time variation common to nearly all sensitive gravimeters, due to slow changes occurring in the springs or mountings of the instrumental systems; this variation is corrected by repeated observations at a base station and in other ways.

drift-barrier lake *See* MORAINAL LAKE

drift current A broad, shallow, slow-moving ocean or lake current.

drift curve Graph of a series of gravity values read at the same station at different times and plotted in terms of instrument reading versus time.

drift deposit Any accumulation of glacial origin; glacial or fluvioglacial deposit.

drift glaciers Small glaciers nourished by blown snow.

drift ice *1.* Floating ice. *2.* Any ice that has drifted from its place of origin.

drifting Opening a drift; driving a drift. *See* DRIFT, *1, 2*

drift map A map showing the distribution of various glacial and fluvioglacial deposits, generally called drift.

drift mine A mine opened by a drift.

drift sheet A sheetlike body of

glacial drift, continuous or discontinuous, deposited during a single glaciation (e.g., Cary drift sheet) or during a closely related succession of glaciations (e.g., Wisconsin drift sheet).

drift theory That theory of the origin of coal which holds that the plant matter constituting coal was washed from its original place of growth and deposited in another locality where coalification then came about.

driller's log A record filled out on tabulated form by the chief of the drilling crew of an oil or gas well drilling rig. This log shows rock character being drilled, drilling progress, drilling tools used, bit size and type, mud weight and condition, mud additives used, as well as a description of operations and personnel on duty each tour, along with any other pertinent or unusual event occurring during the drilling operations.

drilling mud A suspension, generally aqueous, used in rotary drilling and pumped down through the drill pipe to seal off porous zones and to counterbalance the pressure of oil and gas; consists of various substances in a finely divided state among which bentonite and barite are common. Oil may be used as a base instead of water.

drill-stem test A test of the productive capacity of a well when still full of drilling mud. The testing tool is lowered into the hole attached to the drill pipe and placed opposite the formation to be tested. Packers are set to shut off the weight of the drilling mud, and the tool is opened to permit the flow of any formation fluid into the drill pipe, where it can be measured.

driphole *1*. A small hole or niche in clay or rock beneath a point where water drips. *2*. The center hole in a formation built up beneath a water drip.

dripstone Material deposited by dripping solution. A single term to replace stalactite and stalagmite and some other cave deposits.

driven well A well which is sunk by driving a casing, at the end of which there is a drive point, without the aid of any drilling, boring, or jetting device.

drive pipe *1*. A pipe which is driven or forced into a bored hole, to shut off water, or prevent caving. *2*. A thick type of casing fitted at its lower end with a sharp steel shoe, which is employed when heavy driving has to be resorted to for inserting the casing.

drive shoe A protecting end attached to the bottom of drive pipe and casing.

drought Dryness due to lack of rain. An absolute drought is a period of at least 15 consecutive days to none of which is credited 0.254 mm. of rain or more. A partial drought is a period of at least 29 consecutive days, the mean daily rainfall of which does not exceed 0.254 mm. A dry spell is a period of at least 15 consecutive days to none of which is credited 1.016 mm. or more.

drown To submerge land with water, whether by a rise in the level of a lake, ocean, or river, or by a sinking of the land, as the lowering of a coastal region drowns the lower courses of the rivers and connects their valleys into estuaries.

drowned coast The presence of certain long, narrow channels which are largely free from islands suggests that a subsidence of the coast has transformed the lower portions of the old river valleys into tidal estuaries, thus changing a hilly land surface into an archipelago of small islands. A shoreline exhibiting these characteristics is termed a "drowned coast."

drowned glacial erosion coast A

coast having deep estuaries with basin depressions. Fiords and wide glacial troughs like the St. Lawrence, Bay of Fundy, the Skaggerack, and the White Sea are examples.

drowned river mouth When a section of the coast line sinks with reference to the sea, the water invades all the near shore valleys, thus "drowning" them and yielding a drowned river mouth or estuary.

drowned topography Depression whereby the lower courses of most of the rivers are submerged beneath the sea hastens the reduction of what is left above sea level, and by decreasing the slope of the lower courses often causes the building of flood plains at these points. The coasts of Maine, Norway, and Maryland afford excellent examples of such topography.

drum; drumlin [<*Ir., Gael.* druim; also druman, the back, a ridge, summit] *1.* Gravel hills that have an elongated form, are generally steepest toward one side, and rise in every other direction by much more gentle acclivities. *2.* Till is sometimes accumulated in hills of elliptical base and arched profile, known as drumlins, the longer axis often measuring half a mile or more and standing parallel to the direction in which the ice sheet moved, the height reaching 30.5 to 61 m. *3.* A streamlined hill or ridge of glacial drift with long axis paralleling direction of flow of former glacier.

druse *1.* Clusters or aggregates of euhedral or subhedral crystals, commonly incrusting the walls of a cavity. *2.* A cavity of this sort. *3.* (*Insol. residue*) Clusters or aggregates of subhedral crystals with definitely developed faces, commonly incrusted. May be microgeodic. Applied to chert, quartz, oölites, and others. *4.* A hole or bubble in glacial ice filled with a mixture of air and ice crystals.

drusy, *adj.* *1.* A term applied to rocks containing numerous druses. Miarolitic, *q.v.* *2.* A term used to express the appearance or habit of a crystalline aggregate whose surface is covered with a layer of small crystals.

drusy mosaic Crystalline mosaic produced by the deposition of minerals from solution in cavities other than the pores between sedimentary particles.

dry (state of ground, *q.v.*) Condition in which pore space of ground to depths of 7.62 cm. or more is essentially free of water (moisture content is less than permanent wilting percentage, *q.v.*).

dry basin If an interior basin exists in a climate so arid that the superficial flow of water, which constitutes drainage, is only potential and not actual, or else is occasional only and not continuous, it contains no perennial lake and is called a dry basin.

dry-bone ore A miner's term for an earthy, friable carbonate of zinc. Smithsonite, *q.v.* Frequently applied to the hydrated silicate, so-called calamine. Usually found associated in veins or beds is stratified calcareous rocks accompanying sulfides of zinc, iron, and lead. Locally applied to zinc silicate ore.

dry bulk density Natural density.

dry delta Alluvial fan, alluvial cone.

dry firn A dry granular snow which has been compacted by the combined action of sun, wind, and fluctuations in air temperature; also called "dry old snow."

dry hole Somewhat loosely used in oil and gas development, but in general any well that does not produce oil or gas in commercial quantity. A dry hole may flow water, a minor quantity of gas, or may even yield some oil, the oil or gas being in volumes too small to be profitable.

dry ice 1. Solid CO_2. 2. Bare glacier ice with no standing water or slush. 3. Ice with temperature below the freezing point.

dry limestone process A method of controlling air pollution caused by sulfur oxides. The polluted gases are exposed to limestone which combines with oxides of sulfur to form manageable residues.

dry ore An argentiferous ore that does not contain enough lead for smelting purposes.

dry pergelisol Soil material having the requisite mean temperature to be permanently frozen but lacking water.

dry permafrost Ground having the requisite mean temperature to be permanently frozen but lacking ice content, or "dry."

dry valley In England, a valley which, although originally carved out by running water, is now streamless. In the United States, such a valley is referred to as a wind gap.

d-spacing Distance between successive identical planes in a crystal structure that will diffract X rays according to Bragg's law.

DTA Abbreviation for Differential Thermal Analysis.

ductile 1. Capable of being drawn through the opening of a die without breaking and with a reduction of the cross-sectional area, as a ductile wire. 2. *Mineral:* Capable of considerable deformation, especially stretching, without breaking; said of several native metals and occasionally said of some tellurides and sulfides. 3. Pertaining to a substance that readily deforms plastically.

ductility Property of solid material that undergoes more or less plastic deformation before it ruptures.

dug well Well sunk by manual digging, shallow and of large diameter.

Duhem's theorem The state of any closed system is completely defined by the values of any two independent variables, extensive or intensive, provided the initial masses of each component are given. The choice of variables, however, must not conflict with the phase rule.

dull A surface texture of a pebble or grain that lacks luster. This lack is caused by numerous minute irregularities so that reflected light is diffused and scattered.

dumortierite A mineral, $Al_7(BO_3)$-$(SiO_4)_3O_3$. Orthorhombic blue, lavender, greenish blue. Used in manufacture of mullite refractories.

dump A land site where solid waste is disposed of in a manner that does not protect the environment.

dumpy level A leveling instrument in which the telescope is permanently attached to the leveling base, either rigidly or by a hinge that can be manipulated by means of a micrometer screw.

dune 1. *Geol;* A low hill, or bank, of drifted sand. 2. Mounds and ridges of wind-blown or eolian sand are dunes. Once started, a dune becomes an obstacle to blowing sand, and the lodgment of more sand causes the dune to grow. In this way, mounds and ridges of sand, scores and sometimes even hundreds of feet high are built by the wind. 3. A mound, ridge, or hill of wind-blown sand, either bare or covered with vegetation.

dune complex A group of wandering dunes which make up the moving landscape.

dune lake Lake formed as a

result of blocking of the mouths of streams by sand transported along the shore by combined action of winds and waves.

dune ridge A series of parallel foredunes built along the shore of a retreating sea.

dunite An ultramafic rock consisting essentially of olivine, with accessory pyroxene, plagioclase, or chromite.

durability index Term applied to the relative resistance to abrasion exhibited by a sedimentary particle in the course of transportation.

durain Material occurring in megascopic bands in coal characterized by gray to brownish-black color, rough surface, and faintly greasy luster; consists mainly of exinite and inertite.

duration In wave forecasting, the length of time the wind blows in essentially the same direction over the fetch (generating area).

duricrust The case-hardened crust of soil; duricrust is usually formed in arid or semiarid climates, but the plinthites of lateritization have also been called duricrust, as have some silcretes, both of which formed in areas of higher rainfall. Duricrust forms at sites of cementation, whether at the surface, or at the site of illuviation in a soil. Since erosion removes soft soil down to the site of illuviation, and then stops if the cementation is thorough, an old B horizon or part of it may form a superficial crust—a duricrust.

duripan A horizon in a soil characterized by cementation by silica or other minerals.

durite *See* DURAIN

dust Dust consists of dry organic and inorganic matter so finely divided that it may be picked up and carried away by the atmosphere without difficulty.

dust, volcanic Pyroclastic detritus consisting mostly of particles less than 0.25 mm. in diameter, i.e.,

fine volcanic ash. It may be composed of essential, accessory, or accidental material.

dust avalanche Avalanche, *q.v.*, of dry, loose snow.

Dust Bowl Region in southwestern plains states subject to periodic severe drought.

dust hole Dust well.

dust storm *1.* A strong wind carrying large clouds of dust (or silt), common in desert and plain regions; its influence may be felt thousands of miles from the source. China, the United States, Egypt, the Sahara, the Gobi, and numerous other parts of the world are subject to dust storms. *2.* For the development of a dust storm, there are three essentials: (1) an ample supply of fine dust or dry silt, (2) relatively strong winds to stir it up, and (3) a steep lapse rate of temperature in the dust-carrying air. There is a marked distinction between dust storms and sandstorms, *q.v.*

dust tuff An indurated deposit of volcanic dust. Essentially a fine-grained tuff.

dust well *1.* A pit in glacier ice or sea ice produced when small dark particles on the ice surface are heated by sunlight and sink down into the ice. *2.* Cylindrical tubes generally penetrating the ice to a depth of six or eight inches. They range in size from less than one quarter inch to larger than a foot in diameter. *Syn:* CRYOCONITE HOLES, *q.v.*; DUST HOLES

dwarf fauna A fauna characterized by fossils of stunted size. Some depauperate, *q.v.*, faunas are dwarf.

dy A sapropel, composed of organic matter brought into a lake in colloidal form and precipitated there.

dyke English spelling of dike.

dyke swarm *See* DIKE SWARM

dynamic breccia Tectonic breccia. An autoclastic rock clearly showing that it was formed as a result of tectonic movements which crushed and broke the formation. It usually is found in brittle rocks where loading has not been sufficient to render the rocks plastic.

dynamic climatology The explanation of climate according to the effects of atmospheric circulation.

dynamic geology Deals with the causes and processes of geological change.

dynamic metamorphism Dynamometamorphism. Metamorphism produced exclusively or largely by rock deformation, principally folding and faulting.

dynamochemical Refers to chemical activity induced by pressure and heat, particularly within the earth.

dynamometamorphism *See* DYNAMIC METAMORPHISM

dynamothermal metamorphism A common type of regional metamorphism related to large-scale orogeny. Directed pressures and shearing stresses accompanied by confining pressures of 3000–10,000 bars and temperatures of 400°–800° C. are associated with this type metamorphism.

dystrophic lakes Lakes between eutrophic and swamp stages of aging. Such lakes are shallow and have high humus content, high organic matter content, low nutrient availability, and high BOD.

E

earth That planet of the solar system which is third in order of distance from the sun, and fifth in size of the 9 major planets. Earth's equatorial radius is: 6378 km. (3963.5 mi.); polar radius: 6357 km. (3941 mi.); equatorial circumference: 40,075 km. (24,902 mi.).

earth current *1.* Telluric current. Natural electrical currents circulating in the crust of the earth, constituting a worldwide system subject both to spasmodic and periodic variations in intensity and direction which are consistently related to changes noted in other cosmic phenomena, such as the earth's magnetic field, the aurora and solar activity. *2.* A current flowing through a wire the extremities of which are grounded at points on the earth differing in electrical potential. The earth current is due to this difference, which is very generally temporary and often very large.

earth curvature *See* CURVATURE, EARTH

earth fall A landslide.

earthflow A slow flow of earth lubricated with water, occurring as either a low-angle terrace flow or a somewhat steeper but slow hillside flow.

earth hummocks Patterned ground with an essentially circular mesh, a nonsorted appearance, and a form characterized by its three-dimensional, knob-like shape and cover of vegetation.

earth movement Differential movement of the earth's crust; local elevation or subsidence of the land.

earthquake A sudden motion or trembling in the earth caused by the abrupt release of slowly accumulated strain (by faulting or by volcanic activity). *Partial syn:* SEISMIC EVENT; QUAKE. *Syn:* SHOCK; SEISM; MACROSEISM; TEMBLOR

earthquake engineering The study of the behavior of foundations and structures relative to seismic ground motion, and the attempt to mitigate the effect of earthquakes on such structures. *Syn:* ENGINEERING SEISMOLOGY; EARTHQUAKE SEISMOLOGY

earthquake intensity A measure of the effects of an earthquake at a particular place on humans and/or structures. The intensity at a point depends not only upon the strength of the earthquake, or the earthquake magnitude, but also upon the distance from the earthquake to the epicenter and the local geology at the point. *See also* INTENSITY SCALE

earthquake magnitude A measure of earthquake strength or the strain energy released by an earthquake as determined by seismographic observations.

earthquake prediction The estimating of time and place of occurrence of earthquakes based on scientific data.

earthquake swarm A series of minor earthquakes, none of which may be identified as the main shock, occurring in a limited area and time.

earthquake wave Seismic wave.

earthquake zone An area of the earth's crust in which crustal movements and sometimes associated volcanism occur; a seismic area.

earth rotation The turning of earth on its axis in a counterclockwise motion about the north pole. Angular velocity of earth's rotation is 0.00007292 radians per mean solar second.

earth's crust The external part of the earth, accessible to geological investigation.

earth slide Earth slides are those which most nearly resemble snow slides. When a mass of earth begins to slide and still retains a certain cohesion it moves as a block or blocks without other apparent deformation than the change in level: this is the foirage (landslip) of the French authors and resembles the planche (board) of snow, the initial stage of the snowy avalanche.

earth's orbit The elliptical path of earth through space in its annual journey about the sun. The semimajor axis of the ellipse is 92,700,000 miles with an eccentricity of 0.03. The sun occupies one focus of the ellipse.

earth tide The rising and falling of the surface of the solid earth in response to the same forces that produce the tides of the sea. Semidaily earth tides fluctuate between 7 and 15 centimeters.

earth tilting A slight movement or displacement of the surface of the ground as in some forms of earthquakes.

earth tremor A slight earthquake.

easement An incorporeal right existing distinct from the ownership of the soil, consisting of a liberty, privilege, or use of another's land without profit or compensation; a right of way.

East Pacific rise A broad, low volcanic ridge that extends from the southeastern part of the Pacific Ocean to the North American coastline, along Central America to the Gulf of California.

ebb current The movement of the tidal current away from shore or down a tidal stream.

ebb tide A nontechnical term referring to that period of tide between a high water and the succeeding low water; falling tide.

echelon faults Separate faults having parallel but steplike trends; the group having one more or less general direction but with the individuals parallel to each other and at an angle to that direction.

Echinoderma A phylum of exclusively marine, invertebrate animals, nearly all of which have a radial and five-rayed symmetry and a skeleton made of calcium carbonate plates. Many have spines, hence the popular name: spiny-skinned animals.

Echinodermata See ECHINODERMA

echinoid One of a group of invertebrates; a class of the Echinoderma which includes the sea urchins and their close allies

Echinoidea Class of free-moving echinoderms mostly with rigidly plated bodies of spherical to flattened disklike forms; sea urchins, sand dollars.

echogram Graph of the sea-floor shape made by recording water depths by sound-wave measurements.

echo sounder A survey instrument that determines the depth of water by measuring the time required for a sound signal to travel to the bottom and return.

Eckert projection One of a series of six map projections of the entire earth of which the poles are represented, not as a point, but as lines, each 1/2 the length of the equator. The parallels are rectilinear in each, but the meridians may be rectilinear or curved.

eclogite A granular rock composed essentially of garnet (almandine-pyrope) and pyroxene (omphacite).

eclogite facies Metamorphic rocks of gabbroid composition consisting mainly of omphacite (pyroxene) and garnet.

ecological impact The total effect of an environmental change, either natural or man-made, on the ecology of the area.

ecological indicators Species or associations which indicate certain ecological conditions such as soil, type, climate, temperature, or salinity.

ecologic facies Facies determined by nature of environment.

ecology The study of the mutual relationships between organisms and their environment, including the rock substrate.

economic geology Deals with geological materials of practical utility and the application of geology to engineering. Generally used today in the narrow sense to include only the application of geology to mineral materials.

economic mineral Any mineral having a commercial value. *See* ORE

ecosphere The earth considered as one giant ecosystem.

ecosystem An ecologic system; that system composed of interacting organisms and their environments; the result of interaction between biological, geochemical, and geophysical systems. The geobiocoenose, *q.v.*

ecotope Physical environment of a habitat. *See also* BIOTOPE

ecoulement A downhill gliding of a large mass of rock under the influence of gravity, during or as a result of tectonic deformation. Such a movement resulting from ordinary agents of erosion constitutes a landslide, *q.v.*

edaphic A term referring for the soil conditions or types as ecological factors.

eddy A current of air or water running contrary to the main current, especially one moving in a circle; a whirlpool.

eddy current A circular movement of water of comparatively limited area formed on the side of a main current.

edge water The water surrounding or bordering oil or gas in a pool. Edge water usually encroaches on a field after much of the oil and gas has been recovered and the pressure has become greatly reduced.

edge-water drive A process whereby energy for the production of oil is derived principally from the pressure of edge water in the formation.

edgewise conglomerate A conglomerate consisting of small flat pieces of (usually calcareous) rocks packed in such a manner as to lie steeply inclined with reference to the bedding plane of the stratum.

effective permeability The observed permeability of a porous medium to one fluid phase under conditions of physical interaction between this phase and other fluid phases present.

effective porosity *1. Hydrol:* Often used in same sense as specific yield. It is the ratio of the volume of water, oil, or other liquid, which after being saturated with that liquid, it will yield under any specified hydraulic conditions to its own volume. *2.* The property of rock or soil containing intercommunicating interstices, expressed as a per cent of bulk volume occupied by such interstices.

effective size *Soil mech:* The maximum diameter of the smallest 10 per cent of the particles of a sediment. Equals the grain diameter at the 90 percentile in sedimentary petrography.

effective size of grain *Hydrol:* The diameter of the grains in an assumed rock or soil that would transmit water at the same rate as the rock or soil under consideration and that is composed of spherical grains of equal size and arranged in a specified manner.

effervesce To bubble and hiss, as limestone on which acid is poured.

effloresce To change on the surface or throughout to a powder from the loss of water of crystallization on exposure to the air. Efflorescent, *adj.*

efflorescence Surface encrustation, commonly powdery, produced by evaporation.

effluent [<*Lat.* ex, out; fluo, flow] *1.* Flowing forth out of. *Geol:* Flowing out, as lava through fissures in the side of a volcano, as a river from a lake. *2.* Anything that flows forth. *Geog:* A stream flowing out of another or forming the outlet of a lake. *3.* A discharge of pollutants into the environment, partially or completely treated or in its natural state. Generally used in regard to discharges into waters.

effluent seepage Seepage out of the lithosphere.

effusion *1.* The act or process of effusing or pouring out; that which is effused or poured out. *2. Volcanol:* An emission of liquid lava or pyroclastic material from a vent or fissure.

effusive *1.* Pouring out; pouring forth freely. *2. Geol:* A term applied to that class of igneous rocks derived from magmas or magnetic materials poured out or ejected at the earth's surface (i.e., volcanic rocks), as distinct from the so-called intrusive or plutonic rocks that are injected at depth. *See* EXTRUSIVE ROCKS

einkanter A ventifact upon which only one facet has been cut by wind-blown sand. *See* DREIKANTER; VENTIFACT

ejecta; ejectamenta Material thrown out by a volcano, such as ash, lapilli, bombs.

elastic aftereffect Creep recovery, *q.v.*

elastic afterworking Creep recovery, *q.v.*

elastic bitumen *See* ELATERITE

elastic constants Certain mathematical constants that serve to describe the elastic properties of matter.

elastic deformation A nonpermanent deformation, after which a body returns to its original shape when the load is released. Often limited to that deformation in which stress and strain are linearly related in accordance with Hooke's law.

elastic discontinuity Boundary between strata that reflects seismic waves.

elastic flow The part of the deformation from which the specimen recovers is called "elastic flow."

elasticity The property or quality of being elastic, i.e., an elastic body returns to its original form or condition after a displacing force is removed.

elastic limit The maximum stress that a specimen can withstand without undergoing permanent deformation either by solid flow or by rupture when the stress is released. *Syn:* YIELD POINT

elastic medium A material which returns to its original form or condition after a displacing force is removed.

elasticoviscous Material in which instantaneous elastic strain, under stress below the elastic limit, is followed by continuously developed permanent strain under long sustained stress of constant magnitude. *Syn:* VISCOELASTIC

elastic rebound The recovery of elastic strain.

elastic rebound theory Faulting arises from the sudden release of elastic energy which has slowly accumulated in the earth. Just before the rupture, the energy released by the faulting is entirely potential energy stored as elastic strain in the rocks. At the time of rupture the rocks on either side of the fault spring back to a position of little or no strain.

elaterite A massive amorphous dark-brown hydrocarbon ranging from soft and elastic to hard and brittle. It melts in a candle flame without decrepitation, has a conchoidal fracture, and gives a brown streak.

electrical well logging The process of recording the formations traversed by a drill hole, based on the measurements of two basic parameters observable in uncased holes; namely, the spontaneous potential (S.P.) and the resistivity of the formations to the flow of electric currents. The detailed study *in situ* of the formations penetrated by a drill hole, based on measurements made systematically by lowering an apparatus in the hole responding to the following physical factors or parameters: (1) the resistivities of the rocks; (2) their porosity, (3) their electrical anisotropy, (4) their temperature, (5) the resistivity of the drilling muds.

electric log The log of a well or borehole obtained by lowering electrodes in the hole and measuring various electrical properties of the geological formations traversed. Electrical current is introduced by a number of methods.

electric potential Force required to bring a unit charge from infinity to a charged point of the same sign or energy released by opposite movement.

electrochemical series Electromotive force series.

electrodialysis A process that uses electrical current and an arrangement of permeable membranes to separate soluble minerals from water. Often used to desalinize salt or brackish water.

electromagnetic damping Commonly found in seismometers of the induction type. It may be used in mechanical seismographs by employing a copper plate moving between two permanent magnets. Induction seismometers depend upon voltage generated by motion of coil in the magnetic field.

electromagnetic prospecting A geophysical method employing the generation of electromagnetic waves at the earth's surface; when the waves penetrate the earth and impinge on a conducting formation or ore body they induce currents in the conductors which are the source of new waves radiated from the conductors and detected by instruments at the surface.

electromotive force series Metallic elements arranged in the order of their standard electrode potentials and their ability to replace each other in solutions of their salts. In order of decreasing potential they are: K, Na, Ca, Mg, Fe^{++}, Zn, Fe^{+++}, Cu.

electron The elementary particle of mass 9×10^{-28} grams and unit electrical charge (4.80×10^{-10} e.s.u.).

electron capture A type of radioactive transformation in which an electron from one of the inner shells of an atom is captured by the nucleus; especially important in the transformation $K^{40} - A^{40}$.

electronegativity The power of an atom in a molecule to attract electrons to itself. Differences in electronegativity of the two ions in a compound can be related to the type of bond.

electron microprobe An analytical instrument which utilizes a finely focused beam of electrons to excite X-ray emission from selected portions of a sample surface. The composition of the sample at the point of excitation can be determined by analysis of the emitted X-ray spectrum.

electron shell A group of electrons in an atom, all having approximately the same average distance from the nucleus and approximately the same energy.

electron volt The kinetic energy acquired by an electron falling through a potential difference of 1 volt. It is equal to 1.6×10^{-12} erg.

electro-osmosis Press of a solution against electric potential.

electrostatic precipitator An air pollution control device that removes particulate matter by imparting an electrical charge to particles in a gas stream for mechanical collection on an electrode.

electrostatic valence The quotient of the charge on a simple ion divided by its coordination number.

electroviscosity Viscosity of fluids as influenced by electric properties. Viscosity generally increases more in low-conductivity solutions flowing through narrow capillaries than in high-conductivity fluids.

electrum A natural alloy of gold and silver (Au, Ag), varying from pale to deep yellow. *See* GOLD

element *1.* A substance which cannot be decomposed into other substances. *2.* A substance all of whose atoms have the same atomic number. The first definition was accepted until the discovery of radioactivity (1896), and is still useful in a qualitative sense. It is no longer strictly correct, because (a) natural radioactive decay involves the decomposition of one element into others, (b) one element may be converted into another by bombardment with high-speed particles, (c) an element can be separated into its isotopes. The second definition is accurate, but it has little relevance to ordinary chemical reactions or to geologic processes.

element, equant *Struct. petrol:* A fabric element of approximately equal dimensions.

element, fabric *Struct. petrol:* A single crystal or group of crystals which behaves as a unit with respect to the applied forces.

element, linear *Struct. petrol:* A fabric element of rodlike form where one dimension is much greater than the other two.

element, planar *Struct. petrol:* A fabric element having two dimensions conspicuously in excess of the third is described as planar.

elevation *1.* A particular height or altitude above a general level, as, the height of a locality above the level of the sea; of a building, etc.: above the level of the ground. *2.* In the United States, term generally refers to height in feet above mean sea level. *See* ALTITUDE, 2

elevation correction In gravity measurements, the corrections applied to observed gravity values because of differences of station elevation to reduce them to any arbitrary reference or datum level, usually sea level. The corrections consist of (1) the free-air correction, to take care of the vertical decrease of gravity with increase of elevation, and (2) the Bougeur correction, to take care of the attraction of the material between the reference datum and that of the individual station. *Seismol:* In seismic measurements, the corrections applied to observed reflection time values due to differences of station elevation in order to reduce the observations to an arbitrary reference datum or fiducial plane.

ellipsoidal lavas; ellipsoidal basalts *Syn:* PILLOW LAVAS

elutriation Purification by washing and pouring off the lighter matter suspended in water, leaving the heavier portions behind.

eluvial *1.* Formed by the rotting of rock in place to a greater or less depth. *2.* A term applied to materials carried to lower depths by descending vadose waters.

eluviated horizon The uppermost, leached horizon in a soil. If there is leaching the A horizon is leached first, but some soils are leached throughout and others are not leached.

eluviation The movement of soil material from one place to another in solution or in suspension by natural soil-forming processes. Soil horizons that have lost material through eluviation are referred to as eluvial, and those that have received materials as illuvial.

eluvium Atmospheric accumulations *in situ,* or at least only shifted by wind, in distinction to alluvium, which requires the action of water.

emanations, magmatic A combination of volatile and nonvolatile materials given off by a magma at various stages in its history, and with various compositions and densities. The term usually includes both aqueous liquids and gases, and both the pegmatitic and the hydrothermal fluids. *See* MINERALIZER

emanations, volcanic Volatile or nonvolatile materials emitted from volcanoes, fumaroles, or lavas at the earth's surface, usually consisting of a mixture of water vapor and one or more of the other volcanic gases, *q.v.*

embankment *1.* When the combined action of waves and currents extends a barrier into deep water, an embankment is formed, which in many cases becomes of very grand proportions. In the formation of embankments the debris of which they are composed is swept along the surface of the barrier or terrace leading to them, and deposited when deep water is reached. This process continues until the embankment has been built up to the water surface. *2.* Artificial ridge of earth or broken rocks such as a dike or railroad grade across a valley.

embayed Formed into a bay or bays, as an embayed shore.

embayed crystal Crystal penetrated by another generally euhedral crystal.

embayment *1.* An indentation in a shoreline forming an open bay. *2.* The formation of a bay. *3.* Term describing a continental border area that has sagged concurrently with deposition so that an unusually thick section of sediment results. An embayment is similar to a basin of sedimentation or a geosyncline, and some embayments may be one flank of a larger subsiding feature. *4.* Used in a structural sense to designate a reentrant of sedimentary rocks into a crystalline massif.

embouchure [*Fr.*] The mouth of a river, or that part where it enters the sea. *See* DEBOUCHURE

emerald A bright emerald-green variety of beryl, the color being due to the presence of chromium. Used as a gem.

emergence *1.* A term which implies that part of the ocean floor has become dry land but does not imply whether the sea receded or the land rose. *2.* Point at which an underground stream comes to the surface. *Syn:* RE-SURGENCE; RISE. *3. Paleobot:* An outgrowth, consisting of epidermal and cortical tissues lacking vascular tissues; e.g., rose prickles.

emergent form Emergent forms of topography are those built up partly beneath the water, but gradually rising above it. Deltas built into dry land, lakes filled up with silts, turned into marshes, and later into dry land, are examples of this kind.

emery *1.* An abrasive consisting mainly of pulverized corundum. *2.* A granular rock composed of corundum, magnetite, and spinel with or without margarite, chloritoid, etc., resulting from metamorphism of highly aluminous sediments.

emery rock A rock that contains corundum and iron ores. *See* CORUNDOLITE

eminence A mass of high land; a high ground or place.

emplace *1.* To move to a particular position; said of intrusive rocks. *2.* To develop in a particular place; said of ore deposits.

emulsion A colloidal dispersion of one liquid in another.

enantiomorphous *Crystallog:* Similar in form but not superposable; related to each other as the right hand is to the left, hence, one the mirror image of the other.

enantiotropy The relationship between polymorphs that possess a stable transition point and that therefore can be stably interconverted by changes of temperature and/or pressure.

enargite A mineral, Cu_3AsS_4. Orthorhombic. An ore of copper.

encroachment The advancement of water, replacing withdrawn oil or gas in a reservoir.

endellite Hydrated halloysite. A clay mineral, $Al_2Si_2O_5(OH)_4$.

$4H_2O$, which readily dehydrates to halloysite.

endemic Native or confined naturally to a particular and unusually restricted area or region; biologically the site of origin of a taxon. An endemic taxon remains in the site of origin after other taxa have spread beyond that site.

end member *1.* One of two or more relatively simple compounds or substances occurring in a mixture. *2.* One of two or more distinctive forms between which more or less gradual and continuous variation occurs.

end moraine *1.* Moraine marking the terminal position of a valley glacier. *2.* Terminal moraine.

endogene effects Internal effects upon the margin of an intrusive body itself.

endogenetic *1.* Pertaining to rocks resulting from physical and chemical reactions, their origin being due to forces within the material. In general, they are nonclastic, chemical precipitates formed by "solidification, precipitation, or extraction of the mineral matter from the states of igneous fusion, aqueous solution or vaporization." *Cf.* AU-THIGENIC. *2.* Endogenic. A term applied to processes that originate within the earth and to rocks, ore deposits, and landforms which owe their origin to such processes. Contrasted with exogenetic.

endogenous Produced from within; originating from or due to internal causes. *Syn:* ENDOGE-NETIC; ENDOGENIC. Contrasted with exogenous.

endometamorphism; endomorphism The modification produced in an igneous rock due to the partial or complete assimilation of portions of the rocks invaded by its magma; a phase of contact metamorphism in which attention is directed to the changes suffered by the intrusive

body instead of to those produced in the invaded formations.
Adj: ENDOMETAMORPHIC, ENDOMORPHIC

endoskeleton The foreign material, frequently calcium carbonate, secreted by an animal to form a rigid internal support for its soft parts.

endothermic Designating, or pertaining to, a reaction which occurs with an absorption of heat. Opposed to exothermic.

end product Stable element resulting from radioactive decay.

endurance limit Fatigue limit. Limiting stress below which specimens can withstand hundreds of millions of repetitions of stress without fracturing. Considerably less than the rupture strength.

en echelon Parallel structural features that are offset like the edges of shingles on a roof when viewed from the side.

energy The ability to do work or produce motion. Earth's surface receives much of its energy from the sun.

energy coefficient The ratio of the energy in a wave per unit crest length transmitted forward with the wave at a point in shallow water to the energy in a wave per unit crest length transmitted forward with the wave in deep water.

energy resource The natural supply of energy available for use. Several major sources include: Earth's internal heat (geothermal energy), fossil fuels (principally coal, natural gas, and oil), hydropower (rivers, ocean currents, tides, and waves), nuclear energy, solar energy, and wind.

engineering geology The application of the geological sciences to engineering practice for the purpose of assuring that the geologic factors affecting the location, design, construction, operation, and maintenance of

engineering works are recognized and adequately provided for.

englacial Pertaining to the inside of a glacier.

englacial drift *1.* This is regarded as having been the material embraced within the glacial ice, or borne on its surface, and by its melting let loosely down upon the true till, formed beneath the ice. *2.* May be applied to any erratic material that, at any time during its transportation, may be enclosed within the ice even though it be essentially at the bottom of the glacier and may have been actually at the bottom a little while before and may again be at the base a little later on; or it may be applied, less technically but more significantly, to that only which is embedded in the heart of the ice and borne passively along with it free from basal influence until it is at length brought out to the surface of the terminal slope by the agency of ablation.

engulfment If a great reservoir of molten lava, underlying a cone, should be drawn off through a side fissure instead of through the central vent, as sometimes happens, the upper part of the cone might fall by its own weight. This process is known as engulfment.

enrichment The action of natural agencies which increases the metallic content of an ore. Secondary sulfide enrichment refers to the formation of new sulfide minerals which contain a larger percentage of the metals.

enstatite A mineral of the pyroxene group, $MgSiO_3$.

enterolithic structure Small folds resulting from changes in volume due to chemical changes in a rock.

enthalpy Heat content per unit mass.

entisol Soils of frozen ground,

desert sands, in all areas and under all climates, characterized by a distinct lack of soil horizons.

entrenched meander A meander eroded below the surface of the valley in which it was formed; may result from either uplift of a region without tilting or lowering of base level.

entrenched meander valley One whose stream, having inherited a meandering course from a previous erosion cycle, has sunk itself into the rock with little modification of its original course.

entrenched stream A narrow, meandering trench cut in a wide-open, flat-bottomed trough, the trough being sunk well beneath the general surface of the adjacent upland. *See* INTRENCHED STREAM

entropy *1.* A measure of the unavailable energy in a system, i.e., energy that cannot be converted into another form of energy. *2.* A measure of the mixing of different kinds of sediment; high entropy is approach to unmixed sediment of one kind.

entropy function facies map Map showing areal relations of facies based on degrees of intermixing of three end members; does not distinguish between end members.

entry *1. Coal min:* A haulage road, gangway, or airway to the surface. *2.* An underground passage used for haulage or ventilation, or as a manway. If it is driven directly down a steep dip it becomes a slope. *See* PORTAL

entry pressure The minimum capillary pressure that will force the entry of a nonwetting fluid into capillary openings saturated with a wetting fluid. *Syn:* DISPLACEMENT PRESSURE; FOREFRONT PRESSURE

envelope *1.* The outer part of a recumbent fold; especially used to contrast the sedimentary cover of a recumbent anticline with the crystalline cone. *2.* Metamorphic rocks surrounding an igneous intrusion.

environment The sum total of all the external conditions that may act upon an organism or community, to influence its development or existence. For example, the surrounding air, light, moisture, temperature, wind, soil, and other organisms are all parts of the environment.

environmental facies Facies that are controlled entirely by the nature of the environment. These are not three-dimensional bodies of rock or sediments, but areas inferred from a combination of mutually interacting conditions exhibited as distinctive sedimentary types and biologic communities.

environmental geochemistry The distribution and interrelationship of the chemical elements and radioactivity at the earth's surface, including rocks, water, air, and biota.

environmental geology The application of geologic principles and knowledge to problems created by man's occupancy and exploitation of the physical environment.

environmental impact statement A document prepared by industry or a political entity on the environmental impact of its proposals for legislation and other major actions significantly affecting the quality of the human environment. Environmental impact statements are used as tools for decision making and are required by the National Environmental Policy Act.

Eocene Second epoch of the Tertiary Period; Paleocene below and Oligocene above; also the series of strata deposited during that epoch.

Eogene Lower of two Cenozoic subdivisions, consisting of Paleocene, Eocene, and Oligocene. *Cf.* NEOGENE

eolian Aeolian, *Obs. 1.* Applied to deposits arranged by the wind, as the sands and other loose materials along shores, etc. (From Eolus, the god of winds.) Subaerial is often used in much the same sense. *2.* Applied to the erosive action of the wind, and to deposits which are due to the transporting action of the wind.

eolith The most primitive type of man-made stone implements.

eon; aeon A period of existence; an informal age; an infinite space of time. The term is used by some geologists to denote any one of the grand divisions of geological time. P. E. Cloud has defined an eon as one billion years.

Eötvös effect The vertical component of the Coriolis acceleration experienced when measuring gravity on a moving platform. Changes in course, velocity, and position can influence the apparent gravitational attraction by more than 50 milligals.

Eötvös torsion balance A gravity instrument consisting of a pair of masses suspended by a sensitive torsion fiber and so supported that they are displaced both horizontally and vertically from each other. A measurement is made of the rotation of the suspended system about the fiber caused by slight differences in the direction of gravitational force on the two weights.

Eötvös unit The unit of measurement in work with Eötvös torsion balance having the dimensions of acceleration divided by length, for the gradient and differential curvature values. For the gradient, 1 Eötvös unit $(1E.) = 1 \times 10^{-9}$ gal. per horizontal centimeter.

epeirogenesis Epeirogeny.

epeirogenic movement [$<Gr.$ epeiros, a continent] The broad uplift or depression of areas of the land or of the sea bottom, in which the strata are not folded or crumpled, but may be tilted or may retain their original horizontal attitude.

epeirogeny The broad movements of uplift and subsidence which affect the whole or large portions of continental areas or of the oceanic basins.

ephemeral stream A stream or portion of a stream which flows only in direct response to precipitation.

ephemeris time Time as measured by relative changes in the positions of the earth, moon, and stars.

epi- [*Gr.*] A prefix indicating alternation.

epibole *1.* Biostratigraphic unit corresponding to a hemera. *2.* An abundance or peak zone; a thickness of rock containing one or more taxa in more than usual abundance.

epicenter The point on the earth's surface directly above the focus of an earthquake.

epiclastic Textural term applied to mechanically deposited sediments (gravel, sand, mud) consisting of weathered products of older rocks. Detrital material from pre-existent rocks.

epicontinental Situated upon a continental plateau or platform, as, an epicontinental sea.

epicontinental sea *1.* Those shallow portions of the sea which lie upon the continental shelf, and those portions which extend into the interior of the continent with like shallow depths, such as the Baltic Sea and Hudson Bay, may be called epicontinental seas. *2.* Independent intercontinental seas with an abyssal area are mediterraneans (*Mittelmeere*), while the shallow type, without the abyssal area, constitute the true epicontinental sea.

epidote A mineral, $Ca_2(Al,Fe''')_3$-$(SiO_4)_3(OH)$. Monoclinic. A common mineral in metamorphic rocks.

epieugeosyncline Deeply subsiding troughs with limited volcanism associated with rather narrow uplifts and overlying a deformed and intruded eugeosyncline.

epigene 1. Geological processes originating at or near the surface of the earth. 2. A group name for volcanic and sedimentary classes of rocks, i.e., rocks formed at or near the surface of the earth. *Cf.* HYPOGENE. 3. Foreign. Said of forms of crystals not natural to the substances in which they are found. *Cf.* PSEUDOMORPH

epigenesis *Metal:* Changes in the mineral character of a rock resulting from external environmental forces acting at or near the earth's surface; e.g., mineral replacement during metamorphism.

epilimnion In lakes, the layer of water above the thermocline. It is the area where main primary productivity takes place.

epineritic That portion of the marine environment extending from low tide to a depth of water of less than 120 feet (40 meters or 20 fathoms).

epinorm Theoretical calculation of minerals in metamorphic rocks of the epizone as indicated by chemical analyses.

epipedon A surface layer of soil about one foot in thickness.

epipelagic The upper portion of the oceanic province extending from the surface to a depth of about 185 meters; also the animals that occupy it.

epiplankton Those organisms which habitually live upon a floating object to which they are attached or on which they move freely.

epithermal deposit A deposit formed in rocks of shallow depth from low-temperature hydrothermal solutions.

epizone 1. The "upper zone" of metamorphism. In this zone the distinctive physical conditions are moderate temperature, lower hydrostatic pressure, and powerful stress, and the rocks characteristically produced include mylonites, and cataclastic rocks generally, phyllites, chlorite schists, talc-schists, porphyroids, and in part marbles and quartzites. 2. Metamorphic environment characterized by low temperature and hydrostatic pressure, with or without high stress, resulting in chemical and mechanical metamorphism; characterized by hydrous silicates.

epoch A division of geologic time; when capitalized it becomes a formal division of geologic time corresponding to a series of rock and a subdivision of a period.

epsomite Epsom salt. A mineral, $MgSO_4.7H_2O$. Orthorhombic.

epsom salt Epsomite, a chemical, $MgSO_4.7H_2O$.

equal-area net A type of projection of points on a sphere to a flat surface (circle). The areas of every square degree on the projection are equal.

equant, *adj.* A term applied to crystals that have the same or nearly the same diameter in every direction. *Syn:* EQUIDIMENSIONAL

equant element *See* ELEMENT, EQUANT.

equatorial projection One of a group of cylindrical family map projections that has its center point on the equator and its polar axis vertical; e.g., the Mercator projection on which all straight lines are true compass bearings.

equiareal projection Projection from the center of a sphere through a point on its surface to a plane tangent at the south pole so constructed that areas between meridians and parallels on the plane are equal to corresponding ones on the surface of the sphere.

equigranular, *adj.* A textural term applied to rocks whose essential minerals are all of one order of size.

equilibrium Equilibrium exists in any system when the phases of the system do not undergo any change of properties with the passage of time and provided the phases have the same properties when the same conditions, with respect to the variants, are again reached by a different procedure.

equilibrium constant The equilibrium constant of a chemical reaction is the quotient formed by multiplying together the activities of the products, each raised to a power indicated by its coefficient in the equation, and dividing by a similar product of the activities of the reactants.

equiplanation [<*Lat.* aequus, equal, planus, a plain] All physiographic processes which tend to reduce the relief of a region and so cause the topography eventually to become more plainlike in contour, without involving any loss or gain of material, i.e., the amounts of material remain apparently equal, or are not increased or decreased by the plain-producing process or processes.

equipotential surface A surface on which the potential is everywhere constant for the attractive forces concerned.

equivalent *1. Geol:* Corresponding in geologic age or stratigraphic position; said of formations, etc. *2.* Applied to grains of ore or vein-stuff of varying diameters and density, which fall through water at an equal velocity. Usually used in the plural.

equivalent grade In textural classification, refers to arithmetic mean size.

equivalent molecular unit In the Niggli calculation of the molecular norm, the sum of all cations in the formula of a given mineral; thus $Na_2O.Al_2O_3.2SiO =$ 6 nepheline equivalent units, as there are 6 cations in the formula.

equivalent radius A measure of sedimentary particles' size equal to the radius of a spherical particle of the same density which would have the same settling rate as the sedimentary particles.

equivolumnar waves Distortional waves; secondary waves; shear waves; transverse waves; S-waves, *q.v.*

era *Geol:* In general, a large division of geologic time; specifically, a division of geologic time of the highest order, comprising one or more periods. The eras now generally recognized are the Paleozoic, Mesozoic, and Cenozoic.

erathem The largest recognized time-stratigraphic unit, next in rank above system; the rocks formed during an era of geologic time, such as the Mesozoic Erathem composed of the Triassic System, the Jurassic System, and the Cretaceous System.

E ray *See* EXTRAORDINARY RAY

erg A term used in the Sahara, and applicable elsewhere, for a vast region covered deeply with pure sand and occupied by dunes.

Erian Middle Devonian.

Erian orogeny Early Devonian diastrophism.

erodible Capable of being eroded.

erosion The group of processes whereby earthy or rock material is loosened or dissolved and removed from any part of the earth's surface. It includes the processes of weathering, solution, corrasion, and transportation. The mechanical wear and transportation are affected by running water, waves, moving ice, or winds, which use rock fragments to pound or grind other rocks to powder or sand.

erosion scarp A scarp produced by the agents of erosion.

erosion surface A land surface shaped by the disintegrating, dissolving, and wearing action of streams, ice, rain, winds, and other land and atmospheric agencies.

erosion thrust A thrust fault along which the hanging wall moved across an erosion surface.

erratic *1.* Those large water-worn and ice-borne blocks (boulders) which are scattered so generally over the higher and middle latitudes of the northern hemisphere. *2.* A transported rock fragment different from the bedrock on which it lies, either free or as part of a sediment. The term is generally applied to fragments transported by glacier ice or by floating ice.

error, average The mean of all the errors taken without regard to sign.

error, constant A systematic error which is the same in both magnitude and sign through a given series of observations.

error, personal A systematic error caused by an observer's personal habits in making observations, or due to his tendency to react mentally and physically in the same way under similar conditions.

error, systematic An error whose algebraic sign and, to some extent, magnitude bear a fixed relation to some condition or set of conditions.

In its broadest sense, the term systematic error includes constant errors. Under similar conditions systematic errors tend to be repeated; if the conditions do not change, the error will be constant. The inclusive definition is preferred.

Systematic errors are regular, and therefore are subject to *a priori* determination.

eruption, volcanic The emission or ejection of volcanic materials at the earth's surface from a crater of pipe or from a fissure. Central eruptions are those in which volcanic materials are emitted from a central vent or pipe and ordinarily result in the formation of a volcanic cone; fissure eruptions are those in which lava or pyroclastic materials emanate from a relatively narrow fissure or group of fissures, commonly building lava plains and lava plateaus.

eruption cloud A convoluted, rolling mass of partly condensed water vapor, dust, and ash, generally highly charged with electricity, emitted from a volcano during an explosive eruption.

eruptive *1.* Descriptive of igneous rocks that reach the surface of the earth in a molten condition. *2.* Refers to material thrown out by a volcano. *3.* Applies to any igneous rock (not recommended).

erythrite Cobalt bloom. A mineral, $Co_3(AsO_4)_2.8H_2O$. Monoclinic.

escarpment *1.* [<*Fr.* escarper, to cut steeply] A steep face terminating high lands abruptly. *2.* A slope; a steep descent; a declivity. *Geol:* The steep face frequently presented by the abrupt termination of stratified rocks.

escutcheon A diamond-shaped basal expansion, pit, or cavity seen on the aboral surface of a conodont or the dorsal surface of a bivalve (pelecypod).

eskar Esker, *q.v.*

esker Osar; asar; eschar; eskar;

serpent kame. *1.* Eskers or kames, unlike the drumlins, are rudely stratified accumulations of gravel, sand, and waterworn stones. They are of rough fluvial or torrential origin, and occur in long tortuous ridges (serpent kames), mounds, and hummocks. They have the general direction of the drainage, though sometimes not according with the present course of drainage. *2.* Serpentine ridges of gravel and sand. These are often associated with kames, and are taken to mark channels in the decaying ice sheet, through which streams washed much of the finer drift, leaving the coarser gravel between the ice walls.

esker fan Small plains of gravel and sand built at the mouth of sub-glacial tunnels and channels in the ice; associated with an esker or eskerlike chain of deposits made in the ice sheet at the same time.

essential ejecta *1.* Pyroclastic detritus, whether loose or indurated, which is of immediate, juvenile, magmatic origin. *2.* Juvenile ejecta. Fresh magmatic material thrown out in liquid form by a volcano.

essential minerals Those mineral constituents of a rock that are necessary to its classification and nomenclature. An essential constituent is not necessarily a major constituent, for the presence in a rock of minor amounts of such minerals as nepheline, olivine, or quartz may affect its classification. *See* ACCESSORY MINERALS

essexite A plutonic rock composed essentially of plagioclase hornblende, biotite, and titanaugite with subordinate alkali feldspar and nepheline. An alkalic gabbro.

estuarine Of, pertaining to, or formed in an estuary.

estuary Drainage channel adjacent to the sea in which the tide ebbs and flows. Some estuaries are the lower courses of rivers or smaller streams, others are no more than drainage ways that lead seawater into and out of coastal swamps.

etched *1.* Term applied to a rough frosted surface, as of minerals or sand grains. *2.* Pitted or corroded in such manner that a pattern of pits or lines is produced which is related to the crystal or tectonic structure. *See* FROSTING

etch figure A marking, usually minute pits, produced by a solvent on a crystal surface; the form varies with the species and solvent but conforms to the symmetry of the crystal, hence revealing its crystal structure.

etching A roughening of the surface of a sand grain or crystal as a result of a solvent.

ethane A colorless, gaseous compound (C_2H_6), of the paraffin series contained in the gases given off by petroleum and in illuminating gas.

eucrystalline *adj.* A textural term relating to igneous rocks, such as granites, which are well crystallized.

eugeosyncline A geosyncline in which volcanism is associated with clastic sedimentation. A volcanic part of an orthogeosyncline, located away from the craton.

euhedral A crystal completely bounded by its own regularly developed crystal faces. *Syn:* IDIOMORPHIC, ORTHOMORPHIC

eulittoral That part of the littoral zone less than 50 meters in depth.

eupelagic Having purely pelagic qualities.

euphotic zone The layer of water which receives ample sunlight for photosynthesis. It varies with the light extinction coefficient. Usually no deeper than about 60 meters.

euryhaline Tolerant of considerable difference in salinity; generally refers to marine organisms.

eurypterid One of a group of very large, extinct arthropods related to the trilobites.

eustasy Of or pertaining to worldwide sea level.

eustatic, *adj.* Of or pertaining to worldwide sea level.

eutaxitic Applied to a structure of certain volcanic rocks with a streaked or blotched appearance due to the alternation of bands or elongated lenses of different color, composition, or texture; the bands, etc., having been originally ejected as individual portions of magma which were drawn out together in a viscous state and formed a heterogeneous mass by welding.

eutectic Pertaining to a mixture of substances that do not form solid solutions, having its components in such proportions that its melting point is the lowest possible with those components.

eutectic point or **eutectic temperature** The lowest melting temperature obtainable with mixtures of given components, provided that the components do not form solid solutions.

eutectic ratio The ratio of solid phases forming from the eutectic liquid at the eutectic point; it is such as to yield a gross composition for the crystal mixture that is identical with that of the liquid. Most frequently stated in terms of weight per cent.

eutectic texture; eutectoid texture Intergrowths of minerals, either along crystallographic or bleb boundaries, similar to those precipitated from eutectic solutions.

eutrophication The process whereby a body of water becomes highly productive due to the input of large quantities of nutrients.

eutrophic lake A shallow lake that has good primary productivity, usually abundant littoral vegetation, and dense plankton population. Blooms occur when nutrient and physical conditions are right. The bottom water, during later stages of eutrophication, tends to become depleted of oxygen during the summer.

euxinic Refers to bottom water conditions, usually in enclosed or barred (silled, *q.v.*) basins in which waters are so stagnant that all organic matter reaching the bottom is shielded from oxygen and decay, usually resulting in high sulfide content.

evaporite One of the sediments which are deposited from aqueous solution as a result of extensive or total evaporation of the solvent.

evaporite-solution breccia A breccia formed by a removal of the evaporites.

evapotranspiration A term embracing that portion of the precipitation returned to the air through direct evaporation or by transpiration of vegetation, no attempt being made to distinguish between the two.

event, seismic Applied to any definite phase change or amplitude difference on a seismic record; it may be a reflection, a refraction, a diffraction, or a random signal.

everglade A tract of swampy land covered mostly with tall grass; a swamp or inundated tract of low land. Local in the South.

evolution The theory that life on Earth has developed gradually, generally by change and branching of and within species, and generally from simple to complex.

evolutionary momentum The idea that organic function is activity and activity is motion led to the notion that evolution may continue to inadaptive lengths and result in extinction.

exfoliation The breaking- or peeling-off of scales, lamellae, as concentric sheets from bare rock surfaces by the action of either physical or chemical forces.

exhaustible resources *See* RE-SOURCE CONCEPTS

exhaustion *Min:* The complete removal of ore reserves.

exhumed topography Monadnocks, mountains, or other topographic forms buried under younger rocks and exposed again by erosion. *Cf.* MENDIP

exogene Organism or fossil introduced into a biologic or paleontological association from another place.

exogene effects Effects upon the rocks invaded by igneous masses.

exogenetic *1.* Pertaining to a rock made up of fragments of older rocks and owing its origin chiefly to agents acting from without. *2.* Applied to processes originating at or near the surface of the earth, such as weathering and denudation, and to rocks, ore deposits, and landforms which owe their origin to such processes. *Contr. with* ENDOGENETIC; EDOGENIC

exogenic differentiation Chemical differentiation during a cycle of rock weathering and sediment transportation and deposition.

exogeosyncline A parageosyncline that lies along the cratonal border and obtains its sediments from erosion of complementing highlands in the orthogeosynclinal belt that lies outside the craton.

exomorphism Changes produced in country rock as a result of the intense heat and other properties of magma or lava contact; contact metamorphism.

exoskeleton The protection surrounding the soft body of an animal, as the shell of brachiopods and pelecypods, etc.

exosphere Space beyond the earth's atmosphere; begins at a height of about 1000 km.

exothermic Designating, or pertaining to, a reaction which occurs with a liberation of heat. *Ant:* ENDOTHERMIC

exotic Nonendemic; not native but introduced.

expansion fissures *Petrog:* A system of irregularly radiating fissures which ramify through feldspars and other minerals adjacent to olivine crystals that have been replaced by serpentine. Characteristic of norites and gabbros. The alteration of olivine to serpentine involves a considerable increase in volume, and the stresses so produced are relieved by the fissuring of the surrounding minerals.

exploit *1.* To make complete use of; to utilize. *2.* To make research or experiment; to explore.

exploration *1.* The work involved in looking for ore. Often confused with exploitation. *2.* In the petroleum industry, the search for natural accumulations of oil and gas by any geological, geophysical, or other suitable means.

exploratory well A well drilled either in search of a new and as yet undiscovered pool of oil or gas, or with the hope of greatly extending the limits of a pool already partly developed.

explorer's alidade *See* GALE ALIDADE

explosion breccia A deposit of coarse, indurated volcanic debris containing blocks torn from the walls of a volcanic vent and lying in a matrix of comminuted rock.

explosion caldera A caldera resulting primarily from a violent volcanic explosion which blows out a huge mass of rock, leaving a broad, deep basin in its place. Relatively rare and small in size compared to collapse calderas. An example is Bandai-san in Japan.

explosion crater A volcanic crater formed by a violent explosion commonly developed along rift zones on the flanks of large volcanoes and occasionally at the summit of volcanoes. Distinguished from ordinary craters at the top of volcanic cones and from pit craters, which are produced largely by collapse. *Syn:* EXPLOSION PIT

explosion tuff A tuff of which the constituent ash particles have been dropped directly into place after being ejected from a volcanic vent, the term thus distinguishing such tuffs from the more ordinary types which are washed into place.

explosive evolution Relatively rapid adaptive radiation of a group of animals into numerous different lines of descent in a relatively geologic short time.

explosive index Percentage of pyroclastic ejecta among the total products of a volcanic eruption.

explosive radiation *See* EXPLOSIVE EVOLUTION

exposure *Geol:* The condition or fact of being exposed to view, either naturally or artificially; that part of a rock, bed, or formation which is so exposed; an outcrop.

exsolution Unmixing. Solid solutions of some pairs of minerals form only at high temperatures and become unstable at lower temperatures. When these cool slowly, one mineral may separate out of the other at a certain point in the cooling-temperature curve. This is known as unmixing, or exsolution.

exsolved *See* EXSOLUTION

exsudation The scaling off of rock surfaces through growth of salines by capillary action has been called exsudation. It prob-

ably has only local importance as a weathering process.

extended consequent stream Extended consequents are streams of an older type which become extended across the newly emerged coastal plain. These streams do not differ from those originating on the coastal plain, except in their greater volume of water and hence greater erosive power. They will, therefore, cut deeper than the others, becoming the master streams of their respective regions, and directing to a large extent the further development of their drainage system. Streams extended across a dry delta may also be classed here, as are those extended across a plain of glacial deposition.

extension fractures Fractures that form parallel to a compressive force. In a sense they are tension fractures.

extension joints Joints that form parallel to a compressive force. In a sense they are tension fractures.

extensive quantity Thermodynamic quantities such as volume or mass that depend on the total quantity of matter in the system.

external mold The impression, in adjoining rock, of the outer sides of the hard parts of an organism.

external rotation Rotation of small mineral aggregates in metamorphic rocks whose internal relationships are little disturbed.

extinction *1.* A position at which a birefringent substance on the stage of a polarizing microscope with crossed nicols is dark, even though the line of sight is not parallel to the optic axis; also the darkness so obtained. *2.* The disappearance, natural or otherwise, of a taxon of organisms succumbing to changed conditions of environment.

extinction angle The angle

through which a section of an anistropic crystal must be revolved from a known crystallographic plane or direction to the position of maximum darkness (extinction) under the petrographic microscope.

extinction direction One of two perpendicular directions in a crystal parallel to the vibration planes of the microscope polarizer and analyzer.

extralateral right In U.S. mining law, one who locates on the public domain a claim in which a vein comes to an apex is entitled to extralateral rights which pertain to parts of the vein beyond the planes passed through the side lines of his claim, but lying within vertical cross planes passed through the end lines. Also called the Apex Law.

extraordinary ray E ray in a uniaxial crystal. The ray of light which vibrates in a plane containing the optic axis and at an angle with the basal pinacoid.

extreme *1. Adj:* Applied to the highest and the lowest temperatures (or other meteorological element) which had occurred over a very long record for each month and for the year. *2. Adj; N:* Sometimes applied to the extremes of temperatures in individual months; and again to the average of the highest and the lowest temperatures, as the so-called "mean monthly extremes and mean annual extremes."

extrusion *1.* The act or process of thrusting or pushing out; also a form produced by the process; a protrusion. *2. Geol:* The emission of magmatic material (generally said of lavas) at the earth's surface; also the structure or form produced by the process, such as a lava flow, volcanic dome, or certain pyroclastic rocks.

extrusion flow hypothesis A theory which explains the movement of glacier ice out of a basin-shaped bed or down a gentle gradient by assuming that the differential pressures caused by weight of the overlying ice forces the basal portion of the glacier to flow like a plastic substance. In this theory, extrusion flow depends on the thickness of the ice and the steepness of the surface slope. *See* GRAVITY FLOW

extrusive rocks Applied to those igneous rocks derived from magmas or magmatic materials poured out or ejected at the earth's surface, as distinct from the intrusive or plutonic igneous rocks which have solidified from magmas that have been injected into older rocks at depth without reaching the surface. *Syn:* EFFUSIVE ROCKS; VOLCANIC ROCKS

exudation basin A spoon-shaped depression found at the heads of outlet glaciers of the Greenland icecap.

eyed structure Augen structure, *q.v.*

F

fabric *1.* The orientation in space of the elements of which a rock is composed. *2. Petrol:* That factor of the texture of a crystalline rock which depends on the relative sizes, the shapes, and the arrangement of the component crystals.

fabric element A rock component, ranging from an atom or ion to a mineral grain or ground of grains in pebbles, lenses, layers, etc., that acts as a unit in response to deformative forces.

face *V: 1.* To be directed toward, to turn toward; *2.* To be directed toward younger rocks, e.g., overturned strata and structures face downward. *N: 3. Min:* End or surface where work is being or was last done in any level or gently inclined part of a mining operation; *cf.* BACK; *4. Coal min:* Surface of principal jointing in a coal bed. *5.* In structure, original upper surface of a stratum, especially if it has been raised to a vertical or steeply inclined position. *6. Crystallog:* Plane surface of a crystal.

facet *1.* The polished surface of a cut gemstone. *2.* Any nearly plane surface abraded on a rock fragment, e.g., by glaciation. *3.* Asymmetrically scalloped rock surfaces. *Syn:* FLUTE; SOLUTION RIPPLE

faceted spur *1.* In a river valley the spurs between ravines run down and die out at, or near, the river; in a glaciated valley, these are ground away by the longitudinal erosion up to the level of the ice and after its recession terminate in more or less well-defined inverted V-shapes in the wall of the main valley; these spurs are said to be faceted. *2.* The end of a ridge which has been truncated or steeply beveled by stream erosion, glaciation, or faulting.

facies *Petrog: 1.* General appearance or nature of one part of a rock body as contrasted with other parts. *2.* Part of a rock body as differentiated from other parts by appearance or composition. *3.* A kind of rock distinguished from other more or less related kinds. *4. Stratig:* A stratigraphic body as distinguished from other bodies of different appearance or composition. *5.* A subdivision of a stratigraphic unit; *cf.* LITHOFACIES, BIOFACIES, TECTOFACIES

facies, igneous A variety of igneous rock. Especially applied to an igneous rock that in some respects is a departure from the normal or typical rock of the mass to which it belongs. Thus a mass of granite may grade into a porphyritic facies near its borders.

facies contour Line indicating equivalence in lithofacies development, e.g., a particular value of the sand-shale ratio.

facies-departure map Map showing areal relations of facies based on degree of similarity to some particular sedimentary composition.

facies evolution Gradual change in the nature of facies and their relations with a particular area.

facies family Group of closely related and associated facies, e.g., different parts of an organic reef.

facies fauna A group of animals adapted to life on a restricted type of sea floor or other environment.

facies fossil A fossil, usually a species, which is adapted to life in a restricted environment.

facies map A map showing the distribution of different types of sedimentary facies occurring within a designated geologic unit.

facies sequence Succession of vertically related facies.

facies strike Direction indicated by facies contours.

facies suite All broadly related facies, e.g., all marine deposits.

facies tract System of different but genetically interconnected facies; originally included areas of erosion which furnished sediments to areas of deposition.

facing Applied to the original direction of a layer.

faecal pellet *See* FECAL PELLET

fall; falls A cascade, waterfall, or cataract. The flow or descent of one body of water into another.

falling dune Sand blown off a mesa top into a valley or canyon may form a solid wall, sloping at the angle of rest of dry sand, or a fan extending downward from a reentrant in the mesa wall.

fall line *1.* A line characterized by numerous waterfalls, as the edge of a plateau in passing which the streams make a sudden descent. *2.* A large river whose valley is extended across a coastal plain often has low falls or rapids near the inner margin of the plain, which determine the "head of navigation" or uppermost point that can be reached by vessels from the river mouth.

families of igneous rocks Subgroups, under clans, in the classification of igneous rocks, e.g., the syenite family.

family *Paleon:* A taxonomic division used in the classification of animals and plants; a group of closely related genera and a subdivision of an order.

fan An accumulation of debris brought down by a stream descending through a steep ravine and debouching in the plain beneath, where the detrital material spreads out in the shape of a fan, forming a section of a very low cone. *See* ALLUVIAL FAN

fan bay The head of an alluvial fan which extends a considerable distance into a mountain canyon.

fan fold An anticlinal fold in which the two limbs dip toward one another, or a syncline in which the two limbs dip away from one another.

fanglomerate A fanglomerate is composed of heterogeneous materials which were originally deposited in an alluvial fan but which since deposition have been cemented into solid rock.

fan shooting A refraction type of seismic shooting in which a fan of detectors is laid out from a single shot point.

fast breeder reactor A nuclear reactor that, while utilizing U-235, alters U-238 to a fissionable fuel, and therefore produces more nuclear fuel than it consumes.

fast ice A floe of ice extending from the land and fast to it. *Syn:* LAND FLOE; SHORE FLOE; ICE FOOT

fat clay Clay of relatively high plasticity. *Cd.* LEAN CLAY

fathogram A continuous profile of the depth obtained by echo soundings.

fathom A unit of measurement used for soundings. It is equal to 1.83 meters (6 feet).

fathometer The copyrighted trade name for a type of echo sounder.

fatigue limit *See* ENDURANCE LIMIT

fault A fracture or fracture zone along which there has been displacement of the sides relative to one another parallel to the fracture.

fault basin A region depressed relative to the surrounding regions and separated from them by faults.

fault block *1.* A mass bounded on at least two opposite sides by faults; it may be elevated or depressed relatively to the adjoining region, or it may be elevated relatively to the region on one side and depressed relatively to that on the other. *2.* A body of rock bounded by one or more faults.

fault breccia Angular fragments resulting from the crushing, shattering, or shearing of rocks during movement of a fault or from friction between the walls of the fault.

fault complex Intricate system of interconnecting and intersecting faults of the same or different ages.

fault dip The vertical inclination of the fault surface or shear zone, measured from a horizontal plane.

fault embayment A depressed region in a fault zone or between two faults invaded by the sea. The Red Sea and Tomales Bay on the San Andreas fault are examples.

fault escarpment *See* FAULT SCARP

faulting The movement which produces relative displacement of adjacent rock masses along a fracture.

fault line The intersection of a fault surface with the surface of the earth or with any artificial surface of reference.

fault-line scarp A scarp that is the result of differential erosion along a fault line rather than the direct result of the movement along the fault. *See* OBSEQUENT and RESEQUENT FAULT-LINE SCARPS

fault plane A fault surface that is more or less planar.

fault scarp The cliff formed by a fault. Most fault scarps have been modified by erosion since the faulting.

fault set Two or more parallel faults within an area constitute a fault set.

fault splinter A connecting ramp-like structure between opposite ends of two parallel normal faults.

fault spring A surface spring emerging from a fault that brings an aquifer into contact with an impermeable bed.

fault strike The direction of the intersection of the fault surface, or the shear zone, with a horizontal plane.

fault surface The surface along which dislocation has taken place. *Cf.* FAULT PLANE

fault system Consists of two or more fault sets that were formed at the same time.

fault trap An oil trap whose closure results from the presence of one or more faults.

fault trellis drainage pattern A variety of trellis pattern found where a series of parallel faults have brought together alternating bands of strong and weak rock.

fault wedge Wedge-shaped block between two faults.

fault zone A fault, instead of being a single clean fracture, may be a zone hundreds or thousands of feet wide; the fault zone consists of numerous interlacing small faults or a confused zone of gouge, breccia, or mylonite.

fauna The animals of any place and time that lived in mutual association. The limitations of any fauna are relative. They may be interpreted broadly as of a fauna of a continent or of a geologic period, or restrictively as of the fauna of a small area of the sea bottom during a single season. A paleontologic fauna consists only of those animals the remains of which are preserved as fossils.

faunal Pertaining to a natural assemblage of animals.

faunal break An abrupt change from one fossil assemblage to another in a stratigraphic sequence.

faunal province A region characterized by a specific assemblage of animals more or less widely distributed within that region.

faunal succession The observed changes of life forms in definite and recognizable order through geological time.

faunizone 1. Biostratigraphic unit characterized by the presence of a particular fauna that may have either time or environmental significance. Cf. ASSEMBLAGE ZONE. 2. Biostratigraphic unit consisting of various more or less overlapping biozones; has dominantly time-stratigraphic significance.

faunule A diminutive fauna. *Paleontol:* a subdivision of a fauna.

f-axis *Struct. petrol:* Axis of rotation, normal to t, around which a gliding plane may be bent.

fayalite A mineral of the olivine group, $(Fe,Mg)_2SiO_4$.

feather joints A series of joints that branch diagonally from a larger joint or fault.

fecal pellet Excreta, mainly of invertebrates, present especially in modern marine deposits but also as fossils in sedimentary rocks. Most are of simple ovoid form and 1 mm. or less in size. More rarely they are rod-shaped with either longitudinal or transverse sculpturing. Coprolites, *q.v.*, are of similar origin but much larger. *Syn:* FAECAL PELLET

feeder 1. Small vein joining a larger vein. 2. A spring or stream.

3. A blower of gas, as in a coal mine.

feeder channels Channels parallel to shore along which feeder currents flow before converging and forming the neck of a rip current.

fee land That land that is controlled by an individual. Mineral rights are not tied to fee land, but may be sold separately.

feldspar A group of abundant rock-forming minerals of the general formula, $MAl(Al,Si)_3O_8$, where M can be K, Na, Ca, Ba, Rb, Sr, and Fe. Most widespread of any mineral group and may constitute 60% of the Earth's crust, occurring in all types of rock. Orthoclase, microcline, plagioclase, (albite, and anorthite), and celsian. Also called feldspar, feldspath.

feldspathic Containing feldspar as a principal ingredient.

feldspathic sandstone A sandstone containing from 10 to 25% feldspar intermediate between a pure quartzose sandstone and an arkose. Cf. ARKOSIC SANDSTONE

feldspathoids A group of minerals related to feldspars, but containing less silica. Leucite, nepheline, sodalite, cancrinite, melilite. These minerals take the place of feldspars in igenous rocks that are too low in silica to form the corresponding feldspar.

felsenmeer [*Ger.* sea of rock] Above the limit of the growth of trees (tree or timberline) rock destruction goes on with great rapidity, as is indicated by the wild and chaotic confusion of rock pieces.

felsic A mnemonic term derived from (fe) for feldspar, (1) for lenads or feldspathoids, and (s) for silica and applied to light-colored rocks containing an abundance of one or all of these constituents. Also applied to the minerals themselves, the chief felsic minerals being quartz, feldspars, feldspathoids, and muscovite. *Syn: ACID, 4, q.v.; MAFIC, q.v.; SILICIC, q.v.*

felsite An igneous rock with or without phenocrysts, in which either the whole or the groundmass consists of a cryptocrystalline aggregate of felsic minerals, quartz and potassium feldspar being those characteristically developed. When phenocrysts of quartz are present the rock is termed a quartz felsite, or, more commonly, a quartz porphyry.

felsitic, *adj.* A textural term ordinarily applied to dense, light-colored igneous rocks made up of crystals that are too small to be readily distinguished with the unaided eye. In this sense the term is essentially synonymous with microcrystalline.

felty; felted A textural term applied to dense, holocrystalline igneous rocks or to the dense, holocrystalline groundmass of porphyritic igneous rocks consisting of tightly appressed microlites, generally of feldspar, interwoven in irregular, unoriented fashion. If, as is characteristic of many andesites and trachytes, the crowded microlites of feldspar are disposed in a subparallel manner as a result of flow, and their interstices are occupied by micro- or cryptocrystalline material, the texture is called pilotaxitic or trachytic.

femic A mnemonic term derived from (fe) for iron and (m) for magnesium and applied to the group of standard normative minerals in which these elements are an essential component, including the pyroxene and olivine molecules and most of the normative accessory minerals (magnetite, ilmenite, hematite). The corresponding term for the ferromagnesian minerals actually present in a rock is MAFIC, *q.v.*

fence diagram Three or more geologic sections showing the relationship of wells or outcrop sections to formations. To give proper perspective, scales diminish with distance from the foreground. When several sections are used together, they form a fencelike enclosure, hence the name. Similar in some respects to a block diagram, but having the advantage of transparency, which is not possible in a block diagram.

fenster Window. An erosional break through an overthrust sheet or through a large recumbent anticline whereby the rocks beneath the thrust sheet are exposed. The term "window" has sometimes been erroneously used in the United States for areas in which the normal stratigraphic succession has not been disturbed by faulting, but where older strata are exposed along the crest of an anticline.

ferberite *See* WOLFRAMITE

Ferrel's law Coriolis force, *q.v.* A statement of the fact that currents of air or water are deflected by the rotation of the earth to the right in the northern hemisphere and to the left in the southern hemisphere.

ferride A member of a group of elements related to iron, including Ti, V, Cr, Mn, Fe, Co, Ni.

ferroalloy A metal whose chief use is for alloying with iron to produce special quality steel. Ferroalloy metals include manganese, nickel, chromium, tungsten, molybdenum, vanadium, cobalt, and titanium.

ferroan-dolomite Dolomite in which not more than half of the magnesium has been replaced by iron.

ferrogabbro A gabbroic rock in which the pyroxene or olivine or both are exceptionally rich in iron.

ferromagnesian *Petrol:* Containing iron and magnesium. Applied to certain dark silicate minerals, especially amphibole, pyroxene, biotite, and olivine, and to igneous rocks containing them as dominant constituents.

ferromagnetic Refers to those paramagnetic materials having a magnetic permeability considerably greater than one. They are attracted by a magnet.

ferrosilite A mineral of the pyroxene group, FeSiO₃.

ferruginate *1. Adj:* Cemented with iron minerals generally limonite. *2. V:* To stain with iron.

ferruginous Containing iron. Descriptive of rocks of red color but not necessarily abnormal iron content.

festoon A type of cross-lamination resulting from (1) the erosion of plunging troughs having the shape of a quadrant of an elongate ellipsoid, (2) the filling of the troughs by sets of thin laminae conforming in general to the shape of the trough floors, and (3) the partial destruction of the filling laminae by subsequent erosion, producing younger troughs.

fetch *1.* In wave forecasting, the continuous area of water over which the wind blows in essentially a constant direction. Sometimes used synonymously with fetch length. *Also* GENERATING AREA. *2.* In wind setup phenomena, for inclosed bodies of water, the distance between the points of maximum and minimum water surface elevations. This would usually coincide with the longest axis in the general wind direction.

fiard [*Sw.*] *1.* Sea-drowned valleys which are not river estuaries, occur in lowland countries, and are composed of hard rocks. *2.* A glaciated reentrant of the sea with low glaciated sides. Fiards are shorter and shallower than fiords. For example, Norway is a fiord country, while Sweden has few fiords but many fiards.

fibroblastic Nematoblastic.

fiducial mark *1.* An index line or point. A line or point used as a basis of reference. *2. Photogrammetry:* Index marks rigidly connected with the camera lens through the camera body and forming images on the negative which defines the principal point of the photograph.

fiducial time A time on a seismograph record which may be marked to correspond, by employing necessary corrections, to a datum plane in space.

field *1.* A large tract or area of many square kilometers containing valuable minerals. *See* COAL FIELD. *2.* A colliery, or firm of colliery proprietors. *3.* The immediate locality and surroundings of a mine explosion. *4.* A region or space traversed by lines of force, as, gravitational, magnetic, or electric.

field capacity The amount of water held in a soil by capillary action after gravitational water has percolated downward and drained away; expressed as the ratio of the weight of water retained to the weight of dry soil.

field development well Any well drilled within the presently known or proved productive area of a pool (reservoir) as indicated by reasonable interpretation of subsurface data, with the objective of obtaining oil or gas from that pool.

field focus The total area or volume which the source of an earthquake occupied. If a fault is the source, the focus is the local fault surface, and is called the "field" because it is inferred from the area of shaking as observed in the field.

field ice Sea ice, *q.v.*

field intensity Magnetic field strength (H), *q.v.*

figure of the earth The sea-level surface of the earth, the geoid, or a modified surface, the cogeoid, related to the geoid in a defined manner.

filiform texture Threadlike forms consisting of one mineral embedded in another mineral.

fill *1.* The withdrawal of the river current from the outer side of the curve, leaves that bank bordered by quiet water, in which deposition will probably result. Such deposition may be called a fill. *2.* Material deposited or washed into a cave passageway. Fill is generally prefixed by a word describing its dominant grain size—sand fill, silt fill, clay fill, gravel fill. *3.* Any sediment deposited by any agent so as to fill or partly fill a valley, sink, or other depression. *4.* Material used to raise the surface of the land generally in a low area.

fillers Mineral and other minerals used for a specific purpose in a manufactured product but not as essential constituents.

fill terrace A term to comprise the series of terms—alluvial terrace, glacial terrace, and others—which are formed after the rejuvenation of a stream-filled valley or a valley surface made by aggregation.

filltop terrace Fill terrace whose surface is the original depositional surface.

film, intergranular *See* INTER-GRANULAR FILM

filter pressing *1. Geol:* The process of the straining out of liquid when an igneous rock has partly crystallized and then is subjected to pressure by earth movements, etc. *2.* A process of magmatic differentiation wherein a magma having crystallized to a "mush" of interlocking crystals in liquid becomes compressed, permitting the liquid to move toward regions of lower pressure and hence become separated from the crystals.

filter sand Sand suitable for use in filtering the suspended matter from water.

find, *n. 1.* That which is found, particularly something valuable or interesting. *2.* Meteorite found but not observed to fall.

fine gold Almost pure gold. The value of bullion gold depends on its percentage of fineness. *See* FINENESS; FLOAT GOLD

fine-grained soil Soil consisting mostly of clay and silt, more than 50% by weight smaller than 0.074 mm. in diameter.

fineness The proportion of pure silver or gold in jewelry, bullion, or coin, often expressed in parts per thousand. The fineness of U.S. coin is nine-tenths, or 900 fine; that of English gold coin is eleventwelfths or 917 fine, and English silver coin is 925 fine.

fineness factor A measure of average particle size obtained by summing the products of the reciprocal of the size grade midpoints times the frequency of particles in each class expressed as a decimal part of the total frequency.

fines *1.* The fine fraction of a sediment or the product of rock crushing, particularly that which passes through a grading sieve. *2.* The fraction of sand and gravel finer than 0.074 mm. in particle diameter.

fine sand All grains between .25 mm. and .125 mm. in diameter.

finger Minor structure radiating from a major one.

finger lake Long narrow rock basins occupied by lakes.

fiord; fjord A long, deep arm of the sea, occupying a portion of a channel having high steep walls, a bottom made uneven by bosses and sills, and with side streams entering from high-level valleys by cascades or steep rapids.

fire assay The assaying of metallic ores, usually gold and silver, by methods requiring a furnace heat. It commonly involves the processes of scorification, cupellation, etc.

fireclay *1.* Any clay capable of resisting very high heat without passing into a glassy clay, no alkaline substance being present to form a flux. *2.* Formerly used for almost any soft nonbedded clay immediately underlying a carboniferous coal bed, many of which are not refractory. *Cf.* UNTERCLAY

fireclay mineral Mellorite. A poorly crystallized (partly disordered) kaolinite.

firedamp An explosive coal mine gas consisting mainly of methane.

firefountaining The rhythmic eruption of gas-charged lava (normally basaltic) from a volcanic vent, either a localized central vent or a fissure, forming a fountain of molten rock.

fire opal A transparent hyacinth-red opal which may or may not show play of colors.

firn *1.* A name given to snow above the glaciers which is partly consolidated by alternate thawing and freezing, but has not yet become glacier ice. *See* NÉVÉ. *2.* Compacted, granular but still pervious snow with a density usually greater that 0.4 but less than 0.82. By some workers considered to be any snow that has survived one or more ablation seasons. Firn may later become glacial ice. *Syn:* ACCUMULATION AREA. Névé is used as a synonym for firn in both senses.

firn basin Accumulation area of a glacier.

firn field *1.* A mass of firn which is not part of a glacier. *2.* The accumulation areas of a glacier.

firn limit The highest level on a glacier to which the snow cover recedes during the ablation season. Firn limit is preferable to firn line, for a zone of snow patches usually marks the transition from bare ice to a solid snow cover.

firn snow *See* FIRN, 2.

first arrival The primary or first impulse recorded by seismographs. In the refraction method of seismic prospecting quantity observed is the time between the initiation of the seismic wave by an explosion and the first disturbance indicated by a seismic detector at a measured distance from the shot point. Since first arrivals only are considered, the waving causing the disturbance is that which has traveled the minimum time path between shot point and detector. *See* FERMAT'S PRINCIPLE

first law of thermodynamics The first law of thermodynamics introduces the concept of internal energy of a system and expresses the fact that the change of energy of the system is equal to the amount of energy received from the external world. The energy received from the external world is equal to the heat taken in by the system and the work done on the system.

firth A narrow arm of the sea; also the opening of a river into the sea. *Syn:* ESTUARY; FRITH

fishing The operation of attempting to recover a piece of drilling or other equipment broken off or lost from the drilling tools and left in the hole.

fishtail bit A rotary bit used to drill soft formations. The blade is flattened and divided, the divided ends curving away from the direction of rotation. Resembling a fishtail, hence the name.

fissile bedding Term applied to bedding which consists of laminae less than 2 mm. in thickness.

fissility A property of splitting easily along closely spaced parallel planes.

fission 1. The splitting of an atomic nucleus into at least two parts (other atomic nuclei) plus the ejection of two or more neutrons (and occasionally other particles) plus the emission of energy. 2. Separation of calyx of a coral by cleavage. 3. Reproduction of the asexual type in which one polyp divides to form two new ones, as in coelenterates.

fissionable Capable of undergoing fission, usually by the action of neutrons, but also of protons, deuterons, alpha particles, electrons, and gamma radiation.

fissure 1. An extensive crack, break, or fracture in the rocks. A mere joint or crack persisting only for a few inches or a few feet is not usually termed a fissure by geologists or miners, although in a strict physical sense it is one. 2. Where there are well-defined boundaries, very slight evidence of ore within such boundaries is sufficient to prove the existence of a lode. Such boundaries constitute the sides of a fissure.

fissure eruption See ERUPTION, VOLCANIC

fissure polygons Nonsorted polygons.

fissure vein 1. A cleft or crack in the rock material of the earth's crust, filled with mineral matter different from the walls and precipitated therein from aqueous solution, or introduced by sublimation or pneumatolysis. 2. A mineral mass, tabular

in form, as a whole, although frequently irregular in detail, occupying or accompanying a fracture or set of fractures in the inclosing rock; this mineral mass has been formed later than the country rock, either through the filling of open spaces along the latter or through chemical alteration of the adjoining rock.

fiveling Crystal formed by fivefold cyclic twinning.

fix The position on a map of a point of observation obtained by surveying processes. Also, the act of determining such a position.

fixed carbon In the case of coal, coke, and bituminous materials, the solid residue other than ash, obtained by destructive distillation, determined by definite prescribed methods.

fixed ground water Ground water held in saturated material with interstices so small that it is attached to the pore walls, and is usually not available as a source of water for pumping.

flagging *Geophys:* The use by surveyors of flags of cloth or paper to mark instrumental or shot locations.

flaggy Strata from 10 to 100 mm. thick.

flags Thin bedded, hard sandstone or limestone that can be used for flagstones.

flagstone A rock that splits readily into slabs suitable for flagging.

flake A flat fragment from a larger piece.

flamboyant structure The optical continuity of the crystals or grains as disturbed by a divergent structure caused by slight differences in orientation.

flame photometry Spectrum measurement of a substance heated to incandescence in a flame.

flame structure 1. Load cast showing evidence of some horizontal slip. 2. Load cast in which part of an underlying layer has been squeezed irregularly upward into the overlying layer.

flame test The use of the characteristic coloration imparted to a flame to detect the presence of certain elements.

flanking moraine The sidelong or flanking moraines left by lobations or tonguelike projections of an ice sheet.

flanks The limbs of folds. *Syn:* LEGS; SHANKS; BRANCHES; SLOPES

flap A gravity-collapse structure. A bed that has slid down the side of an anticline and bent over so that it is now upside down.

flaser Streaky, parallel layers surrounding the granular lenticular bodies in flaser structure.

flaser gabbro A cataclastic gabbro in which are preserved lenses (phacoids, augen) of undeformed rock. *See* MYLONITE

flaser structure A structure developed in gneisses, gabbros, etc., by dynamic metamorphism. Small lenses of granular material are separated by wavy ribbons and streaks of finely crystalline, foliated material, usually aggregates of parallel scales in wavy or bent lines.

flash box A box in which a light source, electromagnet, and telescope are all mounted in the pendulum apparatus of gravitational recording.

flat 1. A general term meaning smooth, or even; a surface of low relief. *See* TIDAL FLAT; VALLEY FLAT, etc. 2. [*Derbyshire; N. Wales*] A horizontal vein or ore deposit; auxiliary to a main vein; also any horizontal portion of a vein elsewhere not horizontal.

flatiron A triangular-shaped sloping mesa-type of hogback ridge,

often occurring in series on the flank of a mountain.

flat joint In igneous rocks, joint dipping at an angle of 45° or less, randomly oriented with respect to others.

flat lode A lode which varies in inclination from the horizontal to about 15°. *See* FLAT

flatness 1. A measure of the shape of a pebble given by the sum of the long and intermediate diameters of the pebble divided by twice the short diameter. 2. A measure of the shape of a pebble given by the ratio of the radius of curvature of the most convex portion of the flattest face to the mean radius of the pebble.

flat of ore A horizontal ore deposit occupying a bedding plane in the rock. *See* FLAT

flats and pitches 1. In the Upper Mississippi lead and zinc district the term is applied to the near horizontal solution openings in the galena dolomite (flats) and the interconnecting inclined joints or fractures (pitches) in which the ore has been deposited. 2. Applied to certain ore bodies of characteristic form that occur in regions of bedded sedimentary rocks. Such ore bodies have a steplike form with the "flats" following nearly horizontal bedding planes and the "pitches" following steeply dipping joint planes or fractures.

flattening, plane of *Struct. petrol:* The pebbles or grains are flat and perpendicular to the greatest principal stress axis. The plane of schistosity is called a plane of flattening.

flaw Blatt [*Ger.*]. A steep, transverse fault along which the displacement has been parallel to the strike of the fault. That is, a steep, transverse strike-slip fault. *Cf.* TEAR FAULT flaxseed ore An oölitic iron ore in which the oölites have been somewhat flattened parallel to the bedding plane so that they are disk-shaped rather than spherical.

fleckshiefer An argillaceous rock in which there has been incipient production of new minerals as a result of low-grade metamorphism. See SPOTTED SLATE

flexible Bends without breaking and without a tendency to return to its original form.

flexible sandstone Itacolumite, q.v.

flexural slip Movement in relatively competent rocks by sliding of beds past each other.

flexure 1. Syn: FOLD. 2. A broad domical structure.

flexure correction A correction necessary in pendulum observations of gravity. The vibrating pendulum produces oscillations of the receiver case, of the pillar, and of the surface soil. Rather complex coupled vibration phenomena arise and the period of the pendulum itself changes.

flexure fault A growth fault, or contemporaneous fault.

flexure-flow fold A flexural fold resulting from flow within layers in which thickening occurs in hinge areas and thinning takes place along the limbs.

flexure-slip folding A flexure fold in which the mechanism of folding is slip along the bedding planes or along the surface of foliation. There is no change in thickness of individual strata.

flint A dense fine-grained form of silica which is very tough and breaks with a conchoidal fracture and cutting edges. Of various colors, white, yellow, gray, and black. See CHERT

flint clay A flintlike fire clay which when ground up develops no plasticity.

float; floater; float mineral; float ore Terms much used among miners and geologists for pieces of ore or rock which have fallen from veins or strata, or have been separated from the parent vein or strata by weathering agencies. Not usually applied to stream gravels.

float copper 1. In the Lake Superior region, fine scales of metallic copper (especially produced by abrasion in stamping) which do not readily settle in water. 2. Native copper found away from its original rock. Cf. FLOAT ORE

float gold; flour gold Particles of gold so small and thin that they float on and are liable to be carried off by the water.

floating 1. Relations of large sedimentary particles not in contact with each other contained in much finer grained matrix. 2. Relations of quartz sand grains more or less sparingly disseminated in limestone.

floating sand grain Isolated sand grain, particularly in limestone, not in contact with other scattered grains.

float ore Fragments of vein material found on the surface, and usually downstream or downhill from the outcrop.

flocculation In waste water treatment, the process of separating suspended solids by chemical creation of clumps or flocs.

floe Mass of floating ice some 30m. to 8 km. across not fast to any shore, formed by breaking up of the frozen surface of a large body of water.

floeberg A thick mass of floe ice heaped together by the collision of floes with each other or with the shore.

floe till See TILL

flood 1. Any relatively high streamflow which overtops the

natural or artificial banks in any reach of a stream. *2. Sedimentary petrol:* A term implying the occurrence of a particular species so far in excess of all others as to constitute almost a pure concentrate.

flood basalt *See* PLATEAU BASALT

flood basin The flood basins can be defined either as the tracts actually covered by water during the highest known floods or as the flat areas between the sloping low plains on one side and the river lands on the other, occupied by heavy soils and commonly having either no vegetation or a strictly swampy vegetation.

flood classification Floods are classified by recurrence magnitude, as 100-year flood, 50-year flood, etc.

flood control That aspect of flood planning that has to do with the elimination or retention of floodwaters so as to prevent death, injury, and monetary damage.

flood control reservoirs Those artificial bodies of water constructed for the purpose of retaining floodwater so that land below the reservoir will not be flooded.

flood current The movement of the tidal current toward the shore or up a tidal stream.

flood frequency Over a period of years, the average number of times a flood of a given magnitude is likely to occur.

flooding A term used for the drowning out of a well by water, often resulting from drilling too deeply into the sand.

flood maps Maps that depict the extent of former floods or the extent of any particular magnitude of flood.

flood peak The maximum rate of flow attained at a given point during a flood event.

flood plain That portion of a river valley, adjacent to the river channel, which is built of sediments during the present regimen of the stream and which is covered with water when the river overflows its banks at flood stages.

flood-plain meander scar A class of scars which includes any and all features on a flood plain that mark the former course of a stream meander.

flood-plain scrolls Patches of material having curved crescentic shapes originating from deposition along the inside curve of river meanders, and incorporated in large numbers into the flood plain.

flood planning The sum total of those activities of men that tends to decrease or eliminate death, injury, and monetary damage from floods.

flood tide *1.* The flow, or rising toward the shore, is called flood tide, and the falling away, ebb tide. *2.* A nontechnical term referring to that period of tide between low water and the succeeding high water; a rising tide.

floodway A large capacity channel constructed to divert floodwaters safely through and around populated areas.

flood zone [stream] A strip of land on the flood plain bordering a stream that is subject to flooding at equal frequency.

floor *1.* The bed or bottom of the ocean. A comparatively level valley bottom; any low-lying ground surface. *2.* That part of any subterraneous gallery upon which you walk or upon which a tramway is laid.

flora The plants collectively of a given formation, age, or region. *Cf.* FAUNA

florizone A biostratigraphic zone or group of strata characterized by a particular fossil floral assemblage. Analogous to faunizone.

flos ferri An arborescent variety of aragonite.

flotation Method of mineral separation whereby a froth created in water by a variety of reagents floats some finely crushed minerals whereas others sink.

flour copper Very fine scaly native copper that floats on water and is very difficult to save in milling. *See* FLOAT COPPER

flour gold The finest gold dust, much of which will float on water. *See* FLOAT GOLD

flow, gliding That type of solid flow which takes place by the combined mechanisms of translation- and twin-gliding.

flow, pseudoviscous Load recrystallization. The type of solid flow which takes place under a strain and a stress to low to produce gliding flow, producing instead intergranular movement and dimensional orientation for the most part.

flowage folds *1.* Minor folds that are the result of the flowage of rocks toward a synclinal axis, toward which the minor folds are overturned. *2.* Folds in which layers of rock are thinned at crest and thickened at trough.

flow banding A structure of igneous rocks, especially common to silicic lava flows, due to the movement or flow of magmas or lavas and evidenced by the alternation of mineralogically unlike layers.

flow breccia A type of lava flow, usually of silicic composition, in which fragments of solidified or partly solidified lava, produced by explosion or flowage, have become welded together or cemented by the still fluid parts of the same flow.

flow cast A "roll," lobate ridge, or other raised feature produced on the underside of a sand layer by the sand's flowing into a depression in underlying soft hydroplastic sediment. The underlying rock, typically coal or mudstone, preserves no diagnostic structure.

flow cleavage That variety of rock cleavage that is the result of solid flow of the rock. *See* FOLIATION

flow-duration curve A curve secured from an arrangement of daily stremflows in the order of their magnitudes.

flow earth Material on a slope characterized by local derivation and lack of sorting. *Syn:* SOLIFLUCTION MANTLE

flow folding Ptygmatic folding. Folding in beds which offer so little resistance to deformation that they assume any shape impressed upon them by more rigid rocks surrounding them or by the general stress pattern of the deformed zone.

flow gneiss Gneiss whose structure was produced by flowage in an igneous mass before complete solidification.

flowing well Well from which water or oil flows without pumping.

flow layer Rock layer differing mineralogically or structurally from adjacent layers produced by flowage before the complete consolidation of magma.

flow lines *See* FLOW STRUCTURE; FLOW TEXTURE

flowmeter A meter that indicates the rate at which water flows.

flow stage That stage in the consolidation of a magma when it is still sufficiently fluid to flow as a liquid.

flowstone Deposits of calcium carbonate have also accumulated against the walls in many places where water trickles from the rock.

flow stretching Orientation and possible deformation of crystals with long axes in direction of plastic flow in metamorphic rocks.

flow structure Fluidal structure. A structure of igneous rocks, generally but not necessarily restricted to volcanic rocks, in which the stream or flow lines of the magma are revealed by alternating bands or layers of differing composition, crystallinity, or texture, or by a subparallel arrangement of prismatic or tabular crystals.

flow surface Plane separating adjacent flow layers.

flow symmetry Symmetry of movement comparable to the symmetry of equal and interchangeable parts located with reference to a center or one or more axes or planes.

flow texture Fluidal texture. A texture common to the glassy groundmass of extrusive rocks; especially lavas, in which the stream or flow lines of the once molten material are revealed by a subparallel arrangement of prismatic or tabular crystals or microlites.

flow units The nearly contemporaneous subdivisions of a lava flow (usually basaltic) which consists of two or more parts poured one over the other during the course of a single eruption.

fluid *1.* Having particles which move easily and change their relative position without a separation of the mass, and which yield easily to pressure; capable of flowing; liquid or gaseous. *2.* A substance in which the constituent particles are in disordered array, permitting deformation when subjected to stress. The distinction between plastic and fluid is generally in degree of deformation under a given stress. *3.* A substance of properties intermediate between those of a gas and a liquid, *q.v.* The specific definition of fluid in phase equilibrium studies includes all noncrystalline phases, with the exception of the special limiting cases of a liquid and a gas phase, when in equilibrium with each other.

fluid contact The surface in a reservoir separating two regions characterized by predominant differences in fluid saturation.

fluid inclusions *See* INCLUSIONS, FLUID

fluidization Process in which gas passes through loose fine-grained material, mixes with it, and causes it to flow like a liquid; may occur at the time of volcanic eruption as in a glowing avalanche.

fluid pressure Pressure exerted by fluid contained in rock.

flume *1.* A deep, narrow ravine or gorge, with nearly perpendicular walls and a stream forming a series of cascades. *2.* A channel, either natural or man-made, which carries water.

fluorapatite A mineral of the apatite group, $Ca_5(PO_4)_3F$.

fluorbarite Trade name for a fluorite-barite mixture used in glassmaking.

fluorescence Emission of visible light by a substance exposed to ultraviolet light, useful in examining well cuttings for oil shows and in prospecting for some minerals.

fluoride A compound of fluorine with one other element or radical.

fluorine dating A relative dating method for bones of Pleistocene or Holocene age based on the gradual combination of fluorine in ground water with the calcium phosphate of buried bones.

fluorite A mineral CaF_2. Isometric. The principal ore of fluorine.

fluorographic method A method involving exposing soil samples to ultraviolet light and recording the emitted light on a light-sensitive medium. The densities of the recorded sample images are measured by a transmission photometer.

fluorologging A logging technique based on the principle that the rocks overlying an oil accumulation are characterized by anomalously high fluorescent intensities.

fluorspar Commercial name for fluorite-rich rock.

flushing The driving of oil or gas from a trap through the agency of ground water. Flushing may occur wherever there is sufficient water in motion, but particularly near the rim of artesian basins.

flute Asymmetrical scalloped rock surfaces. Drapes of drip or flowstone (commercial cave). *Syn:* FACET; SOLUTION RIPPLE

flute casts Sharp, subconical welts occurring on bottom surfaces of sandstone layers, in which one end is rounded or bulbous and the other flares out to merge gradually with the striated bottom of the sandstone layer.

fluting 1. Smooth gutterlike channels or deep smooth furrows worn in the surface of rocks by glacial action. 2. Fluting is a peculiar method of surface decay by which granite or gneisses are left with a corrugated or fluted surface.

fluvial Of, or pertaining to, rivers; growing or living in streams or ponds; produced by river action, as, a fluvial plain.

fluvial geomorphic cycle The normal cycle of erosion by streams, leading to the formation of a peneplain.

fluviatile Belonging to a river; produced by river action; growing or living in fresh-water rivers.

fluviatile dam Dam formed in a valley of sediment deposited by a tributary.

fluviation All of the numerous activities engaged in and the various processes employed by streams are grouped under fluviation.

fluvioglacial Glaciofluvial. Pertaining to streams flowing from glaciers or to the deposits made by such streams.

flux 1. Substance that reduces the melting point of a mixture. 2. Passage across a physical boundary such as CO_2 from atmosphere to hydrosphere, or across a chemical boundary as CO_2 from atmosphere to organic matter. 3. State of change.

fluxing ore An ore containing an appreciable amount of valuable metal, but smelted mainly because it contains fluxing agents required in the reduction of richer ores.

flux stone Limestone, dolomite, or other rock used in metallurgical processes to lower the fusion temperature of ore.

fly ash All solids, including ash, charred paper, cinders, dust, soot, or other partially incinerated matter, that are carried in a gas stream.

flysch The widespread deposits of sandstones, marls, shales, and clays, which lie on the northern and southern borders of the Alps. Although largely consisting of sandy and calcareous shales (hence the name—in reference to their fissile character) the flysch also contains beds of sandstone and conglomerate.

foam crust An ablation feature on snow which has the appearance of small overlapping waves like sea foam on a beach. Foam crust is the result of evaporation by the sun's rays. Foam crust may develop into ploughshares.

focal sphere Theoretical sphere enclosing the focal region of an earthquake.

focus *Seismol:* The source of a given set of elastic waves. The true center of an earthquake, within which the strain energy is first converted to elastic wave energy.

fold *1.* A bend in strata or any planar structure. *2.* A broad median external undulation or plica that may be situated on either the dorsal or ventral valve of a brachiopod. *3.* Major rounded elevation of a brachiopod shell along longitudinal mid-line, affecting both outer and inner shell surfaces; generally on brachial valve.

fold, flexure A type of fold, in size microscopic to orogenic, in which movement took place normal to the axial line and parallel with the limbs, producing notable shortening.

foldbelt Orogenic belt; a linear region which has been folded and deformed during mountain building.

fold fault An overfold, the middle limb of which is replaced by a fault surface.

fold nappe A recumbent fold, of which the reversed middle limb has been completely sheared out as a result of the great horizontal translation.

fold system Group of folds showing common characteristics and trends and, presumably, of common origin.

foliate A general term for any foliated rock. *See* GNEISS; SCHIST

foliation *1.* The laminated structure resulting from segregation of different minerals into layers parallel to the schistosity. *2.* Foliation is considered synonymous with "flow cleavage," "slaty cleavage," and schistosity by many writers to describe parallel fabrics in metamorphic rocks. Considerable ambiguity attends their current use. *3.* In addition to its rude stratification, the ice of the deeper portions of a glacier often acquires a strati-

form structure which may perhaps best be called foliation to distinguish it from deposition.

foliation, axial-plane *See* AXIAL-PLANE FOLDING

fondo The environment represented by the bottom of a body of water; all the environment not encompassed by unda and clino, *q.v.*

fondoform The subaqueous landform constituting the main floor of the water body.

fondothem A rock unit formed in a fondo, *q.v.*, environment.

food chain The sequence of organisms in which each is food for a higher member of the sequence.

food cycle Production, consumption and decomposition of food and the energy relationships involved.

fool's gold Pyrite, a sulfide of iron, FeS_2.

foot The bottom of a slope, grade, or declivity. A term for the lower part of any elevated landform.

foothill One of the lower subsidiary hills at the foot of a mountain, or of higher hills.

footwall The mass of rock beneath a fault plane, vein, lode, or bed of ore.

foram Abbreviated name for foraminiferan.

Foraminifera Subclass of the Sarcodina; unicellular animals mostly of microscopic size that secrete tests, composed of calcium carbonate, or build them of cemented sedimentary grains, consisting of one to many chambers arranged in a great variety of ways. Most are marine.

force That which changes or tends to produce a change in the state of rest or motion of a body.

ford Passage across a stream where water is not too deep for wading or the movement of land vehicles.

foredeep A long, narrow, crustal

depression, or furrow, bordering a folded orogenic belt or island arc on the convex side, commonly on the oceanward side.

foredune Dune developed along the shoreward face of a beach ridge.

foreland *1.* A promontory; a jutting of high land into the sea. At the beginning of a cycle the waves attack the coast at all points, cutting or nipping back the initial form of the land into a cliff; at a later stage, transportation of material alongshore begins, and the waste from the edge and bottom of the land, together with the river sediment, is built out at certain points in front of the older mainland in deposits of various shapes, which are appropriately grouped together under the general term forelands. *2.* In folded mountain ranges three zones may be distinguished: (1) A rigid, unyielding mass which is not folded, (2) the zone of folding, (3) the zone of diminishing action, where the folding gradually dies away or ends in a fault. The side of the range toward which the overturned folds incline is called the foreland, and may be either the unfolded mass or the zone of diminishing action. *3.* The relatively stable area, lying in shallower water, represented by the continental platform. *4.* The resistant block toward which the geosynclinal sediments move when compressed. *5.* In its structural sense: The region in front of a series of overthrust sheets.

foreland shelf Part of the relatively stable continental region that extends inward from the hingebelt of a geosyncline.

fore limb Steeper dipping side of an asymmetrical anticline produced by lateral thrusting.

fore reef *1.* The steeply dipping

talus slope commonly found on the seaward side of an organic reef. *2.* The high energy side of a reef.

foreset beds The series of inclined layers accumulated as sediment rolls down the steep frontal slope of a delta. *See* BOTTOM-SET BEDS; TOP-SET BEDS

foreshock An earthquake which precedes a larger earthquake within a fairly short time interval (of the order of a few days or weeks), and which originates at or near the focus of the larger earthquake.

foreshore Lower shore zone, between ordinary low and high water levels. *Cf.* BACKSHORE

foresight A sight on a new survey point, made in connection with its determination; or a sight on a previously established point, to close a circuit.

foreslope Slope extending from the outer margin of an organic reef to an arbitrary depth of 6 fathoms.

forest bed Interglacial deposit containing woody remains of trees and other vegetation.

form All the faces of a crystal that have a like position relative to the elements (planes, axes, etc.) of symmetry.

format Informal rock stratigraphic unit bounded by marker horizons believed to be isochronous surfaces that can be traced across facies changes, particularly in the subsurface, and useful for correlations between areas where the stratigraphic section is divided into different formations that do not correspond in time value.

formation *1.* Something naturally formed, commonly differing conspicuously from adjacent objects or material, or being noteworthy for some other reason. *2. Stratig:* The primary unit of formal mapping or description. Most formations possess certain distinctive or combinations of distinctive lithic features. Boundaries are not based on time criteria. Formations may be combined into groups or subdivided into members. *3. Speleol:* Secondary deposit in a cave forming stalactites, stalagmites, or other cave deposits. *4. Paleobot:* A biome, or a climax, both of which represent preferred usage.

formation factor The electrical resistance of a rock saturated with an electrolyte, divided by the resistivity of the electrolyte. There is an inverse linear relationship between the formation factor and the porosity and permeability of the rock. Known also as formation resistivity factor.

formation water Water naturally occurring in sedimentary strata. *Cf.* CONNATE WATER

form contour Topographic contour determined by stereoscopic study of aerial photographs without ground control or by other means not involving conventional surveying.

form energy Potentiality of minerals to develop crystal form within a solid medium.

form genus *1.* A complex of genera which are known to be distinct but which are not or have not as yet been named because all have the same general habit and are therefore included in the single taxon. *2.* Fossil plants are named according to the same set of rules that governs the naming of living plants, but the names usually refer to parts rather than complete organisms. Since the generic name applies to only a leaf, a seed, or some other part, the genus is called a form genus or artificial genus, as contrasted with a natural genus. Animal parts may also constitute form genera in this sense, as in conodonts.

form species A morphotypic spe-

cies, especially where the relationship of its members are not known.

forsterite A mineral of the olivine group, Mg_2SiO_4.

fosse *1.* Depression or unfilled area often found between the terraced ice contact of glacial sand plains and morainal mounds forming a belt within the ice-covered field, as on Nantucket. *2.* A ditch, moat, or trench, specifically between a glacier and a moraine or rock wall.

fossil *1.* The remains or traces of animals or plants which have been preserved by natural causes in the earth's crust exclusive of organisms which have been buried since the beginning of historic time. *2.* Anything dug from the earth. *Obs.*

fossil assemblage Fossils naturally associated in a stratum; they possibly were derived from more than a single fossil community.

fossil community Fossils that represented a living community.

fossil fuel A deposit of organic material containing stored solar energy which can be used as fuel. The most important are coal, natural gas, and oil; oil shale and tar sand have future potential.

fossiliferous Containing organic remains.

fossilize To convert into a fossil; to petrify.

fossil ore Fossiliferous red hematite.

fossil soil A soil developed upon an old land surface and later covered by younger formations.

fossil wax OZOCERITE

founder To sink deeply, generally beneath the sea.

foundry sand Includes those siliceous sands that are used to make the forms of casting metals.

foveolate, *adj.* Pitted. The pit (foveola) may be solitary or not.

fractional crystallization Separation of a magma into two phases, crystals and liquid, possibly followed by a gross separation of the two phases from each other by other processes such as filter pressing, etc.

fractionation Separation of a substance from a mixture, e.g., one isotope from another of the same element.

fracture *1.* The manner of breaking and appearance of a mineral when broken, which is distinctive for certain minerals, as conchoidal fracture. *2.* Breaks in rocks due to intense folding or faulting. *3.* The process of breaking oil-, gas-, or water-bearing strata by injecting a fluid under such pressure as to cause partings in the rock.

fracture cleavage Fracture cleavage is a capacity to part along closely spaced parallel surfaces of fracture or near-fracture, commonly in a single set, but occasionally in intersecting sets. It is closely related to joint structure, but the joints are so closely spaced as to give the rock a distinctive structure not ordinarily to be described in terms of joints.

fractured Fissured. Broken by interconnecting cracks. A common structure in limestone oil reservoirs (Ellenburger, W. Tex.; Hunton, Okla.). Reported to exist in certain shale reservoirs (Mancos, Col.; Monterey, Calif.).

fracture porosity Porosity resulting from presence of openings produced by the breaking or shattering of an otherwise less pervious rock.

fracturing *See* HYDROFRACTURING

fragmental Consisting of broken material, particularly that which has been moved from its place of origin.

fragmental rocks Clastics.

fragmental texture A general term applied to that of rocks composed of fine materials and including sandy, conglomeratic, bouldery, and brecciated materials. A texture of clastic rocks.

fragmentary *1. Geol:* Applied to rock masses composed of the fragments or debris of other rocks; rocks not homogeneous in texture; nearly synonymous with breccias or breccioconglomerates, *q.v.* *2.* Rocks consisting of a congeries of particles which have not grown together but are fragments which have been broken off their parent masses and brought together by some external agency, their coherence being caused either by mechanical compression or by a cement of some other substance.

francolite A mineral of the apatite group containing carbonate and fluorine.

franklinite A mineral, (Fe,"Zn,-Mn")(Fe,"'Mn"')$_2$O$_4$, a member of the spinel group. Isometric. An ore of zinc.

Frasch process A process for mining sulfur in which superheated water is forced into the sulfur deposits for the purpose of melting the sulfur. The molten sulfur is then pumped to the surface. Used extensively in Louisiana and Texas.

free *1.* Native; uncombined with other substances, as, free gold or silver. *2.* Coal is said to be "free" when it is loose and easily mined, or when it will "run" without mining.

free-air anomaly The difference at any point on the earth between the measured gravity and the gravity calculated for the theoretical gravity at sea level and a "free-air" coefficient determined only by the elevation of the station with respect to sea level.

free-air correction *See* FREE-AIR ANOMALY

free energy The capacity of a system to perform work, a

change in free energy being measured by the maximum work obtainable from a given process.

free period (of a seismograph) The time for one complete swing of the seismograph mass when all damping is removed and the earth is quiet.

freezing interval That temperature interval between the solidus and the liquidus for a given composition. *Syn:* CRYSTALLIZATION INTERVAL

freibergite A silver-rich variety of tetrahedrite.

frequency distribution The numerical or quantitative distribution of objects or material in a series of closely related classes generally selected on the basis of some progressively variable physical character.

frequency domain Measurements as a function of frequency or operations in which frequency is the variable, in contrast to the time domain.

fresh water Water with less than 0.2% salinity, e.g., 2000 ppm, not the same as potable (=drinkable) water.

freshwater limestone Underclay limestone. *1.* A limestone formed by direct precipitation in fresh water. *2.* A thin dense nodular, relatively unfossiliferous limestone underlying the coals and closely related to the underclay in the Central Interior Coal Basin.

friable Easily crumbled, as would be the case with rock that is poorly cemented.

fringing reef The coral reefs around other lands or islands rest on the bottom along the shores. They are either fringing or barrier reefs, according to their position. Fringing reefs are attached directly to the shore, whereas barrier reefs, like artifical moles, are separated from the shore by a body of water usually termed a lagoon.

frit *N: 1.* Material of which glass is composed; *2.* Semifused stony mass. *V: 3.* To partly fuse.

front *1. Petrol:* A metamorphic zone of changing mineralization developed outward from a large expanding igneous intrusion; *cf.* BASIC FRONT. *2. Topog:* More or less linear outer slope of a mountain range that rises above a plain or plateau.

frontal moraine Terminal moraine.

front pinacoid Orthopinacoid. *See* PINACOID

frost *1.* A light, feathery deposit of ice caused by the condensation of water vapor, directly in the crystalline form, on terrestrial objects whose temperatures are below freezing, the process being the same as that by which dew is formed, except that the latter occurs only when the temperature of the bedewed object is above freezing. *2.* The occurrence of temperatures below freezing.

frost action The weathering process caused by repeated cycles of freezing and thawing.

frost circles Circular cracks developed by freezing in horizontal thin bedded limestone cut by two sets of joints meeting at right angles; commonly 4.6 to 7.6m. in diameter.

frost crack Opening in soil produced by the development of an ice wedge.

frost-crack polygon Nonsorted polygon produced by low temperature contraction of frozen ground.

frost creep Soil creep resulting from frost action.

frost-heaved mound Stone ring, *q.v.*

frost heaving The lifting of a surface by the internal action of frost.

frost hillocks The marked upward bulging sometimes present in the center of each polygon in cellular soils.

frosting A lusterless, ground-glass, or mat surface imposed on the surface of rounded quartz grains because of innumerable close contacts with other similar grains. Generally believed to be caused by wind action.

frost line The maximum depth to which the ground becomes frozen; it may be given for a particular winter, for the average of several winters, or for the extreme depth ever reached.

frost mound General term for knolls, hummocks, and hills associated with frozen ground; includes earth hummocks, falsens, pingoes.

frost polygons See POLYGON GROUND

frost splitting Breaking of rock by water freezing in its cracks.

frost stirring Frost heaving and thrusting in surface zone of annual freeze and thaw; does not involve mass movement.

frost thrusting Lateral soil movement resulting from freezing.

frost weathering The mechanical disintegration of earth materials brought about by frost action. Frost wedging.

frost wedging Frost weathering.

frozen Said of vein material which adheres closely to the inclosing walls.

frozen ground Tjäle. Ground that has a temperature 0° C. or lower and generally contains a variable amount of water in the form of ice.

fucoid A term commonly applied in the past to any indefinite marking found on a sediment which could not be referred to a described genus. The term was derived from the marine alga, fucus, which it was supposed might leave such a marking if buried under favorable conditions.

fugitive constituents Those substances which were present in the magma before crystallization set in, but were for the greater part lost during the process of crystallization, so that they do not commonly appear as rock constituents.

fulcrum The intersection of the end of a recurved spit with the next succeeding stage in development of a compound recurved spit.

fulgurite [<*Lat.* fulgur, lightning] Little tubes of glassy rock that have been fused from all sorts of other rocks by lightning strokes. They are especially frequent in exposed crags on mountain tops.

Fuller's earth A fine earthy material containing clay (predominantly montmorillonite). Has the property of absorbing coloring matter from oils and fats.

fumarole A hole or vent from which fumes or vapors issue; a spring or geyser which emits steam or gaseous vapor; found usually in volcanic areas.

fundamental strength The maximum stress that a substance can withstand, regardless of time, under given physical conditions without rupturing or plastically deforming continuously.

Fungi Class of thallophytes, multicelled plants that feed on organic matter rather than perform photosynthesis; probably polyphyletic.

funnel pluton Pluton having the general shape of an inverted cone; most consist of layered gabbroic rocks.

furrow 1. [*Ger.* Furche; *Fr.* sillon] A valley or channel-like hollow in the continental border, and more or less at right angles to it. 2. An elongated depression in the earth's crust of a depth excessive in comparison to the ordinary, more or less equidimensional depressions of the ocean floors and the continental platforms.

furrow cast Impression on the lower side of a sedimentary layer of a furrow in the surface of its underlying bed.

fusain Coal material with the appearance and structure of charcoal; friable, sooty, and generally high in ash; consists mainly of fusite.

fusibility scale A list of minerals arranged in the order of their fusibility; von Kobell's scale of fusibility follows: (1) stibnite, (2) natrolite, (3) almandine garnet, (4) actinolite, (5) orthoclase, (6) bronzite.

fusiform, *adj.* Spindle-shaped; narrowed both ways from a swollen middle.

fusion *1.* Act or operation of melting or rendering liquid by heat. *2.* State of being melted or dissolved by heat. *3.* Union or blending of things as if melted together. *4.* The combination of certain light nuclei, such as deuterium and tritium, forming a heavier nucleus and releasing energy.

fusion-pressure curve; fusion curve A pressure vs. temperature plot of the univariant equilibrium crystal liquid in a unary system; sometimes incorrectly assumed to represent the behavior of the same substance in the earth (not a unary system).

fusulinid Any fossil belonging to one of the several genera of the Fusulinidae; a foraminifer shaped like a grain of wheat. Fusulinids are important guide fossils in the Pennsylvanian and Permian systems.

future ore A class of ore whose existence is a reasonable possibility in view of the strength and continuity of geologic-mineralogic relationships and extent of ore bodies already developed, a measure of whose continuity is available as a criterion of what may be expected as mining operations progress. *Syn:* POSSIBLE ORE

G

gabbro *1.* A plutonic rock consisting of calcic plagioclase (commonly labradorite) and clinopyroxene, with or without orthopyroxene and olivine. Apatite and magnetite or ilmenite are common accessories. *2.* Loosely used for any coarse-grained dark igneous rock.

gage height The water-surface elevation referred to some arbitrary datum.

gahnite A mineral, $(Zn,Fe,''Mg)$-Al_2O_4, a member of the spinel group. Isometric.

gal An acceleration of one centimeter per second. A milligal is 0.001 gal.

Gale alidade Explorer's alidade. A light compact alidade, with a low pillar and a reflecting prism through which the ocular may be viewed from above. As used by petroleum geologists it is commonly equipped with the Stebinger drum, *q.v.*

galena Lead glance. A mineral, PbS. Isometric. The principal ore of lead.

gallery *1.* A subsidiary passage in a cave at a higher level than the main passage. *2.* A horizontal or nearly horizontal underground passage either natural or artificial.

galvanometer An instrument for measuring a small electric current or for detecting its presence or direction by means of the movement of a magnetic needle or of a wire or coil in a magnetic field.

gamma Common unit of magnetic intensity, equal to 10^{-5} oersted.

gamma radiation Emission by radioactive substances of quanta of energy corresponding to X rays and visible light but with a much shorter wave length than light; may be detected by gamma-ray Geiger counters.

gamma ray Rays or quanta of energy emitted by radioactive substances corresponding to X rays and visible light but having a much shorter wave length than light.

gamma-ray well logging A method of logging boreholes by observing the natural radioactivity of rocks through which the hole passes. It was developed for logging holes which can not be logged electrically because they are cased.

gamma structure *1.* Thrust sheet with underlying low-angle thrust lane steepening abruptly downward. *2.* Overthrusting or overfolding in one direction only.

gangue The nonmetalliferous or nonvaluable metalliferous minerals in the ore; veinstone or lode filling. The mineral associated with the ore in a vein.

ganister *1.* A highly refractory siliceous sedimentary rock used for furnace linings. *2.* A mixture of ground quartz and fire clay, used in lining Bessemer converters. *3.* A local name for a fine close-grained siliceous clay that occurs under certain coal beds in Derbyshire, Yorkshire, and the North of England.

gap *1.* In Pennsylvania, any deep sharp notch in a mountain ridge. Water gaps are those notches or passes which penetrate to the bases of the mountains, and give passage to the larger streams. *2.* Any deep notch, ravine, or opening between hills or in a ridge or mountain chain. *3.* In faulting, the horizontal separation, *q.v.*, can be measured parallel to the strike of the fault. Gap is that component of this separation measured parallel to the strike of the disrupted index plane (bed, vein, dike, etc.). Overlap is defined in the same way; however, gap is used when it is possible to walk at right angles to the strike of the disrupted index plane and miss it completely. Overlap is used when under similar conditions one would cross the index plane twice in certain places. *4.* A steep-sided furrow which cuts transversely across a ridge or rise.

garnet A mineral group, formula $A_3B_2(SiO_4)_3$, where A=Ca,Mg,-F^{++}, and Mn^{++}; and B=Al, Fe^{++}, Mn^{++} and Cr. Isometric. The principal end members are almandine ($Fe''Al$), pyrope (Mg-Al), spessartite (MnAl), grossularite (CaAl), andradite (Ca-Fe'''), and uvarovite (CaCr). Garnet from contact metamorphosed limestones is usually grossularite-andradite, from pegmatites usually spessartite, from schists usually almandine-pyrope, from kimberlites usually pyrope. Used as a gem and as an abrasive.

garnierite Name given to various poorly defined hydrated magnesium nickel silicates. An ore of nickel.

gas Naturally occurring gaseous hydrocarbon produced in association with petroleum or as marsh gas.

gas cap Free gas occurring above oil in a reservoir, and present whenever more gas is available than will dissolve in the associated oil under existing pressure and temperature.

gas coal Bituminous coal that is suitable for the manufacture of flammable gas because it contains about 33–38% volatile matter. *Cf.* HIGH-VOLATILE BITUMINOUS COAL. *See also* COAL GAS. *Syn:* GAS FLAME COAL

gas-cut mud Drilling mud returned from the bottom of a drill hole, characterized by a fluffy texture, gas bubbles, and reduced density due to the retention of entrained natural gas rising from the strata traversed by the drill.

gas cycling (recycling) A secondary-recovery process involving injection into the reservoir of the gas or a portion of the gas produced with the oil from that reservoir. When pressure is maintained, gas cycling may be one of the means employed.

gaseous transfer The process whereby a magma differentiates by separation of a gaseous phase which then moves relative to the magma.

gas field A tract or district yielding natural gas.

gash fractures Open gashes diagonal to the fault or fault zone; they are tension fractures.

gasification Conversion of coal to synthetic natural gas, which is cleaner and more easily transported than coal.

gas-oil ratio *1.* The quantity of gas produced with oil from an oil well, usually expressed as the number of cubic feet of gas produced per barrel of oil. *Abbrev:* GOR. *2.* Reservoir gas-oil ratio.

gas phase Any chemical substance in the form of gas as contrasted with liquid or solid form is said to be in the gas phase.

gas pool An area of subsurface

formations, under adequate closure, which will yield gas in economic quantities. *Syn:* GAS FIELD

gas sand An oil sand or sandstone formation containing a large quantity of natural gas.

gas streaming A process of differentiation in which the formation of a gas phase at a late stage in the crystallization results in partial explusion, by the escaping gas bubbles, of residual liquid from among the network of crystals.

gastrolith Gizzard (literally stomach) stone. Highly polished, well-rounded pebbles associated with saurian skeletons. Believed to have been gizzard stones.

gastropod A member of the phylum Mollusca, class Gastropoda; usually with a calcareous exoskeleton or shell, which is asymmetrically coiled and without internal chambers or partitions. *Syn:* SNAIL

Gastropoda A class of the phylum Mollusca; commonly known as gastropods, *q.v.*, or snails.

gas well *1.* A deep boring, from which natural gas is discharged. *2.* As used in oil and gas leases, a well having such a pressure and volume of gas, and, taking into account its proximity to market, as can be utilized commercially.

gather A collection of data sets which have some element in common, as a common-depth-point gather or a common-offset gather.

gauge pressure The pressure as read on an ordinary spring or Bourdon type gauge, usually expressed in pounds per square inch. It is the absolute pressure less that exerted by the atmosphere. Abbreviation: psig.

gauss The unit of magnetic field intensity, equal to 1 dyne per unit pole. The preferred term for this unit is the oersted.

Gaussian curve Normal distribution curve.

geanticline A broad uplift, generally referring to the land mass from which sediments in a geosyncline are derived. Originally used as a synonym of anticlinorium and the opposite of synclinorium.

geest Material derived from rock decay *in situ*.

Geiger counter An instrument which detects gamma rays given off by radioactive substances, consisting of a discharge tube which responds to the ionization produced by the rays in a gas which fills the tube. A Geiger-Mueller counter.

Geiger-Mueller counter An ionization chamber with its vacuum and applied potential so adjusted that a gamma ray or other ionizing particle through it causes a momentary current to flow. The surges of current can be amplified and counted so as to measure the intensity of radioactivity in the neighborhood of the chamber.

gel A jellylike material formed by coagulation of a colloidal dispersion.

gem A general term for any precious or semiprecious stone, as, diamond, ruby, topaz, etc., especially when cut or polished for ornamental purposes.

gemology [*Am.*] **gemmology** [*Eng.*] The study of gems.

gene The specific factor for an inherited characteristic, now thought to originate in deoxyribonuclei acid (DNA) in the chromosomes.

gene complex The interacting system of all the genetic factors of an organism.

genera Plural of genus, *q.v.*

generative folds Folds that increase in amplitude in successive beds with the accompaniment of an increase in stratigraphic thickness toward the axial region.

generic Pertaining to a genus.

generic name Name of a genus, consisting of one word that is capitalized; in contrast, a specific name generally is not capitalized.

genetic Pertaining to relationships due to a common origin.

genetic drift Gradual change with time in the genetic composition of a continuing population resulting from the random selection or elimination of genetic features. The effect is unrelated to the benefits or detriments of the genes involved.

genetics The science that deals with the processes of heritable characteristics or features from generation to generation.

genotype The genetic constitution of an organism or a species in contrast to its observable characteristics (the latter is the phenotype).

genus A group of species believed to have descended from a common direct ancestor that are similar enough to constitute a useful unit at this level of taxonomy.

geo *1*. In Iceland, a narrow inlet walled in by steep cliffs. *2*. An element in many compound words of Greek origin, meaning the earth.

geocentric Pertaining to, or measured from, the earth's center; having or relating to the earth as a center.

geochemical anomaly A concentration of one or more elements in rock, soil, sediment, vegetation, or water markedly different from the normal concentration in the surroundings. Sometimes applied also to abnormal concentrations of hydrocarbons in soils.

geochemical cycle The sequence of stages in the migration of elements during geologic changes. Rankama and Sahama distinguish a major cycle, proceeding from magma to igneous rocks to sediments to sedimentary rocks to metamorphic rocks and possibly through migmatites back to magma, and a minor or exogenic cycle proceeding from sediments to sedimentary rocks to weathered material and back to sediments again.

geochemical exploration Exploration or prospecting methods depending on chemical analysis of the rocks or soil, or of soil gas, or of plants.

geochemical facies *1*. Area characterized by particular physiochemical conditions influencing the production and accumulation of sediment; *2*. Facies defined on the basis of trace element occurrences, associations and ratios.

geochemical prospecting *1*. The search for concealed deposits of metallic ores by analyzing soils, surface waters, and/or organisms for abnormal concentrations of metals. *2*. The search for petroleum accumulations by analyzing soil gases for hydrocarbons.

geochemistry The study of (a) the relative and absolute abundances of the elements and of the atomic species (isotopes) in the earth, and (b) the distribution and migration of the individual elements in the various parts of the earth (the atmosphere, hydrosphere, crust, etc.), and in minerals and rocks, with the object of discovering principles governing this distribution and migration. Geochemistry may be defined very broadly to include all parts of geology that involve chemical changes or may be focused more narrowly on the distribution of the elements.

geochron Time interval corresponding to a rock stratigraphic unit; may vary from place to place depending on age of rock unit.

geochronologic sequence The placing of the rocks of the earth's crust in a systematic framework of variously ranked time-stratigraphic units to show their relative position and age with respect to earth history as a whole.

geochronologic unit Unit of geologic time, e.g., period, epoch.

geochronology The study of time in relationship to the history of the earth, or a system of dating developed for this purpose. Absolute chronology (sometimes called absolute age) involves dating of geologic events in years. Relative chronology involves the system of successive eras, periods, and epochs used in geology and paleontology. Literally the science of earth time.

geochronometry Measurement of geologic time.

geode Hollow, globular bodies, varying in size from an inch to a foot or more in size, and characteristic of certain limestone beds, while rarely found in shales. Significant features are (1) subspherical shape, (2) a hollow interior, (3) a clay film between the geode wall and the enclosing limestone matrix, (4) an outer chalcedonic layer, (5) an interior drusy lining of inward projecting crystals, and (6) evidence of expansion or growth.

geodepression A long, narrow depression, not necessarily filled by sediments.

geodesic coordinates Coordinates used in the study of the geometry of metric curved spaces in the neighborhood of a point O, called the origin, any point in the neighborhood of which may be identified by the geodesic through it and O and by the arc distance along the geodesic.

geodesic line A line of shortest distance between any two points

on any mathematically defined surface. Also termed a geodesic.

geodesy; geodetics The investigation of any scientific questions connected with the shape and dimensions of the earth. This is the function of a geodetic survey.

geodetic coordinates See GEODESIC COORDINATES

geodetic datum A datum consisting of 5 quantities: the latitude and the longitude of an initial point, the azimuth of a line from this point, and two constants necessary to define the terrestrial spheroid.

It forms the basis for the computation of horizontal control surveys in which the curvature of the earth is considered.

geodetic line The shortest-distance line between any two given points on the surface of the spheroid.

geodynamics That branch of science that deals with the forces and processes of the earth's interior.

geofracture Master fracture of great age, separating blocks of the earth's crust, that has influenced later tectonic activity. *Cf.* GEOSUTURE

geognosy *1.* The science which treats of the solid body of the earth as a whole and of the different occurrences of minerals and rocks of which it is composed and of the origin of these and their relations to one another. *2.* A term invented to express absolute knowledge of the earth, in contradistinction to geology, which embraces both the facts and our reasonings respecting them. That part of geology treating of the materials of the earth and its general exterior and interior constitution.

geographical cycle Cycle of erosion, *q.v.* Every land form passes through a comparatively systematic series of changes from its youth, when its form is defined chiefly by constructional processes, past its maturity, when the processes of subaerial sculpture have carved a great variety of mouldings and channelings, toward its old age, in which the accomplishment of the full measure of denudation reduces the mass essentially to base level, however high it may have been originally. *See* GEOMORPHIC CYCLE

geographic center The geographic center of an area on the earth has been defined as that point on which the area would balance if it were a plate of uniform thickness. In other words, it is the center of gravity of that plate.

geographic latitude Latitude measured in angle between axis of the earth and plane of the horizon at a point on the earth's surface. *Cf.* GEOCENTRIC LATITUDE

geographic longitude The angle between the line perpendicular to the standard spheroid at the observers' position and the plane of the prime meridian.

geographic province Large region characterized by similar geographic features.

geography The science that treats of the surface of the earth, including its form and development, the phenomena that take place thereon, and the plants, animals, and peoples that inhabit it, considered in relation to the earth's surface; also a book or treatise on the above subject.

geoid The figure of the earth considered as a mean sea-level surface extended continuously through the continents.

geologic; geological The generally preferred usage is as follows: geologic data; geologic investigation or survey; geological organization, survey, or society; geological era; geological time.

geologic age The age of a fossil or geologic event referred to the geologic time scale. Age may be expressed in "years before present" as determined by radiometric dating methods, or as relative age as determined by geologic methods.

geological horizon A term used to denote a particular level in the rocks or a particular thickness of rocks at a particular level.

geologic column A composite vertical chronologic diagram that shows the subdivisions of all or part of geologic time and the sequence of stratigraphic units present in a particular region or locality.

geologic hazard A geologic condition, either natural or man-made, that poses a potential danger to life and property. Examples: earthquake, landslides, flooding, faulting, beach erosion, land subsidence, pollution, waste disposal, and foundation and footing failures.

geologic high Sometimes used in oil fields to indicate a later geological formation regardless of elevation; opposed to geologic low, which refers to earlier formations. *Cf.* TOPOGRAPHIC HIGH

geologic history The history of planet Earth throughout geologic time including a record of life and associated physical and chemical changes affecting the planet from its formation to the present.

geologic map A map showing surface distribution of rock varieties, age relationshps, and structural features.

geologic province Large region characterized by similar geologic history and development.

geologic range Stratigraphic range; the distribution of any taxonomic group of oganisms through geologic time.

geologic thermometer A term applied to known temperature limits within which certain minerals or mineral aggregates must have formed; based on the thermal data relating to the fusion points of rocks and minerals, and the inversion- or transition-points of allotropic modifications of rock-forming compounds, and in general to the equilibrium conditions and stability ranges under different conditions of pressure for various minerals, allotropes, solid-solutions, eutectics, and other mineral aggregates.

geologic thermometry Measurement or estimation, by direct or indirect methods, of the maximum, minimum, or actual temperatures at which geological processes occur or have occurred in the past.

geologic time *1.* All of the time since the origin of the earth. *See also* TIME STRATIGRAPHIC for a discussion of its division.

geologic time scale An arbitrary chronologic arrangement of geologic events commonly presented in a vertical chart form with the oldest event and time unit at the bottom and the youngest at the top.

geologic time unit Time unit corresponding with a time-stratigraphic unit, e.g., period, epoch, age.

geologist One versed in geology, or engaged in geological study or investigation.

geology *1.* The science which treats of the earth, the rocks of which it is composed, and the changes which it has undergone or is undergoing. *2.* Earth science including physical geology and geophysics; the history of the earth, stratigraphy, and paleontology; mineralogy, petrology; and engineering, mining, and petroleum geology.

geomagnetic Pertaining to the magnetic field of the earth.

geomagnetic poles Poles of the earth's magnetic field located

about 6437 km. above its surface. One pole is at 78½° N., 69° W., and the other pole at 78½° S., 111° E. They do not correspond to the surface magnetic poles.

geomagnetic reversal A change in the polarity of the earth's magnetic field. The present condition is termed *normal polarity* and the opposite condition is called *reversed polarity*.

geomagnetism The earth's magnetic field and its associated phenomena; also, the branch of geophysics which studies the magnetic phenomena exhibited by the earth and its atmosphere.

geomechanics That branch of geology dealing with the response of natural earth materials to the application of deforming forces and combining the basic fundamentals of structural geology.

geomorphic Of, or pertaining to, the figure of the earth or the form of its surface; resembling the earth.

geomorphic cycle Geographical cycle; cycle of erosion, *q.v.*

geomorphogeny That part of geomorphology which treats of the origin and development of the earth's surface features.

geomorphology That branch of both physiography and geology which deals with the form of the earth, the general configuration of its surface, and the changes that take place in the evolution of landforms.

geopetal Any rock feature that indicates the relation of top to bottom at the time of formation of the rock.

geophone A detector, placed on or in the ground in seismic work, which responds to the ground motion at the point of its location. *Syn:* SEISMOMETER; SEISMOGRAPH; GEOTECTOR; PICKUP; JUG; TORTUGA

geophysical The form geophysical is used to the exclusion of any other that might be derived from the term geophysics.

geophysical survey The exploration of an area in which geophysical properties and relationships unique to the area are mapped by one or more geophysical methods.

geophysicist One who studies or practices the science of the physics of the earth.

geophysics The science of the earth with respect to its structure, composition, and development.

Geophysics is a branch of experimental physics dealing with the earth, including its atmosphere and hydrosphere. It includes the sciences of dynamical geology and physical geography, and makes use of geodesy, geology, seismology, meteorology, oceanography, magnetism, and other earth sciences in collecting and interpreting earth data.

geophysics, applied Geologic exploration or prospecting using the instruments and applying the methods of physics and engineering; exploration by observation of seismic or electrical phenomena or of the earth's gravitational or magnetic fields or thermal distribution.

geosphere *1.* The solid portion of the earth, synonymous with the lithosphere. *2.* Inclusive term for the earth and its hydrosphere and atmosphere. *3.* Any one of the so-called spheres which occur as concentric layers within the earth.

geostrophic Pertaining to deflective force due to rotation of the earth.

geosuture A boundary zone between major contrasting tectonic units of the earth's crust. In many places a fault which probably extends through the entire thickness of the crust. *Cf.* GEOFRACTURE

geosynclinal *See* GEOSYNCLINE

geosynclinal cycle *See* TECTONIC CYCLE

geosynclinal prism The load of sediments which accumulates in the downwarped area of a geosyncline.

geosyncline *1.* Large generally linear trough that subsided deeply throughout a long period of time in which a thick succession of stratified sediments and possibly extrusive volcanic rocks commonly accumulated. The strata of many geosynclines have been folded into mountains. Many different kinds have been differentiated and named. *2.* The area of such a trough. *3.* A stratigraphic surface that subsided in such a trough.

geotectocline A tectocline filled with sediment. *See* GEOSYNCLINE

geotectogene *See* TECTOGENE

geotectonic Pertaining to the form, arrangement, and structure of the rock masses composing the earth's crust. Structural.

geothermal energy The internal energy of the earth, available to man as heat from heated rocks, water, etc.

geothermal gradient The change in temperature of the earth with depth, expressed either in degrees per unit depth, or in units of depth per degree.

geothermic; geothermal Of, or pertaining to, the heat of the earth's interior.

geothermic degree The average depth within the earth's crust corresponding to an increase of one degree in temperature.

gerontic Pertaining to the old age stage of an individual.

geyser Literally, a roarer; intermittent hot springs or fountains. Columns of water are thrown out at intervals with great force, often rising to between 30 and 60 meters; and after the jet of water ceases, a column of steam rushes out with a thundering roar, and the eruption then ceases for an interval.

geyser basin A broad topographic basin in which a number of geysers are grouped together.

geyserite Siliceous sinter, *q.v.* A general term for siliceous deposits, usually opaline silica, formed around thermal springs and geysers, whether loose, compact, or concretionary.

geyser pipe The narrow tube or well of a geyser extending downward from the surface pool, which is ordinarily contained in the top of a sinter mound.

geyser pool The comparatively shallow pool of heated water ordinarily contained in a sinter crater or sinter mound at the top of a geyser pipe or orifice.

ghost Faint indication of a structure such as a crystal or fossil more or less obliterated by diagenesis or replacement.

giantism *See* GIGANTISM

Gibbs free energy A thermodynamic potential, associated with the variables pressure and temperature. In an irreversible thermodynamic process at constant temperature and pressure the Gibbs free energy of a system decreases; in a reversible process the Gibbs free energy remains constant.

gibbsite A mineral, $Al(OH)_3$. Monoclinic. A principal constituent of many bauxites.

gigantism A type of evolution in which overspecialization is shown as an increase in the size of an animal or one or more of its parts.

gilsonite One of the varieties of natural asphalt having a black color, brilliant luster, brown streak, and conchoidal fracture.

girdle *1. Struct. petrol:* The pattern shown by many petrofabric diagrams whereby the points are concentrated in a band of varying width, commonly normal to a fabric axis. *2.* A marginal band (in a chiton) of uniform width, differentiated from the central area which, on the back of the animal, consists of eight articulating plates or valves.

girdle, ac (also **ab, bc**) *Struct. petrol:* A girdle of points in a petrofabric diagram having a trend parallel with the plane of the a and c fabric axes (or with the a and b axes, or with the b and c axes). *Syn:* GIRDLE MAXIMUM

Gish-Rooney method An artificial-current conductive direct-current method of measuring ground resistivity which avoids polarization by continually reversing the current with a set of commutators.

gizzard stone *See* GASTROLITH

glacial *Geol:* Pertaining to, characteristic of, produced or deposited by, or derived from a glacier.

glacial advance *1.* Increase in the area and thickness of a glacier. *2.* A time interval marked by such increase.

glacial boulder A boulder that has been transported by a glacier.

glacial canyon Glacial canyons are characterized by several peculiar features: (1) They are U-shaped rather than V-shaped in cross profile; (2) small tributary gorges usually enter at levels considerably above the canyon bottoms; (3) in longitudinal profile the canyon bottoms are irregularly terraced; (4) the canyons are sometimes locally expanded into amphitheaters; (5) the canyon bottom is not always obdurate rock.

glacial cycle The phrase glacial cycle is here reserved for the ideal case of so long a continuance of glaciation under fixed climatic conditions, except for changes of climate with change of altitude due to degradation, that glacial erosion would be carried to its completion, truncating all the higher mountains at the snowline, and therefore causing snowfall to replace rainfall, and normal erosion to replace glacial erosion.

glacial drift Sediment (a) in transport in glaciers, (b) deposited by glaciers, and (c) predominantly of glacial origin, made in the sea or in bodies of glacial meltwater. *See* DRIFT

glacial epoch The Pleistocene Epoch, the earlier of the two epochs comprised in the Quaternary Period; characterized by the extensive glaciation of regions now free from ice.

glacial erosion Reduction of the earth's surface as a result of the presence or passage of a glacier.

glacial erratic Erratic.

glacial flow Glacier flow.

glacial geology The study of features resulting from glacial erosion and deposition. Contrast with glaciology, the study of the physics, form, and regimens of glaciers.

glacial groove A large furrow cut by the abrading action of rock fragments contained in a glacier.

glacial ice Ice that is flowing or that shows evidence of having flowed.

glacial lake *1.* A sheet of water owing its existence to the effects of the glacial period. They are of two classes, those excavated in the rock, and those produced by the irregular deposit of heaps of drift. *2.* Lake fed by glacial meltwater. *3.* Lake lying against or on a glacier.

glacial lobe *1.* One of the lobate protrusions of the margin of an ice sheet, sometimes a score or more miles in width, as where the ice has been free to spread out in depressions along its margin. *2.* A tonguelike projection from the main mass of a continental glacier.

glacial markings Small features, etched in rock as the result of glacial abrasion, such as chatter-marks and striations.

glacial maximum *1.* The position of greatest advance of a glacier. *2.* The time of greatest advance of a glacier.

glacial mill A vertical or steeply inclined opening in glacial ice through which melt water can pour. Water may have a vertical motion. *Syn:* MOULIN

glacial plain Plains constructed by the direct action of the ice itself.

glacial polish The smooth, even surface produced on bedrock by the movement of abrasive-laden glacial ice.

glacial recession Reduction in area and thickness of a glacier. A time interval marked by such reduction. Glacial retreat, *q.v.*

glacial retreat A glacier is said to retreat when its front recedes. The ice may be actually moving forward toward this front, but the rate of backward melting at the front, if it exceeds the rate of forward movement, will cause the position of the front to recede.

glacial scour lake *See* FINGER LAKE

glacial scratches *See* GLACIAL STRIAE

glacial stairway Glaciated valley whose floor rises in a series of irregular steplike benches.

glacial striae (*sing.* stria) *1.* Usually straight, more or less regular scratches, commonly parallel in sets, on smoothed surfaces of rocks, due to glacial abrasion; glacial scratches. 2. Curved, crooked, and intermittent gouges, of irregular depth and width, and rough definition, or certain rock surfaces, sometimes due to abrasion by icebergs. *See* GLACIAL STRIATION

glacial striation *1.* Fine-cut lines on the surface of the bedrock which were inscribed by the overriding ice. *2.* A scratch or small groove cut into a surface of the bedrock or mantle by rock fragments carried in a glacier.

glacial theory The theory that large elevated portions of the temperate and frigid zones were covered during the early Quaternary, and perhaps during some earlier epochs, by slowly moving ice sheets and glaciers, that transported vast masses of drift to lower latitudes, assisted by icebergs drifting along the coast. No longer a theory, but accepted as fact.

glaciate *1.* To cover with glacier ice. *2.* To subject to the action of a glacier.

glaciated *1.* Said of a country which has been scoured and worn down by glacial action, or strewn with ice-laid drift. *2.* Covered by and subjected to the action of a glacier.

glaciation Alteration of the earth's solid surface through erosion and deposition by glacier ice.

glaciation limit The lowest level at which glaciers can develop.

glacier A mass of ice with definite lateral limits, with motion in a definite direction, and originating from the compacting of snow by pressure.

glacier bands *1.* Any of several phenomena which appear to the observer as a series of bands on a glacier surface. *See* ALASKAN BANDS; ICE BANDS; OGIVES; PARALLEL BANDS. *2.* The ice layers or lenses causing this banded appearance.

glacier burst The sudden release of a reservoir of water which has been impounded within or by a glacier.

glacieret; hanging glacier *1.* Small alpine glaciers are sometimes called glacierets, or, if visible high in the sides of mountain valleys, hanging glaciers. *2.* A very small glacier on a mountain slope or in a cirque. Sometimes called cirque glacier.

glacier flood *See* GLACIER BURST

glacier flow The motion that exists within the body of a glacier. Glacier flow may exist without movement of the glacier front. *See* EXTRUSION FLOW HYPOTHESIS; GRAVITY FLOW; STREAMING FLOW

glacier ice A body of ice developed from snow which becomes large enough to move from its place of accumulation.

glacierization In British usage, approximately equivalent to glaciation, or gradual covering of land by glaciers.

glacier lake A lake produced by the damming of a natural drainage by a glacier, as a lake ponded by a glacier advancing across a valley.

glacier milk The white, turbid glacier meltwater containing rock flour in suspension.

glacier table A large block of stone supported by a column or pedestal of ice on the surface of a glacier.

glacier tongue *1.* An outlet glacier of an ice sheet or icecap. *2.* The lower part of a valley glacier.

glacier well Moulin; glacier mill. All streams on the surface of a glacier eventually encounter the marginal crevasses and plunge down in foaming cascades, producing well-known "glacier wells" or "glacier mills."

glacier wind *1.* A cold wind blowing out of apertures in a glacier front. This wind is caused by the difference in density between the cold air inside the ice caves and the warmer outside air. 2. Any wind blowing off a glacier, usually a katabatic wind, *q.v.*

glacio A combining form frequently used with other words to denote formation by or relationship to glaciers.

glaciofluvial Fluvioglacial. Pertaining to streams flowing from glaciers or to the deposits made by such streams.

glacio-isostacy Balance and level in areas influenced by the weight of glacial ice.

glaciology The study of all aspects of snow and ice or the science that treats all processes associated with solid existing water.

glance A term used to designate various minerals having a splendent luster, as silver glance, lead glance, etc.

glance copper Chalcocite.

glass A state of matter intermediate between the close-packed, highly ordered array of a crystal, and the poorly packed, highly disordered array of a gas. Most glasses are supercooled liquids, i.e., metastable, but there is no break in the change in properties between the metastable and stable states. The distinction between glass and liquid is solely on the basis of viscosity, and is not necessarily related, except indirectly, to the difference between metastable and stable states.

glass, volcanic *See* VOLCANIC GLASS

glass sand An extremely pure silica sand useful for making glass and pottery.

glass sponge Popular term for a class of Porifera (sponges) in which the skeletal framework consists of six-rayed spicules of silica. *Syn:* HYALOSPONGE

glauberite A mineral, $Na_2Ca(SO_4)_2$. Monoclinic.

glauconite A green mineral, closely related to the micas and essentially a hydrous potassium iron silicate. Commonly occurs in sedimentary rocks of marine origin. Also used as a name for a rock of high glauconite content.

glauconitic sandstone A quartz sandstone or an arkosic sandstone rich in glauconite grains.

glaucophane A mineral of the amphibole group, $Na_2(Mg,Fe)_3$-$Al_2Si_8O_{22}(OH)_2$.

G layer The seismic region of the earth's interior below 5160 km.; the inner core.

glei; gley A soil horizon in which the material is bluish gray or blue-gray, more or less sticky, compact, and often structureless. It is developed under the influence of excessive moisture.

glide direction The direction of gliding along glide planes in a mineral.

glide line *Struct. petrol:* In single-crystal deformation, the possible direction, or directions, of movement in a glide plane; in a tectonite the direction of movement in an s-surface indicated either from field observation or from interpretation of preferred orientation of the fabric.

glide plane; slip plane *Struct. geol:* In single-crystal deformation, a lattice plane on which translation- or twin-gliding takes place in a tectonite, an s-surface characterized (in a statistical sense) by a preferred orientation of one or more fabric elements which indicate movement in the s-surface.

gliding, *n.* Slip or movement along certain lattice planes in crystalline substances, produced by deformation and characterized either as translation-gliding or twin-gliding, *q.v.*

glimmer [*Ger.*] Mica.

glint In Norway, a boundary. An escarpment, particularly one produced by the outcrop of a dipping resistant formation.

glint lake A lake whose basin is excavated in hard rock where a glacier is held in check by an escarpment. An illustration is Torne Träsk in northern Lapland.

global tectonics Tectonics on a global scale, such as tectonic processes related to very large-scale movement of material within the earth. *Specif.* new global tectonics. *Cf.* MEGATECTONICS

globigerina ooze A calcareous marine deposit formed in deep water (but less than 3660 m. consisting chiefly of calcareous shells of Foraminifera, especially *Globigerina* spp. These are surface-dwelling forms, which reach the bottom only after death.

Globigerina ooze contains more than 30% $CaCO_3$, of which the greater part consists of pelagic Foraminifera. The carbonate content averages 64% but ranges from 30 to 97%.

globular *Petrol:* A textural term synonymous with spherulitic, *q.v.*

globular map projection A map projection representing a hemisphere, on which the equator and a central geographic meridian are represented by straight lines intersecting at right angles; these lines are divided into equal parts. All meridians, except the central one, are represented by circular arcs connecting points of equal division on the equator with the poles. The parallels, except the equator, are circular arcs dividing the central and extreme outer meridians into equal parts. The extreme outer meridian limits the projection and is a full circle.

Glomar Challenger A research ship specially designed to obtain long sediment cores by drilling into the ocean floor. It can drill in water more than 6000 meters deep. The Deep Sea Drilling Proj-

ect, using the *Glomar Challenger* to obtain cores throughout the ocean, began in the later 1960s and has extended into the 1970s.

glory hole *1.* A large open pit from which ore is or has been extracted. *See* MILLING. *2.* An opening through which to observe the interior of a furnace.

glowing avalanche Volcanic eruption feature similar to a glowing cloud (nuée ardente) but propelled by gravity rather than by the auto-explosion of magma.

glowing avalanche deposit Chaotically brecciated material that should not be confused with welded tuff.

glowing cloud Nuée ardente.

gneiss A coarse-grained rock in which bands rich in granular minerals alternate with bands in which schistose minerals predominate. *See* AUGEN GNEISS; CHARNOCKITE; COMPOSITE GNEISS; GRANITE GNEISS; GRANULITE; INJECTION GNEISS; ORTHOGNEISS; PARAGNEISS; SCHIST

gneissic; gneissoid, *adj.* Having the appearance or character of gneiss, *q.v.*

gneissose *1.* Resembling gneiss. *2.* Having composite structure of alternating schistose and granulose bands and lenses which differ in mineral composition and texture.

goethite A mineral, $FeO(OH)$. Orthorhombic. Most limonite is impure goethite.

gold A mineral, Au, that forms a complete solid-solution series with silver, and partial solutions with other metals. Isometric. Sp. Gr.=19.3 when pure. The natural gold-silver alloy is electrum.

gold dust Fine particles of gold, such as are obtained in placer mining. An impure dust is sometimes called commercial dust.

Goldschmidt's mineralogical phase rule Under natural rock-forming conditions, the probability of finding a system with a variance (degrees of freedom) of less than two (temperature and pressure) is small. Any given natural mineral assemblage, igneous or metamorphic, seems to be the stable one over a range of temperature and pressure. Thus, with a variance of two, the phase rule is reduced to a special case, $P=C$, in which the maximum number of phases possible is equal to the number of components.

goldstone See AVENTURINE

Gondwana; Gondwanaland Ancient continent including Australia, Antarctica, Africa, South America, and India south of the Ganges River, plus smaller islands; fragmented and drifted apart since early Triassic Time.

goniatite An ammonoid cephalopod, characteristic of the Devonian and Carboniferous, in which the sutures have developed undigitated lobes and saddles that give an angular appearance to the suture line.

goniometer An instrument for measuring the angles between crystal faces.

gorge [Fr.] 1. As most generally used in English, the word gorge means a narrow passage, with precipitous, rocky sides, enclosed among the mountains. A ravine need not be enclosed; and this word would hardly be applied to a mere depression in the soil, as ravine might be. 2. A canyon; a rugged and deep ravine or gulch. 3. A jam; as an ice gorge.

gossan; gozzan A yellow to reddish deposit of hydrated oxides of iron produced near surface by the oxidation and leaching of sulfide minerals. A visible guide to potential sulfide ore deposits at depth.

gouge N: 1. A layer of soft material along the wall of a vein, favoring the miner, by enabling him after "gouging" it out with a pick, to attack the solid vein from the side; 2. Finely abraded material occurring between the walls of a fault, the result of grinding movement; 3. In Nova Scotia, a narrow band of gold-bearing slate next the vein, which can be extracted by a thin, long-pointed stick. V: 4. To work a mine without plan or system; 5. To contract the face of a mine working by neglecting to keep the sides cut away.

graben A block, generally long compared to its width, that has been downthrown along faults relative to the rocks on either side.

grab sample 1. Sample of rock or sediment taken more or less indiscriminately at any place. 2. Subaqueous sample of bottom sediment obtained by an instrument with movable jaws that close after being dropped to the bottom.

grade 1. Continuous descending curve of a stream channel just steep enough for current to flow and transport its load of sediment. 2. Measure of inclination expressed in per cent. 3. Varying inclination, e.g., of a road. 4. Expression of relative quality, e.g., high grade or low grade. 5. Biolog: A particular stage of evolution of an animal in which one or more features have undergone a chronocline.

graded Geol: Brought to or established at grade, through the action of running water carrying a load of sediment, by eroding or degrading at some places and depositing or aggrading in other places.

graded bedding Diadactic structure, q.v. A type of stratification each stratum of which displays a gradation in grain size from coarse below to fine above.

graded profile See PROFILE OF EQUILIBRIUM

graded sediment Geol: A sediment consisting chiefly of grains of the same size range. *Engin:* A sediment having a uniform or equable distribution of particles from coarse to fine.

graded shoreline A shoreline that has been straightened by the building of bars across embayments and by the cutting back of headlands. The coast consists of alternately steep and low reaches.

graded slope Unbroken slopes worn or built to the least inclination at which the waste supplied by weathering can be urged onward.

graded stream A graded stream is one in which, over a period of years, slope is delicately adjusted to provide, with available discharge and with prevailing channel characteristics, just the velocity required for the transportation of the load supplied from the drainage basin

grade level Where the slopes are steep, erosion will occur; and near the sea, where the gradients are gentle, deposition of the water-borne sediments will take place, so that eventually the whole course of the stream will be reduced to a uniform gradient. When this condition is reached the stream is said to have found its grade level.

grade scale A subdivision of an essentially continuous scale of particle sizes into a series of size classes. *See* WENTWORTH SCALE

grade scale, Atterberg A decimal grade scale for particle size, with 2 mm. as the reference point, and involving the fixed ratio 10. Subdivisions are the geometric means of the grade limits: 0.2, 0.6, 2.0, 6.3, 20.0.

grade scale, Phi A logarithmic transformation of the Wentworth grade scale based on the negative logarithm to the base 2 of the particle diameter.

grade scale, Tyler standard A scale for sizing particles based on the square root of 2 used as specifications for sieve mesh. Alternate class limits closely approximate the class limits on the Udden grade scale, and the intermediate limits are the geometric means of the Udden scale values: 0.50, 0.71, 1.00, 1.41, 2.00.

grade scale, Udden A grade scale for particle size, with 1 mm. as the reference point, and involving the fixed ratio 2 to $\frac{1}{2}$, depending on whether the scale is increasing or decreasing, as $\frac{1}{4}$, $\frac{1}{2}$, 1, 2, 4.

grade scale, Wentworth An extended version of the Udden grade scale with descriptive class terms for various sizes of sedimentary particles.

gradient *1.* Slope, particularly of a stream or a land surface. Measurements are expressed in per cent, feet per mile or degrees. *2.* Change in value of one variable with respect to another variable, especially vertical or horizontal distance, e.g., gravity, temperature, magnetic intensity, electric potential, etc.

grading Degree of mixing of size classes in sedimentary material. Well graded implies more or less uniform distribution from coarse to fine; poorly graded implies uniformity in size or lack of continuous distribution. *Cf.* SORTING

grading factor The coefficient of sorting of a clastic sediment. Perfect sorting has a grading factor of 1.0.

gradiometer Any instrument that measures grades or slopes. An instrument for measuring the gradient of any physical quantity, such as the magnetic or gravitational field. A form of torsion balance with 3 weights sensitive to gravity gradients but not to curvatures.

graduation Division, *q.v.*

grahamite A hydrocarbon resembling albertite in its jet-black luster. It is soluble in carbon disulfide and chloroform but not in alcohol, and is fusible. Occurs in veinlike masses. Specific gravity 1.145. Has conchoidal fracture and is brittle.

grain *Geol: 1.* The particles or discrete crystals which comprise a rock or sediment; *2.* The individual particles which form settled snow, firn, and glacier ice. Glacier grains are single crystals of ice, but snow and firn grains are agglomerations of many snow crystals; *3.* A direction of splitting in rock, less pronounced than the rift and usually at right angles to it. *4. Coal min:* In England, the lines of structure or parting in the rocks parallel with the main gangways and hence crossing the breasts. *5.* A unit of weight equal to 0.0648 part of a gram, 0.000143 part of an avoirdupois pound, and 0.04167 part of a pennyweight. Derived from the weight of a grain of wheat.

grain growth Solid state enlargement of some crystals at the expense of others producing a coarser texture in an essentially monomineralic rock like limestone; commonly termed recrystallization.

grain plane A plane of parting in metamorphic rocks at right angles to the cleavage.

grain size A term relating to the size of mineral particles that make up a rock or sediment.

grainstone A limestone in which there is no mud and in which the allochems are in contact and self-supporting; a biosparite, *q.v.*

granite *1.* A plutonic rock consisting essentially of alkali feldspar and quartz. Sodic plagioclase, usually oligoclase is commonly present in small amounts and muscovite, biotite, hornblende, or rarely pyroxene may be mafic constituents. *2. Seismol:* A rock in which velocity of the compressional wave lies somewhat between 5.5 and 6.2 km./sec. *3.* Loosely used for any light-colored, coarse-grained igneous rock.

granite gneiss *1.* A coarsely crystalline, banded metamorphic rock of granitic composition. *2.* A primary igneous gneiss of granitic composition. *See* AUGEN GNEISS; GNEISS; ORTHOGNEISS

granite porphyry *See* QUARTZ PORPHYRY

granite tectonics The structural features of plutons and their relationships.

granitic *1.* Of, pertaining to, or composed of, granite or granite-like rock. *2.* A textural term applied to coarse- and medium-grained, granular igneous rocks in which all or nearly all of the mineral constituents are anhedral (xenomorphic) and of approximately the same size. In this sense the term is synonymous with hypidiomorphic-granular or hypautomorphic-granular. *See* HYPIDIOMORPHIC

granitization; granitisation A term used in somewhat different connotations by different authors, but in general, referring to the production of a granitic rock from sediments by an unspecified process. Some would limit the term to the production of granite in place, without the formation of a "notable" amount of liquids; others would include all granitic rocks formed from sediments by any process, regardless of the amount of liquid formed or any evidence of movement. The precise mechanism, frequency, and magnitude of the process are still in dispute.

granitoid A term applied to the texture of holocrystalline igneous or metasomatic rocks such as granites in which the constituents are mostly anhedral or xenomorphic and of uniform size.

granoblastic *1.* The texture of metamorphic rocks composed of equidimensional elements. *2.* A term applied to secondary texture due to diagenetic change either by crystallization or recrystallization in the solid state, in which the grains are of equal size (equigranular).

granodiorite A plutonic rock consisting of quartz, calcic oligoclase or andesine, and orthoclase, with biotite, hornblende, or pyroxene as mafic constituents. Granodiorite is intermediate between quartz monzonite and quartz diorite and contains at least twice as much plagioclase as orthoclase.

granophyre An intrusive silicic rock featuring an intergrowth texture of quartz and alkali feldspar.

granophyric A textural term applied to generally fine-grained intergrowths of quartz and alkali feldspar in igneous rocks.

granular *1.* A textural term applied to holocrystalline rocks in which most of the mineral grains are equant or equidimensional. *2.* Also applied to sedimentary rock made up of grains or granules. *3.* *Paleobot:* Covered with very small grains; minutely or finely mealy.

granular disintegration A type of mechanical weathering, in regions of great extremes of temperature, involving the breaking of rock into small fragments or into the component crystal units without chemical decay, due to varying coefficients of expansion.

granularity *Petrol:* The feature of rock texture relating to the size of the constituent grains or crystals. Expressed by such terms as fine-, medium-, or coarse-grained; phanerocrystalline, mi-

crocrystalline, etc. Essentially synonymous with grain size.

granule *1.* Rounded rock fragments larger than very coarse sand grains but smaller than pebbles. *2.* Granular mineral products used primarily to form a protective and decorative coating on composition roofing.

granulite *1.* A metamorphic rock composed of even-sized, interlocking granular minerals. *2.* A metamorphic belonging to a high-temperature facies characterized by the presence of mica and hornblende. Coarse and fine bands alternate and produce a regular planar schistosity. *3.* In French literature, the term has been used as a synonym for muscovite granite. *See* CHARNOCKITE, GNEISS; QUARTZITE

granulite facies Gneissic rocks produced by deep-seated high-grade regional metamorphism.

granulometric Refers to size measurements of sedimentary particles.

graphic A rock texture resulting from the regular intergrowth of quartz and feldspar. The quartz is commonly cuneiform, resembling runic inscriptions on the background of feldspar.

graphic log A graphic record, usually in the form of a strip, on which the formations penetrated in drilling are drawn to a uniform vertical scale. In addition to lithology, such logs usually indicate the points at which oil, gas, or water was found, and the lengths of casing used. Usually, also, conventional colors and symbols are used in order to abbreviate the record.

graphite Plumbago; black lead. A mineral, native carbon, dimorphous with diamond. Hexagonal. Black to steel-gray, very soft (H=1).

graptolite Extinct colonial organism that produced chitinous enclosing and supporting structures; generally considered to be related to primitive chordates; one of the Graptolithina.

Graptolithina Class of extinct colonial animals referred to the hemichordates; they constructed variously formed and arranged series of tiny chitinous thecae.

grassland soil Prairie, chestnut, reddish chestnut, brown, and chernozem soils.

graticule *1.* A network of lines representing geographic parallels and meridians forming a map projection. *2.* A template, divided into appropriately designed blocks or cells, for graphically integrating a quantity such as gravity. Graticules are much used in computing terrain corrections and the gravitational or magnetic attraction of irregular masses.

grating *Opt:* A system of close equidistant and parallel lines or bars, used for producing spectra by diffraction. *Applied geophys:* A graticule.

gravel *1.* Accumulation of rounded waterworn pebbles. The word gravel is generally applied when the size of the pebbles does not much exceed that of an ordinary hen's egg. The finer varieties are called sand while the coarser varieties are called shingle. *2.* An accumulation of rounded rock or mineral pieces larger than 2 mm. in diameter. Divided into granule, pebble, cobble, and boulder gravel. *3.* Consists of rock grains or fragments with a diameter range of from 76 mm. (3 in.) to 4.76 mm. (retention on a No. 4 sieve). The individual grains are usually more or less rounded. *4.* Accumulation of uncemented pebbles. Pebble gravel. May or may not include interstitial sand ranging from 50 to 70% of total mass. *Cf.* BOULDER, COBBLE, and GRANULE GRAVEL

gravimeter *1.* An instrument for measuring variations in the magnitude of the earth's gravitational field; a gravity meter. *2.* An instrument for determining specific gravities, particularly of liquids. *See* HYDROMETER

gravitational constant The constant γ in the law of universal gravitation. Its value is $6.673\pm.033\times10^{-8}$ cm.^{3}/gm. sec.^{2}

gravitational gliding Extensive sliding of strata down the slope of an uplifted area producing low angle overthrust faults, recumbent folds, or nappes.

gravitational separation The separation of oil, gas, and water in a reservoir rock in accordance with their relative gravities.

gravitational theory One of the migration theories which assume oil and gas to move because of their buoyancy or lower specific gravities relative to that of the associated water.

gravitational water Water which exists in the large pores of the soil and which the force of gravity will remove from the soil when conditions for free drainage exist. *Syn:* FREE WATER

gravity; gravitation *1.* The resultant effect upon any body of matter in the Universe of the inverse square law attraction between it and all other matter lying within the frame of reference and of any centrifugal force which may act on the body because of its motion in any orbit. *2.* The resultant force on any body of matter at or near the earth's surface due to the attraction by the earth and to its rotation about its axis. *3.* The force exerted by the earth and by its rotation on unit mass or the acceleration imparted to a freely falling body in the absence of frictional forces.

gravity, Baumé Specific gravity of a fluid in the relation which the weight of a unit volume of the fluid bears to that of the same volume of water. Antoine Baumé devised a system in which fluids with a specific gravity of 1.00 (that of water) were at 10°. The Baumé scale for oil was replaced in 1921 by the API Gravity scale. The Baumé scale is defined by the equation:

$$\text{Degrees Baumé} = \frac{140}{\text{Sp. Gr. 60° F}} - 130.$$

See API GRAVITY

gravity anomaly Difference between theoretical calculated and observed terrestrial gravity; excess observed gravity is positive and deficiency is negatively anomalous. *See* BOUGUER, FREE-AIR, and ISOSTATIC ANOMALY

gravity compaction Compaction of sediment resulting from overburden pressure.

gravity fault *1.* A fault along which the hanging wall has moved down relative to the footwall. *See* NORMAL FAULT. *2.* Sometimes restricted to those faults that are the result of withdrawal of support, either below or on the side.

gravity flow A type of glacier movement in which the flow of the ice is caused by the downslope component of gravity in an ice mass resting on a sloping floor.

gravity gradiometer An instrument for measuring the gradient of gravity.

gravity instruments Devices for measuring the differences in the gravity force or acceleration at two or more points. They are of three principal types (a) A static type in which a linear or angular displacement is observed or nulled by an opposing force (*see* GRAVIMETERS); (b) A dynamic type in which the period of oscillation is a function of gravity and is the quantity directly observed; (c) A gradient measuring type, e.g., Eötvös torsion balance.

gravity meter *See* GRAVIMETER

gravity of oils Specific gravity, expressed usually in degrees API. *See* API GRAVITY

gravity sliding Downslope shearing movement of part of the earth's crust on the flank of a large uplifted area.

gravity tectonics Tectonics in which the dominant propelling mechanism is downslope gliding under the influence of gravity.

gravity unit; G unit One-tenth of a milligal (prospecting but not scientific usage).

gravity wave A wave whose velocity of propagation is controlled primarily by gravity. Water waves of a length greater than 5.08 centimeters are considered gravity waves.

gray copper ore Tetrahedrite; tennantite, *q.v.*

gray podzol Gray soil with leached A zone and thin dark humus zone (Ao); B horizon may be thin.

graywacke; greywacke A type of sandstone marked by: (1) large detrital quartz and feldspars (phenocysts) set in a (2) prominent to dominant "clay" matrix (and hence absence of infiltration or mineral cement) which may on low-grade metamorphism (diagenesis) be converted to chlorite and sericite and partially replaced by carbonate, (3) a dark color, (4) generally tough and well indurated, (5) extreme angularity of the detrital components (microbreccia), (6) presence in smaller or larger quantities of rock fragments, mainly chert, quartzite, slate or phyllite, and (7) certain macroscopic structures (graded bedding, intraformational conglomerates of shale or slate chips, slip bedding, etc.) and (8) certain rock associations.

greasy Applied to the luster of minerals. Having the luster of oily glass, as nepheline.

great circle The line of intersection of the surface of a sphere and any plane which passes through the center of the sphere. The shortest distance between any two points on a sphere is along the arc of a great circle connecting the two points.

green belt An area restricted from being used for buildings and houses; it often serves as a separating buffer between pollution sources and concentrations of population.

green earth *1.* Glauconite, found in cavities of amygdaloids and other eruptive rocks, and used as a pigment by artists. *2.* Chlorite; a variety of talc.

greenhouse effect The thermal effect resulting when comparatively short wavelength solar radiation penetrates the atmosphere rather freely, only to be largely absorbed near and at the earth's surface, whereas the resulting long wavelength terrestrial radiation thus formed passes upward with great difficulty. The effect is because the absorption bands of water vapor, ozone, and carbon dioxide are more prominent in the wavelengths occupied by terrestrial radiation than the short wavelengths of solar radiation. Hence the lower atmosphere is almost perfectly transparent to incoming radiation, but partially opaque to outgoing long-wave radiation. The effect is increased when the atmospheric content of carbon dioxide, particulates, water vapor, or ozone is increased, and vice versa.

green marble A commercial term for serpentine.

green mud A fine-grained terrigenous mud or oceanic ooze found near the edge of the continental shelf, and similar to other terrigenous muds except for the greenish color and, perhaps, less organic matter. It occurs at depths of 91.5 to 2287.5 meters. A deep-sea terrigenous deposit characterized by the presence of a considerable proportion of glauconite and $CaCO_3$ in variable amounts up to 50%.

greenockite A mineral, CdS. Hexagonal.

greensand Glauconitic sand; less commonly, any glauconitic sediment.

greenschist A metamorphosed basic igneous rock which owes its color and schistosity to abundant chlorite. *See* GREENSTONE; OPHIOLITE.

greenschist facies Metamorphic rocks produced under low temperature conditions.

greenstone An old field term applied to altered basic igneous rocks which owe their color to the presence of chlorite, hornblende, and epidote. *See* GREENSCHIST; OPHIOLITE

greisen *1.* A pneumatolytically altered granitic rock composed largely of quartz, mica, and topaz. The mica is usually muscovite or lithium mica, and tourmaline, fluorite, rutile, cassiterite, and wolframite are common accessories. *2.* A granitoid but often somewhat cellular rock composed of quartz and muscovite or some related mica, rich in fluorine. It is the characteristic mother rock of the ore of tin, cassiterite, and is in most cases a result of the contact mineralizers. *3.* A coarse-grained tin-bearing rock which contains muscovite, quartz, topaz, or tourmaline.

Grenville A provincial series of the Precambrian of Canada and New York.

Grenville orogeny A name that is

widely used for major plutonic, metamorphic, and probably deformational event during the Precambrian, dated radiometrically as between 880 and 1000 m.y. ago which affected a broad province along the southeastern border of the Canadian Shield. Originally, the name Grenville was used for a metasedimetary series in the southern part of the province, and the name Laurentian was used for the associated plutonic rocks. Pertinent objections have been raised to use of Grenville for the orogeny, the province, and for its northwestern structural "front," but these uses will be continued until generally acceptable alternatives are proposed.

grid *1.* A systematic array of points or lines: (a) At or along which field observations are made. (b) For which computations are made. *2.* The control electrode in thermionic tubes.

gridiron twinning *See* CROSSED TWINNING

grinding The crushing of small grains through the effect of continued contact and pressure by larger pebbles.

grinding pebbles Pebbles, usually of chert or quartz, used for grinding in ball mills, etc. where contamination with iron must be avoided.

grindstone *1.* A tough sandstone of fine and even grain, composed almost entirely of quartz, mostly in angular grains. It must have sufficient cementing material to hold the grains together but not enough to fill the pores and cause the surface to wear smooth. *2.* A large circular stone made from sandstone and used quite extensively for the sharpening of many different tools and instruments.

grit *1.* Sand, especially coarse sand. *2.* Coarse-grained sand-

stone. *3.* Sandstone with angular grains. *4.* Sandstone with grains of varying size producing a rough surface. *5.* Sandstone suitable for grindstones.

groin; groyne A shore-protection and improvement structure (built usually to trap littoral drift or retard erosion of the shore).

groove casts Rounded or sharp crested rectilinear ridges, a few mm. high and many cm. long, occurring on undersurfaces of sandstone layers lying on mudstone.

grossularite A mineral of the garnet group $Ca_3Al_2(SiO_4)_3$. Grossular.

grotto A hole eroded in the wall of a cave by seepage or lateral stream erosion. A small cave.

ground cover Grasses or other plants grown to keep soil from being blown or washed away.

ground ice Anchor ice. *1.* Ice formed about stones over the bottoms of streams or lakes. *2.* Bodies of more or less clear ice in permanently frozen ground. Deposits which are evidently only temporary features are excluded under this definition, and the term is not applied to deposits which seem to be on top of the ground. Stagnant earth-covered glaciers appear to fall about in the dividing line of this definition. If their glacial origin is evident, they would be excluded.

ground-ice mound Pingo.

ground-ice wedges Ground ice occurring chiefly in a network of more or less vertical wedges, surrounding isolated bodies of frozen ground.

ground magnetometer A magnetometer primarily suitable for making observations of magnetic field intensity on the surface of the earth.

groundmass The material between the phenocrysts in a porphyritic igneous rock. It includes the basis or base as well as the smaller crystals of the rock. *Syn:* MATRIX

ground moraine The material deposited from a glacier on the ground surface over which the glacier has moved. It is bordered by lateral and/or end moraines.

ground motion The displacement of the ground due to the passage of elastic waves arising from earthquakes, explosions, seismic shots, machinery, wind, traffic, and other causes.

ground resistivity The resistivity of soil and rock materials. *See* RESISTIVITY

ground roll Seismic surface waves generated during reflection shooting; predominantly Rayleigh waves.

ground swell A long high ocean swell. Also, this swell as it rises to prominent height in shallow water. Not usually so high or dangerous as blind rollers.

ground water Phreatic water, *q.v.* That part of the subsurface water which is in the zone of saturation.

ground-water barrier A body of material which is impermeable or has only low permeability and which occurs below the surface in such a position that it impedes the horizontal movement of ground water and consequently causes a pronounced difference in the level of the water table on opposite sides of it. *Syn:* GROUND-WATER DAM

ground-water divide A line on a water table on each side of which the water table slopes downward in a direction away from the line.

ground-water flow That portion of the precipitation which has been absorbed by the ground and has become part of the ground water, alternately being discharged as spring and seepage

water into the stream channels and leaving no drainage as runoff.

ground-water level Ground-water surface.

ground-water reservoir Aquifer, *q.v.*

ground-water runoff That part of the runoff which consists of water that since its last precipitation has existed as ground water.

ground-water surface This level, below which the rock and subsoil (down to unknown depths) are full of water, is known as the ground-water level, ground-water surface, or water table.

group *1. General:* An association of any kind based upon some feature of similarity or relationship. *Stratig: 2.* Lithostratigraphic unit consisting of two or more formations; *3.* More or less informally recognized succession of strata too thick or inclusive to be considered a formation; *4.* Subdivision of a series. *Obs. Cf.* STAGE

group velocity *1. Oceanog:* The velocity at which a wave group travels. In deep water, it is equal to half the velocity of the individual waves within the group. *2. Geophys:* The velocity of individual wave crests in a dispersive, continuous medium in which several frequencies are apparent. In a medium in which the velocity varies with frequency the wave train changes its shape as it progresses, in which case the group velocity may differ materially from the phase velocity.

growth fabric Crystal arrangement determined by growth from a plane surface such as the wall of a vein.

growth twinning Twinning resulting from change in lattice orientation during the growth of a crystal. *Cf.* DEFORMATION TWINNING

groyne *See* GROIN

grunerite A mineral of the amphibole group, $(Fe,Mg)_7Si_8O_{22}(OH)_2$.

grus An accumulation of angular, coarse-grained fragments resulting from the granular disintegration of crystalline rocks (esp. granite) generally in an arid or semiarid region.

Guadalupian The third of four provincial series of the Permian in the United States.

guano 1. Applied to deposits of the excrement of bats, birds, or other animals. 2. A substance found in great amounts on some coasts or islands frequented by marine birds; it is composed chiefly of their excrement, and is rich in phosphates and nitrogenous matter. It is extensively used for soil fertilizer.

guide fossil Any kind of category (species, genus, etc.) of fossil useful in the identification of a stratigraphic unit. The best guide fossil of a time-stratigraphic unit should have short stratigraphic range, wide geographic distribution, broad ecologic tolerance, and occur abundantly. *Syn:* INDEX FOSSIL

gulch A small ravine; a small, shallow canyon with smoothly inclined slopes and steep sides. Local in Far West.

gulf 1. (1) A portion of the sea partially enclosed by a more or less extensive sweep of the coast. The distinction between gulf and bay is not always clearly marked, but in general a bay is wider in proportion to its amount of recession than a gulf; the latter term is applied to long landlocked portions of sea opening through a strait, which are never called bays. (2) A deep hollow, chasm, abyss. (3) A profound depth (in river or ocean); the deep. 2. A large deposit of ore in a lode. 3. Elongated karst valley, steep-sided, level floor of alluvium; a "window" of an underground drainage system, generally formed by the merging of several collapse sinks. *Syn:* UVALA (in a limited sense)

Gulfian Upper Cretaceous.

gulf-type (Hoyt) **gravimeter** A meter consisting of a mass suspended at the end of a spring, the latter so designed that its extension will cause the mass to rotate. By this means the linear displacement of the spring is converted into an angular deflection which is more easily measured. The design also minimizes the sensitivity to seismic disturbances and the basic instrument is therefore well suited for underwater observations.

gulf-type (Vacquier) **magnetometer** A flux-gate or saturable reactor type of recording magnetometer. Used primarily in aircraft and there includes means for keeping the measuring element aligned in the direction of maximum intensity (i.e., total field). In this case it records variations in the total field regardless of variations in its direction. Sometimes used in the sense of including the equipment for establishing the position of the aircraft as well as the magnetometer itself.

gully 1. A small ravine. 2. Any erosion channel so deep that it cannot be crossed by a wheeled vehicle or eliminated by plowing. 3. (*Oceanog.*) The extension of a trough or basin which penetrates the land or a submarine elevation either with a uniform or a gradually diminishing depth, or which is bounded on one side by land and on the other by a submarine elevation, may be called a gully [*Ger:* Rinne; *Fr:* chenal], if long and narrow.

gully erosion Removal of soil by running water, with formation of channels that cannot be smoothed out completely by normal cultivation.

gully gravure A process or more exactly a recurrent cycle of processes by which the steep slopes of hills and mountains retreat; in the cycle, gullying is recurrent in time but shifts laterally in place. The slope retreats by repeated scoring or graving. Each groove is a gully, and each new groove is parallel to earlier grooves. In time all parts of the slope are grooved. The slope retreats by repeated incisions, each of about the same depth and so disposed as to reduce rather than emphasize inequalities.

gumbo *1.* A name current in western and southern states for those soils that yield a sticky mud when wet. In southwest Missouri a puttylike clay associated with lead and zinc deposits. In Texas, a clay encountered in drilling for oil and sulfur. *2.* The stratified portion of the lower till of the Mississippi Valley.

gumbotil *1.* Leached deoxidized clay containing siliceous stones; the product of thorough chemical decomposition of clay-rich till. *2.* Gray to dark-colored, thoroughly leached, nonlaminated, deoxidized clay, very sticky and breaking with starchlike fracture when wet, very hard and tenacious when dry. Chiefly the result of weathering of drift.

G unit Gravity unit, *q.v.*

Günz The oldest of the four classical glacial stages of Europe, sometimes called the First Glacial Stage. It is now known that there were earlier glaciations in the Pleistocene.

gut *1.* A narrow passage such as a strait or inlet. *2.* A channel in otherwise shallower water, generally formed by water in motion.

Gutenberg discontinuity The seismic-velocity discontinuity marking the mantle-core boundary, at which the velocities of P-waves are reduced and S-waves disappear. It probably reflects the change from a solid to a liquid phase.

guyot (gē'yō) Tablemount, *q.v.*

Gymnospermae Class of the Spermatophyta or Pteropsida; plants whose seeds are not enclosed in an ovary.

gyprock *1.* A rock composed chiefly of gypsum. *2.* A driller's term for a rock of any kind in which he has trouble in making holes.

gypsum Alabaster; selenite; satin spar. A mineral, $CaSO_4.2H_2O$. Monoclinic. A common mineral of evaporites. Used in the manufacture of plaster of Paris.

gypsum flower Curved, twisted crystal growths of gypsum resembling flowers.

gypsum test plate An accessory to the polarizing microscope which gives a full wavelength path difference (retardation).

gyre A great closed ringlike system of ocean currents which rotates in circular motion in each of the major ocean basins. Gyres located in the northern hemisphere turn clockwise while those in the southern hemisphere rotate counterclockwise.

gyttja *1.* A sapropelic black mud in which the organic matter is more or less determinable. *2.* A natural solid hydrocarbon, tasmanite.

H

habit 1. *Crystallog:* The characteristic shape, as determined by the crystal faces developed and their shapes and relative proportions. 2. *Biol:* The mode of life of an organism in contrast to its habitat, *q.v.*

habitat The environment in which the life needs of a plant or animal are supplied.

hachure A short line used in drawing and engraving, especially in shading and denoting different surfaces as in map drawing to represent slopes of the ground.

hackly Showing jagged points in fracture.

hadal That depth zone of the oceans below 6500 meters.

hade 1. The angle of inclination of a vein, fault, or lode measured from the vertical. 2. To deviate from the vertical; said of a vein, fault, or lode.

Haeckel's law Recapitulation in the sense that each individual recapitulates the complete morphology of some ancestral stage.

half life The time period in which half the initial number of atoms of a radioactive element disintegrate into atoms of the element into which they change directly.

halide A compound characterized by a halogen such as chlorine, iodine, or bromine as the anion.

halite Rock salt. A mineral NaCl. Isometric. A common mineral of evaporites.

halloysite A clay mineral related to kaolinite and with essentially the same chemical composition, $Al_2Si_2O_5(OH)_4.2H_2O$. Crystals observed under the electron microscope are slender hollow tubes. Water of hydration is irreversibly lost under atmospheric dehydration. The level of hydration should be specified as a prefix, e.g., fully hydrated halloysite.

halmyrolysis; halmyrosis 1. A group name for the processes by which ions are removed from solution in sea water. 2. The chemical rearrangements and replacements that occur while the sediment is still on the sea floor. 3. Submarine weathering of sediment or rock.

halo A crescent or ring partly or entirely surrounding a central area characterized by values of opposite sign. Encountered principally in magnetic and geochemical surveys.

halogen Any one of the elements in Group VIIb of the periodic table (F,Cl,Br,I).

hammada Plateau in a desert region whose rocky surface has been denuded by wind erosion.

hammock A hummock. Local in Southeast and on the Gulf Coast.

hand level A small leveling instrument in which the spirit level is so mounted that the observer may view the bubble at the same time that he observes an object through the telescope.

hand specimen *Petrol:* A piece of rock trimmed to a size, usually 2.5×7.5×10.0 cm. for megascopic study and for preservation in a reference or study collection.

hanging 1. The hanging wall; the rock on the upper side of a mineral vein or deposit. 2.

Sticking or wedging of part of the charge in a blast furnace. *See* HANG, 2

hanging glacier *1.* A comparatively small glacier on an incline so steep that the forefoot of the ice constantly breaks off and falls downward. *2.* A small glacier terminating on a steep slope or at the lower end of a hanging valley. *See* GLACIERET

hanging side; hanging wall; hanger The wall or side above the ore body.

hanging valley *1.* A valley the floor of which is notably higher than the level of the valley or shore to which it leads. *2.* A tributary valley whose floor is higher than the floor of the trunk valley in the area of junction.

hanging wall The mass of rock above a fault plane, vein, lode, or bed of ore.

hard *1.* Containing certain mineral salts in solution, especially calcium carbonate; said of water having more than eight or ten grains of such matter to the gallon. *2.* In ceramics, requiring great heat: said of muffle colors in porcelain decoration. *3.* Solid, compact, difficult to break or scratch. *See* HARDNESS SCALE

hard coal Anthracite.

hardness *1.* Resistance to scratching or abrasion. The brittle hardness of the mineralogist differs from the penetration (ductile) hardness of the metallurgist. *2.* Quality of water that prevents lathering because of calcium and magnesium salts which form insoluble soaps.

hardness scale The empirical scale by which the hardness of a mineral is determined as compared with a standard. The Mohs scale is as follows: 1. Talc; 2. Gypsum; 3. Calcite; 4. Fluorite; 5. Apatite; 6. Orthoclase; 7. Quartz; 8. Topaz; 9. Corundum; 10. Diamond.

hardpan *1.* A hard impervious layer, composed chiefly of clay, cemented by relatively insoluble materials, does not become plastic when mixed with water, and definitely limits the downward movement of water and roots. It can be shattered by explosives. *2. Placer min:* Applied to layers of gravel occurring a few feet below the surface and cemented by limonite.

hard radiation Radiation of high energy.

hard rock Rock which requires drilling and blasting for its economical removal. Loosely used to distinguish igneous and metamorphic from sedimentary rock.

hard rock geology Geology of igneous and metamorphic rocks.

hard water Water containing dissolved salts of calcium and magnesium; it reacts with soap to produce insoluble precipitates that prevent lathering.

Harker diagram *See* VARIATION DIAGRAM

harmonic folding Folding in which, with depth, there are no sudden changes in the form of the folds. Contrasts with disharmonic folding, *q.v.*

harmonic relations Parallelism in beds or planes as inclinations vary.

harpolith Large, sickle-shaped intrusions injected into previously deformed strata, then, with the host rocks, stretched horizontally in the direction of maximum orogenic displacement.

Hartman's law The acute angle between two sets of intersecting shear planes is bisected by the greatest principal stress axis, whereas the obtuse angle is bisected by the least principal stress axis.

harzburgite A peridotite composed of olivine and subordinate orthopyroxene with accessory opaque oxides. *Syn:* SAXONITE

hausmannite A mineral, Mn_3O_4.

Hawaiian eruption A type of volcanic eruption in which great quantities of extremely fluid basaltic lava are poured out, mainly issuing in lava fountains from fissures on the flanks of a volcano. Explosive phenomena are rare, but much spatter and scoria are piled into cones and mounds along the vents. Characteristic of shield volcanoes.

Hawaiian-type bomb Pancake-shaped bomb.

haystack Rounded, conical-shaped hill developed as a result of solution. *Syn:* COCKPIT; HUM; PEPINO; MOGOTES

head *1.* A comparatively high promontory with either a cliff or steep face. It extends into a large body of water, such as a sea or lake. An unnamed head is usually called a headland. *2.* Pressure of a fluid upon a unit area due to the height at which the surface of the fluid stands above the point where the pressure is determined. Usually expressed as pounds per square inch, and sometimes as actual feet of head or fluid column. *3.* The section of surf rip current which has widened out seaward of the breakers. *4.* Those features and structures in consolidated and unconsolidated soils which are clearly the result of slow flow, from higher to lower ground, while oversaturated with water from melting snow or ice, rain, or lines of springs or seepages.

head erosion Headward erosion; headwater erosion, *q.v.*

headland *1.* Any projection of the land into the sea; generally applied to a cape or promontory of some boldness and elevation. *See* CAPE; HEAD; PROMONTORY; TONGUE. *2.* In soil conservation: (1) The source of

a stream. (2) The water upstream from a structure.

headwall The steep, wall-like cliff at the back of a cirque.

headward erosion Head erosion; headwater erosion. A valley is lengthened at its upper end and is cut back by the water which flows in at its head, the direction being determined by the greatest column of water which enters it. This is called headward erosion.

headwater erosion Head erosion; headward erosion, *q.v.*

head wave A critically refracted seismic wave.

heat budget Amount of heat seasonally absorbed and given off by a body of water.

heat budget, annual (of a lake) The amount of heat necessary to raise its water from the minimum temperature of winter to the maximum summer temperature.

heat capacity The amount of heat required to raise the temperature of the system one degree absolute. Heat capacity at constant pressure refers to the heat required to raise the system one degree with the system held at a constant pressure. Heat capacity is an extensive quantity.

heat conductivity Thermal conductivity, *q.v.*

heat content Enthalpy, *q.v.*

heat flow Dissipation of heat coming from within the earth by conduction or radiation at the surface; average about 1.2×10^{-6} cal./cm.^{2}/sec.

heat gradient Change in temperature of the earth with depth; approximately 30° C. per kilometer in the upper part of the crust.

heave *1.* In faulting, the horizontal component of the dip separation; that is, the apparent horizontal component of displacement of a disrupted index plane on a vertical cross section, the strike of which is perpendicular to the strike of the fault. 2. A rising of the floor of a mine caused by its being too soft to resist the weight on the pillars.

heaving shale A condition in which shale squeezes into a drill hole. Often this occurs adjacent to faulted traps where the shale is under considerable pressure, though gas pressure within the formation and the nature of clay minerals in the shale have an important contributing effect.

heavy liquids *Petrol:* A general term applied to a group of heavy organic liquids, inorganic solutions, and fused salts (heavy melts) used for the determination of the specific gravity of mineral particles, or for the separation of minerals having, respectively, lower and higher specific gravities (or densities) than the liquid used. Examples are bromoform, methylene iodide, and clerici solution.

heavy metals Metallic elements with high molecular weights, generally toxic in low concentrations to plant and animal life. Such metals are often residual in the environment and exhibit biological accumulation. Examples include mercury, chromium, cadmium, arsenic, and lead.

heavy minerals The accessory detrital minerals of a sedimentary rock, of high specific gravity, which are separated in the laboratory from minerals of lesser specific gravity by means of liquids of high density, such as bromoform. *Igneous petrol:* Mafic minerals.

heavy spar Barite.

hectare Metric unit of land measure, 10,000 square meters; 2.471 acres.

hectorite A clay mineral of the montmorillonite (smectite group), containing magnesium and lithium.

hedenbergite *See* PYROXENE

hedreocraton Persistent craton that strongly influenced later continental development.

height of instrument (H.I.) *1.* In spirit leveling: The height of the line of sight of a leveling instrument above the adopted datum. *2.* Stadia surveying: The height of the center of the telescope (horizontal axis) of transit or telescopic alidade above the ground or station mark. *3.* Trigonometrical leveling: The height of the center of the theodolite (horizontal axis) above the ground or station mark. *4.* The elevation of the center of the surveying telescope.

height of wave The vertical distance between a crest and the preceding trough. *See* SIGNIFICANT WAVE HEIGHT

helictite A distorted twiglike lateral projection of calcium carbonate, found in caves, etc. *Cf.* STALACTITE; STALAGMITE

heliotrope An instrument composed of one or more plane mirrors, so mounted and arranged that a beam of sunlight may be reflected by it in any desired direction.

Placed over a survey station, a heliotrope is used to direct a beam of sunlight toward a distant survey station, where it can be observed with a theodolite. It provides an excellent target in observing horizontal directions, such targets having been observed on at distances of over 241 kilometers and approaching 322 kilometers.

Helmholtz free energy A thermodynamic potential that is a function of temperature and volume. This potential is useful in determining the course of constant volume isothermal processes.

hematite A mineral, Fe_2O_3. Hexagonal rhombohedral. The principal ore of iron.

hemera The time interval corresponding to the acme of a taxon. This is not a good time unit because an acme (EPIBOLE, *q.v.*) may be expected to vary in time value from place to place.

Hemichordata Subdivision of the Protochordata or Chordata consisting of animals with a notochord only in the preoral region and with three primary segments of the coelom in the adult.

hemidome That form in a crystal composed of two parallel domatic planes in the triclinic, or of two parallel orthodomatic planes in the monoclinic system of crystallization. *Obs.*

hemihedral *Crystallog:* Having a lower grade of symmetry than, and only half as many faces as, the corresponding form of full or normal symmetry for the system. HEMIMORPHIC; TETARTO-HEDRAL

hemimorphic *Crystallog:* Having no transverse plane of symmetry and no center of symmetry, and composed of forms belonging to only one end of the axis of symmetry.

hemimorphite Calamine. A mineral, $Zn_4Si_2O_7(OH).H_2O$. Orthorhombic. A common secondary mineral; an ore of zinc.

hemiprism A form produced by two parallel planes cutting the two lateral axes in the triclinic system of crystallization *Obs.*

Henry's law The amount of a gas absorbed by a given volume of a liquid at a given temperature is directly proportional to the pressure of the gas. It is applicable only to dilute solutions.

heptane A liquid hydrocarbon of the paraffin series, formula C_7H_{16}.

Hercynian orogeny The Late Paleozoic orogenic era of Europe, extending through the Carboniferous and Permian; synonymous with the Variscan orogeny.

hercynite A mineral, a member of the spinel group, $(Fe,Mg)Al_2O_4$. Isometric.

heredity *1.* The transmission of characters from parents to offspring; the science which deals with this process is genetics. *2.* That organic relation between successive generations that secures persistence of characters and yet allows new ones to emerge.

herringbone cross-lamination Under conditions of frequently shifting current direction, as along the littoral zone, thin layers of sand are cross-laminated in opposite direction in alternating layers, giving rise to a "herringbone structure" if viewed in transverse section.

herringbone texture A long narrow spine, with more or less symmetrically arranged parallel zones of different composition on either side of it.

Hertz Unit of frequency. Hertz=hz=cycles per second.

heteroblastic A term applied to designate rocks in which the essential constituents are of two distinct orders of magnitude.

heterodesmic Bonded in more than one way; said of atoms in crystals.

heterogeneous *1.* A characteristic of a medium or a field of force which signifies that the medium has properties which vary with position within it. *Ant:* HOMOGENEOUS. *2.* Differing in kind; having unlike qualities; possessed of different characteristics; opposed to homogeneous. *3. Paleobot:* Lacking uniformity in kind of part or organ.

heterogeneous equilibria More than one homogeneous system interacting. *Syn:* PHASE EQUILIBRIA, *q.v.*, each phase being a homogeneous system within itself.

heteromorphism The phenomenon whereby two magmas of

identical chemical composition may crystallize into two different mineral aggregates as a result of different cooling histories.

heterozygous Having differently constituted paired chromosomes; a heterozygous individual will not exhibit recessive characters.

heulandite A member of the zeolite mineral group, $(Na,Ca)_{4-6}Al_6$ $(Al,Si)_4Si_{26}O_{72}.24H_2O)$. Monoclinic.

hexacoral Coral with hexagonal symmetry.

Hexacoralla One of the three main types of corals constituting a subclass distinguished by hexagonal symmetry.

hexagonal system The crystal system that has three equal axes intersecting at 120° and lying in one plane, and a fourth unequal axis perpendicular to the other three.

hexahedron The cube.

hexane A liquid hydrocarbon of the paraffin series, formula C_6H_{14}.

hexoctahedron Isometric crystal form bounded by 48 faces each cutting the three crystallographic axes at different distances.

hiatus That part of a lacuna that is represented by nondeposition.

high, structural A term applied to the upper or higher part of a dome or other anticlinal structure. *See* STRUCTURE

high-angle fault A fault with a dip greater than 45°.

high-energy environment Environment characterized by considerable current and/or wave action that prevents the settling of fine-grained sediment.

high-grade *1.* Rich ore. *2.* To steal or pilfer ore or gold, as from a mine by a miner. *See* HIGH-GRADING

high-grade ore Rich ore. *Cf.* LOW-GRADE ORE

high-grading Larceny of small particles of ore or gold by employees in a mine.

highland *1.* Includes all mountains, the heights of the land which are greater than hills, and all elevated masses of land, which are called plateaus, or tablelands. *2.* Elevated or mountainous land; an elevated region or country.

high-oblique photograph An oblique photograph which shows the horizon.

high plain *1.* An extensive area of relatively level land not situated near sea level. *2.* (Plural) In the United States, plains extending along the east side of the Rocky Mountains.

high-quartz Quartz formed at temperatures of more than 573° C. consisting of SiO_4 tetrahedra arranged in a more symmetrical lattice than those formed at lower temperatures; inversion is reversible. Beta quartz.

high-rank graywacke Feldspathic graywacke.

high-rank metamorphism Metamorphism accomplished under conditions of high temperature and pressure. *See* METAMORPHIC GRADE

high-speed layer A layer, usually sedimentary, in which the speed of wave propagation is greater than it is in at least one adjacent layer.

high tide Maximum height reached by each rising tide.

high-volatile A bituminous coal Nonagglomerating bituminous coal having less than 69% of fixed carbon (dry, mineral-matter-free) and more than 31% of volatile matter (dry, mineral-matter-free) and 14,000 or more B.t.u. (moist, mineral-matter-free).

high-volatile B bituminous coal Nonagglomerating bituminous coal having 13,000 or more, and less than 14,000 B.t.u. (moist, mineral-matter-free).

high-volatile C bituminous coal

Either agglomerating or non-weathering bituminous coal having 11,000 or more, and less than 13,000 B.t.u. (moist, mineral-matter-free).

high-water line In a strict interpretation the intersection of the plane of mean high tide with the shore. The shoreline delineated on the nautical charts of the U. S. Coast and Geodetic Survey is an approximation of the high tide line.

hill *1.* A prominence smaller than a mountain. Some hills, like some mountains, are volcanic heaps, and many, like mountains, are produced by dissection of plateaus and plains; but none are the direct result of uplift. *2.* In general, the term hill is properly restricted to more or less abrupt elevations of less than 305 meters, all altitudes exceeding this being mountains. *3.* A mass of material rising above the level of the surrounding country and culminating in a well-marked crest or summit. *4.* An arch or high place in a mine. *5.* In Scotland, the surface at a mine. *6.* In the North of England and the Midlands, an underground inclined plane.

hill creep; hillside creep Soil and any superficial material may move slowly downhill under the influence of gravity. Near the surface, the dip of strata may be modified by this creep.

hillock A small hill.

hinge Line along which maximum curvature of a fold occurs; it lies at the intersection of the axial plane and an s-surface.

hingebelt Zone of downfolding separating a frontal geosyncline from the continental foreland.

hinge fault *1.* A fault in which the movement is an angular or rotational one on one side of an axis normal to the fault plane. *2.* A fault along one side of which the wall rocks have rotated as a block with respect to the rocks on the other wall.

hinge line *1.* The line which separates the region in which a beach has been upwarped from that in which it is still horizontal. First use in this sense attributed to Frank Leverett. *2.* A node line drawn from an extinct outlet through points where a beach splits vertically because of uplifts of the outlet. *3.* A line in the plane of a hinge fault separating the part of the fault along which thrust or reverse movement has occurred from that showing normal movement. *4.* A line of abrupt flexure, usually applied to relatively gentle tiltings due to loading attending advance and retreat of continental glaciers. *5.* The line along which articulation takes place in a brachiopod.

hinterland *1.* That zone containing the beach flanks and the area inland from the coast line to a distance of five miles. *2.* The region lying behind the coast district. *3.* A subjective term referring to the relatively undisturbed terrain on the back of a folded mountain range, i.e., the side away from which the thrusting and recumbent folding appears to have taken place. *4.* The actively moving block which forces the geosynclinal sediments toward the foreland. If both blocks move equally, the distinction between foreland and hinterland breaks down. *5. Geog:* The land which lies behind a seaport or seaboard and supplies the bulk of the exports, and in which are distributed the bulk of the imports of that seaport or seaboard.

histogram Multiple-bar diagram showing relative abundances of specimens, materials, or other quantitative determinations divided into a number of regularly arranged classes.

histosol In soil classification, an order characterized by more than half organic matter in its upper 80 cm. or organic matter filling interstices.

hogback *1.* A sharp anticlinal, ridge, decreasing in height at both ends until it runs out. *2.* A ridge produced by highly tilted strata. *3.* A name applied in the Rocky Mountain region to a sharp-crested ridge formed by a hard bed of rock that dips rather steeply downward. *4.* In England, a sharp rise in the floor of a coal seam.

hog wallow Faintly billowing surface with low coalescent mounds 15 to 25 cm. higher than the depressions.

hollow *1.* A small ravine; a low tract of land encompassed by hills or mountains. *2.* A large sink. (Colloquial)

Holmes' classification A classification of igneous rocks based primarily upon the degree of saturation, and secondarily on other aspects of the mineralogical composition.

holoblast Newly grown minerals.

Holocene Recent; that period of time (an epoch) since the last ice age (Wisconsin in North America; Würm in Europe); also the series of strata deposited during that epoch.

holocoen The interactions between environmental factors and organisms, equivalent to ecosystem.

holocrystalline Applied to rocks consisting entirely of crystallized minerals and no glass. The minerals may or may not have crystal boundaries, and the rocks may be granular or porphyritic.

holocrystalline-porphyritic, *adj.* A term originated by Harry Rosenbusch for the texture of

depositional (one or more holocrystalline groundmass.

holohedral *Crystallog:* Belonging to the highest symmetry class of its crystal system.

holohyaline, *adj.* Applied to rocks consisting entirely of glass.

hological approach Analysis of a system (e.g., as to heat budget) where the system is taken as a whole and treated as a "black box" in which only input and output are considered.

holomictic lakes Lakes where bottom and surface waters are mixed periodically.

holoplankton Animals which live their complete life cycle in the floating state. *Syn:* PERMANENT PLANKTON

holosome Intertongued stratigraphic unit that may be either depositional (one or more holostromes) or hiatal (one or more hiatuses).

holostrome Stratigraphic unit consisting of beds laid down in a complete transgressive-regressive sequence including strata that may later have been removed by erosion.

holosymmetric Holohedral.

Holothuroidea One of a subdivision of the phylum Echinodermata; a free living animal with elongated, more or less cylindrical body. *Syn:* SEA CUCUMBER

holotype The single specimen chosen by the original author of a species; it is the name bearer for that species and any concept of the species must include it. A specimen so selected by a later author is no longer called a holotype, but is termed the lectotype, *q.v.,* or neotype, *q.v.*

homeoblastic A term used instead of equigranular and applied to metamorphic rocks to indicate that the texture so described is due to recrystallization.

homeomorph One of a group of

animals or taxa that are homeo-morphous.

homeomorphous Having homeo-morphy, *q.v.*

homeomorphy Like appearance for any reason of either individual organisms or taxa.

homeostasis The tendency of a system to remain at or return to normal, after or during an outside stress.

homeotype *See* HOMOTYPE

homoaxial folding Folding along parallel axes.

homoclinal shifting Because of the lack of homogeneity of struc-ture in a cuesta or homoclinal ridge the law of equal declivities does not apply. Erosion is very slow on the gentle dip slope of resistant rock; but on the steep-er obsequent slope, or escarp-ment, it is rapid. The divide formed by the crest line of the cuesta or monoclinal ridge is thus forced to migrate toward the dip, and as the general level of the surface is lowered the subsequent streams and the val-ley lowlands migrate also in the same direction. The process is termed homoclinal shifting.

homoclinal valley Monoclinal val-ley, *q.v.*

homocline *1. Geol.* A group of inclined beds of the same dip, which may be either monoclinal, one limb of a fold, or isoclinal, but whose actual relations are not determinate. Used in a more restricted sense than a monocline in that it applies to small or fragmentary areas. *2.* A structural condition in which the beds dip uniformly in one direction. *Cf.* MONOCLINE

homoeotype A specimen com-pared by a competent observer with the holotype, lectotype, or other primary type of a species.

homogeneous *1.* Of the same kind or nature; consisting of similar parts, or of elements of a like nature, opposed to hetero-geneous. *2. Geochem:* Alike in all parts; consisting of a single phase; as, homogeneous system, homogeneous mixture, homoge-neous equilibrium. *3.* A charac-teristic of a medium which sig-nifies that the physical property of every element of volume has the same value regardless of its location. *4.* Consisting through-out of identical or closely simi-lar material which may be a single substance or a mixture whose proportions and properties do not vary.

homogeneous deformation Affine deformation. Homogeneity or inhomogeneity of deformation is a mathematical concept ex-pressed in terms of affinity or nonaffinity. Affinity means the relation between two figures in the same plane that correspond to each other, point to point, and straight line to straight line. Geometrically, an affine trans-formation is one in which similar figures remain similar figures, similarly situated. The mathemat-ical expression for such a trans-formation is: $y=ax+b$ where a is not equal to zero.

homogeneous equilibria Those equilibria between various atom-ic, ionic, or molecular species within a homogeneous phase, typically in liquid or gas phases.

homogeneous phase Any phase whose chemical composition and physical state are uniform throughout, neglecting irregular-ities on an atomic scale. It may be gaseous, liquid, or crystalline.

homogenous Homologous.

homologue An organ, animal, or taxon that possesses homology.

homology Resemblance due to in-heritance from a common ances-tor.

homonym Any one of two or more identical names used to identify different organisms or objects. In taxonomy, only the oldest is a valid name. Commonly used to identify a junior homonym.

homoplastic Having homoplasy.

homoplasy Resemblance not due to inheritance from a common ancestry; includes chance similarity, parallelism, convergence, mimicry, and analogy.

homopolar bond *See* BOND, HOMOPOLAR

homopycnal inflow Homopycnal (equally dense) inflow occurs where a sediment-laden stream enters a basin filled with fluid of comparable density, as when a stream enters a fresh-water lake. The resulting delta is the classical type with top-, fore-, and bottom-set beds.

homoseism Lines drawn through all points simultaneously affected by an earthquake shock.

homotaxial The similar arrangement of lithofacies of biofacies regardless of age. A sandstone-shale-limestone-shale cycle of Eocene age is homotaxial to a like cycle of the Cambrian.

homotype A specimen compared by a competent observer with the holotype, lectotype, or other primary type of a species.

"honeycomb" coral A compound coral belonging to a Paleozoic family which has prismatic corallites so arranged as to resemble the cells of a honeycomb.

hoodoos Pillars developed by erosion of horizontal strata of varying hardness in regions where most rainfall is concentrated during a short period of the year. *See* EARTH PILLARS; PILLAR

hook The end of a spit turned toward the shore, owing to a deflection of the current that built it, or to the opposing action of two or more currents. *See* RECURVED SPIT

Hooke's law Stress is proportional to strain.

hook valley Barbed tributaries, *q.v.*

horizon *1.* A particular level, with or without thickness, e.g., fossils can be collected from a horizon, therefore that horizon has thickness. *2.* The subdivision of the soil solum or profile.

horizontal Any direction at a point which lies in the plane tangent to the gravity equipotential surface at the point.

horizontal axis (theodolite or transit) The axis about which the telescope of a theodolite or transit rotates when moved vertically.

horizontal classification That taxonomic classification in which species are arranged into superspecific units that are of the same age. Frequently results in polyphyletic, *q.v.*, higher taxonomic categories.

horizontal displacement Strike slip.

horizontal fault A fault, the dip of which is zero. *Cf.* VERTICAL FAULT

horizontal separation In faulting the distance between the two parts of a disrupted index plane (bed, vein, dike, etc.) measured in any specified horizontal direction.

horizontal slip In faulting the horizontal component of the net slip.

horn A high pyramidal peak with steep sides formed by the intersecting walls of three or more cirques, as, the Matterhorn.

hornblende A mineral of the amphibole group, $(Ca,Na)_{2-3}(Mg,Fe^{+2},Fe^{+3},Al)_5(Al,Si)_8O_{22}(OH)_2$.

hornblendite A plutonic rock composed essentially of hornblende.

horn coral A solitary coral, conical in shape, and generally belonging to the subclass Rugosa.

Syn: CUP CORAL; SIMPLE CORAL; SOLITARY CORAL

hornfels A fine-grained non-schistose metamorphic rock resulting from contact metamorphism. Large crystals may be present and may represent either porphyroblasts or relic phenocrysts. This word, however, is often employed in so wide a sense as to lose any precise meaning, and some writers have affronted the English language by using hornfels as a verb. *See* CALC-SILICATE HORNFELS; FLECKSCHIEFER; MACULOSE; PORCELANITE; SPILOSITE; SPOTTED SLATE

hornfels facies Pyroxene-hornfels facies.

hornito A small mound of driblet or spatter built on the back of a lava flow (generally pahoehoe) and formed by the gradual accumulation of clots of lava ejected through an opening in the roof of an underlying lava tube. *Syn:* DRIBLET CONE

horn quicksilver Mercurial horn ore; calomel, *q.v.*

horn silver Cerargyrite.

horse *Min:* 1. A large block of unmineralized rock included in a vein. 2. Rock occupying a channel cut into a coal bed. 3. Ridge of limestone rising from beneath residual phosphate deposit in Tennessee. 4. In structure, a large block of displaced wall rock caught along a fault, particularly a high-angle normal fault.

horse latitudes The belts of calms, light winds and fine, clear weather between the trade wind belts and the prevailing westerly winds of higher latitudes. The belts move north and south after the sun in a similar way to the doldrums.

horseshoe lake *See* OXBOW

horsetail ore Ore in fractures which diverge from a major fracture.

horst 1. A mass of earth-crust which is limited by faults and which stands in relief with respect to its surroundings. 2. A block of the earth's crust, generally long compared to its width that has been uplifted along faults relative to the rocks on either side. 3. A pendant with the connection portion smaller than the body. A knobby ledge of limestone beneath a thin mantle of soil.

host rock A rock body, or wall rock enclosing an epigenetic mineral deposit.

Hotchkiss superdip A modification of the dip needle which is used as a magnetometer. It measures variations in the total field. *See* DIPPING COMPASS; DIP NEEDLE

hot spot Localized melting region in the mantle below the base of the lithosphere, of the order of a few hundred kilometers in diameter and persistent over at least a few tens of millions of years, and whose existence is inferred from volcanic activity above it.

hot spring A thermal spring whose water has a higher temperature than that of the human body (above 98° F.).

hourglass valley A valley locally constricted, as where it cuts through a fault scarp or a former drainage divide, so that it is much narrower than on either side.

Hubble constant Relates the rate of nebular recession to distance from a center.

Hudsonian orogeny A time of plutonism, metamorphism, and deformation during the Precambrian in the Canadian Shield, dated radiometrically as between 1640 and 1820 m.y. ago.

huebnerite *See* WOLFRAMITE

huerfano [*Sp.* Pronounced "warefa-no."] 1. Tejon (disk-shaped) and huerfano (orphan) are used for circumscribed eminences. The latter is applied especially to solitary eminences standing far away from kindred masses. 2. A hill or mountain of older rock, entirely surrounded, but not covered, by any kind of later sedimentary rock. Used occasionally in southwestern United States. *Syn:* LOST MOUNTAIN; ISLAND HILL

Hughes balance A low-frequency (500–2000 cycles) induction balance used in early treasure or pipe finders.

hum *1.* A residual standing above a recently eroded limestone surface. *2.* A rounded, conical-shaped hill resulting from solution. *Syn:* COCKPIT; HAYSTACK; PEPINO; MOGOTES

humic Derived from plants, carbonaceous. *Cf.* BITUMINOUS

humic acid *1.* Any of the various complex organic acids supposedly formed by the partial decay of organic matter. An indefinite term, but widely used. *2.* Black, acidic, organic matter extracted from soils, low-rank coals, and other decayed plant substances by alkalis.

humidity The condition of the atmosphere in respect to water vapor.

humidity, relative The ratio of the actual amount of water vapor present in the portion of the atmosphere under consideration to the quantity which would be there if it were saturated.

hummock *1.* A mound or knoll. Ice is said to be hummocky when it consists of the fragments and dislocated parts of a previously existing sheet, refrozen together, and thus producing a rugged surface. *2.* A small elevation; hillock. *3.* A pile or ridge of ice on an ice field. *4.* A more or less elevated piece of ground rising out of a swamp, often densely wooded; an area of deep,

rich soil, usually covered by hardwood vegetation; a small elevation; a hillock.

hummocky moraine Area of glacial knob and kettle topography; may have been produced by either live or stagnant ice.

humus Substance of organic origin that is fairly but not entirely resistant to further bacterial decay. It is black, has a higher carbon content, commonly 52–58%, and lower nitrogen content than the original material. It may accumulate subaerially in soil or subaqueously in sediment.

humus layer The top portion of the soil which owes its characteristic features to its content of humus. The humus may be incorporated or unincorporated in the mineral soil.

Huronian The next to the oldest provincial Precambrian Series of the Canadian Shield.

Huygens' principle An important principle of wave propagation involving the concept that every point on an advancing wave front may be regarded as the source of a wavelet and that a later wave front is the envelope which may be drawn tangent to all the wavelets.

hyacinth A transparent red, or brownish, variety of zircon, sometimes used as a gem.

hyaline Transparent, like glass.

hyalite A variety of opal which occurs in clear globular or botryoidal forms resembling drops of melted glass.

hyalocrystalline, *adj.* A textural term applied to porphyritic rocks in which phenocrysts and groundmass are equal or nearly equal in amount; the ratio of crystals to groundmass being between 5 to 3 and 3 to 5.

hyalo-ophitic, *adj.* An intersertal texture in which the basis is more abundant and not so much separated by the crystals as in the usual intersertal texture.

hyalopilitic, *adj.* A textural term applied to igneous rocks consisting of needlelike microlites in a glassy groundmass. Pilotaxitic is the same texture without glass. When the amount of glass is decreased and the number of crystals is increased, the texture becomes intersertal, although that texture includes also rocks in which the interstitial material is not glass.

hybrid rocks Rocks of supposed heterogeneous origin. The term was originally used to refer to igneous rocks formed by the mixing of two compositionally contrasted primary world-wide magmas; it is now used in a more general sense to include all igneous rocks formed through the mixing of materials from several sources, as by assimilation of solid igneous rocks by later intrusions from the same source, or assimilation of country rocks. In this latter sense synonymous with contaminated rocks. Hybridization, *n.*

hydatogenic *Geol:* Derived from or modified by substances in a liquid condition. Said of the genesis of ores and other minerals. Contrasted with pneumatogenic.

hydrargillite Gibbsite, *q.v.*

hydrate A compound formed by the union of water with some other substance, and represented as actually containing water; e.g., gypsum, $CaSO_4.2H_2O$. Less properly, a hydroxide; e.g., calcium hydrate, $Ca(OH)_2$.

hydration The chemical combination of water with another substance.

hydraulic *1.* Of or pertaining to fluids in motion; conveying or acting, by water; operated or moved by means of water, as, hydraulic mining. *2.* Hardening or setting under water, as, hydraulic cement.

hydraulic cement Cement formerly produced by burning impure limestone (waterlime) which contains proper proportions of alumina and silica; now supplanted by Portland cement.

hydraulic conductivity Ratio of flow velocity to driving force for viscous flow under saturated conditions of a specified liquid in a porous medium.

hydraulic discharge Discharge of ground water in the liquid state directly from the zone of saturation upon the land or into a body of surface water through springs or artificial openings such as wells.

hydraulic fracturing A general term, for which there are numerous trade or service names, for the fracturing of rock in an oil or gas reservoir by pumping a fluid under high pressure into the well. The purpose is to produce artificial openings in the rock in order to increase permeability.

hydraulic gradient *1.* Pressure gradient. As applied to an aquifer it is the rate of change of pressure head per unit of distance of flow at a given point and in a given direction. *2.* As applied to streams, the slope of the energy grade line, or slope of line representing the sum of kinetic and potential energy along the channel length. It is equal to the slope of the water surface in steady, uniform flow. *3.* A vector point function equal to the decrease in hydraulic head per unit distance in direction of greatest decrease in rate.

hydraulic jump *1.* In fluid flow, a change in flow conditions accompanied by a stationary, abrupt turbulent rise in water level in the direction of flow. *2.* A type of stationary wave.

hydraulic lime A variety of calcined limestone which, when pulverized, absorbs water without swelling or heating and affords a paste or cement that hardens under water.

hydraulic limestone A limestone which contains some silica and alumina, and which yields a quicklime that will set or form a firm, strong mass under water, as in hydraulic cements.

hydraulic mining A method of mining in which a bank of gold-bearing earth or gravel is washed away by a powerful jet of water and carried into sluices, where the gold separates from the earth by its specific gravity.

hydraulic profile The vertical section of the piezometric surface of an aquifer.

hydraulic radius *Hydraul.* and *Geomorph:* The ratio of the cross-sectional area of a stream to its wetted perimeter. *Symbol: R. Syn:* HYDRAULIC MEAN DEPTH

hydraulics That branch of science or of engineering which treats of water or other fluid in motion, its action in rivers and canals, the works and machinery for conducting or raising it, its use in driving machinery, etc.

hydrobiotite A clay mineral composed of mixed layers of biotite and vermiculite.

hydrocarbon A compound containing only the two elements carbon and hydrogen.

hydrochemical prospecting Prospecting guided by the trace element content of ground and surface water.

hydroclastic Clastic through the agency of water; said of fragmental rocks deposited by water.

hydrodynamic Of or relating to the force or pressure of water or other fluids.

hydroelectric The use of water under the force of gravity to turn turbines to produce electricity.

hydroelectric reservoir A reservoir for which the primary use of water in storage is for the production of electricity.

hydroexplosion A general term for a volcanic explosion caused by the generation of steam from any body of water. It includes phreatic, phreatomagmatic, submarine, and littoral explosions.

hydrofracturing Process of increasing the permeability of strata near a well by pumping in water and sand under high pressure. Hydraulic pressure opens cracks and bedding planes and sand introduced into these serves to keep them open when pressure is reduced.

hydrogenation theory Because petroleum contains approximately 5% more hydrogen than is usually found in marine organic matter, one theory for the origin of petroleum is that hydrogen—given off by rocks through different agencies—combined with organic matter contained in beds relatively nearer the surface. The hydrogen might also have been of organic origin.

hydrogen bond Force uniting hydrogen atoms to other atoms of greater electronegativity.

hydrogen escape The concept that much hydrogen has been lost from the atmosphere by escape from the atmosphere into space.

hydrogenesis A process of natural condensation of moisture in the air spaces in the surface soil or rock.

hydrogen ion concentration (pH) A measure of the acidity of a solution. Usually given as the "pH" of the solution, which is the logarithm, to the base 10, of the reciprocal of the concentration of hydrogen ions in grams (H^+) per liter. The actual occurrence of the hydrogen ions as hydrated ("hydronium," H_3O^+) ions in the liquid is immaterial.

hydrogen sulfide (H_2S) A malodorous gas made up of hydrogen and sulfur with the characteristic odor of rotten eggs. It is emitted in the natural decomposition of organic matter and is also the natural accompaniment of advanced stage of eutrophication.

hydrogeochemistry Geochemistry of water.

hydrograph A graph showing stage, flow, velocity, or other property of water with respect to time.

hydrography *1.* The description of the sea, lakes, rivers, and other aqueous portions of the earth's surface. *2.* That branch of geography which treats of the waters that form part of the surface of the terraqueous globe, as streams, rivers, lakes, seas, and the great oceans.

hydrolith An aqueous rock that is chemically precipitated, such as rock salt or gypsum.

hydrologic cycle The complete cycle of phenomena through which water passes, commencing as atmospheric water vapor, passing into liquid and solid form as precipitation, thence along or into the ground surface, and finally again returning to the form of atmospheric water vapor by means of evaporation and transpiration. *Syn:* WATER CYCLE

hydrology The science that relates to the water of the earth.

hydrolysate; hydrolyzate Sediments consisting partly of chemically undecomposed, finely ground rock powder and partly of insoluble matter derived from hydrolytic decomposition during weathering. Bauxites, clays, shales, etc., belong to this class.

hydrolysis Chemical process in which a salt combines with water to form an acid and a base.

hydrolyzate Hydroxide resulting from hydrolysis; may be precipitated as sediment.

hydrometamorphism The alteration of rocks by the addition, subtraction, or exchange of material brought or carried in solution by water, without the influence of high temperature or pressure.

hydrometer A tubular device made of glass with the lower end weighted, graduated in specific gravity, degrees API, or other units, designed to determine the gravity of liquids by the depth to which the hydrometer sinks when immersed.

hydromica A term of rather loose usage applied to illite and sericite. *See* ILLITE

hydromorphic soils Various bog and marsh soils formed under local conditions.

hydromuscovite Illite; hydromica.

hydrophilic Having strong affinity for water; said of colloids which swell in water and are not easily coagulated.

hydrophobic Lacking strong affinity for water; said of colloids whose particles are not highly hydrated and which coagulate easily.

hydrophone A pressure-sensitive commonly magnetostrictive or piezoelectric device used to detect P-waves in marine seismic surveys.

hydroplasticity Plasticity that results from the presence of pore water in sediments.

hydropower The energy of moving water, especially rivers, but also ocean currents, tides, and waves.

hydroscope An instrument for detecting moisture, especially in the air.

hydrosol A colloidal system in which water is the dispersion medium.

hydrosphere *1.* The water portion of the earth, as distinguished from the solid part which is called the lithosphere, *q.v.* 2. In a more inclusive sense, the water vapor in the atmosphere, the sea, the rivers and the ground waters. 3. The liquid and solid water that rests on the lithosphere, including the solid, liquid, and gaseous materials that are suspended or dissolved in the water.

hydrostatic head The height of a vertical column of water, the weight of which, if of unit cross section, is equal to the hydrostatic pressure at a point.

hydrostatic level Static level. That level which for a given point in an aquifer, passes through the top of a column of water that can be supported by the hydrostatic pressure of the water at that point.

hydrostatic pressure The pressure exerted by the water at any given point in a body of water at rest.

hydrostatics That branch of physics which relates to the pressure and equilibrium of liquids, as, water, mercury, etc.; the principles of statics applied to water and other liquids.

hydrostatic stress State of stress in which all principal stresses are equal.

hydrothermal An adjective applied to heated or hot aqueous-rich solutions, to the processes in which they are concerned, and to the rocks, ore deposits, and alteration products produced by them. Hydrothermal solutions are of diverse sources, including magmatic, meteoric, and connate waters.

hydrothermal alteration Those phase changes resulting from the interaction of hydrothermal stage fluids ("hydrothermal solutions") with preexisting solid phases, such as kaolinization of feldspars, etc. Also includes the chemical and mineralogical changes in rocks brought about by the addition or removal of materials through the medium of hydrothermal fluids, e.g., silicification.

hydrothermal deposit A mineral deposit formed in rock at high temperatures and pressures from hydrothermal fluids.

hydrothermal stage A stage in the normal sequence of crystallization of a magma containing volatiles, at which time the residual fluid is strongly enriched in water and other volatiles. The exact limits of the stage are variously defined by different authors, in terms of either phase assemblage (one vs. two liquids, gas vs. liquid), temperature, composition, or vapor pressure. Most definitions place it as the last stage of igneous activity, presumably coming at a later time (and hence lower temperature) than the pegmatitic stage.

hydrothermal synthesis Mineral synthesis in the presence of water at elevated temperatures.

hydrothermal water Warm water ascending from a deeper zone within the earth; needs not be exclusively magmatic.

hydroxide A compound of an element with the radical or ion, OH, as, sodium hydroxide, NaOH.

hydroxylapatite *See* APATITE

hydrozincite A mineral, $Zn_5(CO_3)_2(OH)_6$, occurring in massive fibrous, earthy, or compact incrustations as a result of alteration of primary zinc minerals. Zinc bloom.

Hydrozoa A class of Coelenterata. An aquatic, usually marine, animal of small size, somewhat more highly specialized than the sponges; usually colonial; some build a skeletal deposit of calcium carbonate; rarely preserved as fossils.

hygrometer An instrument or apparatus for measuring the degree of moisture of the atmosphere.

hygroscopic Having the property of readily absorbing moisture from the atmosphere.

hygroscopic coefficient The amount or percentage of water absorbed and held by a material in a saturated atmosphere.

hygroscopic water Water which is so tightly held by the attraction of soil particles that it cannot be removed except as a gas, by raising the temperature above the boiling point of water.

hypabyssal A general adjective applied to minor intrusions such as sills and dikes, and to the rocks that compose them, which have crystallized under conditions intermediate between the plutonic and extrusive classes, being distinguished from these types in some cases by texture and in others only by mode of occurrence.

hypabyssal rocks Igneous rocks that have risen from the depths as magma but solidified mainly as minor intrusions such as dikes and sills before reaching the surface.

hyperfusible Any substance, such as water, occurring in the end-stage magmatic fluids that serves to lower melting ranges. *Syn:* HYPERFUSIBLE COMPONENT

hyperpycnal inflow Hyperpycnal (more dense) inflow occurs where the sediment-laden fluid flows down the side of the basin and then along the bottom as a turbidity current, with vertical mixing inhibited because the dense fluid seeks to remain at the lowest possible level. Delta formation by such flow is most frequent at the mouth of submarine canyons.

hypersthene A mineral of the pyroxene group, $(Mg,Fe)SiO_3$. Orthorhombic.

hypidiomorphic 1. A general term applied to those minerals of igneous rocks that are bounded only in part by their characteristic crystal faces, i.e., crystal forms intermediate between allotriomorphic (anhedral) and idiomorphic (euhedral). *Syn:* SUBHEDRAL; HYPAUTOMORPHIC. 2. A textural term applied to granular plutonic rocks in which there are few idiomorphic minerals, most of the constituents being hypidiomorphic or subhedral. The texture of such rocks is said to be hypidiomorphic-granular or hypautomorphic-granular.

hypocrystalline Applied to igneous rocks that consist partly of crystals and partly of glass. *Syn:* HYPOHYALINE

hypogene, *adj.* 1. Used by Sir Charles Lyell and intended as a group name for plutonic and metamorphic classes of rocks, i.e., rocks formed within the earth. 2. Used by Sir Archibald Geikie for geological processes originating within the earth. 3. Applied to mineral or ore deposits formed by generally ascending waters. Contrasted with supergene, *q.v.*

hypolimnion That part of a lake below the thermocline.

hyponym When a name is first applied to a group, the group should be so described, or illustrated, or referred to preserved plants, that its identity can be recognized by other botanists. A name not so described or identified is a hyponym and nonvalid for the group for which it was intended.

hypopycnal inflow Hypopycnal (less dense) inflow occurs where sediment-laden fluid moves out over the surface of denser fluid filling the basin, as in the case of a stream discharging into the ocean. If the magnitude of discharge is small, a lunate bar forms off the outlet; if moderate to large, a cuspate, arcuate, or bird-foot type of delta will form.

hypothermal Originating at relatively high temperatures (300°–500° C.).

hypothermal deposits Deposits formed at high temperatures and pressures in and along openings in rocks by deposition from fluids derived from consolidating igneous rocks.

hypothetical reserve A term used by the United States Geological Survey for those reserves not yet known, but which should accrue from knowledge of general geological conditions and increased technology. Such concepts are hazardous and planning should not consider them.

hypotype A described or figured specimen used in extending or correcting the knowledge of, or in other publication of, a species.

hypsithermal interval Postglacial warm interval extending from about 7000 to 600 B.C. responsible for the last 6-foot (1.8288 meters) eustatic rise of seal level. *Syn:* POSTGLACIAL CLIMATIC OPTIMUM; THERMAL MAXIMUM

hypsographic map A topographic map on which elevations are referred to a sea-level datum.

hypsometric Relating to elevation above a datum, usually sea level.

hysteresis *1.* A lag in the return of an elastically deformed specimen to its original shape after the load has been released. *2.* An effect, involving energy loss, found to varying degrees in magnetic, electric, and elastic media when they are subjected to variation by a cyclical applied force. In such media the polarization or stress is not a single valued function of the applied force, or, stated in another way, the state of the medium depends on its previous history as well as the instantaneous value of the applied force.

ice The solid state of water, specifically the dense substance formed by the freezing of liquid water or by the recrystallization of fallen snow. Ice formed by the sublimation of water vapor is usually known as snow or hoar, *q.v. See* GLACIER ICE; SEA ICE; GROUND ICE

ice age A glacial period or part of a glacial period; most frequently refers to the last glacial period.

iceberg *1.* [*Dan.* berg, mountain] Large masses of ice broken off from the lower ends of the huge glaciers, which in high altitudes force their way down to and into the sea. They float nine-tenths in the water and one-tenth out, so that for every meter of ice above water there are nine meters below. *2.* A glacier on a seacoast often stretches out its icy foot into the ocean, and, when this part is finally broken off, by the movement of the sea or otherwise, it becomes an iceberg. *3.* A mass of land ice (including shelf ice) that has broken away from its parent formation on or near the coast and either floats in the sea or is stranded on a shoal.

icecap A dome-shaped or plate-like cover of perennial ice and snow covering all of the summit area of a mountain mass so that no peaks emerge through it, or covering a flat landmass such as an Arctic island, spreading due to its own weight outward in all directions, and having an area of less than 50,000 sq. km.

ice cascade When the slope of a glacier bed increases suddenly, an ice cascade is developed, but an ice cascade has little in common with the rapids or falls of rivers. Ice fall.

ice-contact forms Stratified drift bodies such as kames, kame terraces, and eskers, deposited in contact with melting glacier ice.

ice dammed lake Whenever a continental glacier, either in advancing its front or in retiring, lies across the lines of drainage upon their downstream side, water is impounded along the ice front so as to form ice dam lakes.

iced firn A mixture of ice and firn.

ice drift Loose floating ice.

ice face Ice cliff.

ice fall That portion of a glacier which flows down a steep gradient, resulting in a zone of crevasses and seracs, *q.v.*

ice foot In polar seas the land is often bordered by a fringe of ice, called the ice foot. *See* FAST ICE

ice front *1.* The terminus of a glacier. *2.* The seaward-facing cliff of shelf ice.

ice island *1.* A large tabular ice mass, several tens of meters thick and several square kilometers in area, floating in the Arctic ice pack. Ice islands may have an undulating surface, and streams and ponds in summer. *2.* Any tabular iceberg. *3.* Ice-island iceberg. *4.* Floeberg. *5.* An island completely covered by ice and snow.

ice jam *1.* Fragments of broken river ice lodged in a narrow portion of the river channel. *2.*

Large fragments of lake ice thawed loose in early spring and shoved against the shore.

Iceland spar Transparent cleavage fragments of calcite, which, owing to its strong double refraction, are used for optical purposes.

ice layer Approximately horizontal layer of ground ice, sometimes lenticular in form.

ice lobe Glacial or glacier lobe, q.v.

ice mountain An iceberg.

ice pack Pack ice. A term applied to a phenomenon which occurs in Arctic regions, when the ice first formed on the surface of the sea becomes broken up and separate portions piled or packed on top of each other, forming masses very difficult to navigate.

ice pan A large, flat piece of sea ice, protruding several centimeters to a meter above the water, usually composed of winter ice up to one year old.

ice plateau *1.* An ice-covered highland area whose upper surface is relatively level, and whose sides slope steeply to lowlands or the ocean. *2.* Any ice sheet with a level or gently rounded surface.

ice pole The area around which the more consolidated part of the Arctic ice pack is located. In 1959 its location was about 84° N. Lat. and 160° W. Long. *Syn:* POLE OF INACCESSIBILITY

ice push The expansion ice formed on a lake or embayment of the sea which accompanies a rise in temperature. As the ice expands horizontally it pushes unconsolidated debris on the shore into ramparts or irregular ridges.

ice-push ridge Lake rampart; walled lake.

ice rafting Transport of rock particles and other materials by floating ice.

ice-scoured plain A region reduced to the condition of a plain by actual ice scour.

ice sheet A glacier forming in continuous cover over a land surface, with the ice moving outward in many directions. Continental glaciers, icecaps, and some highland glaciers are examples of ice sheets.

ice shelf Floating ice permanently attached to a landmass, as the Ross Ice Shelf in the Antarctic. Now the preferred term for ice barrier, ice front, and shelf ice, q.v.

ice tongue *1.* That portion of a valley glacier below the firn line. *2.* Outlet glaciers from an icecap. *3.* Glacial lobes, q.v.

ice vein Ice wedge.

ice wedge; ground-ice wedge Vertical, wedge-shaped vein of ground ice.

ice-wedge polygon Large-scale polygonal feature commonly outlined by shallow trenches underlain by ice wedges.

ichnofossil A trace fossil, q.v.

ichnology The study of trace fossils, those inorganic evidences of animals, such as burrows, footprints, trails, etc.

ichor, *n.* A term proposed for a granitic juice or liquor, capable of granitizing rocks, and derived from a granitic magma. *Syn:* EMANATION; MINERALIZER; RESIDUAL MAGMA

icicle A pendant, somewhat conical, shaft of ice formed by the freezing of dripping water.

Iddings' classification A classification of igneous rocks by J. P. Iddings (1913) which attempts to correlate the mineralogical classifications of Harry Rosenbusch and Ferdinand Zirkel with the chemicomineralogical C.I.P.W. or norm classification system.

iddingsite A reddish-brown mixture of silicate minerals formed by the alteration of olivine.

ideal gas Perfect gas. One which

obeys the gas laws perfectly, particularly Boyle's law. The term implies infinitely small gas particles and no interactions between particles.

ideal solutions A solution in which the forces between molecules are identical throughout. In nonideal solutions (containing several different types of molecules or ions) the interactions between unlike types are different from that between identical types, and the behavior will be nonideal.

identified resource A specific site and volume of resource in which the locality, quality, and quantity have been demonstrated and are known.

identified subeconomic resource A specific resource body that is known, but not now economic.

idioblastic An idioblastic mineral is one which shows crystal faces against an adjacent mineral.

idiogenous Said of deposits contemporaneous in origin with the rocks in which they occur, i.e., primary deposits which are constituents of the rocks in which they occur.

idiogeosyncline *1.* An intermontane trough in which sediments accumulated. *2.* Marginal basin, short lived and weakly folded.

idiomorphic Euhedral; automorphic, *q.v.*

idocrase Vesuvianite. A mineral, $Ca_{10}Al_4(Mg,Fe)_2(SiO_4)_5(Si_2O_7)_2(OH)_4$. Tetragonal. Common in contact-metamorphosed limestones.

igneous, *adj. Petrol:* Formed by solidification from a molten or partially molten state. Said of the rocks of one of the two great classes into which all rocks are divided, and contrasted with sedimentary. Rocks formed in this manner have also been called plutonic rocks, and are often divided for convenience into plutonic and volcanic rocks, but there is no clear line between the two.

igneous breccia *1.* Breccia consisting of igneous rock. *2.* Breccia produced by igneous action, includes pyroclastic, flow and contact breccias.

igneous complex Intimately associated and roughly contemporaneous igneous rocks.

igneous emanations Gases and liquids which are given off during igneous activity.

igneous facies *See* FACIES, IGNEOUS

igneous rocks Formed by solidification of hot mobile material termed magma.

igneous rock series An assemblage of igneous rocks in a single district and belonging to a single period of igneous activity, characterized by a certain community of chemical, mineralogical, and occasionally also textural properties. *See* CONSANGUINITY

ignimbrite *1.* A silicic volcanic rock forming thick, massive, compact, lavalike sheets that cover a wide area in the central part of North Island, New Zealand. The rock is chiefly a fine-grained rhyolitic tuff formed mainly of glass particles (shards) in which crystals of feldspar, quartz, and occasionally hypersthene or hornblende are embedded. The glass particles are firmly "welded" and bend around the crystals, and evidently were of a viscous nature when they were deposited. The deposits are believed to have been produced by the eruption of dense clouds of incandescent volcanic glass in a semimolten or viscous state from groups of fissures. *Syn:* WELDED TUFF. *2.* The deposit of a fiery cloud or pyroclastic flow, extensive and generally thick with well-developed prismatic jointing.

Illinoisan The third of four classical glaciations in the Pleistocene of North America.

illite A group of clay minerals abundant in sedimentary rocks and intermediate in composition between muscovite and montmorillonite, $K_y(Al,Mg,Fe)^2(Al_y,Si_{4-y})O_{10}(OH)_2$, $y<1$. Characterized by a 10-Angstrom X-ray diffraction spacing and no tendency to expand with water or other solvating agents. Glimmertone, hydromica, illidromica.

illuvial horizon The horizon of deposition in a soil.

illuviation The deposition in an underlying layer of soil (usually the B horizon) of mineral or organic matter.

ilmenite A mineral, $FeTiO_3$. Hexagonal rhombohedral. The principal ore of titanium.

ilvaite Lievrite. A mineral, $CaFe''_2Fe'''(SiO_4)_2(OH)$. Orthorhombic.

imbibition The tendency of granular rock or any porous medium to "imbibe" a fluid, usually water, under the force of capillary attraction, and in the absence of any pressure.

imbricate The shingling or overlapping effect of stream flow upon flat pebbles in the stream bed. The pebbles are inclined so that the upper edge of each individual is inclined in the direction of the current.

imbricate structure *1.* In general refers to tabular masses that overlap one another as shingles on a roof. *2. Struct. geol:* A series of thrust sheets dipping in the same direction, sometimes called shingle-block structure. Schuppen structure [*Ger.*].

immature soil *1.* A soil in which erosion exceeds the rate at which the soil develops downward. *2.* Soil that has not reached maturity as shown by lack of zonation, e.g., young alluvial soils.

immiscibility or **miscibility gap** A break in an otherwise completely miscible series between two materials, either liquid or crystalline.

immiscible Said of two or more liquid or solid solutions that are not capable of being mixed, as oil and water. The two (or more) liquids can still be in equilibrium with each other.

impact The wearing away of smaller rocks or fragments through the effect of definite blows of relatively larger fragments on the relatively smaller rocks or fragments.

impact bomb Porous mass of impactite formed by splattering; may enclose pebbles, etc.

impact crater A crater formed by the impact of an unspecified projectile, especially craters formed on the Earth or Moon surface where the nature of the impacting body is unknown.

impactite Vesicular glassy to finely crystalline material produced where a meteor has struck the earth; consists of meteoric material and slag.

impact law A formula expressing the relationship of large particle diameter and fluid density to the settling velocity of large particles in a given liquid medium.

impact slag Glassy material produced mainly by the melting of local sediment or rock where a meteorite has struck the earth.

impedance In electric circuits or measurements, the generalized resistance or opposition to current flow in other than pure direct currents; commonly includes complex components of resistance, inductance, and capacity.

impermeable; impervious *Hydrol:* Having a texture that does not permit water to move through it perceptibly under the head differences ordinarily found in subsurface water.

impervious Impermeable, *q.v.* Impassable; applied to strata such as clays, shales, etc., which will not permit the penetration of water, petroleum, or natural gas.

impoundment A body of water, such as a pond, confined by a dam, dike, floodgate, or other barrier; also the process of.

impregnated Containing metallic minerals, scattered or diffused through the mass. Properly used in referring to country rock containing minerals similar to that in the vein.

impression *1.* The form or shape left on a soft surface by materials which have come in contact with it. Usually occurs as a negative or concave feature on the top of a bed. A cast of it will sometimes be found on the base of the overlying bed. *See* IMPRINT. *2.* Results from the burial of plant parts in soils which subsequently harden into rock, in much the same manner as an imprint is left when a leaf is pressed into the surface of wet cement which is allowed to "set." Impressions thus preserve the external structural features of plant parts.

imprint *1.* The impression made on a soft surface of mud or sand by organic or inorganic materials which have come in contact with it. *Syn:* IMPRESSION. *2.* Evidence of deformative movement within rocks.

Impsonite A black asphaltic pyrobitumen with a hackly fracture and high fixed carbon content. Impsonite is derived from the metamorphosis of petroleum.

impulse A short-period force or action; in seismograph prospecting, the effect of an explosive or mechanical source of seismic waves.

Inarticulata Class of brachiopods whose shells lack articulating teeth and sockets, mostly with chitino-phosphatic shells.

inceptisol Soils on new volcanic deposits, or soils of recently deglaciated areas, or other soils that are so young they have only a slight horizon development.

incidence, angle of The angle between the normal to the wave front and the normal to a reflecting or refracting surface.

incineration The controlled process by which solid, liquid, or gaseous combustible wastes are burned. The residue produced contains little or no combustible material.

incinerator An engineered apparatus used to burn waste substances and in which all the combustion factors—temperature, retention time, turbulence, and combustion air—can be controlled.

incise Cut down into, as a river cuts into a plateau.

incised or **entrenched meander** A deep tortuous valley cut by a rejuvenated stream, the meandering course having been acquired in a former cycle.

inclination *1. Geol:* The dip of a bed, fault, vein, or other tabular body measured from the horizontal. *2. Geophys:* The angle between any direction and the vertical. *See* DIP; MAGNETIC INCLINATION

inclined contact Applied to a contact plane of oil or gas with underlying water, in which the plane slopes or is inclined. Usually due to the fact that the water is in motion because of a hydraulic gradient across the bed.

inclined extinction Oblique extinction.

inclined fold Fold whose axial plane is not vertical.

inclinometer *1.* A dipping compass. *2.* An instrument for measuring inclination or slope, as of the ground or of an embankment; clinometer.

inclosed meanders Stream meanders that are more or less

closely bordered or inclosed by the valley walls.

included gas Gases in isolated interstices in either the zone of aeration or the zone of saturation.

inclusion *1. Petrol:* A crystal or fragment of another substance or a minute cavity filled with gas or liquid enclosed in a crystal. *2.* A fragment of older rock enclosed in an igneous rock; a xenolith.

inclusions, fluid During the crystallization of minerals, or during recrystallization following fracturing, small portions of the fluids present may become trapped within the mineral grains. Most of these are small (0.001–0.01 mm.), and frequently contain a small bubble of a gaseous phase in addition to the liquid (usually hydrous) phase.

incompetent bed *1. Geol:* Not combining sufficient firmness and flexibility to transmit a thrust and to lift a load by bending; consequently, admitting only the deformation of flowage: said of strata and rock structure. *See* COMPETENT, *1. 2.* Competent and incompetent are relative terms. An incompetent bed is one that is relatively weak and thus cannot transmit pressure for any distance. *3.* A bed which, because it lacks strength or cohesiveness, is unable to lift its own or the weight above it without breaking, when it undergoes such movement as folding.

incompetent folding *See* FLOW FOLDING

incompressibility modulus Modulus of volume elasticity; modulus of elasticity, *q.v.*

incongruent melting Melting accompanied by decomposition or by reaction with the liquid, so that one solid phase is converted into another; melting to give a liquid different in composition from the original solid. For example, orthoclase melts incongruently to give leucite and a liquid richer in silica than the original orthoclase.

incongruent solution Dissolution accompanied by decomposition or by reaction with the liquid, so that one solid phase is converted into another; dissolution to give dissolved material in different proportions from those in the original solid.

incongruous drag folds Drag folds that do not have the characteristics of congruous drag folds, *q.v.*

incretion Cylindrical concretions with a hollow core.

incrustation A crust or hard coating of anything upon or within a body, as, a deposit of lime inside a steam boiler.

index contour Certain contour lines (usually every fifth) accentuated by use of a line heavier than the intervening lines and whose elevation or other value is indicated by figures along its course.

index ellipsoid Ellipsoid whose axes are proportional in length to the indices of refraction in different directions. *Syn:* INDICATRIX

index fossil *1.* A fossil characteristic of an assemblage zone and so far as is known restricted to it. *2.* A fossil, the name of which designates a biostratigraphic zone. *3.* Less desirable—a fossil that is a guide to a particular stratigraphic level. *Cf.* GUIDE FOSSIL

index map A map which shows the location of collections of related data, whether in the form of other maps, or statistical tables, or descriptions.

index mineral *Metamorph. petrol:* A mineral whose first appearance (in passing from low to higher grades of metamorphism) marks the outer limit of the zone in question.

index of refraction A characterizing number which expresses the ratio of the velocity of light *in vacuo* to the velocity of light in the substance. The conventional symbol is *n*.

index plane A surface of any bed, dike, or vein, which may be regarded as a plane and used as a base for measurement of fault movements. *Syn:* INDEX HORIZON

index zone Zone recognizable by paleontologic or lithologic characters that can be traced laterally and identifies a reference position in a stratigraphic section.

indicated ore Ore for which tonnage and grade are computed partly from specific measurements and partly from projection for a reasonable distance on geologic evidence.

indicated reserves Reserves or resources for which tonnage and grade are computed partly from specific measurements, samples, or production data and partly from projection for a reasonable distance on geologic evidence.

indicator, *n.* An erratic whose place of origin in the bedrock is known.

indicator plant *Geobot. prospecting:* Plant or tree that grows exclusively or preferentially on soil rich in a given metal or other element.

indicator vein A vein which is not metalliferous itself, but, if followed, leads to ore deposits.

indicatrix Index ellipsoid.

indigenous *1.* Originating in a specific place; *in situ. 2.* Designating rocks, minerals, or ores originating in place, as opposed to those transported from a distance.

induced magnetization Impermanent magnetization produced by an applied magnetic field; it is reversible like that in a soft iron magnet.

induction *1.* The process by which a magnetizable body becomes magnetized by merely placing it in a magnetic field. *2.* The process by which a body becomes electrified by merely placing it in an electric field. *3.* The process by which electric currents are initiated in a conductor by merely placing it in an electromagnetic field.

induction log A continuous record of the conductivity of strata traversed by a borehole as a function of depth.

indurated Rendered hard; confined in geological use to masses hardened by heat, baked, etc., as distinguished from hard or compact in natural structure.

In modern usage the term is applied to rocks hardened not only by heat, but also by pressure and cementation.

induration The process of hardening of sediments or other rock aggregates through cementation, pressure, heat, or other cause.

inequigranular A textural term applied to rocks whose essential minerals are of different orders of size, e.g., porphyries.

inexhaustible resource *See* RESOURCE CONCEPTS

infant stream A stream which has just begun the work of tearing down and carrying away the upland.

inferred reserves Reserves or resources for which quantitative estimates are based largely on broad knowledge of the geologic character of the deposit and for which there are few, if any, samples or measurements. Estimates of inferred reserves or resources should include a statement of specific limits within which the inferred material may lie.

infiltration *1.* The flow of a fluid into a substance through pores or small openings. It connotes flow into a substance in contradistinction to the word percolation, which connotes flow through a porous substance. *2.* The flow or movement of water through the soil surface into the ground. *3.* The deposition of mineral matter among the grains or pores of a rock by the permeation or percolation of water carrying it in solution. *4.* The material filling a vein as though deposited from a solution in water.

infiltration capacity The maximum rate at which the soil, when in a given condition, can absorb falling rain.

infiltration rate Maximum rate at which soil can absorb rain or shallow impounded water.

infiltration velocity Volume of water moving downward into soil per unit of area and time.

influent *1.* A tributary stream or river; affluent. *2.* A stream or stretch of a stream is influent with respect to ground water if it contributes water to the zone of saturation. The upper surface of such a stream stands higher than the water table or other piezometric surface of the aquifer to which it contributes.

infraglacial Applied to deposits formed and accumulated underneath or in the bottom parts of glaciers and ice sheets, and to the geological action of the ice upon rocks over which it flows.

infralittoral *Geol:* Below the region of littoral deposits.

infraneritic Pertaining to the marine environment in water from 120 feet (20 fathoms, 40 meters) to 600 feet (100 fathoms, 195 meters) deep.

infrared That portion of the electromagnetic spectrum with wavelengths of from 0.7 to about 1.0 micrometer, just beyond the red end of the visible spectrum.

infrastructure Structure produced at a deep crustal level, in a plutonic environment, under condi-

tions of elevated temperature and pressure, which is characterized by plastic folding and the emplacement of granite and other migmatitic and magmatic rocks.

ingrown meander A meander deepened as a result of rejuvenation of a stream course, as, from down-valley tilting.

inherent ash Ash derived from mineral constituents of vegetable material in coal rather than from accompanying sediment.

initial dip The angle of slope of bedding surfaces at the time of deposition, the contacts between layers usually being approximately parallel with the surface of deposition unless subsequently altered by differential compaction or other deformational processes.

initial open flow A term applied to the rate of flow of a gas well immediately following completion.

initial production The volume or quantity of gas or oil initially produced by a well in a certain interval of time, usually 24 hours.

injected igneous body An intrusive igneous body that is entirely inclosed by the invaded formations, except along the relatively narrow feeding channel. Examples are dikes, sills, laccoliths, phacoliths, etc. Contrasted with subjacent igneous body.

injection complex General term for ancient rocks, mostly plutonic, underlying the oldest sedimentary formations in the eastern United States.

injection folding Deformation in a plastic layer between more competent layers resulting from differential changes in thickness.

injection gneiss A gneiss whose banding is wholly or partly due to lit-par-lit injection of granitic magma. *See* COMPOSITE GNEISS; GNEISS; MIGMATITE

injection metamorphism Metamorphism accompanied by intimate injection of sheets and streaks of liquid magma in zones near plutonic rocks.

injection well Well into which water or gas is pumped in an oil field to promote secondary recovery or to maintain subsurface pressure.

inland ice Informal designation for the interior zone of a large ice cap or continental glacier.

inland seas Seas that are largely surrounded by land or shallow water so that communication with the open ocean is restricted to one or a few straits.

inlet A short, narrow waterway connecting a bay, lagoon, or similar body of water with a large parent body of water. An arm of the sea (or other body of water) that is long compared to its width and that may extend a considerable distance inland. *See* TIDAL INLET

inlier A more or less circular or elliptical area of older rocks surrounded by younger strata. Often the result of erosion of the crest of an anticline. *Ant:* OUTLIER

inner core Central part of earth's core beginning at depth of about 5000 km., probably solid.

inosilicate A class or structural type of silicate characterized by the linkage of Si-O tetrahedra into linear chains by the sharing of oxygen. Single chain minerals, e.g., pyroxenes, are characterized by the radical (SiO_3). Double chain minerals, e.g., amphiboles, are characterized by the radical (Si_4O_{11}).

in place Said of rock occupying, relative to surrounding masses, the position that it had when formed. If an ore body is continuous to the extent that it may maintain that character, then it is "in place." *See* IN SITU

inselberg Prominent steep-sided residual hills and mountains rising abruptly from plains make a landscape type rather common in Africa. The residuals are generally bare and rocky, large and small, isolated and in hill and mountain groups, and they are surrounded by lowland surfaces of erosion that are generally true plains, as distinguished from peneplains.

insequent *Geol:* developed on the present surface, but not consequent on nor controlled by the structure; said of streams, drainage, and dissection of a certain type. A type of drainage in which young streams flowing on a nearly level plain wander irregularly.

insequent stream *1.* Streams the courses of which are not due to (consequent upon) determinable factors are called insequent streams. [in (con) sequent=insequent].

inshore In beach terminology, the zone of variable width extending from the shore face through the breaker zone.

inshore current Any current in or landward of the breaker zone.

in situ In its natural position or place. *Geol:* Said specifically of a rock, soil, or fossil when in the situation in which it was originally formed or deposited. *See* IN PLACE

insolation *Meteor:* Received solar radiation, as by the earth; also rate of delivery of all direct solar energy per unit of horizontal surface.

insoluble residue Siliceous residue. Material remaining after a specimen has been dissolved in hydrochloric or acetic acid, chiefly composed of shale, chert, quartz, other siliceous material and various insoluble detrital minerals.

instability *See* METASTABLE

insulated stream A stream or reach of a stream is insulated with respect to ground water if it neither contributes water to the zone of saturation nor receives water from it. It is separated from the zone of saturation by an impermeable bed.

intake Recharge, *q.v.*

intake area (of an aquifer) An area where water is absorbed which eventually reaches a part of an aquifer that is in the zone of saturation.

integrated drainage The drainage of any initial surface becomes "integrated" as the local undrained depressions, common to all such surfaces, become incorporated into one master drainage system.

intensity (of an earthquake) A number describing the effects of an earthquake on man, on structures built by him, and on the earth's surface. The number is rated on the basis of an earthquake intensity scale.

intensity scale A standard of relative measurement of earthquake intensity. Three such systems are: the Mercalli scale, the modified Mercalli scale, and the Rossi-Forel scale. The Richter scale is a measurement of magnitude rather than of intensity.

intensive variable A thermodynamic variable that is independent of the total amount of matter in the system, such as temperature and pressure.

interbed A typically thin bed of rock material alternating with contrasting thicker beds of rock.

interbedded Occurring between beds, or lying in a bed parallel to other beds of a different material; interstratified.

intercalate, *v.* To insert among others, as, a bed or stratum of lava between layers of other material; to interstratify.

intercepts *Crystallog:* Distances cut off on axes of reference by planes.

intercept time Delay time, *q.v.*

interface Contact surface separating two different substances.

interfacial angle *Crystallog:* The angle between two crystal faces.

interfacial tension The force tending to reduce the area of contact between two liquids or between a liquid and a solid.

interference *1.* The masking of a desired signal by others arriving at very nearly the same time. *2.* The vector sum of the displacements (or velocities or accelerations) of wave trans-arriving at a point from the same source by different paths. Also used to describe the same result when waves arrive from two or more sources.

interference colors In crystal optics, the colors displayed by a birefringent crystal in crossed, polarized light. Thickness and orientation of the mineral and the nature of the light are factors that affect the colors and their intensity. The pattern or figure that a crystal displays in polarized light under the conoscope is frequently used in the optical identification of minerals. The figure consists of colored rings and curves combined with black bars.

interference figure A system of colored rings and curves combined with black bars and curves seen when a mineral grain is examined in a certain way under the microscope or other suitable optical instrument. The interference figure is due to birefringence, *q.v.* and is one of the most useful optical aids in identifying minerals.

interference ripple mark *1.* Results when two sets of symmetrical ripples are formed by two systems of waves crossing nearly at right angles. The cell-like pattern of some interference ripple marks led Edward Hitchcock to regard them as "tadpole nests." *2.* Ripples formed during a sin-

gle phase of current action by a complex current. *See* OSCILLATION CROSS RIPPLE MARK

interfinger To grade or pass from one material into another through a series of interlocking or overlapping wedge-shaped layers.

interfluve The district between adjacent streams flowing in the same general direction.

interfolding Simultaneous development of differently oriented folds that do not cross or interfere.

interformational conglomerate Those gravels and their indurated equivalents that often are present within a formation of which the constituents have a source external to the formation.

interglacial Pertaining to the time between glaciations. An interglacial age.

intergranular A textural term applied to volcanic rocks in which there is an aggregation of grains of augite, not in parallel optical continuity (as in subophitic texture), between a network of feldspar laths which may be diverse, subradial, or subparallel. Distinguished from an intersertal texture by the absence of interstitial glass, or other substances which may fill the interstices between the feldspar laths. Characteristic of certain basaltic and doleritic rocks. *Cf.* GRANULITIC

intergrowth *Petrol.* and *Mineral:* Commonly applied to a state of interlocking of different crystals due to simultaneous crystallization.

interior The country extending indefinitely inland of the hinterland.

interior basin *1.* A basin situated within the relatively stable area of a continent. *2. Geog.* An undrained basin.

interior salt domes Domes in east Texas, northern Louisiana, southern Arkansas and Missis-sippi located at a distance from the Gulf Coast salt domes. Three general groupings are: (1) in the Tyler Basin, northeast Texas, (2) on the east flank of the Sabine arch in northern Louisiana, (3) in southern Alabama, eastern Louisiana, and south-central Mississippi. *See* SALT DOME; PIERCEMENT DOME

interior sea A body of water, usually marine, lying upon a continental platform and largely surrounded by land.

interior valley A large steep-sided isolated solution valley, generally floored with alluvium. (Jamaica) *Syn:* POLJE

interlobate Situated between lobes. *Geol:* Lying between adjacent glacial lobes, as deposits.

intermediate focus earthquake Earthquake whose focus is at a depth of between 65 and 300 km.

intermediate rock An igneous rock containing between 52 and 66% SiO_2.

intermittent stream *1.* Stream which flows but part of the time, as, after a rainstorm, during wet weather, or during part of the year. *See* PERENNIAL STREAMS. *2.* One which flows only at certain times when it receives water from springs (spring fed) or from some surface source (surface fed) such as melting snow in mountainous areas.

intermontane Lying between mountains.

intermontane area Structural and topographic basin enclosed by diverging and converging mountain ranges.

intermontane glaciers Glaciers produced by the confluence of several valley glaciers and occupying a trough between separate mountain ranges.

intermontane troughs Subsiding areas in an island arc region which lie among the positive elements of the area.

intermountain Area between mountains that mark the margins of an orogen. [*Ger.*] Zwischengebirge. *Syn:* INTERMONTANE AREA

internal energy *See* FIRST LAW OF THERMODYNAMICS

internal mold *See* MOLD, INTERNAL

internal rotation Rotation of individual small particles within a rock such as might be accomplished by flowage.

internal waves Waves that occur within a fluid whose density changes with depth, either abruptly at a sharp surface of discontinuity (an interface), or gradually.

International Active Sun Years An international cooperative program for the scientific investigation of solarterrestrial phenomena during periods of maximum sunspot activity.

International Geophysical Year An international cooperative program conducted from July 1, 1957, to December 31, 1958, for the scientific investigation of geophysical phenomena.

International Hydrologic Decade An international cooperative program conducted over the ten-year period 1965–74, for the purpose of training hydrologists and setting up networks for measuring hydrologic data.

International Years of the Quiet Sun An international cooperative program for the scientific investigation of solarterrestrial phenomena during periods of minimum sunspot activity.

interpretive log A sample log based on rotary cuttings in which the geologist has attempted to portray only the rock cut by the bit at the level indicated, ignoring the admixed material from a higher level.

interrupted profile The break or

interruption in a normal stream profile where the head of the second-cycle valley after rejuvenation touches the first-cycle valley. *See* NICKPOINT

interrupted stream *1.* One which contains (1) perennial reaches with intervening intermittent or ephemeral reaches or (2) intermittent reaches with intervening ephemeral reaches. *2.* An interrupted stream flows at some places and not at others; and an intermittent stream flows at some times and not at others. *See* LOST RIVER

interrupted water table One that has a pronounced descent along a ground-water dam.

intersertal A texture of igneous rocks wherein a base or mesostasis of glass and small crystals fills the interstices between unoriented feldspar laths, the base forming a relatively small proportion of the rock.

interstadial; interstade Refers to short nonglacial periods within a glacial stage.

interstate waters According to law, waters defined as: (1) rivers, lakes, and other waters that flow across or form a part of state or international boundaries; (2) waters of the Great Lakes; (3) coastal waters—whose scope has been defined to include ocean waters seaward to the territorial limits and waters along the coastline (including inland streams) influenced by the tide.

interstice Pore. *1.* Void, *q.v.*

interstitial A term referring to the spaces between particles (e.g., the spaces between the sand grains).

interstitial deposits Deposits that fill the pores of rocks. Frequently used in place of impregnation deposits.

interstitial matrix Fine sedimentary material occurring between coarser grains; maximum size 0.02 mm.

interstitial solid solution Occurrence of foreign atoms or ions in the lattice interstices of a crystal.

interstratified Interbedded; strata laid between or alternating with other strata. *See* INTERBEDDED, INTERCALATED

intertongued lithofacies Stratigraphic body, distinguished from others by its gross lithologic character, that intertongues with its neighbor rather than passing into it by gradual lithologic change. *Cf.* STATISTICAL LITHOFACIES; LITHOSOME

interval *1.* The vertical distance between strata or units of reference. *2.* Contour interval is the vertical distance between two successive contour lines on a topographic, structure, or other contour map.

interval velocity The ratio of any distance interval to the corresponding time interval in: (1) a time distance curve in refraction shooting; (2) a well velocity survey (borehole survey).

interzonal soil Immature soil.

intraclast A component of a limestone consisting of a torn-up, rounded, and reworked fragment of a weakly consolidated penecontemporaneous sediment that has been redeposited to form a new sediment.

intracratonal geosyncline *See* PARAGEOSYNCLINE

intracratonic Situated within a stable continental region.

intracratonic basin Ovate structurally depressed area of considerable size within a continent; autogeosyncline.

intracyclothem Cyclic sequence of strata resulting from the splitting of a cyclothem.

intrafacies A minor stratigraphic facies occurring within a major one.

intraformational Formed by, existing in, or characterizing the interior of a geological formation.

intraformational conglomerate A conglomerate the clasts of which are derived from the formation of which the conglomerate is a part.

intrageosyncline Geosyncline situated within a continent.

intragranular movements In rock deformation, displacements that take place within the individual crystals by movement along glide planes.

intramontane Situated or acting within a mountain.

Intra-Pacific province *See* ATLANTIC SERIES

intrastratal solution Chemical attrition which acts on the constituents of a rock after deposition.

intratelluric Applied to the period of crystallization of a magma prior to its effusion as a lava, and represented in many volcanic rocks by phenocrysts formed under comparatively deep-seated conditions. Crystals such as these, belonging to an earlier generation than the groundmass, are also described as intratelluric.

intrazonal soil Soil which is unique to local conditions because those local conditions override the more widespread effect of climate.

intrenched meander *See* ENTRENCHED MEANDER

intrenched stream A stream flowing in a gorge or narrow valley that has been cut as a result of rejuvenation of the stream. *See* ENTRENCHED STREAM

intrinsic ash Inherent ash.

intrusion *1.* A body of igneous rock that invades older rock. The invading rock may be a plastic solid or magma that pushes its way into the older rock. Some magmas may be emplaced by magmatic stoping, *q.v.* Intrusion should not be used for bodies of igneous-looking rock that are the result of metasomatic replacement. 2. The process of formation of an intrusion.

intrusion displacement Faulting coincident with the intrusion of an igneous rock.

intrusive *Petrol:* Having, while fluid, penetrated into or between other rocks, but solidifying before reaching the surface. Said of plutonic igneous rocks and contrasted with effusive or extrusive.

intrusive contact A contact between an igneous rock and some other rock indicating that the igneous rock is the younger. The younger rock may send dikes into the older rock, have inclusions of the older rock, or be chilled against the older rock.

intrusive rock A rock that consolidated from magma beneath the surface of the earth.

invar An alloy of nickel and steel having a very low coefficient of thermal expansion. Invar is used in the construction of some leveling rods and first-order leveling instruments. Invar is also used in the construction of pendulums.

invariant equilibrium A phase assemblage having zero degrees of freedom, i.e., neither temperature, pressure, nor composition may be varied without loss of one or more phases.

inverse zoning In plagioclase, the change by which crystals become more calcic in outer parts. *Syn:* REVERSED ZONING

inversion *1.* The folding back of strata upon themselves, as by the overturning of a fold in such a manner that the order of succession appears to be reversed. 2. A change of phase, generally from one solid to another of different structure, as quartz inverting to tridymite. In the strict sense both phases must have the same chemical composition (i.e., they are polymorphs), but the term is

widely used in petrology for phase changes involving minor changes in composition, as biotite "inverting" to chlorite. Transformation has a similar meaning. 3. An atmospheric condition where a layer of cool air is trapped by a layer of warm air so that it cannot rise. Inversions spread polluted air horizontally rather than vertically so that contaminating substances cannot be widely dispersed. An inversion of several days can cause an air pollution episode.

inversion point *1.* The temperature at which one polymorphic form of a substance, in equilibrium with vapor, reversibly changes into another under invariant conditions. 2. The temperature at which one polymorphic form of a substance inverts reversibly into another under univariant conditions, the pressure being specified. 3. Loosely used to indicate the lowest temperature of which a monotropic phase inverts at an appreciable rate into a stable phase, or at which a given phase dissociates at an appreciable rate, under given conditions. 4. A point of maximum or minimum on the curve expressing the variation of any physical quantity, such as volume, velocity, or electrical resistance with change of condition, or at which the quantity changes (algebraic) sign. 5. A single point at which different phases are capable of existing together at equilibrium. *Syn:* TRANSITION POINT

invertebrate *1.* Without a backbone or spinal column. 2. Of, or pertaining to, the Invertebrata. *3.* Of, or pertaining to, all the phyla of animals exclusive of the Chordata or animals with notochord or backbones.

inverted limb The overturned limb of a fold.

inverted plunge A plunge of a

fold such that the younger rocks plunge beneath the older rocks.

involute Refers to coiled shells in which there is considerable overlap of older whorls by younger whorls.

involution 1. *Struct. geol:* Refolding of large nappes such as those in the Alps. Two nappes may be refolded together after they formed; or one nappe may penetrate the underside or upper side of an older nappe thus causing the older nappe to wrap around the younger nappe. 2. *Glacial geol:* Used as a synonym of brodel and cryoturbation, *q.v.*

iodate A salt or ester of iodic acid; a compound containing the radical IO_3.

iodide A compound of iodine with one other more positive element or radical.

ion exchange Reversible exchange of ions contained in a crystal for different ions in solution without destruction of crystal structure or disturbance of electrical neutrality. The process is accomplished by diffusion and occurs typically in crystals possessing one or two dimensional channel-ways where ions are relatively weakly bonded. Also occurs in resins consisting of three dimensional hydrocarbon networks to which are attached many ionizable groups. *Syn:* BASE EXCHANGE

ionic substitution The partial or complete proxying of one or more types of ions for one or more other types of ions in a given structural site in a crystal lattice.

ionization chamber A device roughly similar to a Geiger counter that reveals the presence of ionizing radiation.

ionization constant *See* DISSOCIATION CONSTANT

ionization potential The voltage required to drive an electron

completely out of an atom or molecule, leaving a positive ion but without imparting any kinetic energy to the electron.

ionosphere The highest layer of the earth's atmosphere in which ionization takes place. It lies above the stratosphere; its lower limit is about 56.3 km. high in daytime and about 96.6 km. during nights. The ionosphere reflects radio signals.

iridescence The exhibition of colored reflections from the surface of a mineral; a play of colors. Labradorite and some other feldspars show it. The tarnish on the surface of coal, chalcopyrite, etc., is iridescent.

iridosmine A natural alloy of iridium and osmium. Hexagonal.

iron A heavy magnetic malleable and ductile chemically active mineral. The native metallic element Fe (iron).

iron bacteria Usually refers to bacteria that cause the precipitation of iron oxide through their metabolic processes. However, in certain acid environments with relatively low Eh there are bacteria that reduce iron, also.

iron hat *See* GOSSAN

iron meteorite Meteorite consisting of iron and nickel.

iron pan A type of hardpan, *q.v.*, in which a considerable amount of iron oxide is present.

ironstone 1. Any rock containing a substantial proportion of an iron compound from which the metal may be smelted commercially. 2. An iron-rich sedimentary rock, either deposited directly as a ferrugenous sediment or resulting from chemical replacement.

irreversible processes Any process which proceeds in one direction spontaneously, without external interference.

irrotational wave Compressional wave, or P wave.

isallobar An imaginary line or

a line upon a chart connecting the places of equal change of atmospheric pressure within a specified time.

isanomaly; isonomaly A line on a map connecting points of equal anomaly. Used especially for maps showing magnetic anomalies.

isinglass Mica in thin transparent sheets.

island *1.* A tract of land, usually of moderate extent, surrounded by water; distinguished from a continent or the mainland, as, an island in the sea, an island in a river. *2.* A body of land extending above and completely surrounded by water at the mean high-water stage. An area of dry land entirely surrounded by water or a swamp; an area of swamp entirely surrounded by open water.

island arc Curved chain of islands, like the Aleutians, generally convex toward the open ocean, margined by a deep submarine trench and enclosing a deep sea basin.

island shelf The zone around an island or island group, extending from the low-water line to the depths at which there is a marked increase of slope to greater depths. Conventionally its edge is taken at 100 fathoms (or 200 meters).

island slope The declivity from the outer edge of an island shelf into great depths.

isobaric surface A surface in the atmosphere, every point of which has the same barometric pressure.

isobath *1.* Line on a marine map or chart joining points of equal depth usually in fathoms below mean sea level. *2.* A line on a land surface all points of which are the same vertical distance above the upper or lower sur-

face of an aquifer may be called an isobath of the specified surface, or merely a line of equal depth to the surface.

isocal A line constructed on a map, somewhat similar to a contour line, but connecting points of equal calorific value of the coal in a bed.

isocarb *1.* A line constructed on a map, somewhat similar to a contour line, but connecting points of equal content of fixed carbon of coal in the bed. The fixed carbon is computed on ash- and moisture-free basis. *2.* A line passing through points whose carbon ratios, *q.v.*, are equal.

isochemical series A series of rocks having essentially identical chemical compositions.

isochore *1.* A line drawn through points of equal interval between two beds or other planes. *2.* A line connecting points of constant volume, as in P-V-T phase diagrams.

isochore map *1.* Convergence map. A map indicating by means of isochores, the varying interval (convergence) between two designated stratigraphic planes. Differs from isopach map in that it may express the variations in many units and the effects of one or more unconformities, whereas the isopach map expresses variation within a single unit. Current usage shows little distinction between the terms. *2.* A map showing, by contours, the thickness of a pay section in a pool. The map is a basis for estimating reservoir content.

isochron *1.* A line on the surface of the earth connecting points at which a characteristic time or interval has the same value. *2.* In seismic surveying, a contour line passing through points at which difference between arrival times from two reflecting surfaces is equal.

isochroneity Equivalent in time.

isochronous surface A time plane within a body of sediment or sedimentary rocks.

isoclinal *Geol:* Dipping in the same direction; hence, an isoclinal.

isoclinal fold Carinate fold. A fold the limbs of which have parallel dips; may be an anticline or a syncline. The fold may be vertical, overturned, or recumbent.

isocline *Geol:* A series of isoclinal strata. An anticline or syncline so closely folded that the rock beds of the two sides or limbs have the same dip. *See* ISOCLINAL FOLD. *Syn:* ISOCLINIC LINE

isoclinic line Isocline; isodip line. A line joining points at which the magnetic inclination (dip) is the same.

isocon Contour line indicating equal concentration, e.g., salinity.

isodimorphism Relations of two substances that are both similarly isomorphous and dimorphous.

isodynamic line Any joining points of equal magnetic intensity. Applicable to the total intensity or the vertical, horizontal, north-south or east-west components.

isofacial, *adj.* Isograde. Said of all rocks belonging to the same facies.

isofacies map A map showing, by suitable patterns, the distribution of one or more facies within a designated stratigraphic unit.

isogal In gravity prospecting, a contour line of equal gravity values (after gal, the common unit of gravity measurement; one gal=1 cm./sec./sec.).

isogam In magnetic prospecting, a contour line of equal magnetic values (after gamma the common unit of magnetic measurements; one gamma=10^{-5} oersted).

isogeotherm A line or curved surface beneath the earth's surface through points having the same mean temperature. Also called isogeothermal lines.

isogonic line A line joining points of equal magnetic declination.

isograd *1.* A line connecting those rocks comprising the same facies. *2.* A line connecting similar temperature-pressure values. *3.* A line marking the boundary between two facies. *4.* A line of equal grade of metamorphism drawn on a map to distinguish metamorphic zones defined by index minerals.

isogram Contour line.

isogyre In crystal optics, a black or shadowy part of the interference figure that is produced by extinction and indicates the emergence of those components of light having equal vibrational directions.

isohyetal Marking equality of rainfall.

isohyetal line A line on a land or water surface all points along which receive the same amount of precipitation.

isoline map General term for any map on which some variable feature is contoured, e.g., isopach map.

isolithic lines Lines drawn on a paleolithologic map connecting points of similar lithology and separating rocks of differing characteristics.

isolith map One which portrays (by isolith lines) variations in aggregate thickness of a given lithologic facies as measured perpendicular to the bedding at selected points. These may be outcrops or drill holes.

isomagnetic Designating or pertaining to lines connecting points of equal magnetic force.

isomer One of two or more compounds containing the same elements united in the same proportion by weight, but differing in properties because of a difference in structure; usually applied to compounds having the same molecular weight, as distinguished from polymers.

isomeric Composed of the same elements united in the same proportion by weight, but differing in one or more properties owing to difference in structure.

isomesiac The same medium, e.g., the environment of marine, lacustrian, or terrestrial deposition.

isometric projection A division of descriptive geometry in which a three-dimensional object is projected on to one plane. *Cf.* NORMAL PROJECTION.

isometric system *Crystallog:* That system of crystals in which the forms are referred to three equal mutually perpendicular axes. *See* CUBIC SYSTEM

isomorphous Two or more different chemical compounds that can crystallize in intermediate compositions with intermediate physical properties.

isopach; isopachous line; isopachyte [*Brit.*] A line on a map drawn through points of equal thickness of a designated unit.

isopachous *adj.* Of equal thickness. Describing maps, charts, etc., in which the descriptive effect is indicated by lines drawn through points of equal thickness.

isopachous strike Direction indicated by isopachs.

isopiestic level Level of uniform pressure at the base of isostatically balanced areas of the earth's crust; depth of compensation.

isopiestic line A contour of the piezometric surface of an aquifer. It is an imaginary line all points along which have the same static level.

isopleth *1*. A line, on a map or chart, drawn through points of equal size or abundance. *2*. A line of constant composition, as in a binary temperature vs. composition plot.

isoporic line A line drawn through points whose annual change in magnetic declination is equal.

isopycnic Of equal density.

isorads Lines joining points of equal radioactivity, drawn from Geiger- or scintillation-counter data to form an isorad map.

isoseismic line An imaginary line connecting all points on the surface of the earth where an earthquake shock is of the same intensity.

isostasy Theoretical balance of all large portions of the earth's crust as though they were floating on a denser underlying layer; thus areas of less dense crustal material rise topographically above areas of more dense material.

isostatic Subject to equal pressure from every side; being in hydrostatic equilibrium.

isostatic adjustment Isostatic compensation, *q.v.*

isostatic anomaly *1*. The difference between the observed value of gravity at a point after applying to it the isostatic correction and the normal value of gravity at the point. *2*. Anomaly on a map of observed gravity anomalies after applying the isostatic correction. Negative isostatic anomalies indicate undercompensation, implying a tendency to rise; positive isostatic anomalies connote overcompensation and a tendency to sink.

isostatic compensation Isostatic

adjustment. *1.* An equilibrium condition in which elevated masses such as continents and mountains are compensated by a mass deficiency in the crust beneath them. The compensation for depressed areas is by a mass excess. *2.* The process in which lateral transport at the earth's surface by processes such as erosion and deposition is compensated by lateral movements in a subcrustal layer.

isostatic correction The adjustment made to values of gravity or to deflections of the vertical observed at a point to take account of the assumed mass deficiency under topographic features for which a topographic correction is also made.

isostratification map Contour map showing the number or thickness of beds in a stratigraphic unit.

isostructural Two or more chemical compounds with similar crystal structures, but with little tendency to show isomorphism.

isotherm Any line connecting points of equal temperature. *See* ISOGEOTHERM

isothermal Having equal degrees of heat. Isothermal relationships between pressure and volume of a gas or other fluid are those resulting when the temperature is constant, and when heat is added or subtracted by an outside substance or body. *See* ADIABATIC

isothermal plane or **section** In ternary systems, a horizontal section cut through the solid temperature-composition model at a given temperature and showing phase relationships at that temperature.

isotope dilution An analytical technique involving addition of a known amount of an isotopic mixture of abnormal composition to the unknown amount of

an element of normal or known isotopic composition. *See* SPIKE, ISOTOPIC

isotopes Elements having an identical number of protons in their nuclei, but differing in the number of their neutrons. Isotopes have the same atomic number, differing atomic weights, and almost but not quite the same chemical properties. Different isotopes of the same element have different radioactive behavior.

isotopic *1.* Refers to chemical isotopes. *2.* Formed in the same sedimentary basin or geologic province.

isotopic fractionation Process resulting in the relative enrichment of one isotope in a mixture.

isotropic Having the same properties in all directions; most commonly used for optical properties. Typical of amorphous substances and of crystals of the isometric system. Contrasts with anisotropic, *q.v.*

isotropic fabric Random orientation in space of the elements of which a rock is composed.

isotropic symmetry Symmetry in which parts are equal and interchangeable in all directions from a center, characterized by a sphere.

isotropy Condition of having the same properties in all directions. In an isotropic elastic medium, the velocities of propagation of elastic waves are independent of direction.

isthmus A narrow strip of land, bordered on both sides by water, that connects two larger bodies of land.

itabirite A Brazilian term for sedimentary iron formation, or taconite, composed chiefly of hematite and silica.

itacolumite Flexible sandstone; articulite. *1.* A fine-grained micaceous, thin-bedded sandstone, thin slabs of which have a certain degree of flexibility. 2. A schistose and flexible quartzite containing mica, chlorite, and talc. *See* QUARTZITE

iteration Repeated occurrence of

similar evolutionary trends in successive offshoots of a group.

iterative evolution *1.* Repeated development of new forms from a conservative stock. *2.* Repeated and independent evolution.

J

Jacob's staff A single straight rod, pointed and iron shod at the bottom, and having a socket at the top; used instead of a tripod for supporting a compass.

jade; jadeite Nephrite. A hard and extremely tough material of varying composition, greenish white to deep green in color, used in making jewelry and carved ornaments. Both jadeite, a variety of pyroxene (nearly $NaAl$ Si_2O_6), and nephrite, a variety of amphibole, are properly called jade.

jarosite A mineral, $KFe_3(SO_4)_2$ $(OH)_6$. Hexagonal rhombohedral.

jaspagate A jasper agate, usually applied to jasper agate in which the jasper predominates.

jasper Red, brown, green, impure, slightly translucent cryptocrystalline quartz with a dull fracture, abundant enough on Lake Superior and elsewhere to be a rock.

jasperoid *1.* A rock consisting essentially of cryptocrystalline, chalcedonic, or phenocrystalline silica, which has formed by the replacement of some other material, ordinarily calcite or dolomite. *2.* Silicified limestone.

jaspilite A rock consisting essentially of red jasper and iron oxides in alternating bands. *See* ITABIRITE; TACONITE

jelly opal A transparent orangered opal which may or may not show an opalescent play of colors.

jet *1.* A dense black lignite, taking a good polish. Sometimes used for jewelry. *2.* A black marble.

jetty *1.* In the United States: On open seacoasts, a structure extending into a body of water, and designed to prevent shoaling of a channel by littoral materials, and to direct and confine the stream or tidal flow. Jetties are built at the mouth of a river or tidal inlet to help deepen and stabilize a channel. *2.* In Great Britain, jetty is synonymous with wharf or pier.

jig Device for concentrating minerals. Crushed ore is fed into a box containing water whose level is rapidly raised and lowered by action of a piston causing heavier minerals to sink to the bottom from which they are drawn off.

Johannsen's classification A mineralogical classification of igneous rocks in which a rock is characterized by a number, the Johannsen number, consisting of three or four digits, each one of which has a specific mineralogical significance.

JOIDES Joint Oceanographic Institutions for Deep Earth Sampling. A program to obtain cores of sediments in the deep oceans.

join The line (or plane) drawn between any two (or three) composition points.

joint Fracture in rock, generally more or less vertical or transverse to bedding, along which no appreciable movement has occurred.

joint set A group of more or less parallel joints.

joint system Consists of two or more joint sets, *q.v.*, or any group of joints with a characteristic pattern, such as a radiating pattern, a concentric pattern, etc.

Jolly balance A toroidal spring balance used primarily for measuring specific gravity (density) by weighing a specimen when immersed in air and again when immersed in a liquid of known density (usually water, if the specimen is insoluble in H_2O).

jug A colloquial equivalent of detector, geophone, etc.

Jura A mountain system in northern Europe; geologically refers either to the orogeny that formed those mountains or to the middle period of the Mesozoic. *See* JURASSIC

Jurassic *Geol.:* The middle of three periods comprising the Mesozoic Era. Also the system of strata deposited during that period.

juvenile Coming to the surface for the first time; fresh, new in origin; applied chiefly to gases and waters.

juvenile gases Gases from the interior of the earth which are new and have never been at the surface of the earth.

juvenile water Water that is derived from the interior of the earth and has not previously existed as atmospheric or surface water.

K

Kansan Second of four classical glacial stages of the Pleistocene of North America.

kainite A mineral, $KCl \cdot MgSO_4 \cdot 3H_2O$. Monoclinic.

Kainozoic Cenozoic.

kame *1.* A conical hill or short irregular ridge of gravel or sand deposited in contact with glacier ice. *2.* Kames is a Scotch term applied to assemblages of short, conical, often steep hills, built of stratified materials and interlocking and blending in the most diversified manner. *3.* A mound composed chiefly of gravel or sand, whose form is the result of original deposition modified by settling during the melting of glacier ice against or upon which the sediment accumulated. *4.* A hill of stratified drift deposited, usually as a steep alluvial fan, against the edge of an ice sheet by debouching streams of sediment-laden meltwater. *5.* A low, steep-sided hill of stratified drift, formed in contact with glacier ice.

kame-and-kettle topography Surface formed by a kame complex interspersed with kettles. *Obs.*

kame complex A series of interconnecting kames.

kame field A group of kames, including in places kettles and eskers.

kame terrace *1.* A terracelike body of stratified drift deposited between a glacier and an adjacent valley wall. *2.* A terrace of glacial sand and gravel, deposited between a valley ice lobe (generally stagnant) and the bound-ing rock slope of the valley. *3.* Remnant of a depositional valley surface built in contact with glacial ice.

K/Ar age Radioactive age based on determination of the potassium 40-argon 40 ratio.

kaolin *1.* A rock composed essentially of clay minerals of the kaolinite group, most commonly kaolinite, but also halloysite, endellite, dickite, nacrite, etc. *2.* China clay; porcelain clay. A clay, mainly hydrous aluminum silicate, from which porcelain may be made. *See* KAOLINITE

kaolinite A common clay mineral. Two-layer hydrous aluminum silicate having the general formula $Al_2(Si_2O_5)(OH)_4$. It consists of sheets of tetrahedrally coordinated silicon joined by an oxygen shared with octahedrally coordinated aluminum. Essentially, there is no isomorphous substitution. The mineral characteristic of the rock kaolin. *See* FIRE CLAY MINERAL

karat A unit used to describe the fineness of gold alloys. One part gold and 23 parts of other metals=1 karat. 24-karat gold is pure. *Symbol:* k.

karren The furrows that occur from solution by rain wash on blocks of extremely pure limestone in karst areas. Such furrows range from three centimeters wide and 15 centimeters long to 6 meters wide and 30 meters long. They are oriented downslope in parallel rows.

karst A type of topography that is

formed over limestone, dolomite, or gypsum by dissolving or solution, and that is characterized by closed depressions or sinkholes, caves, and underground drainage.

karst plain A plain on which sinkholes, uvala, subterranean drainage, and other karst features are developed.

katabatic wind A wind that flows down slopes that are cooled by radiation, the direction of flow being controlled orographically. Such winds are the result of downward convection of cooled air. *Syn:* MOUNTAIN WIND, CANYON WIND; GRAVITY WIND

katamorphism Metamorphism at or near the surface of the lithosphere, forming simple minerals from complex ones.

katatectic layer Layer of solution residue, generally consisting of gypsum and/or anhydrite, in salt dome caprock.

katatectic surface Contact of two katatectic layers.

katazone The deepest zone in the depth zone classification of metamorphic rocks. This zone would be characterized by high temperatures and pressures. The distinctive minerals include sillimanite, enstatite, hypersthene, orthoclase, cordierite, etc.

Kazanian Fourth of five ages of the Permian Period; also the stage of strata deposited during that age.

K-bentonite A clay material having the geologic appearance of a bentonite, but consisting of a clay mineral interlayer-mixture of montmorillonite and illite, a high proportion of the interlayer cations being potassium. Metabentonite.

K-capture The capturing by an atomic nucleus of an orbital electron from the same atom. The electron usually captured is in the K-orbit. In geology significant only in K^{40}, of which about 10% decays by K-capture to Ar^{40}.

Keewatin The oldest provincial Precambrian series of rocks on the Canadian Shield.

Kelly; Kelly joint In rotary drilling, the square, grooved, or hexagonal member supported at the upper end by the swivel and passing through the rotary table, with the lower end screwed into and supporting the drill pipe. It transmits the rotary motion of the table to the drill pipe and can be raised or lowered through bushings in the table.

keratophyre A name originally applied to trachytic rocks containing highly sodic feldspars, but now more generally applied to all salic lavas and dike rocks characterized by containing albite or albite oligoclase, chlorite, epidote, and calcite.

kernite A mineral, $Na_2B_4O_7$.-$4H_2O$. Monoclinic. An important ore of boron.

kerogen The solid, bituminous mineraloid substance in oil shales which yields oil when the shales undergo destructive distillation.

kerogen shale Oil shale.

kerosene sand A tar sand in Australia from which many of the lighter volatiles have not yet escaped.

kettle A depression in drift, made by the wasting away of a detached mass of glacier ice that had been either wholly or partly buried in the drift. Sölle [*Ger.*].

kettle basin; kettle hole *1.* Bowl-shaped depressions, usually 9.2 to 15.3 meters deep and 30 to 153 meters deep in larger diameter. Each depression, according to the accepted explanation, was the resting place, and often the burial place of a huge mass of ice that became detached during the melting; the final melting away of the ice left a hole where the ice lay. *2.* A steep-sided hole or depression in rock, sand, or gravel, having a shape more or less like the interior of a kettle.

Keuper Uppermost of three provincial series of the Triassic of northern Europe.

Keweenawan The youngest of the provincial Precambrian series on the Canadian Shield.

key *1*. A low island near the coast; used especially on the coasts of regions where Spanish is or formerly was spoken, as, the Florida Keys. *2*. A low insular bank of sand, coral, etc., as one of the islets off the southern coast of Florida. *See* CAY

key bed *1*. A bed with sufficiently distinctive characteristics to make it easily identifiable in correlation. *2*. A bed the top or bottom of which is used as a datum in making structure contour maps. *See* MARKER BED, *1* and *2*

key horizon The top or bottom of a bed or formation or a particular layer of fauna or flora that is so distinctive as to be of great help in stratigraphy and structure.

K-feldspar Potash or potassium-bearing feldspar, $KAlSi_3O_8$, Orthoclase, sanidine, or microcline.

kick Arrival of seismic wave.

kidney ore A variety of hematite, occurring in compact kidney-shaped masses.

kieselguhr German name for diatomaceous earth.

kieserite A mineral, $MgSO_4.H_2O$. Monoclinic.

kilkenny coal Anthracite.

Killarney Revolution Supposed post-Proterozoic orogeny.

kilometer A length of one thousand meters, equal to 3280.8 feet, or 0.621 of a mile. The chief unit for long distances in the metric system.

kimberlite A variety of mica peridotite consisting essentially of olivine, phlogopite, and sub-ordinate melilite, with minor pyroxene, apatite, perovskite, and opaque oxides. Some examples contain diamonds.

Kinderhookian Lowest of four ages of the Mississippian Period of North America; also the stage of strata deposited during that age.

kindred, *n. Petrol:* A group of igneous rocks which show consanguineous chemical and mineral characters, and which appear to be genetically related. *Cf.* SERIES; SUITE; CLAN; BRANCH. *See* CONSANGUINITY

kinetic metamorphism The deformation of rocks without accompanying chemical reconstitution.

kingdom The largest taxonomic division of organisms. Conventionally two kingdoms, plants and animals, are recognized. Some biologists favor the recognition of one or more additional kingdoms for the reception of intermediate or ancestral organisms. *See* PROTISTA

klint A calcareous reef or bioherm, more resistant to erosion than the rocks that enclose it, and which thereby form ridges or hills. Plural is klintar.

klintite Biohermal limestone, particularly the massive core.

klippe An isolated block of rocks separated from the underlying rocks by a fault that normally, but not necessarily, has a gentle dip. Generally, the rocks above the fault are the older. A klippe may be an erosional remnant of an extensive overthrust sheet or it may have moved into place by gravity sliding.

knickpoint Points of abrupt change in the longitudinal profile of stream valleys.

knob A rounded hill or mountain, especially an isolated one. Local in the South.

knob-and-basin topography Knob-

and-kettle topography; kame-and-kettle topography.

knoll *1.* A submerged elevation of rounded shape rising from the ocean floor, but less prominent than a seamount. *2.* A small rounded hill.

knot *1.* The unit of speed used in navigation. It is equal to 1 nautical mile (6080.20 feet, 1853 meters) per hour. *2.* The meeting point of two or more mountain chains.

kratogen An early variation of craton, *q.v.*

kreep An acronym for a basaltic lunar rock type first found in Apollo 12 fines and breccias and characterized by unusually high contents of potassium (K), rare-earth elements (REE), phosphorous (P), and other trace elements in comparison to other lunar rock types. The material, which is found in a variety of crystalline and glassy (shock-melted?) rock types, is distinctly different from the iron-rich mare basalts. The term nonmare basalt is equivalent.

K section Circular cross section through a strain ellipsoid; there are two such sections, designated K_1 and K_2.

kugel [*Ger.* ball, sphere, globe, bulb] *Geol:* A general term applied to those igneous rocks characterized by spheroidal structures, such as orbicular granite, corsite, and especially kugelminette.

Kullenberg corer Device for sampling sediments of the deep sea bottom.

kyanite Cyanite; disthene. A mineral, Al_2SiO_5, trimorphous with sillimanite and andalusite. Triclinic.

L

L *Earthquake seismol:* A phase designation referring to surface waves without respect to type (Love or Rayleigh). From "Undae longae," or "Long waves," as on early earthquake records, the surface waves were distinguished by their long periods relative to the preliminary waves (P and S). LQ and LR refer to Love (Querwellen) waves and Rayleigh waves, respectively; in modern usage, Q and R are preferred to LQ and LR.

labile Applied to particles in a rock that decompose easily.

labile stage The temperature range in which nucleation and/or growth of crystals takes place readily.

labradorite A feldspar mineral of the plagioclase series having nearly equal proportions of calcium and sodium.

laccolith A concordant, intrusive body that has domed up the overlying rocks and also has a floor that is generally horizontal but may be convex downward.

lacuna *1.* A cavity, hole, or gap. *2.* An unrecorded stratigraphic record at an erosion surface, consisting of (a) the record destroyed by erosion=EROSIONAL VACUITY, *q.v.,* and (b) the record never represented by strata—HIATUS, *q.v.*

lacustrine *1.* Pertaining to, produced by, or formed in a lake or lakes; e.g., "lacustrine sands" deposited on the bottom of a lake, or a "lacustrine terrace" formed along the margin of a lake. *2.* Growing in or inhabiting lakes;

e.g., "lacustrine fauna." *3.* Said of a region characterized by lakes; e.g., a "lacustrine desert" containing the remnants of numerous Pleistocene lakes that are now dry. *Cf.* LIMNIC. *Syn:* LACUSTRAL; LACUSTRIAN

ladder lode; ladder veins Roughly parallel fractures normal to the walls of a dike that have been filled with gangue or ore.

lag *1.* Any time delay. *2.* The phase angle by which the current is behind the e.m.f. in an induction circuit. *3.* The time delay between the breaking of the bridgewire in a detonating cap and the resulting explosion. Sufficiently small in modern caps to be negligible. *4.* The time delay between the arrival of a signal in a piece of equipment and the response, such as the making or breaking of a circuit by a relay.

lag gravel Residual accumulations of coarser particles from which the finer material has been blown away; similar to "desert pavement," but more restricted in its implications as to the extent and continuity of the accumulation. *See* DESERT PAVEMENT; PEBBLE ARMOR

lagoon *1.* Body of shallow water, particularly one possessing a restricted connection with the sea. *2.* Water body within an atoll or behind barrier reefs or islands.

lagoon cycle Refers to the filling of a lagoon by sediments from the land, atmosphere, and sea and the eventual erosion and destruction of these deposits by wave action.

laguna [*Sp.*] *1.* Shallow ephemeral lakes, mostly found in the lower parts of the bolsons, and fed by streams whose sources are in the neighboring mountains, and which flow only during time of storm. *2.* A lake or pond. *3.* A pseudokarst feature; large shallow sinks with clay bottoms; developed in silts and sands of south-central United States.

lahar *1.* Landslide or mudflow of pyroclastic material on the flank of a volcano. *2.* Deposit produced by such a landslide. Lahars are described as wet if they are mixed with water derived from heavy rains, escaping from a crater lake or produced by melting snow. Dry lahera may result from tremors of a cone or by accumulating material becoming unstable on a steep slope. If the material retains much heat, they are termed hot lahars.

lake *1.* Any standing body of inland water, generally of considerable size. *2.* A pool of other more or less fluid substance, as oil or asphalt. *3.* A pigment formed by absorbing animal, vegetal, or coal-tar coloring matter from an aqueous solution by means of metallic bases.

lake pitch Asphalt from the Pitch Lake, Trinidad. It is richer than the land pitch in bituminous matter; soluble in petroleum spirit.

lake rampart A ridge of shore materials (sand, gravel, or driftwood) along a lake shore formed by the shoreward movement of lake ice, also called ice rampart.

lake terrace In small lakes, lowering of the lake level generally cuts short the shoreline cycle before it has reached an advanced stage. During an interval of stationary water level a map or line of low cliffs is cut, and in front of this there is a narrow shelf, partly cut and partly built, which remains as a lake terrace when the water level falls.

Lamarckism The idea that changes acquired or developed by individuals during their lifetimes are transmitted to their offspring; the inheritance of acquired characters. Similar to kinetogenesis of Cope; Recent believers in Lamarckism were the Russians Michurin and his once politically powerful follower, T. D. Lysenko.

Lambert azmuthal equal-area map projection An azmuthal map projection in which the pole of the projection is the center of the mapped area. The azimuths of great circles radiating from the center of the map are represented by a scale varying in distance from the center. Distortion is greatest near the periphery.

Lambert conformal conic map projection A map projection on which all geographic meridians are represented by straight lines which meet in a common point outside the limits of the map, and the geographic parallels are represented by a series of arcs or circles having this common point for a center. Meridians and parallels intersect in right angles, and angles on the earth are correctly represented on the projection. This projection may have one standard parallel along which the scale is held exact, or there may be two such standard parallels, both maintaining exact scale. At any point on the map, the scale is the same in every direction. It changes along the meridians, and is constant along each parallel. Where there are two standard parallels, the scale between those parallels is too small; beyond them, too large.

lamella *Biol:* One of the layers of a cell wall; a thin layer in a shell, etc., like a leaf in a book.

lamellar Composed of thin layers, plates, scales, or lamellae;

disposed in layers like the leaves of a book.

lamellar flow Movement of liquid whereby successive layers glide over one another like cards in a sheared pack of playing cards. Contrasted with turbulent flow.

Lamellibranchiata Pelecypoda, *q.v.*; Bivalvia.

lamina *1.* Unit layer or sheet of a sediment in which the stratification planes are one centimeter or less apart. Laminae need not be parallel to bedding. When they are at an angle to the bedding planes the term "cross-lamination" may be used. *2.* A layer in a sedimentary rock less than 1 cm. in thickness that is visually separable from other layers above and below, the separation being determined by a discrete change in lithology, a sharp break in physical lithology, or by both. *3. Paleontol:* In some corals the sheetlike structure formed by the juxtaposition of two layers of skeletal material in septa and the column, i.e., the axial structure. *4.* A blade or expanded portion.

laminar flow *1.* That type of flow in which the stream lines (or stream surfaces) remain distinct from one another (except for molecular mixing) over their entire length. Under laminar flow the head loss is proportional to the first power of the velocity. It is typical of ground-water movement under most conditions. *2.* A flow of current without turbulence. A smooth flow at relatively slow velocity in which the fluid elements follow paths that are straight and are parallel to the channel walls. *See* LAMELLAR FLOW

laminated quartz Vein quartz characterized by slabs or films of other material. Laminated quartz is a general term including book structure and ribbon structure.

lamination *1.* The layering or bedding less than 1 cm. in thickness in a sedimentary rock. *2.* The more or less distinct alternation of material, which differ one from the other in grain size or composition.

lamprophyre A group name applied to dark dike rocks in which dark minerals occur both as phenocrysts and in the groundmass and light minerals occur only in the groundmass. They differ from normal rocks in which light and dark minerals occur both as phenocrysts and in the groundmass. The essential constituents of lamprophyres are biotite, hornblende, or pyroxene or combinations of the three, and feldspar or feldspathoids. Olivine is present in some varieties.

lamp shell *1.* Brachiopod, *q.v.* *2.* A terebratuloid brachiopod, the most characteristic Mesozoic brachiopod shell form, an ovate, curved hinge type which is aptly compared with an ancient Roman pottery lamp.

land bridge A landmass or chain of islands. Some scientists believe such connections between continents permitted animal migration. There is no modern evidence for the existence of extensive "bridges" in the past.

landform One of the multitudinous features that taken together make up the surface of the earth. It includes all broad features, such as plain, plateau, and mountain, and also all the minor features, such as hill, valley, slope, canyon, arroyo, and alluvial fan.

landmass A part of the continental crust lying above sea level, considered as a unit without regard to size or relief.

landslide Landslip, *q.v. 1.* The perceptible downward sliding or falling of a relatively dry mass of earth, rock, or mixture of the two. *2.* Earth and rock which becomes loosened from a hillside by moisture or snow, and slides or falls down the slope. *See* SLIDE

landslip Landslide, *q.v.*

land-tied island *See* TOMBOLO

land use The sum total or particular parts of man's use of the land.

land-use planning The sum of man's activities that plan for the use of the land; it may be good, bad, or indifferent.

langbeinite A mineral, K_2Mg_2 $(SO_4)_3$. Isometric. Occurs in potassium salt deposits; mined as a source of K_2SO_4.

lapidary A skilled workman who cuts and polishes gems or other stones.

lapies *See* KARREN

lapilli Essential, accessory, and accidental volcanic ejecta ranging mostly from 4 mm. to 32 mm. in diameter.

lapis lazuli A rock composed primarily of the mineral lazurite, a translucent rich blue stone used as a gem. Frequently flecked with small crystals of pyrite.

lap-out map Map showing the areal distribution of formations immediately overlying an unconformity.

lapse rate Temperature gradient, *q.v.*

Laramide orogeny In broad sense the diastrophic movements beginning, perhaps in the Jurassic, certainly in the Lower Cretaceous, and continuing until the Lower Eocene.

Larsen variation diagram Weight per cent of each oxide constituent in a rock analysis is plotted as the ordinate against the "abscissa position," defined as 1/3 $SiO_2+K_2O-FeO-MgO-CaO$; a smooth curve is drawn through the points representing a given constituent for a series of analyses.

larvikite A nepheline-bearing syenite with abundant rhombic phenocrysts of feldspar. Titanau-

gite, barkevikite, and lepidomelane are minor constituents and apatite, opaque oxides, zircon, and olivine are accessories.

LASA Large aperture seismic array. A geophone array in Montana set up to detect nuclear explosions and distinguish them from earthquakes.

late magmatic minerals Those formed during the later stages of magmatic activity, principally those formed between the main stage of crystallization and the pegmatitic stage. More or less synonymous with "reaction" minerals.

lateral accretion Wherever a stream meanders, it digs away its outer bank while the inner is building up the water level by the deposition of material brought there by rolling or pushing along the bottom.

lateral erosion The action of the stream in impinging on one side of its channel and undermining the bank at that point, so that masses of material tumble down to be ultimately disintegrated; at the same time the channel keeps shifting toward the bank which is being undercut.

lateral migration Movement of oil or gas through permeable zones parallel to the stratification; a better term is parallel migration.

lateral moraine [<*Fr.* moraine laterale] *1.* Certain aggregations of drift which are left by a valley glacier after melting. *2.* An elongate body of drift, commonly thin, lying on the surface of a glacier in a valley, at or near the lateral margin of the glacier. *3.* An end moraine built along the lateral margin of a glacier lobe occupying a valley. First used by Louis Agassiz.

lateral planation The reduction of the land in interstream areas to a plane parallel to the stream profile, effected by the lateral swinging of the stream against its banks.

lateral secretion The theory that the contents of a vein or lode are derived from the adjacent wall rock.

laterite *1.* A latosol, *q.v.*, 2. Lateritic latosol, *q.v.* 3. The hard horizon of illuviation in a lateritic latosol. *See also* PLINTHITE

lateritic latosol Red and reddish soil leached of soluble minerals and of alumina and silica, but retaining oxides and hydroxides of iron.

late stage effects *See* LATE MAGMATIC MINERALS

lath Long thin mineral crystal.

latite The extrusive equivalent of monzonite and a variety of trachyandesite in which potash feldspar and plagioclase are present either as normative or modal minerals in nearly equal amounts. Potash feldspar is often concealed in the fine-grained crystalline or glassy groundmass and thus a chemical analysis is often necessary for correct classification. Augite or hornblende is usually present and sometimes biotite plus accessory apatite and opaque oxides.

latitude *1.* Distance on the earth's surface from the equator, measured in degrees of the meridian. 2. In plane surveying, the perpendicular distance in a horizontal plane of a point from an east-west axis of reference. *See* DEPARTURE

latosol Refers to those highly leached soils rich in iron, alumina, or silica, that once formed under very humid climate with relatively high temperature. Tropical woodland soils is a synonym, but many latosols are now in desert areas, although they originated under a different climatic regime.

lattice drainage *See* RECTANGULAR DRAINAGE

lattice energy For an ionic crystal, the energy absorbed when a mole of the crystal is dispersed into infinitely separated ions.

lattice orientation *Struct. petrol:* A preferred orientation that is shown by the lattice of the mineral. This is recognized by optical studies using the petrographic microscope or by X-ray studies.

Laué X-ray diffraction pattern produced by a single crystal with fixed orientation when exposed to a small unfiltered beam of radiation.

Laué camera A single-crystal X-ray diffraction apparatus involving a stationary crystal and film and utilizing white X radiations.

Laurasia Hypothetical continent in the northern hemisphere which supposedly broke up at about the end of the Carboniferous Period to form the present northern continents.

laurdalite A variety of nepheline syenite with alkalic feldspar of rhombic form. Also contains biotite, pyroxene or amphibole and accessory apatite and sodalite.

Laurentian Gneissic granite of the Lake Superior area.

lava Fluid rock such as that which issues from a volcano or a fissure in the earth's surface; also the same material solidified by cooling.

lava blisters Small, hollow, steep-sided swellings raised on the surface of some pahoehoe lava flows and formed by gas bubbles puffing up the viscous crust of the flow. *Syn:* TUMULUS

lava cascade A cascade of fluid, incandescent lava, formed when a lava river passes over a cliff or over a precipitous part of its course.

lava cavern In the production of an aa surface, if the crust is of sufficient strength not to be broken by the underflow the lava beneath may flow out and leave a cavern.

lava cone A volcanic cone built

entirely or largely of lava flows, usually basaltic lavas that were very mobile at the time of eruption, and resembling a miniature shield volcano.

lava-dam lake   A lake ponded behind a lava flow which has obstructed a stream valley.

lava dome; cumulo dome *1.* If the eruptions at a volcanic vent are exclusively of the explosive type, the material of the mountain which results is throughout tuff or cinder, and the volcano is described as a cinder cone. If, on the other hand, the vent at every eruption exudes lava, a mountain of solid rock results which is a lava dome. *2.* Lava domes are the greater masses of lava which, in the form of many individual flows, have issued from central vents in the proper directions to build a dome-shaped pile of lava. The world type is Mauna Loa.

lava field   A wide expanse of lava flows, usually many square miles in extent, covering level or nearly level terrain and generally bearing clusters of cinder cones.

lava flow   A lateral, surficial outpouring of molten lava from a vent or a fissure; also, the solidified body of rock that is so formed.

lava fountain; fire fountain   A rhythmic jetlike eruption of lava issuing vertically from a central volcanic vent or from a fissure.

lava lake   A lake of fluid molten lava, usually basaltic, and ordinarily contained in a summit crater or in a pit crater on the flanks of a shield volcano.

lava pit; fire pit   A pit crater, usually developed at the summit or on the flanks of a shield volcano, or on the floor of a caldera containing an active or congealed lava lake.

lava plain   A broad stretch of level or nearly level land, usually many hundreds of square kilometers in extent, underlain by a relatively thin succession of lava flows, most of which are basaltic and the product of the fissure eruption.

lava plateau   A broad, elevated tableland or flat-topped highland, usually many hundreds or thousands of square kilometers in extent, underlain by a thick succession of lava flows, most of which are tholeiitic basalts and the product of fissure eruption.

lava shield   Shield volcano.

lava tube   Many lava flows develop hard crusts by the cooling and solidification of the upper surface. Later the supply of lava for the particular flow may cease and the liquid lava may drain out leaving a long tubular opening under the crust. *Cf.* LAVA CAVERN

lava tunnel   A lava cavern or lava tube, *q.v.*, open at the ends.

law [science]   A formal statement of the invariable and regular manner in which natural phenomena occur under given conditions; e.g., the "law of superposition" or a "law of thermodynamics."

law of constancy of interfacial angles   The statement in crystallography that the angles between corresponding faces on different crystals of one substance are constant. *Syn:* CONSTANCY OF INTERFACIAL ANGLES

law of crosscutting relationships   A stratigraphic principle whereby relative ages of rocks can be established; a rock (esp. an igneous rock) is younger than any other rock across which it cuts.

law of equal declivities   Where homogeneous rocks are maturely dissected by consequent streams, all hillside slopes of the valleys cut by the streams tend to develop at the same slope angle, thereby producing symmetric profiles of ridges, spurs, and valleys.

law of faunal assemblages   A general law of geology: Similar assemblages of fossil organisms (faunas and floras) indicate similar geologic ages for the rocks that contain them.

law of faunal succession A general law of geology: Fossil organisms (faunas and floras) succeed one another in a definite and recognizable order, each geologic formation having a different total aspect of life from that in the formations above it and below it; or, the age of rocks can be determined from their fossil content.

law of homonymy A principle in taxonomy stating that any name that is a junior homonym of another name must be rejected and replaced. *See also* HOMONYMY

law of nature A generalization of science, representing an intrinsic orderliness of natural phenomena or their necessary conformity to reason. *Syn:* NATURAL LAW

law of original continuity A general law of geology: A waterlaid stratum, at the time it was formed, must continue laterally in all directions until it thins out as a result of nondeposition or until it abuts against the edge of the original basin of deposition.

law of original horizontality A general law of geology: Waterlaid sediments are deposited in strata that are horizontal or nearly horizontal, and parallel or nearly parallel to the earth's surface.

law of priority The valid name of a genus or species can be only that name under which it was first designated.

law of reflection The angle between the reflected ray (normal to the wave front) and the normal to the reflecting surface is the same as the angle between this normal and the incident ray, provided the reflected wave is of the same type (travels with the same velocity) as the incident wave.

law of refraction Snell's law.

When a wave crosses a boundary the wave normal changes direction in such a manner that the sine of the angle of incidence between wave normal and boundary normal divided by the velocity in the first medium equals the sine of the angle of refraction divided by the velocity in the second medium.

law of stream gradients A general law expressing the inverse geometric relation between stream order and the mean stream gradient of a given order in a given drainage basin.

law of superposition A general law upon which all geologic chronology is based: In any sequence of sedimentary strata (or of extrusive igneous rocks) that has not been subsequently disturbed by overthrusting or overturning, the youngest stratum is at the top and the oldest at the base, the older strata being successively covered or overlain by younger and younger layers; or, each bed is younger than the bed beneath, but older than the bed above it.

law of universal gravitation Newton's law of gravitation. Every particle of mass attracts every other particle with a force inversely proportional to the square of the distance between them. $F = \gamma m_1 . m_2 / r^2$.

layer A bed or stratum of rock.

layer depth The depth of the mixed layer of ocean water to the top of the thermocline.

lazulite A mineral, $(Mg,Fe)Al_2(PO_4)_2(OH)_2$. Monoclinic.

lazurite A mineral, $(Na,Ca)_8(Al,Si)_{12}O_{24}(S,SO_4)$. The principal constituent of lapis lazuli. A blue, isometric, translucent mineral.

leach To wash or drain by percolation. To dissolve minerals or metals out of the ore, as by the use of cyanide or chlorine solutions, acids, or water.

leachate A solution obtained by leaching, as in the downward percolation of meteoric water through soil or solid waste and containing soluble substances, such as a landfill.

lead *1.* An indication of interesting geological conditions which might suggest a mineral deposit. *2.* Amount of time by which one seismic event precedes another, such as amount by which an actual arrival precedes that expected in the absence of interesting structural situation, thus suggesting some anomalous situation.

lead glance Galena.

lead-uranium ratio The ratio of amount of lead to amount of uranium in a rock or mineral, or of amounts of various isotopes of the two elements; used in computing the geologic age of the rock or mineral.

league *1.* A unit of linear measure. A land league=3 statute miles, or 15,840 feet (4.8280 km.). A nautical league=3 geographical miles or 18,240.78 feet (5.5597 km.). *2.* An area embraced in a square 5000 varas on each side. It contains 4428.40 acres, 6.919 square miles (17.92 square kilometers). Term used chiefly in Texas land descriptions.

lean Applied to poor ores, or those containing a lower proportion of metal than is usually worked.

lean clay Clay of relatively low plasticity. *Cf.* FAT CLAY

lease Contract between landowner and another granting the latter right to search for and produce oil or mineral substances upon payment of an agreed rental, bonus, and/or royalty.

least squares *See* METHOD OF LEAST SQUARES

least-time path Minimum time path.

Le Chatelier's rule If conditions of a system, initially at equilibrium, are changed, the equilibrium will shift in such a direction as to tend to restore the original conditions.

lectotype Subsequently designated type upon which a species is based providing the holotype had never been designated by the original author.

ledge *1.* A bed of several beds as in a quarry or natural outcrop, particularly those projecting in a steplike manner. *2.* The surface of such a projecting bed. *3. Min:* Projecting outcrop or vein, commonly quartz, that is supposed to be mineralized; also any narrow zone of mineralized rock. *4.* In northern Michigan, bedrock.

lee *1.* Shelter, or the part or side sheltered or turned away from the wind. *2.* Chiefly nautical: The quarter or region toward which the wind blows.

Lee configuration An electrical resistivity measuring method using two current electrodes and three equispaced potential electrodes.

lee side *Geol:* That side of glaciated rocks that looks away from the quarter whence the ice moves, or moved, as indicated by rough and weathered surfaces; opposed to shock side or stoss side. Also used for the lee sides of dunes.

left-handed separation Left-lateral separation.

left-lateral fault Commonly used as a general fault term to imply a strike-slip fault with left-lateral separation.

left-lateral separation A lateral fault along which, in planned view, the apparent movement of the side opposite the observer is to the left.

left strike-slip fault A strike-slip fault in which the movement is to the left as one crosses the fault, that is the ground on the opposite side has moved toward the left.

leg *1.* A prop of timber supporting the end of a stull, or cap of a set of timber. 2. A single cycle in a wave train on a seismogram.

legend Explanation of the symbols and patterns shown on a map or diagram.

Lemberg solution An aqueous solution of logwood extract and $AlCl_6$ which produces a violet stain on calcite but leaves dolomite unchanged.

lens A body of ore or rock thick in the middle and thin at the edges; similar to a double convex lens. See LENTICULAR

lens, sand A body of sand with the general form of a lens, thick in the central part and thinning toward the edges.

lensing The thinning out of a stratum in one or more directions.

lenticular Shaped approximately like a double convex lens. When a mass of rock thins out from the center to a thin edge all around, it is said to be lenticular in form. See LENS

lentil 1. Lens-shaped rock body. 2. A minor rock-stratigraphic unit, a subdivision of a formation similar in rank to a member, having relatively small geographic extent and presumably wedging out in all directions.

Leonardian The second of four provincial epochs of the Permian of North America; also the series of strata deposited during that epoch.

lepidoblastic A term applied to that type of flaky schistosity due to an abundance of minerals like micas and chlorites with a general parallel arrangement.

lepidolite A mineral of the mica group, $K(Mg,Li,Al)_2(AlSi_3)O_{10}$ $(OH,F)_2$, rose, lilac, or gray. Monoclinic.

lepidomelane A mineral of the biotite series, characterized by high iron content.

leptothermal Deposits intermediate between Waldemar Lindgren's mesothermal and epithermal zones.

leucite A mineral, $KAlSi_2O_6$. Pseudo-isometric. Found in K-rich volcanic rocks.

leucitohedron Trapezohedron.

leucocratic A term applied to light-colored rocks, especially igneous rocks, containing between 0 and 30% of dark minerals, i.e., rocks whose color index is between 0 and 30. See MESOCRATIC; MELANOCRATIC

levee An artificial or a natural bank confining a stream to its channel or, if artificial, limiting the area of flooding; also a landing place, pier, or quay.

levee delta Deltas occasionally take the form of long narrow ridges upon one or both sides of a stream, resembling the natural levees in the "goosefoot" of the Mississippi. Normally the point where a tributary valley joins a larger one is marked by a notch in the wall of the latter, but in some cases a bisected spur appears instead.

level, spirit A small closed vessel of transparent material (glass), having the inside surface of its upper part curved (circular) in form; the vessel is nearly filled with a fluid of low viscosity (alcohol or ether), enough free space having been left for a bubble (blister) of air which will always assume a position at the top of the vessel.

leveling In surveying, the operation of ascertaining the comparative levels of different points of land, for the purpose of laying out a grade, etc., by sighting through a leveling instrument at one point to a leveling staff at another point.

leveling instrument A surveyor's level bearing a telescope.

level of zero amplitude The level to which seasonal change of temperature extends into permafrost. Below this level the temperature gradient of permafrost is more or less stable the year around. An abbreviation of level of zero annual amplitude.

level rod A graduated rod used in measuring the distance between points on the ground and the line of sight of a leveling instrument.

level surface A surface which at every point is perpendicular to the plumb line or the direction in which gravity acts.

Lg *Earthquake seismol:* A phase designation given to a slow, short period Love wave which is found to travel only along nonoceanic paths. The subscript g refers to the possible importance of the granitic layer for their propagation.

Lias; Liassic The lower most of three epochs of the Jurassic Period; also the series of strata deposited during that epoch.

Liassic *See* LIAS

licks An American term given to boggy grounds affording salt springs, because the cattle go down to lick the salt there. In Brazil these are called carrieros.

liesegang rings Rings or bands resulting from rhythmic precipitation in a gel.

life cycle The phases, changes, or stages an organism passes through during its lifetime. *Cf.* ONTOGENY

light, polarized Light in which the vibrations are in one plane. Light is usually polarized with Nicol prisms (or modifications of them) or with manufactured materials such as Polaroid.

light-colored mineral Light mineral.

light minerals *Geol:* Applied to those rock-forming minerals that have a specific gravity less than 2.8, including such materials as quartz, calcite, feldspars, feldspathoids, and some micas. Also applied to the rock-forming min-

erals that are light in color, these minerals generally being the same as those that are classified as light on the basis of weight.

light ruby silver Proustite.

lignite *1.* A brownish-black coal in which the alteration of vegetal material has proceeded further than in peat but not so far as sub-bituminous coal. *2.* Consolidated lignitic coal having less than 8300 B.t.u. (moist, mineral-matter-free).

liman *1.* Shallow lagoon or embayment with muddy bottom. *2.* Area of mud or slime deposited near a river's mouth.

liman coast [<*Lat.* limus, mud] An alluvial coast usually with many lagoons.

limb *1.* One of the two parts of an anticline or syncline on either side of the axis. *2.* The graduated margin of an arc or circle in an instrument for measuring angles.

limburgite Glassy nepheline basalt. Phenocrysts of titanaugite, olivine, and opaque oxides in glassy groundmass.

lime Calcium oxide, CaO. Loosely used for calcium hydroxide, calcium carbonate, and even for calcium in deplorable expressions such as carbonate of lime or lime feldspar.

lime feldspar Misnomer for calcium feldspar. Anorthite. *See* PLAGIOCLASE

limestone *1.* A bedded sedimentary deposit consisting chiefly of calcium carbonate ($CaCO_3$) which yields lime when burned. Limestone is the most important and widely distributed of the carbonate rocks and is the consolidated equivalent of limy mud, calcareous sand, or shell fragments. *2.* A general term for that class of rocks which contain at least 80 per cent of the carbonates of calcium or magnesium. The suitability of the rock for the manufacture of lime is not an essential characteristic.

limestone reefs Aggregates of calcareous skeletons and other structures of plants and animals growing upward from submerged continental or island basements to the level of the ocean.

limnetic; limnic Pertaining to fresh water.

limnetic zone In fresh water, the water zone from the surface to the light compensation level.

limnobios The life of the fresh-water environment.

limnology The scientific study of fresh waters, especially that of ponds and lakes. In its broadest sense it deals with all physical, chemical, meteorological, and biological conditions pertaining to such a body of water.

limonite A field term for a group of brown, amorphous, naturally occurring, hydrous ferric oxides. May consist of the minerals goethite, hematite, and lepidocrocite.

Lindgren's volume law A principle, pointed out by Waldemar Lindgren, that replacement occurs on approximately a volume-by-volume basis.

lineage *1.* A series of genera and species which form an evolutionary series, each one being ancestral to its successor in the geological sequence; a line of evolution. *2.* A broad belt of descent embracing series of communities each of which may include a wide range of forms that can be described as so many morphological species; a bundle of lines of descent. *Syn:* PLEXUS

lineament *1. Lunar:* A conspicuous linear feature on the surface of the moon; e.g., a rill, wrinkle ridge, crater chain, ray, and fault. Also, a less distinctive lunar feature, such as an elongated valley, a mountain ridge, and a straight

section of a crater wall. *2. Photo:* Any line, on an aerial photograph, that is structurally controlled, including any alignment of separate photographic images, such as stream beds, trees, or bushes that are so controlled. *3. Tect:* Straight or gently curved, lengthy features of the earth's surface, frequently expressed topographically as depressions or lines of depression. Their meaning has been much debated and some certainly express valid structural features such as faults, aligned volcanoes, and zones of intense jointing, but the meaning of others is obscure.

linear *1. N:* A straight or gently curved physiographic feature on the earth's surface. *2. Adj:* In paleobotany, long and narrow, the sides parallel or nearly so, as blades of most grasses.

linear cleavage A property of metamorphic rocks whereby they break in long pencil-like fragments; results from two intersecting cleavages or from linear parallelism of platy or prismatic minerals.

linear element *See* ELEMENT, LINEAR

linear system A system whose output is directly proportional to its input.

lineation Any linear structure within or on a rock resulting from primary flowage in igneous rock or secondary flowage in metamorphic rock shown by rotation of mineral grains or other bodies, intersection of planes, slippage along gliding planes, and growth of crystals.

line of bearing The direction of the strike, or outcrop.

line of dip A line of greatest inclination of a stratum to the horizontal.

line of section On a map, a line indicating the position of a profile section or cross section.

lingulid One of an ancient family of brachiopods with simple shells, dating from the Cambrian and persisting practically without change to the present.

linguoid One of the modifications of the simple normal type of current ripple shows a highly irregular pattern with a wide range in the variety of forms.

link A unit of linear measure, one one-hundredth of a chain, and equivalent to 7.92 inches.

linked vein A steplike vein in which the ore follows one fissure for a short distance, then passes by a cross fissure to another nearly parallel, and so on.

Linnaean Conforming to the principles of binomial nomenclature as advocated by Carl von Linné, who Latinized his name to Carolus Linnaeus.

Lipalian A theoretical geologic period immediately antedating the Cambrian. Unknown anywhere. No equivalent to Sinian, which is known, and is considered a system that lies between the Cambrian and Precambrian by many, but not all, Russian authors.

liquefied natural gas (LNG) Natural gas that has been cooled to about −160° C. for shipment or storage as a liquid. Liquefaction greatly reduces the volume of the gas, and thus reduces the cost of shipment and storage, even though high-pressure cryogenic containers must be used.

liquefied petroleum gas (LPG) Also known as "bottled gas," liquefied petroleum gas consists primarily of propanes and butanes recovered from natural gas and in the refining of petroleum.

liquid 1. Flowing freely like water; fluid, not solid. 2. Characterized by free movement of the constituent molecules among themselves, but without the tendency to separate from one another characteristic of gases. 3. As generally used, a fluid of high density and low compressibility. The specific recognition of a liquid as distinct from a gas of the same composition requires the simultaneous presence of both phases at equilibrium. See GAS; FLUID; VAPOR

liquid flow Movement of a liquid, generally one of low viscosity, involving either or both laminar and turbulent flow.

liquid immiscibility A process of magmatic differentiation involving the separation of the magma into two or more immiscible liquid phases, which are then separated from each other by gravity or other processes.

liquid limit A property of soil used in civil engineering. Water content of a soil-water mixture, at which the consistency changes from the plastic to the liquid state as determined by the standardized tests. Syn: ATTERBERG LIMIT, q.v.

liquidus The locus of points in a temperature-composition diagram representing the maximum solubility (saturation) of a solid component or phase in the liquid phase. In a binary system it is a line, in a ternary system it is a curved surface, and in a quaternary system it is a volume.

lith-; litho- [Gr.] 1. A prefix meaning stone or stonelike. 2. -lith A suffix meaning rock or rocklike.

lithic; lithologic Refers to sediments and rocks in which rock fragments are more important proportionally than feldspar grains.

lithic tuff 1. Tuffs that consist dominantly of crystalline rock fragments are called lithic tuffs. This is in contrast to the crystal tuff, q.v. Formed from quickly cooled volcanic materials, giving the rock a fine-grained structure and a crystal fabric. 2. An indurated deposit of volcanic ash in which the fragments are composed of previously formed rocks, e.g., accidental particles of sedimentary rock, accessory pieces of earlier lavas in the same cone, or small bits of nes lava (essential ejecta) that first solidify in the vent and are then blown out.

lithification *1.* That complex of processes that converts a newly deposited sediment into an indurated rock. It may occur shortly after deposition—may even be concurrent with it—or it may occur long after deposition. *Cf.* INDURATION; CEMENTATION. *2.* A type of coal-bed termination wherein the disappearance takes place because of a lateral increase in impurities resulting in a gradual change into bituminous shale or other rock.

lithify To turn to rock; to petrify; to crystallize as from a magma; to consolidate, such as the process of induration of a loose sediment. *Cf.* INDURATION

lithium An element of atomic number 3. It is a proposed fuel for fusion energy.

lithofacies The rock record of any sedimentary environment, including both physical and organic characters,

lithofacies map A map showing the areal variation in overall aspect of the lithology of a stratigraphic unit. Emphasis may be placed on a dominant, average, or specific lithologic aspect of the unit in question.

lithofraction Breaking of rock fragments during transportation in streams or by wave action on beaches. *Cf.* SPLITTING

lithographic limestone; lithographic stone An exceedingly fine-grained limestone used for lithography.

lithographic texture A term used to denote grain size in calcareous sedimentary rocks. The grain size corresponds to that of clay, or less than 1/256 mm.

lithoidal A term applied to dense, fine-grained, crystalline igneous rocks, to devitrified glasses, or to a groundmass of crystalline material as distinguished from glassy varieties. *Syn:* STONY. *See* FELSITIC

lithology *1.* The physical character of a rock, generally as determined megascopically or with the aid of a low-power magnifier. *2.* The microscopic study and description of rocks. *Obs.* PETROGRAPHY

lithophile elements Elements enriched in the silicate crust. Elements with a greater free energy of oxidation, per gram atom of oxygen, than iron; they concentrate in the stony matter or slag crust of the earth, as oxides and more often as oxysalts, especially silicates.

lithophysae Hollow, bubblelike structures composed of concentric shells of finely crystalline alkali feldspar, quartz, and other minerals. Found in certain silicic volcanic rocks, such as rhyolite and obsidian.

lithosol One of a group of azonal soils having no clearly expressed soil morphology and consisting of a freshly and imperfectly weathered mass of rock fragments; largely confined to steep hillsides and sand dunes.

lithosphere *1.* In plate tectonics: a layer of strength relative to the underlying asthenosphere for deformations at geologic rates, which includes the crust and part of the upper mantle and is of the order of 100 km. in thickness. *2.* In geochemistry and general geology: the silicate shell of the earth; includes mantle and crust; part of the classification lithosphere, hydrosphere, atmosphere, biosphere, *q.v.*

lithostatic pressure The equal,

all-sided pressure in the crust of the earth due to the weight of the overlying rocks.

lithostratigraphic unit Unit consisting of stratified, mainly sedimentary, rocks grouped on the basis of lithologic rather than biologic characters or time value. Rock-stratigraphic unit. *Cf.* BIO-STRATIGRAPHIC UNIT

lithostratigraphy *1.* Stratigraphy based only on the physical and petrographic features of rocks. *2.* Recognition and interpretation of the physical characters of sedimentary rocks.

lithotype *1.* Rock defined on the basis of certain selected physical characters. *2.* One of the four microscopically recognizable constituents of bonded coal; vitrain, clarain, durain, fusain.

lithozone Stratigraphic zone defined by lithology.

lit-par-lit [*Fr.* leaf-by-leaf] *Petrol:* Used to designate the intimate penetration of bedded, schistose, or other foliate rocks by innumerable narrow sheets and tongues of granitic rock, usually granitic igneous rocks.

littoral *1.* Belonging to, inhabiting, or taking place on or near the shore. *2.* The benthonic environment between the limits of high and low tides.

littoral benthal Benthonic animals dwelling in the littoral, *q.v.*

littoral cone An adventive or accidental ash or tuff cone built on the stagnant surface of an aa lava flow and formed by the explosion of the lava when it runs into a body of water, usually the sea. Such cones are the result of steam explosions that hurl into the air large amounts of ash, lapilli, and small bombs derived from the new lava.

littoral current Longshore current. Generated by waves breaking at an angle to the shoreline, which move usually parallel to, and adjacent to the shoreline within the surf zone.

littoral deposits Deposits of littoral drift located between high and low water lines.

littoral drift *1.* Applied to the movement along the coast of gravel, sand, and other material composing the bars and beaches. *2.* The material moved in the littoral zone under the influence of waves and currents.

littoral shelf Shallow, near shore, terracelike part of submerged lake bed produced by wave erosion and deposition.

littoral zone *1.* Strictly, zone bounded by high and low tide levels. *2.* Loosely, zone related to the shore, extending to some arbitrary shallow depth of water.

living fossil A modern animal that has descended from a very ancient stock with comparatively little change.

L-joints Primary flat joints. Horizontal or nearly horizontal joints that are found in igneous rocks and related to the intrusion of magma.

Llandeillian The fourth of six epochs of the Ordovician Period; also the series of strata deposited during that epoch.

Llandoverian The lowermost of three epochs of the Silurian Period; also the series of strata deposited during that epoch.

llano [*Sp.*] *Topog:* *1.* An extensive plain with or without vegetation. *2.* The term is an exact equivalent of the English word plain, and by Spanish-speaking persons is so used. Generally the term is applied to the vast treeless plains of South America.

Llanvirnian The third of six epochs of the Ordovician Period; also the series of strata deposited during that epoch.

load *1*. In erosion and corrasion the material which is transported may be called the "load." The load is transported by two methods: a portion floats with the water, and another portion is driven along the bottom. *2*. The sediment moved by a stream, whether in suspension or at the bottom, is its load. *3*. The quantity of material actually transported by a current. This is usually somewhat less than the actual capacity of the current.

load cast Roll or other irregularity at the base of an overlying stratum, commonly sandstone, projecting into an underlying stratum, commonly shale or clay, produced by differential settling and compaction. *Cf.* FLOW CAST

loaded stream A stream is loaded when it has all the sediment it can carry; it is but partly loaded when it is carrying less than it might.

loam A soil composed of a mixture of clay, silt, sand, and organic matter.

lobe *1*. The tongue of land within the meander. When the lobe lies between two meanders and is connected with the mainland by a narrow passage, the narrow passage is the neck. The cutting action of the river narrows the neck until finally the river breaks through and forms a new channel or a cutoff. *2*. A projection of a glacial margin or of a body of glacial drift beyond the main mass of ice or drift. *See* GLACIAL LOBE. *3*. In a cephalopod, undulations of the suture line which are convex toward the apex and concave toward the aperture of the shell. *4. Paleobot:* Any part or segment of an organ; specifically a part of a petal or calyx or leaf that represents a division to about the middle.

local currents Natural earth currents of local origin such as those arising from the oxidation of sulfide deposits. A term used in electrical prospecting.

local metamorphism Contact metamorphism.

Locke hand level Hand level with fixed bubble tube that can be used only for horizontal sighting.

lode *Eco. geol:* A mineral deposit consisting of a zone of veins; a mineral deposit in consolidated rock as opposed to placer deposits. *streams:* A local English term for a channel or water course, usually partly artificial and embanked above the surrounding country.

lode mining claim A mining claim including a lode, fissure, or fissure vein. In the United States the maximum length along the lode or vein is 457.5 meters and the maximum width is 183 meters.

lodestone; loadstone A piece of magnetite possessing polarity like a magnetic needle.

lode tin Tin ore (cassiterite) occurring in veins, as distinguished from stream tin or placer tin.

lodgement till Till deposited beneath a moving glacier, characterized by compact fissile structure and stones oriented with their long axes parallel to the direction of flow.

loess A homogeneous, nonstratified, unindurated deposit consisting predominantly of silt, with subordinate amounts of very fine sand and/or clay; a rude vertical parting is common at many places.

loess doll Calcareous concretion occurring in loess.

loessification Development of loess from swampy terrace sediments by weathering and downslope creep.

loess kindchen A spheroidal or irregular nodule of calcium carbonate found in loess. *Syn:* LOESS PUPPEN; LOESS MÄNNCHEN

Loewinson-Lessing classification A chemical system of classification of igneous rocks.

log *1.* A graphic presentation of the lithologic and/or stratigraphic units traversed by a borehole. *2.* The similar presentation of the variation of some physical property in a borehole with depth, such as resistivity, self-potential, gamma-ray intensity or velocity. *3.* The record of formations penetrated, drilling progress, record of depth of water, oil, gas, or other minerals, the record of size and length of pipe used, and other written or recorded facts having to do with drilling a well.

Logan's Line A structural discontinuity along the northwestern edge of the northern Appalachians, between complexly deformed geosynclinal rocks on the southeast and undisturbed cratonic and shield rocks on the northwest. It is interpreted by many geologists as having been formed during the Taconic orogeny of Early Paleozoic time.

lognormal distribution A distribution in which the logarithm of a parameter is normally distributed.

log strip Long narrow paper strip on which a well log may be plotted.

longitude *1.* Distance east or west on the earth's surface, measured by the angle which the meridian through a place makes with some standard meridian, as, that of Greenwich or Paris. *2.* A coordinate distance, linear or angular, from a north-south reference line. *3. Geod:* The angle between the plane of the geodetic meridian and the plane of an initial meridian, arbitrarily chosen.

longitude correction The east-west corrections made to observed magnetic intensities by subtracting the earth's normal field.

longitudinal dune A very general term for various types of linear dune ridges, commonly more or less symmetrical in cross profile, which are known or inferred to extend parallel to the direction of the dominant dune-building winds.

longitudinal fault A fault whose strike is parallel with the general structure.

longitudinal joint Joint extending parallel to flow lines in igneous rock, steeply dipping and best developed where flow lines are horizontal.

longitudinal stream A stream which runs parallel to the strike of the rocks.

longitudinal valley A valley having a direction the same as the strike. *See* SUBSEQUENT STREAM

longitudinal wave Primary wave; pressure wave; compressional wave; dilatational wave; irrotational wave. An elastic wave in which the displacements are in the direction of wave propagation.

long limb That side of an asymmetrical fold which extends farther than the other before inclination is reversed, generally the more gently dipping side.

long range order Orderliness of arrangement in crystal structure extending indefinitely beyond neighboring atoms or molecules; the normal or ideal state in a crystalline solid.

longshore bar Refers to slightly submerged sand ridges which extend generally parallel with the shore line and are submerged at least by high tides.

longshore drift The material transported by longshore drifting or longshore currents.

longwall A system of working a seam of coal in which the whole seam is taken out and no pillars left, excepting the shaft pillars, and sometimes the main-road pillars.

loop A pattern of field observations which begin and end at the same point with a number of intervening observations. Such a pattern is useful in correcting for drift in gravity-meter observations, diurnal variation in magnetometer surveys, and to detect faults or other cause of misclosure in seismic dip shooting.

lopolith A large floored intrusive that is centrally sunken into the form of a basin.

lorac A precision radio surveying technique in which two or more fixed transmitters emit continuous waves and in the resulting standing wave pattern the position of a mobile receiver is determined by measuring with it the phase difference of the waves emanating from two of the transmitters.

loran A pulse-type electronic navigation system for measuring distance differences with respect to fixed transmitters of known geographic position.

lost river *Geol:* A river that, by a secular increase in aridity, at first periodically in the driest season, and at last permanently, has lost its trunk, its detached tributaries losing themselves in the arid ground. River in a karst region which drains into an underground channel.

louderback Outlier of lava flow forming dip slope in block faulted region; proves topography was produced by faulting rather than by erosion alone.

Love wave Q-wave. *1.* A transverse wave propagated along the boundary of two elastic media which both have rigidity, i.e., both media must be capable of propagating transverse waves. *2.* A surface seismic wave in which the particles of an elastic medium

vibrate transverse to the direction of the wave's travel, with no vertical component.

low *1.* A region of the atmosphere where the barometric pressure is below normal, usually surrounded by closed isobars with the point of minimum pressure in the center. Not strictly synonymous with cyclone, *q.v.,* since lows often appear on the weather map without a well-defined cyclonic wind circulation. Moreover, an extra-tropical cyclone, *q.v.,* always has a frontal structure; a low need not have. *2.* Former stream channels in coal beds. They do not extend downward through the entire thickness of coal. They are now filled with sandstone, clay, shale. *See* WASHOUT. *3.* Minimum (gravity or magnetic).

low, structural An area in which the beds are structurally lower than in neighboring areas; a syncline or structural depression; sometimes also applied to saddles between local "highs" along the crests of anticlines.

low-angle fault A fault dipping less than 45°.

low-energy environment Environment characterized by general lack of wave and/or current action; very fine-grained sediment is permitted to settle.

low-grade ore Ore which is relatively poor in the metal for which it is mined. *Cf.* HIGH-GRADE ORE

lowlands The lowlands include the extended plains or country lying not far above tide level.

low-oblique photograph An oblique aerial photograph with the entire picture below the horizon.

low-quartz The low-temperature polymorph of SiO_2 stable at temperatures below 573°C. Alphaquartz.

low-rank graywacke Nonfeldspathic graywacke.

low-rank metamorphism Metamorphism accomplished under conditions of low to moderate temperature and pressure. *See* METAMORPHIC GRADE

low tide; low water Minimum height reached by each falling tide; the mean value of all low waters over a considerable period.

low-velocity correction Weathering correction.

low-volatile bituminous coal Nonagglomerating bituminous coal having 78% or more, and less than 86%, of fixed carbon (dry, mineral-matter-free) and 22% or less, and more than 14% of volatile matter (dry, mineral-matter-free).

Luder's line Slip bands. The dark and light bands inclined at about 45° that appear in rods, especially those made of metal, subjected to tension or compression. These lines (actually surfaces) are zones within which the maximum deformation of the grains takes place.

Ludlovian Uppermost of three epochs of the Silurian Period; also the series of strata deposited during that epoch.

luminescence The emission of light by a substance that has received energy or electromagnetic radiation of a different wavelength from an external stimulus.

lumping Practice of ignoring minor differences in the recognition or definition of species and genera.

lunar *1.* Pertaining to or occurring on the moon, such as "lunar probe" designed to pass close to the moon, or "lunar dust" consisting of fine-grained material produced by meteoritic bombardment. *2.* Resembling the surface of the moon, such as the "lunar landscape" of certain glaciers.

lunar cracks The crescent-shaped cracks that more or less parallel the crown of a landslip; they

sometimes precede the landslide and are a warning of impending movement.

lunar crater Crater [lunar].

lunar geology A science that applies geologic principles and techniques to the study of the moon, esp. its composition and the origin of its surface features. *See also* SELENOLOGY

lunarite A general term for light-toned, brightly reflecting surface rocks of the lunar highlands (terrae).

lunar playa A relatively small (as much as a few kilometers long), level area on the moon's surface, occupying a low place in the ejecta blankets surrounding lunar craters such as Tycho and Copernicus. It is believed to be either a fallback deposit or a small lava flow.

lunar regolith A thin, gray layer on the surface of the moon, perhaps several meters deep, consisting of partly cemented or loosely compacted fragmental material ranging in size from microscopic particles to blocks more than a meter in diameter. It is believed to be formed by repeated meteoritic and secondary fragment impact over a long period of time. *Syn:* LUNAR SOIL; SOIL

lunar soil Lunar regolith.

lunate bar A crescentic-shaped bar commonly found off the entrance to a harbor.

luster The character of the light reflected by minerals; it constitutes one of the means of distinguishing them.

lutaceous Argillaceous, *q.v.* Descriptive of a fine-grained texture and particularly, but not entirely, applicable to silts and clays and their derivatives.

lutite A general name used for consolidated rocks composed of mud (silts and/or clays) and of the various associated materials which, when mixed with water, form mud; e.g., shale, mudstone, and calcilutite.

lutyte Variety of lutite.

lysimeter Structure containing a mass of soil and so designed as to permit the measurement of water draining through the soil.

lysocline The depth or level in the ocean at which the solubility of calcium carbonate increases significantly.

maar A crater formed by violent explosion not accompanied by igneous extrusion, commonly occupied by a small circular lake.

Maastrichtian The uppermost age of the Cretaceous Period; also the stage of strata deposited during that age.

macadam effect Cementation of calcareous fragments resulting from wetting, partial solution, and deposition of cement by evaporation.

macro- [<*Gr.* makro-] Prefix meaning large, long; visibly large.

macroaxis The b-axis (long) in orthorhombic and triclinic crystals.

macroclastic Composed of fragments visible without magnification.

macrocrystalline *1.* Applied to the texture of holocrystalline igneous rocks in which the constituents are distinguishable with the naked eye. Opposed to microcrystalline. *2.* In recrystallized sediments, the texture of a rock with grains or crystals over .75 mm. in diameter.

macrodome Crystal form whose faces are parallel to the macro- or b-axis in orthorhombic and triclinic systems.

macrofacies Suite of genetically related facies.

macrofossil Fossils large enough to be studied without the aid of optical magnification.

macropinacoid *See* PINACOID

macropolyschematic A term applied to a body of rock or mineral deposit whose fabric consists of macroscopically different domains, i.e., a coarsely mixed fabric.

macroscopic *See* MEGASCOPIC

macrostructure A structural feature of rocks that can be discerned by the unaided eye, or with the help of a simple magnifier.

maculose A term given to the group of contact-metamorphosed rocks represented by spotted slates, to denote their spotted or knotted character. May be applied to either the rocks or their structures. *See* SPOTTED SLATE

Maestrichtian Misspelling of Maastrichtian, *q.v.*

mafic *Petrol:* Subsilicic; basic. Pertaining to or composed dominantly of the magnesian rock-forming silicates; said of some igneous rocks and their constituent minerals. Contrasted with felsic. In general, synonymous with "dark minerals," as usually used.

maghemite γ -Fe_2O_3. A mineral, Fe_2O_3, dimorphous with hematite. Strongly magnetic.

magma Naturally occurring mobile rock material, generated within the earth and capable of intrusion and extrusion, from which igneous rocks are thought to have been derived through solidification and related processes. It may or may not contain suspended solids (such as crystals and rock fragments) and/or gas phases. *Adj:* magmatic.

magma chamber A large reservoir in the earth's crust occupied by a body of magma.

magmagranite Granite produced by crystallization of a magma.

magmatic *Petrol:* Of, pertaining to, or derived from magma.

magmatic deposits Certain kinds of mineral deposits form integral parts of igneous rock masses and permit the inference that they have originated, in their present form, by processes of differentiation and cooling in molten magmas.

magmatic differentiation *Petrol:* The process by which different types of igneous rocks are derived from a single parent magma, or by which different parts of a single molten mass assume different compositions and textures as it solidifies. Also applied to ores produced by this process. *Syn:* MAGMATIC SEGREGATION

magmatic segregation *See* MAGMATIC DIFFERENTIATION

magmatic stoping A process of igneous intrusion whereby a magma gradually eats its way upward by breaking off blocks of the country rock. As originally proposed, the hypothesis assumed that these blocks sank downward.

magmatic water Water that exists in, or which is derived from, molten igneous rock or magma.

magmatism Development and movement of magma within the earth. *Cf.* VOLCANISM

magmatist *1.* One who believes that much granite has crystallized from a mobile magma whatever the origin of that material may have been. *Cf.* TRANSFORMIST. *2.* One who believes that much granite is a primary igneous rock produced by differentiation from basaltic magma.

magma type A categorization of a magma according to any given scheme of classification, e.g., "nonporphyritic central magmatype."

magnafacies That lithofacies representing the total duration of deposition under the same lithotope during time, regardless of age, as long as the lithofacies is continuous.

magnesia Magnesium oxide, MgO. Loosely, magnesium carbonate.

magnesian limestone Any limestone containing more than 20% of magnesia is a magnesian limestone. Often used, however, as specifying the limestones of Permian age.

magnesian marble Applied to both dolomitic marbles and marbles with magnesian silicates. *See* OPHICALCITE; PREDAZZITE

magnesioferrite A mineral of the spinel group, $(Mg,Fe'')Fe''_2O_4$. Isometric. Strongly magnetic.

magnesite A mineral, $MgCO_3$. Hexagonal rhombohedral, usually massive to compact and earthy.

magnet *1.* Any body which orients itself in definite direction when suitably suspended in any magnetic field such as that of the earth. *2.* Any shaped mass of ferromagnetic material which has been permanently magnetized.

magnetic anomaly Any departure from the normal magnetic field of the earth as a whole. Comparable to a topographic feature on a topographic map.

magnetic bearing A bearing measured clockwise from magnetic north at the point of observation.

magnetic compass An instrument having a freely pivoted magnetic needle that aligns with the Earth's magnetic field such that one end of the needle points to the magnetic north.

magnetic declination The acute angle between the direction of the magnetic and geographic meridians. In nautical and aeronautical navigation the term magnetic variation is preferred.

magnetic dip Magnetic inclination, *q.v.*

magnetic dip pole Place where

the horizontal component of the earth's magnetic field vanishes; not the same as the geomagnetic pole; it changes in position relatively rapidly.

magnetic direction The position of the magnetic field in a rock sample which has been oriented to the present magnetic field of the earth.

magnetic equator The line on the surface of the earth where the magnetic needle remains horizontal, or does not dip, i.e., where the magnetic lines of force are horizontal. *Syn:* ACLINIC LINE

magnetic field *1.* The space through which the force or influence of a magnet is exerted. *2.* The space about a conductor carrying an electric current in which, as it may be shown, magnetic force is also exerted.

magnetic field strength (H) Field intensity. The force exerted on a unit pole at any point is the field strength at that point.

magnetic flux Through any surface element it is the product of the component of the intensity normal to the surface element and its surface area. It is the average intensity over unit area of an equipotential surface.

magnetic force The force, attractive or repulsive, exerted between two magnetic poles; the force which produces or changes magnetization.

magnetic inclination; magnetic dip The acute angle between the vertical and the direction of the earth's total magnetic field in the magnetic meridian plane.

magnetic intensity Magnetic field strength, *q.v.*

magnetic line of force In general, a curved line which at each point has the direction of the force which would be exerted on a unit pole at that point. It is a concept of rather limited utility in geophysical prospecting.

magnetic meridian Magnetic north.

magnetic permeability The ratio of the magnetic induction to the inducing field strength.

magnetic polarity reversal The phenomenon of a switch in position of the north and south magnetic poles of the earth. Paleomagnetic studies reveal that such reversals have occurred numerous times in the geologic past.

magnetic polarization The magnetic moment per unit volume.

magnetic pole *1.* Either of several points on the earth's surface where the lines of magnetic force are vertical; and end of the axis of the earth's magnetic polarity, not coinciding with a geographical or geomagnetic pole (*q.v.*) and slowly changing its position. The north magnetic pole is in northern Canada. *2.* One of two points near opposite ends of a magnet toward which the magnetic lines of force are oriented and concentrated; if the magnet is permitted to rotate in all directions by use of a central pivot, one pole will point in the direction of the earth's magnetic pole near the North Pole and the pole at this end is defined as the "north seeking" or positive pole of the magnet; the other pole is the "south seeking" or negative pole.

magnetic pyrites Pyrrhotite.

magnetic storm A considerable variation of the earth's magnetic field, with time, occurring over extensive areas. The variations are greater, more irregular and more rapid than the diurnal variations. In areas where they are frequent they present a serious obstacle to effective magnetic surveys.

magnetic susceptibility A measure of the degree to which a substance is attracted to a magnet; the ratio of the intensity of magnetization to the magnetic field strength in a magnetic circuit.

magnetic unit in prospecting The gamma $(\gamma)=10^{-5}$ oersteds.

magnetism That peculiar property possessed by certain bodies (as iron and steel) whereby, under certain circumstances, they naturally attract or repel one another according to determinate laws.

magnetite Magnetic iron ore. A mineral, of the spinel group, $Fe''Fe_2''O_4$. Isometric, black, commonly in octahedrons. An important ore of iron.

magnetization The magnetic state of a body defined by a vector having a direction and a magnitude called the magnetic moment per unit volume; this vector may be considered as representing the number and degree of orientation of the elementary magnetic dipoles of the body.

magnetohydrodynamic (MHD) generator An expansion engine in which hot, partially ionized gases are forced through a magnetic field. Movement of the electrically conducting gas through the field generates an electric current that is collected by electrodes lining the expansion chamber. MHD is thus an efficient way to generate electricity from the combustion of fuels without going through an intermediary steam turbine.

magnetometer An instrument used for measuring magnetic intensity; in ground magnetic prospecting usually an instrument for measuring the vertical intensity; in airborne magnetic prospecting usually an instrument for measuring the total intensity. Also, instruments used in magnetic observatories for measuring various components of the magnetic field.

magnetometry Measurement of the earth's magnetic field.

magnetosphere The region surrounding the earth within which the magnetic field is confined. Due to influence exerted by the solar wind, the magnetosphere is somewhat teardrop-shaped with the tail trailing into space away from the Sun.

magnetostriction The effect of stress on magnetization.

magnitude (of an earthquake) A quantity characteristic of the total energy released by an earthquake, as contrasted to "intensity," which describes its effects at a particular place. The Richter magnitude scale is related to the logarithm of an observed displacement on a calibrated instrument and its distance from the epicenter. Each step of one magnitude represents a 10-fold increase in observed amplitude.

malachite A mineral, $Cu_2(CO_3)$-$(OH)_2$. Monoclinic, commonly green, botryoidal. A typical alteration product in the oxidized zone of copper sulfide deposits.

Malm The uppermost of three epochs of the Jurassic; also the series of strata deposited during that epoch.

malpais A term used in the southwestern U.S. for a region of rough and barren lava flows. The connotation of the term varies according to the locality.

Malthusian principle The concept that all animals, including man, potentially outbreed the food supply; conversely, the food supply is the primary limiting factor on population. Thusly, if allowed a free breeding range, most populations maintain themselves at the point of starvation.

mammal A warm-blooded vertebrate animal that brings forth its young alive and suckles them.

Mammalia Class of vertebrates; warm-blooded animals clothed in hair which are viviparous and nurse their young. Jur.-Rec.

mammillary *Mineral:* Forming smoothly rounded masses resembling breasts or portions of spheres. Said of the shape of some mineral aggregates, as, malachite or limonite. Similar to but on a larger scale than botryoidal.

manganese hydrate Psilomelane.

manganese nodule Irregular potatolike mass of manganese-rich material that occurs on the ocean floor. Where concentrated, these nodules have potential value from the cobalt, copper, and nickel they contain, providing the problems of their removal from the ocean depths can be solved.

manganite A mineral, $MnO(OH)$. Monoclinic.

mantle *1.* Layer of the earth between crust and core; bounded above by Mohorovičić discontinuity at depth of about 35 km. below the continents and about 10 km. below the oceans; bounded below by the Wiechert-Gutenberg discontinuity about 2900 km. below the surface of the earth; believed to consist of ultrabasic material. *2.* Mantle rock; regolith.

mantle rock Regolith.

map A representation on a plane surface, at an established scale, of the physical features (natural, artificial, or both) of a part or the whole of the earth's surface or of any desired surface or subsurface data, by means of signs and symbols, and with the means of orientation indicated.

map projection An orderly system of lines on a plane representing a corresponding system of imaginary lines on an adopted terrestrial or celestial datum surface. Also, the mathematical concept of such a system.

map scale The ratio of the distance between two points shown on a map and the actual distance between the points on the earth's surface. Scale is commonly expressed as a representative fraction (R.F.), as 1/1000.

map series A group of maps having some common unifying characteristic and uniform format; e.g., Geologic Quadrangle Map Series.

marble A metamorphic rock composed essentially of calcite and/or dolomite. *See* CALC-SILICATE MARBLE; MAGNESIAN MARBLE

marcasite White iron pyrites. A mineral, FeS_2, the orthorhombic dimorph of pyrite.

mare One of the several dark, low-lying, level, relatively smooth, plains-like areas of considerable extent on the surface of the moon, having fewer large craters than the highlands, and composed of mafic or ultramafic volcanic rock.

mare basin A large, approximately circular or elliptical topographic depression in the lunar surface, filled or partly filled with mare material.

mare material Dark, relatively smooth, heavily cratered igneous rock (chiefly of mafic or ultramafic composition) underlying the lunar maria.

mare ridge Wrinkle ridge.

marginal fissure A fracture bordering an igneous intrusion which has become filled with magma.

marginal sea A semienclosed sea adjacent to a continent, floored by submerged continental mass.

marginal thrust or **upthrust** Thrust fault along the margin of an intrusive that dips toward the intrusive.

marginal trench Narrow steepsided depression extending parallel to a continental margin and generally more than 1830 meters deeper than the general level of the adjacent ocean floor from which it is separated by an outer ridge 183 to 915 meters high.

marigram A graphic record of the rise and fall of the tide.

marine Of, or belonging to, or caused by the sea.

marine abrasion Erosion of a bedrock surface by the to-and-fro movements of an overlying layer of sand under the influence of waves.

marine-built terrace A terrace seaward of a marine-cut terrace which consists of materials removed in the cutting of the marine-cut terrace.

marine cave A cave formed by wave action; generally at sea level and affected by tides. *Syn:* SEA CAVE

marine-cut terrace Plain of marine abrasion, marine denudation, marine erosion, submarine denudation. Shore platform; wave-cut plain; wave-cut terrace.

marine denudation *1.* Used to signify the wearing or scouring action of water, or any chemical process affecting the floor of the ocean. *2.* Used to describe the action of the sea in breaking up and destroying the solid land.

marine plane The final stage of shore development.

marine terrace A narrow elevated seaward-sloping wave-cut platform exposed by uplift along a seacoast.

marine transgression The spread or advance of the sea over coastal land areas, as a result of rising sea level or subsidence of the land.

marker band Identifiable thin bed occurring at a particular stratigraphic position throughout a considerable area.

marker bed; marker horizon *1.* A bed which accounts for a characteristic segment of a seismic refraction time-distance curve and which can be followed over reasonably extensive areas. *2.* A bed which yields characteristic reflections over a more or less extensive area. *See* HORIZON. *3.* A stratigraphic bed selected for use in preparing structure maps, paleogeology maps, and others which emphasize the nature or attitude of a plane or surface.

Generally selected for lithologic characteristics, but biologic factors and unconformities may control. *Syn:* KEY BED. *See* HORIZON, *1.*

marl *1.* A calcareous clay, or intimate mixture of clay and particles of calcite or dolomite, usually fragments of shells. Marl in America is chiefly applied to incoherent sands, but abroad, compact, impure limestones are also called marls. *2.* Marl is an old term of considerable range of usage. In coastal-plain geology of the United States it has been used as a name for little indurated sedimentary deposits of a wide range of composition among which are slightly to richly calcareous clays and silts; fine-grained calcareous sands; clays, silts, and sands containing glauconite; and unconsolidated shell deposits. In the interior of the United States the name is used for calcareous deposits of lakes in which the percentage of calcium carbonate may range from 90 to less than 30%. *3.* A calcareous clayey rock soft enough to weather to a slope instead of a cliff.

marlstone; marlite An indurated mixture of clay materials and calcium carbonate (rarely dolomite), normally containing 25 to 75% clay.

marly Resembling marl; abounding with marl.

marmatite A dark iron-rich variety of sphalerite.

marsh *1.* Marshes proper are shallow lakes, the waters of which are either stagnant or actuated by a very feeble current; they are, at least in the temperate zone, filled with rushes, reeds, and sedge and are often bordered by trees which love to plunge their roots into the muddy soil. In the tropical zone a large number of marshes are completely hidden by a multitude of plants or forests of trees, between the crowded trunks of which the black and stagnant water can only here and there be seen. *2.* A tract of low, wet ground, usually miry and covered with rank vegetation. It may, at times, be sufficiently dry to permit tillage or hay-cutting, but requires drainage to make it permanently arable. It may be very small and situated high on a mountain, or of great extent and adjacent to the sea. *Cf.* BOG; SWAMP. *3.* In biological usage: A herbaceous plant-dominated ecosystem in which the rooting medium is inundated for long periods if not continually.

Marshall line Andesite line.

marsh gas Methane (CH_4), the chief constituent of natural gas. Results also from the partial decay of plants in swamps. In the miner's language, synonymous with fire damp.

martite The mineral hematite, Fe_2O_3, occurring in black octahedral crystals. A pseudomorph after magnetite.

massif [*Fr.*] *1. Geol.* and *Phys. geog:* A mountainous mass or a group of connected heights, whether isolated or forming a part of a larger mountain system. A massif is more or less clearly marked off by valleys. *2.* Isolated mountains rarely occur; they are either ranged together in a chain, or form some irregular combination, so as to constitute a group (massif) about the central summit. *3.* A mountainous mass which breaks up into peaks towards the summit, and has relatively uniform characteristics. *4.* Body of intrusive igneous or metamorphic rock at least 10 to 20 miles in diameter occurring as a structurally resistant mass in an uplifted area that may have been a mountain core.

massive *1. Petrol:* (a) Of homogeneous structure, without stratification, flow-banding, foliation,

schistosity, and the like; said of the structure of some rocks. Often, but incorrectly, used as synonymous with igneous and eruptive. (b) Occurring in thick beds, free from minor joints and lamination; said of some stratified rocks. *2. Mineral:* Without definite crystalline structure; amorphous; not a very good usage. *3.* Applied to shale to indicate it is very hard to split. *4.* A term properly applied to strata more than 100 mm. in thickness.

massive oölith Interior of granular, smooth, or chalk-textured material comprising nearly the entire mass of the spheroid.

mass movement Unit movement of a portion of the land surface as in creep, landslide, or slip.

mass properties The properties of a sediment as an aggregate, including porosity, permeability, density, color, etc.

mass spectrometer An instrument for separation and measurement of isotopic species by their mass.

mass susceptibility; specific susceptibility The magnetic susceptibility per unit mass, therefore equal to the magnetic susceptibility divided by the density.

mass transport The net transfer of water by wave action in the direction of wave travel. *See* ORBIT

mass wasting *1.* The slow downslope movement of rock debris. *2.* A general term for a variety of processes by which large masses of earth material are moved by gravity either slowly or quickly from one place to another.

master fracture. *1.* Zone of disturbance peripheral to a continent extending to a depth of several hundred miles and including the foci of deep-seated earthquakes. *2.* Great faults or fault systems generally involving much lateral movement that have played a prominent part in the development of earth structure.

master river One of the larger and dominating rivers of the drainage system of the land sculpture and base-leveling of any area of the earth's surface.

matrix 1. In a rock in which certain grains are much larger than the others, the grains of smaller size comprise the matrix. The groundmass of porphyritic igneous rocks. 2. The natural material in which any metal, fossil, pebble, crystal, etc., is embedded.

matterhorn A sharp, hornlike or pyramid-shaped mountain peak, somewhat resembling the Swiss peak of that name. See ARÊTE

mature 1. Having reached the maximum vigor and efficiency of action or the maximum development and accentuation of form. Said of streams, the sculpture of land by erosion, and the resultant topography. Cf. YOUNG; OLD. 2. Said also of sediment that is a long and complete cycle or two or more cycles removed from the original crystalline parent rock, so that all or nearly all weatherable material is absent.

mature river A river in the third and most perfect stage of development.

mature soil A soil with well-developed characteristics produced by the natural processes of soil formation, and in equilibrium with its environment. Generally consists of several differently characterized zones. See SOIL PROFILE

mature valley Mature valleys are wider, deeper, and have gentler gradients and more and larger tributaries than young valleys. In early maturity they are roughly U-shaped, instead of V-shaped, as before. In late maturity they have conspicuous flats.

maturity 1. Maturity may be said to last through the period of greatest diversity of form, or maximum topographic differentiation, but until about three-quarters of the original mass are carried away. During maturity no vestige of plain surface remains. 2. That stage in the development of streams or in land sculpture at which the process is going on with maximum vigor and efficiency or the maximum development and accentuation has been reached. Cf. YOUTH; OLD AGE

maturity index The measure of the progress of a sediment in the direction of chemical stability. A sediment may be said to be mature when it contains only the most stable mineral species and is relatively deficient in the more mobile oxides.

maximum 1. Struct. petro: A place on a point diagram or contour diagram, q.v., where there is a concentration of points. There may be several maxima within one diagram. 2. **gravity; magnetic; radioactivity** An anomaly in which the intensity has greater values at a central area compared to its immediate environment. Syn: HIGH

maxwell The practical, electromagnetic unit of magnetic flux in the c.g.s. system. Equal to one gauss-cm.^{2}.

M-boundary or **M-discontinuity** Mohorovičić discontinuity.

M crust Crustal layer, probably basaltic and about 8.3 km. thick, immediately above the Mohorovičić discontinuity in which plastic deformation occurs only as the result of tectonic disturbance.

meander 1. One of a series of somewhat regular and looplike bends in the course of a stream, developed when the stream is flowing at grade, through lateral shifting of its course toward the convex sides of the original curves. 2. A land-survey traverse along the bank of a permanent natural body of water.

meander belt 1. The part of a flood plain between two lines tangent to the outer bends of all the meanders. 2. The zone within which channel migration occurs. Within such a zone channel migration is indicated by abandoned channels, accretion topography, oxbow lakes. This is an engineering usage but is not recommended for geology.

meander core The central hill encircled by the meander.

meander cusp The eroded edge of an alluvial terrace made by a meandering stream.

meander cutoff The shortened channel resulting when a stream cuts through a meander neck.

meandering stream A characteristic habit of mature rivers; may be defined as winding freely on a broad flood plain, in rather regular river-developed curves.

meander line A surveyed line, usually irregular, but not a boundary line. A traverse line.

meander neck Strip of land between adjacent loops of a meandering stream.

meander scar Crescentic cuts in the upland bordering a stream, which were cut by lateral planation on the outer part of the meander loop.

meander scroll 1. Point bar. 2. Lake in well-defined portion of an abandoned river channel, commonly an oxbow.

meander spur A projection of high land into the concave part of a meander that has been undercut.

mean refractive index The mean of the values of the index of refraction for the extreme red and the extreme violet rays.

mean sea level The average

height of the sea for all stages of the tide.

Mean sea level is obtained by averaging observed hourly heights of the sea on the open coast or in adjacent waters having free access to the sea, the average being taken over a considerable period of time.

mean stress Algebraic average of three principal stresses.

measured reserves Reserves or resources for which tonnage is computed from dimensions revealed in outcrops, trenches, workings, and drill holes and for which the grade is computed from the results of detailed sampling.

mechanical analysis An analysis of the particle-size distribution of a sediment.

mechanical origin When the origin of a rock is effected by an external force, as, water flowing in streams, currents, etc., it is used in contradistinction to rocks which may be said to have a chemical origin, or whose particles have been consolidated by a chemical force.

mechanical seismograph A seismic detector in which, except for the mirror and beam of light, all amplification of the ground motion is accomplished by mechanical means. Now little used but employed extensively in earliest seismic prospecting.

mechanical twinning Deformation twinning, q.v.

medial moraine 1. Whenever two mountain glaciers bearing lateral moraines unite, the lateral moraines belonging to the two margins which coalesce give rise to a medial moraine. 2. An elongate body of drift formed by the joining of adjacent lateral moraines below the juncture of two valley glaciers. First used by Louis Agassiz.

median particle diameter A

measure of average particle size obtained graphically by locating the diameter associated with the midpoint of the particle-size distribution.

Medinian Lowermost of three Silurian series.

mediterranean *1.* If the sea penetrates into the interior of any continent, it forms there a mediterranean or inland sea, surrounded almost on all sides by land and leaving only a narrow opening into the ocean. *2.* Mesogeosyncline, *q.v.*

Mediterranean series, province, or **suite** A third great group of igneous rock magmas, high in potassium, and typified by certain igneous rocks in the Mediterranean area. *See* ATLANTIC and PACIFIC SERIES

medium *1.* The physical substance or environment as defined by its pertinent physical properties, e.g., a magnetic medium; one that is magnetizable. *2.* A substance on which a force acts or through which an effect is transmitted.

medium-volatile bituminous coal Nonagglomerating bituminous coal having 69% or more, and less than 78%, of fixed carbon (dry, mineral-matter-free), 31% or less, and more than 22%, of volatile matter (dry, mineral-matter-free).

meerschaum The massive form of the mineral sepiolite.

megacyclothem A rhythm of larger order than a single cyclothem, which includes distinct cycles of cyclothems.

megafauna Animals, living or fossil, which are large enough to be seen and studied with the naked eye.

megaflora Plants, living or fossil, which are large enough to be seen and studied with the naked eye.

megafossil Fossil plants or animals which are large enough to be seen with the naked eye.

megalospheric; megaspheric One of two different kinds of test which occur in many species of Foraminifera because of alternation of sexual and asexual modes of reproduction; the type of test which is small in overall size but which has a large proloculus.

megaripple Large gentle ripple-like feature composed of sand in very shallow marine situations; wave lengths reach 100 m. and amplitude about ½ m.; may be formed by tidal currents.

megascopic Macroscopic. A term applied to observations made with the unaided eye, as opposed to microscopic, made with the aid of the microscope. The term macroscopic is synonymous, but megascopic is preferred.

megashear A transcurrent fault of very large displacement, in excess of 100 kilometers.

megatectonics The tectonics of the very large structural features of the earth, or of the whole earth.

Meinesz zone Relatively long narrow continuous zone of marked negative gravity anomalies.

Meinzer unit The permeability unit most generally applicable to ground-water work and expressed in units of gallon, day, and square foot.

meizoseismal Of, or pertaining to, the maximum destructive force of an earthquake.

meizoseismal curve A curved line connecting the points of the maximum destructive energy of an earthquake shock around its epicentrum.

mélange A heterogeneous medley or mixture of rock materials; specif. a mappable body of deformed rocks consisting of a pervasively sheared, fine-grained, commonly pelitic matrix, thoroughly mixed with angular and poorly sorted inclusions of native and exotic tectonic fragments, blocks, or slabs (of diverse origins and geologic ages) that may be as much as several kilometers in length. Examples include argille scagliose and wildflysch.

melanic Dark-colored; refers to igneous rocks with color index between 30 and 90.

melanite A black variety of andradite garnet.

melanocratic Applied to dark-colored rocks, especially igneous rocks, containing between 60 and 100% of dark minerals, i.e., rocks whose color index is between 60 and 100. *See* LEUCO-CRATIC; MESOCRATIC

melanterite Copperas. A mineral, $FeSO_4 7H_2O$, an alteration product of iron sulfides. Monoclinic.

melilite A mineral group, of general formula $(Na,Ca)_2(Mg,Al)$-$(Al,Si)_2O_7$, the most common end-members including $Ca_2MgSi_2O_7$ (akermanite) and $Ca_2Al_2SiO_7$ (gehlenite). Tetragonal.

mellorite *See* FIRE-CLAY MINERAL

melting point That temperature at which a single pure solid phase changes to a liquid or a liquid plus another solid phase, upon the addition of heat at a specific pressure.

melt water; meltwater Water resulting from the melting of snow or of glacier ice.

member A division of a formation differentiated by separate or distinct lithology or complex of lithologies.

mendip A hill on a coastal plain which at one time was an offshore island.

Meramecan The third of four epochs of the Mississippian Period of North America; also the series of strata deposited during that epoch.

Mercator map projection A conformal map projection of the so-called cylindrical type. The equator is represented by a straight line true to scale; the geographic meridians are represented by parallel straight lines perpendicular to the line representing the equator; they are spaced according to their distance apart at the equator. The geographic parallels are represented by a second system of straight lines perpendicular to the family of lines representing the meridians, and therefore parallel with the equator.

mercury A heavy metal, highly toxic if breathed or ingested. Mercury is residual in the environment, showing biological accumulation in all aquatic organisms, especially fish and shellfish. Chronic exposure to airborne mercury can have serious effects on the central nervous system.

meridian A north-south line from which longitudes (or departures) and azimuths are reckoned; or a plane, normal to the geoid or spheroid, defining such a line.

meridian, magnetic The vertical plane in which a freely suspended symmetrically magnetized needle, influenced by no transient artificial magnetic disturbance, will come to rest. Also, a curve on the earth's surface tangent to such a plane at each place it touches.

meridian, principal (United States public-land surveys) A line extending north and south along the astronomical meridian passing through the initial point, along which township, section, and quarter-section corners are established. The principal meridian is the line from which is initiated the survey of the township boundaries along the parallels.

merocrystalline Hypocrystalline.

merohedral Merosymmetric.

Merostomata Class of arthropods including eurypterids and horseshoe crabs.

merostome A type of arthropod,

e.g., the horseshoe crab, the eurypterid, etc.

merosymmetric *Crystallog:* Having only part of the maximum symmetry of the crystal system concerned.

mesa [*Sp.*] A tableland; a flat-topped mountain or other elevation bounded on at least one side by a steep cliff; a plateau terminating on one or more sides in a steep cliff. Local in Southwest.

meseta [*Sp.*] A tableland.

mesh *1.* One of the openings or spaces in a screen. The value of the mesh is usually given as the number of openings per linear inch. This gives no recognition to the diameter of the wire, so that the mesh number does not always have a definite relation to the size of the hole. *2.* The unit component of patterned ground, e.g., circle, polygon, or intermediate form, not step or stripe.

mesh texture *See* RETICULATE

mesocratic Applied to rocks, especially igneous rocks, containing between 30 and 60% of dark minerals, i.e., rocks whose color index is intermediate between leucocratic and melanocratic.

mesocrystalline, *adj.* The texture of recrystallized rocks with crystal or grain size ranging from 0.20 to 0.75 mm.

mesogene Refers to an environment of mingled hypogene and supergene fluids.

mesogeosyncline A deep complex geosyncline occurring between two closely adjacent continents.

mesohaline *See* BRACKISH

mesolittoral zone Shore zone between high and low tides. *Cf.* LITTORAL

mesonorm Theoretical calculation of minerals in metamorphic rocks of the mesozone as indicated by chemical analyses.

mesosiderite A stony-iron meteorite consisting of nickel-iron en-

closing patches of stony matter composed of hypersthene and anorthite; olivine is also present, but generally as separately enclosed crystals, usually of fairly large size.

Mesozoic One of the grand divisions or eras of geologic time, following the Paleozoic and succeeded by the Cenozoic Era; comprises the Triassic, Jurassic, and Cretaceous periods. Also the erathem of strata formed during that era.

mesozone *1.* The "middle zone" of metamorphism. In this zone the distinctive physical conditions are high temperature and hydrostatic pressure and intense stress, and the rocks characteristically produced include mica schists, garnetiferous and staurolite schists, hornblende schists, amphibolite, and various types of crystalline limestones, quartzites, and gneisses. *2.* The intermediate metamorphic zone in the depth zone classification of metamorphic rock, characterized by minerals such as kyanite, staurolite, almandine, zoisite, etc.

metabentonite Metamorphosed, altered or somewhat indurated bentonite characterized by minerals that are not normally found in bentonites. Does not possess the swelling and absorptive properties of bentonite.

metacinnabar A mineral, HgS, dimorphous with cinnabar. Isometric. Black. A minor ore of mercury.

metacryst Any large crystal developed in a metamorphic rock by recrystallization, such as garnet or staurolite in mica schists. *Syn:* PORPHYROBLAST

metadiorite *1.* Metamorphosed diorite. *2.* Metamorphosed gabbro or diabase. *3.* Metamorphosed sedimentary rock in which both the minerals and texture of a diorite have been produced.

metagranite Granite produced by

metamorphism without even partial remelting.

metahalloysite *See* HALLOYSITE

metal *1.* Any of a class of substances that typically are fusible and opaque, are good conductors of electricity, and show a peculiar metallic luster, as, gold, bronze, aluminum, etc. Most metals are also malleable and comparatively heavy, and all except mercury are solid at ordinary temperatures. Metals constitute over three-fourths of the recognized elements. They form oxides and hydroxides that are basic, and they may exist in solution as positive ions. *2.* Ore from which a metal is derived. *3. Coal min:* In northern England, indurated clay or shale. *See* BIND. *4.* Cast iron, more particularly while melted. *5.* Broken stone for road surfaces or for railway ballast. *6.* Molten glass. *7.* Railway rails. *8.* Copper regulus or matte obtained in the English process.

metallic *1.* Of or belonging to metals, containing metals, more particularly the valuable metals that are the object of mining. *2.* Applied to minerals having the luster of a metal, as, gold, copper, etc.

metallic luster   A luster characteristic of metals in a compact state, and shown also by some other substances, as certain minerals.

metallogenetic province An area characterized by a particular assemblage of mineral deposits or by one or more characteristic types of mineralization. A metallogenetic province may contain more than one episode of mineralization, or metallogenic epoch.

metallogenic epoch Metallogenetic epoch; a unit of geologic time favorable for the deposition of ores or characterized by a particular assemblage of mineral deposits. Several metallogenic epochs may be represented within a single area, or metallogenic province.

metallogenic mineral Ore mineral.

metallogenic province Metallogenetic province.

metallogeny The study of the genesis of mineral deposits, with emphasis on their relationship in space and time to regional petrographic and tectonic features of the earth's crust.

metalloid *1.* An alkali metal, as sodium, or an alkaline-earth metal, as calcium; so called by Sir Humphry Davy because they were not supposed to be well-defined metals. *2.* Certain elements, as arsenic, antimony, that share the properties of metals and nonmentals. *3.* Having the appearance of a metal.

metallurgy The science and art of preparing metals for use from their ores by separating them from mechanical mixture and chemical combination. It includes various processes, as smelting, amalgamation, electrolytic refining, etc.

metamict A mineral, containing radioactive elements, in which varied degrees of lattice disruption have taken place as a result of radiation damage, while at the same time retaining the original external morphology.

metamorphic differentiation Segregation of certain minerals into lenses and bands accomplished by metamorphism.

metamorphic facies Group of metamorphic rocks characterized by particular mineral associations indicating origin under restricted temperature-pressure conditions.

metamorphic grade The grade or rank of metamorphism depends upon the extent to which the metamorphic rock differs from the original rock from which it was derived. If a shale is converted to slate or phyllite the metamorphism is low grade, if it is converted to mica schist with garnet, staurolite, or kyanite the metamorphism is middle grade, and if it is converted to mica schist with garnet and sillimanite the metamorphism is high grade.

metamorphic rank Metamorphic grade.

metamorphic rock Includes all those rocks which have formed in the solid state in response to pronounced changes of temperature, pressure, and chemical environment, which take place, in general, below the shells of weathering and cementation.

metamorphic water Water that is or has been associated with rocks during their metamorphism.

metamorphism Process by which consolidated rocks are altered in composition, texture, or internal structure by conditions and forces not resulting simply from burial and the weight of subsequently accumulated overburden. Pressure, heat, and the introduction of new chemical substances are the principal causes, and the resulting changes, which generally include the development of new minerals, are a thermodynamic response to a greatly altered environment. Diagenesis has been considered to be incipient metamorphism. *See* CONTACT, THERMAL, DYNAMIC, REGIONAL, HIGH RANK, and LOW RANK METAMORPHISM

metamorphosis *Biol:* Process of more or less complete bodily reorganization of an animal that transforms a larva into a succeeding stage of development and growth; common in many invertebrates.

metargillite Rock characterized by weak metamorphic reconstitution, no recrystallization, and without slaty cleavage or foliation.

metasediments Partly metamorphosed sedimentary rocks.

metasomatic *Geol:* Characteristic of, pertaining to, produced by, or occurring during metasomatosis. The term is especially used in connection with the origin of ore deposits. The corresponding noun is metasomatosis, but "replacement" is a good English equivalent.

metasomatism *1.* (Replacement) The process of practically simultaneous capillary solution and deposition by which a new mineral of partly or wholly differing chemical composition may grow in the body of an old mineral or mineral aggregate. *2.* The processes by which one mineral is replaced by another of different chemical composition owing to reactions set up by the introduction of material from external sources. *3.* A practically simultaneous solution and deposition, through small openings, usually submicroscopic, and mainly by hypogene water solutions, by which a new mineral of partly or wholly differing composition may grow in the body of an old mineral or mineral aggregate.

metasomatite A rock in which one substance has completely replaced another, or certain ones completely replace others, producing an entirely new composition.

metasome An individual mineral developed in another mineral.

metastable, *adj. 1.* Stable with respect to small disturbances, but capable of reaction with evolution of energy if disturbed sufficiently. E.g., a mixture of hydrogen and oxygen is metastable, because it is unchanged by jarring or by mild warming, but explodes violently if touched by a flame. *2.* Any phase is said to be metastable when it exists in the temperature range in which another phase of lower vapor pressure is stable. It is not necessary that a vapor phase be present.

metastasis Lateral shifting of the

earth's crust as in the movement of continental masses.

metastasy Metastasis.

metavolcanics Partly metamorphosed volcanic rocks.

Metazoa The large group of animals that have bodies consisting not only of more than one cell, but in which the cells are arranged in two layers, as in the gastrula stage, and are embryologically known as endoderm and ectoderm, q.v.

meteor Originally a general term for any atmospheric phenomenon, and still sometimes used in this sense, particularly in such terms as hydrometeor, optical meteor, etc. Now more commonly restricted to astronomical meteors—sometimes called shooting or falling stars—which are relatively small bodies of matter traveling through interplanetary space, and which are heated to incandescence by friction when they enter the atmosphere, and are either wholly or partially consumed; in the latter case they reach the earth's surface as meteorites.

meteor crater Topographic depression formed by the impact of a large meteor.

meteoric Of or belonging to a meteor, or to the atmosphere. Thus meteoric erosion implies that caused by rain, wind, or other weathering forces of the atmosphere.

meteoric water That which occurs in or is derived from the atmosphere.

meteorite Naturally occurring mass of matter that has fallen to the earth's surface from outer space.

meteorology The science dealing with the atmosphere and its phenomena, especially as relating to weather.

meter; metre A unit of length equivalent in the United States to 39.37 inches exactly.

methane A gaseous hydrocarbon, formula CH_4; the simplest member of the paraffin series. See MARSH GAS

methylene iodide CH_2I_2. In pure form, a straw-colored liquid with a specific gravity of 3.32 at 18° C., a m.p. between 5° and 6° C., a decomposition point of 180° C., and a refractive index of 1.74.

mev Abbreviation for million electron volts.

mgd Millions of gallons per day. Mgd is commonly used to express rate of flow.

mg/l Milligrams per liter; approximately equal to one ppm, q.v.

miargyrite A mineral, $AgSbS_2$. Monoclinic.

miarolitic *Petrol:* Applied to small angular cavities in plutonic rocks, especially common in granites, into which small crystals of the rock-forming mineral project. Also said of igneous rocks containing such cavities; characteristic of, pertaining to, or occurring in such cavities.

mica A mineral group, consisting of phyllosilicates with sheetlike structures. Mostly monoclinic, characterized by very perfect basal cleavage. The general formula of the group is $(K,Na,Ca)(Mg, Fe, Li, Al)_{2-3}(Al, Si)_4O_{10}(OH,F)_2$. See BIOTITE; LEPIDOLITE; MUSCOVITE; PARAGONITE; PHLOGOPITE; ROSCOELITE

mica book Mica crystal that splits readily into thin elastic layers.

micelle A unit of structure built up from complex molecules in colloids.

micrite A limestone with very fine subcrystalline texture, such as comprises most of a sublithographic limestone.

microchemical tests Chemical tests made on minute objects under a microscope. The form, color, and optical properties of the minute crystals are also used.

microcline A mineral, a member of the feldspar group. Composition $KAlSi_3O_8$. Triclinic; dimorphous with orthoclase. A common mineral of granite rocks.

microcrystalline Applied to a rock in which the individual crystals can only be seen as such under the microscope.

microevolution Changes within potentially continuous populations; it probably differs from macroevolution, q.v., only in degree.

microfabric Refers to the microscopic physical constitution of rock.

microfacies The composition, structure, or appearance of a rock or mineral as seen in thin section under the microscope.

microfauna Animals, living or fossil, that are normally so small that their details must be studied under a microscope.

microflora Living or fossil animals too small to be seen with the naked eye. The term is sometimes used synonymously with palynoflora but is more frequently applied to living microscopic algae and fungi. A very localized or small group of plants: plants occupying a very small habitat.

microfold A fold so small that it is observable only in thin section under the petrographic microscope. Generally shown by the parallel arrangement of some elongated crystals.

microfoliation A foliation in rocks distinguished only with the aid of the microscope.

microfossils 1. Fossil remains of organisms whose average representatives are microscopic in size. 2. Fossil remains of microscopic organisms (as in 1 above) and also the fossil remains of skeletal elements of organisms commonly occurring in detached form and requiring microscopic study for identification. 3. Fossil remains as in 1 and 2 and also

the depionic and neanic stages of megafossils, as well as dwarfed forms.

microgranular See MICROCRYSTALLINE

microlite 1. A mineral, $(Ca,Na)_2(Ta,Nb)O_6(OH,F)$. Isometric. An ore of tantalum. Cf. PYROCHLORE. 2. A minute crystal.

microlitic A textural term applied to porphyritic igneous rocks whose groundmasses consist of an aggregate of differently oriented or parallelly oriented microlites in a base that is generally glassy. Hyalopilitic, pilotaxitic, and trachytic textures are all included under this term. When the microlites are not in parallel arrangement, the term felty is synonymous.

microlog (Schlumberger) A resistivity log in borehole surveying obtained with a device consisting of closely spaced electrodes, the arrangement of which is basically the same but in miniature, as the normal and lateral devices in the regular electric survey (ES), q.v. It is designed to measure the resistivity of a small volume of rock next to the borehole.

micrometeorite A very small meteorite or meteoritic particle generally less than one millimeter in diameter.

micromillimeter 1/1,000,000 mm., abbreviation $M\mu$.

micron A unit of length equal to the one one-millionth of a meter and usually denoted by the symbol μ.

micropaleontology Paleontology dealing with microfossils. Syn: MICROGEOLOGY

microperthite A fine-grained intergrowth of potassic and sodic feldspar visible only with the aid of the microscope. See PERTHITE

microphotograph Greatly enlarged photograph made through a microscope.

microplankton Plankton ranging

in size from 60 microns to 1 millimeter, including most phytoplankton.

micropoikilitic Microscopically poikilitic, a microscopic shiller structure. *See* POIKILITIC

microporphyritic Microscopically porphyritic. Essentially synonymous with microphyric.

microscopic *1.* Of, pertaining to, or conducted with the microscope or microscopy; as, a microscopic examination. *2.* Like a microscope; able to see very minute objects. *3.* So small or fine as to be invisible or not clearly distinguished without the use of a microscope. Hence, loosely, very small; minute. Microscopically, *adv.*

microseismometer; microseismograph An apparatus for indicating the direction, duration, and intensity of microseisms.

microseisms More or less persistent feeble earth tremors due to natural causes such as winds, strong ocean wave motion, etc.

microspheric Refers to one form of dimorphic Foraminifera which begins with a small initial chamber and grows to a relatively large size. *Cf.* MEGALOSPHERIC

microstructure A structural feature of rocks that can be discerned only with the aid of the microscope. *Ant:* MACROSTRUCTURE

microstylolite A type of microscopic grain boundary relationship indicating differential solution between two mineral grains and characterized by fine interpenetrating teeth; boundary often marked by a little opaque material concentrated along it. *Cf.* STYLOLITES

microtectonics That phase of structural geology dealing with small features, especially those that must be investigated under the microscope.

Mid-Atlantic ridge A volcanic mountain range that extends es-

sentially north-south along the center of the Atlantic Ocean.

middle limb The overturned limb shared by a pair of adjacent overturned anticline and syncline.

Mid-Indian ridge A volcanic mountain range that extends from the South Pacific Ocean through the Indian Ocean, intersecting the Gulf of Aden between Saudi Arabia and Africa.

mid-ocean canyon Steep-walled, flat-floored continuous depression up to 5 miles wide and 600 feet deep that traverses an abyssal plain; such canyons commonly lead into or out of abyssal gaps.

mid-oceanic islands Isolated features of the earth's relief which rise from the deep-sea floor.

mid-oceanic ridge Mid-ocean ridge.

mid-ocean ridge A continuous, seismic, median mountain range extending through the North and South Atlantic oceans, the Indian Ocean, and the South Pacific Ocean. It is a broad fractured swell with a central rift valley and usually extremely rugged topography, one to three km. in elevation, about 1500 km. in width, and over 84,000 km. in length. According to the hypothesis of sea floor spreading, the mid-ocean ridge is the source of new crustal material. *Syn:* MID-OCEAN RISE; MID-OCEANIC RIDGE

mid-ocean rift Rift valley.

mid-ocean rise Mid-ocean ridge.

migma *1.* A term, originally proposed by M. Reinhard in 1934, which has been variously used, but in general signifies a material having many if not all characteristics in common with magma, but derived from the granitization process. Differentiated from magma only on a genetic basis. *2.* Viscous, domed masses (diapirs) rising in orogenic zones, in a way to leave it very uncertain whether there is a magma or not.

migmatite Rock consisting of

composite igneous or igneous-looking and/or metamorphic materials. *Cf.* COMPOSITE GNEISS; INJECTION GNEISS

migration *1.* The movement of oil, gas, or water through porous and permeable rock. Parallel (longitudinal) migration is movement parallel to the bedding plane. Transverse migration is movement across the bedding planes. *2.* Plotting of seismic reflection time information to allow for the fact that reflecting points do not lie vertically beneath the point of observation. Involves both reflector dip and variation of velocity. Often neglects effects perpendicular to the seismic line.

migratory dune A dune, such as the barchan, which undergoes a translocation more or less as a unit under continued wind action.

Milankovitch curve That curve drawn for any latitude and any geologic time that represents the amount of solar energy received by that latitude; computed originally from an analysis of the solar and earth aberrations.

military geology The application of the principles of geology and related earth sciences, such as soil science, botany, and climatology, to the solution of military problems such as terrain analysis, location of construction materials, foundation conditions, water supply, cross-country movement, airfield siting, road construction and maintenance, siting of underground installations, and determining the effect of ground upon mine detectors.

Miller indices A set of three or four symbols used to define the orientation of a crystal face or internal crystal plain. The indices are determined by expressing the reciprocals of the intercept of the face or plane on the crystallographic axes, and reducing, if necessary, to the lowest integers, retaining the same ratio. General

indices (h, k, l), specific indices, e.g., (201).

millerite A mineral, nickel sulfide, NiS. Hexagonal rhombohedral.

millidarcy The customary unit of measurement of permeability. One one-thousandth of a darcy, *q.v. See* PERMEABILITY

milling ore *1.* A dry ore that can be amalgamated or treated by leaching and other processes; usually these ores are low-grade, free or nearly so, from base metals. *2.* Any ore that contains sufficient valuable minerals to be treated by any milling process.

millstone A hard, tough stone used for grinding cereals, cement rocks, and other materials. Usually a coarse-grained sandstone or fine quartz-conglomerate.

Mima mound; pimple mound *See* PIMPLE PLAIN

mimetic Imitative. Applied to crystals which, by a twinning or malformation resemble simple forms of a higher grade of symmetry.

mimetite A mineral, $Pb_5(AsO_4)_3Cl$, commonly containing some Ca and phosphate. Hexagonal.

minable Capable of being mined.

Mindel The second of four classical glaciations in Europe; there are now known to have been earlier glaciations than the "first" of these four.

Mindel-Riss The term applied in the Alps to the second interglacial stage of the Pleistocene Epoch, after the Mindel glacial stage and before the Riss. *See also* YARMOUTH

mine *1.* In general, any excavation for minerals. More strictly, subterranean workings, as distinguished from quarries, placers, and hydraulic mines, and surface or open works. In a military sense, a mine is a subterranean gallery run under an enemy's works to be subsequently exploded. *2.* Any deposit of mineral or ore suitable for extraction, as, an ore deposit. *3.* The term mine, as used by quarrymen, is applied to underground workings having a roof of undisturbed rock. It is used in contrast with the open-pit quarry. *4.* To dig a mine; to get ore, metal, coal, or precious stones out of the earth; to dig in the earth for minerals; to work in a mine.

mineral A naturally formed chemical element or compound having a definite range in chemical composition, and usually a characteristic crystal form.

mineral cycle *1.* The biochemical cycle, *q.v.,* in which mineral is transferred from soil to plant life to animal life, and returned to the soil. *2.* Properly used, the above cycle plus the crustal cycle, *q.v.*

mineral deposit Any valuable mass of ore.

mineral facies; metamorphic facies All rocks that have originated under temperature-pressure conditions so similar that a definite chemical composition has resulted in the same set of minerals regardless of the manner of crystallization or recrystallization.

mineralization *1.* The process of replacing the organic constituents of a body by inorganic fossilization. *2.* The addition of inorganic substances to a body. *3.* The act or process of mineralizing. *See* MINERALIZE. *4.* The process of converting or being converted into a mineral, as, a metal into an oxide, sulfide, etc.

mineralize *1.* To petrify. *2.* To impregnate or supply with minerals. *3.* To promote the formation of minerals.

mineralized zone A mineral-bearing belt or area extending across or through a district. It is usually distinguished from a vein or lode as being wide, the mineralization extending in some cases hundreds of feet from a fissure or contact plane. *Cf.* CONTACT DEPOSIT. *See* ZONE

mineralizer Mineralizing agent. Substances, especially water and other gases, which, when present in solution in magmas, lower the liquidus temperature and the viscosity, aid crystallization, and permit the formation of minerals containing them. Hydrothermal fluids are presumed to be formed by the concentration of such mineralizers.

mineralogical phase rule See GOLDSCHMIDT'S MINERALOGICAL PHASE RULE

mineralogy The science of the study of minerals.

mineraloid Used to designate materials that are commonly considered to be minerals, but are amorphous, hence excluded by some definitions. Example: allophane.

mineral oil; naphtha A limpid or yellowish liquid, lighter than water, and consisting of hydrocarbons. Petroleum is heavier than naphtha, and dark green in color when crude. Both exude from the rocks, but naphtha can be distilled from petroleum. *See* PETROLEUM

mineral rights Under the law of the United States, mineral rights do not go with the land, necessarily. In other words, some fee land does not have mineral rights, those rights being retained by the government. Fee land that originally had the mineral rights can be sold separately from the mineral rights and vice versa. The state of Texas has given the federal government certain land, but retained the mineral rights.

mineral spring A spring whose water contains large quantities of mineral salt, either those commonly occurring in the locality or of a rare or uncommon character.

mineral water Mineral waters are those which contain some mineral salt generally in sufficient quantities to affect the taste.

mineral wool A substance outwardly resembling wool, presenting a mass of fine interlaced filaments, made by subjecting furnace slag or certain rocks while molten to a strong blast.

miner's inch A measure of water flow equal to 1.5 cubic feet per minute.

minimum *1.* The lowest observed value of temperature, pressure, or other weather element during any given period. *2.* gravity or magnetic low. An anomaly in which the intensity has values below the level of its more or less immediate environment.

minimum duration The time necessary for steady-state wave conditions to develop for a given wind velocity over a given fetch length.

minimum pendulum A pendulum used in gravity measurements so designed that its period is at a minimum with respect to changes of its effective length. Among factors which may tend to change this length are temperature, creep, or knife-edge wear.

minimum time path *Geophys:* Brachistochronic path; least-time path. A Fermat path between two points along which the time of travel is a true minimum.

mining claim That portion of the public mineral lands which a miner, for mining purposes, takes and holds in accordance with mining laws.

mining engineer One versed in, or one who follows, as a calling or profession, the business of mining engineering.

mining engineering That branch of engineering dealing with the excavation and working of mines. It includes much of civil, mechanical, electrical, and metallurgical engineering.

mining width The minimum width necessary for the extraction of the ore regardless of the actual width of ore-bearing rock.

minor elements Trace elements.

Miocene The fourth of the five epochs into which the Tertiary Period is divided. Also the series of strata deposited during that epoch.

miogeosyncline A geosyncline in which volcanism is not associated with sedimentation. The nonvolcanic aspect of an orthogeosyncline, located near the craton.

miohaline *See* BRACKISH

mirabilite Glauber's salt. A mineral, $Na_2SO_4.10H_2O$. Monoclinic.

mire A small muddy marsh or bog; wet, spongy earth; soft, deep mud.

miscibility The property enabling two or more liquids to mix and form one phase when brought together. *Ant:* IMMISCIBILITY

miscibility gap *See* IMMISCIBILITY GAP

mispickel Arsenopyrite.

Mississippian The fifth of seven periods into which the Paleozoic is divided in the United States and some other parts of North America. Approximately equivalent to the Lower Carboniferous of the rest of the world. Also the system of rocks formed during that period.

Missourian Upper Middle Pennsylvanian.

mix crystals Solid solution.

mixed base Term applied to a crude oil in which both the paraffin and the naphthene hydrocarbons are present but neither group dominates.

mixed layer A water layer mixed by wind, wave, or thermal action.

mixed-layer clay minerals Clay crystallites each of which contains two or more differently behaving layer types. Commonly, the differences among layer types are hydration level.

mixed-layer crystal Crystal consisting of either regularly or randomly stratified intergrowths of different related minerals. Such interlayering of clay minerals is common.

mixed liquor A mixture of activated sludge and water containing organic matter undergoing activated sludge treatment in the aeration tank.

mobile belt A portion of the crust of the earth, generally long compared to its width and many scores of miles wide, that is more mobile, as evidenced by geosynclines, folds, and faults, than the adjoining stable blocks of the crust.

mobile shelf 1. Area that alternated between continental and shallow marine conditions and later was strongly folded and faulted. 2. Part of continental platform that subsided intermittently to produce a basin.

mobilization Any process whereby solid rock is sufficiently liquefied to permit it to flow. See RHEOMORPHISM

mode The actual mineral composition of an unaltered igneous rock; contrasted with norm, q.v.

modulus A number or quantity that measures a force or function. See BULK MODULUS; YOUNG'S MODULUS

modulus of compression Bulk modulus. A measure of the stress-strain ratio under simple hydrostatic pressure; defined as $K=\dfrac{FV}{A\triangle V}$; where F/A=pressure, V=volume, and $\triangle V$=change in volume.

modulus of elasticity Incompres-

sibility modulus; modulus of volume elasticity. The ratio of any stress to the resulting strain. The principal moduli of interest in seismology are: (1) Young's modulus for tension; (2) rigidity modulus for shear; (3) bulk modulus for hydrostatic compression or dilation.

modulus of rigidity Shear modulus. A measure of the stress-strain ratio for simple shear defined as $n=\dfrac{FL}{A\triangle L}$, where F/A= tangential force per unit area, L=distance between shear planes, and $\triangle L$=the shear strain produced.

modulus of rupture The measure of the force which must be applied longitudinally in order to produce rupture.

modulus of volume elasticity Incompressibility modulus; modulus of elasticity, q.v.

mofette A fissure or other opening, occurring in a region of recent volcanic activity, from which water vapor and carbon dioxide are emitted. A type of fumarole, q.v.

mogotes Huge castlelike residual masses of limestone between flat-floored valleys; they are riddled with cavernous passages.

Moho Abbreviation for Mohorovičić discontinuity.

mohole Proposed deep borehole to penetrate into the earth's mantle below the Mohorovičić discontinuity.

Mohorovičić discontinuity Seismic discontinuity situated about 35 km. below the continents and about 10 km. below the oceans which separates the earth's crust and mantle.

Mohr's salt Ferrous ammonium sulfate.

Mohs' scale of hardness The hardness scale consisting of 10 minerals from talc softest—

through gypsum, calcite, fluorite, apatite, orthoclase, quartz, topaz, corundum, to the hardest—diamond.

moisture Water diffused in the atmosphere or the ground, including soil water.

moisture equivalent (of soils) The ratio of (1) the weight of water which the soil, after saturation, will retain against a centrifugal force 1000 times the force of gravity, to (2) the weight of the soil when dry. The ratio is stated as a percentage.

molarity Concentration of a solution expressed as moles of solute per liter of solution. Symbol M; thus 0.5M HCl is a solution of HCl containing half a mole of HCl per liter.

molasse *1.* A descriptive term for a paralic (partly marine, partly continental or deltaic) sedimentary facies consisting of a very thick sequence of soft, ungraded, cross-bedded, fossiliferous conglomerates, sandstones, shales, and marls, characterized by primary sedimentary structures and sometimes by coal and carbonate deposits. It is more clastic and less rhythmic than the preceding flysch facies. *2.* An extensive, postorogenic sedimentary formation representing the totality of the molasse facies resulting from the wearing down of elevated mountain ranges during and immediately succeeding the main paroxysmal (diastrophic) phase of an orogeny, and deposited considerably in front of the preceding flysch; specif. the Molasse strata, mainly of Miocene and partly of Oligocene age, deposited on the Swiss Plain and Alpine foreland of southern Germany subsequent to the rising of the Alps.

mold The impression left in the surrounding rock by a shell or other organic structure.

mold, external A mold which shows the form and markings of the outer part of the original shell or organism. Also spelled mould.

mold, internal A mold which shows the form and markings of the inner surfaces of a shell or organism.

mold, natural The cavity left after solution of the original shell or organism, bounded by the external impression (external mold) and the surface of the internal filling (core or steinkern).

molding sand A mixture of sand and loam used by founders in making sand molds.

molecular norm Niggli's classification system, *q.v.*

molecular replacement The petrifaction of an organic substance at such a gradual molecule by molecule rate that the very finest details of the original structure are preserved. *Syn:* HISTOMETABASIS

mole fraction The mole fraction of a given component in a phase is equal to the number of moles of the component in question, divided by the total number of moles of all components in the phase. Mole fractions are thus useful in defining the composition of a phase.

mollisol A group of soils approximately equivalent to soils of the grassland, excluding vertisols; characterized with organic enriched A horizon, accreted B horizon, and a dark surface layer; little or no leaching and adequate to more than adequate accretion for nutrient purposes.

Mollusca A phylum of unsegmented, nonhydroporic, coelomate invertebrate animals that includes the gastropods pelecypods, cephalopods, etc. Of great importance paleontologically.

mollusk One of the Mollusca. Also spelled mollusc.

Mollweide projection An equal-area map projection showing the entire surface of the earth. The entire projection is contained within an ellipse whose major axis (the equator, representing 360° of longitude) is twice the length of the minor axis (the central meridian). The central hemisphere is a circle and all meridians are ellipses.

molybdate A salt or ester of molybdic acid; a compound containing the radical MoO_4^{--}.

molybdenite A mineral, MoS_2. The principal ore of molybdenum. Hexagonal.

monadnock 1. By long continued erosion a land surface may be reduced to an almost level plain, but there may still be a few hills, which, having as yet escaped final destruction, rise conspicuously above the plain. These are monadnocks. The term connotes nothing in regard to form or structure; it means merely a residual of an old topography standing about a plain of subaerial erosion. 2. A residual rock, hill, or mountain standing above a peneplain.

monazite A mineral, (Ce,La)-PO_4, commonly containing thorium. The principal ore of the rare earths and thorium. Monoclinic.

monobasic acid An acid containing one replaceable hydrogen atom per molecule, e.g., HCl, HNO_3.

monoclinal 1. Adjective derived from monocline, q.v. 2. Term signifying sameness in direction of dip. 3. Applied to strata that dip for an indefinite or unknown length in one direction and which do not apparently form sides of ascertained anticlines or synclines. 4. Having the beds sloping in only one direction.

monoclinal scarp A scarp resulting from a steep downward flexure between an upland block and a tectonic basin.

monoclinal shifting (of streams) Flowing along the strike of dipping beds, streams do not usually sink their channels vertically, but shift them down dip at the same time that they are deepened. This process is known as monoclinal shifting of streams.

monocline 1. Strata that dip for an indefinite or unknown length in one direction, and which do not apparently form sides of ascertained anticlines or synclines. 2. Beds inclined in a single direction. 3. A one-limbed flexure in strata which are usually flat-lying except in the flexure itself. 4. A steplike bend in otherwise horizontal or gently dipping beds.

monoclinic symmetry *Struct. petrol:* May refer to either the movement or the fabric. Monoclinic symmetry of movement is analogous to the sliding of cards over one another in one direction; in this movement there is one plane of symmetry parallel to <u>a</u>, the direction of tectonic transport. In monoclinic symmetry of fabric there is one plane of symmetry parallel to <u>a</u>, the direction of tectonic transport.

monoclinic system One of the six crystal systems characterized by either the single twofold axes of symmetry, a single plan of symmetry, or a combination of the two. Crystals belonging to this system are referred to three unequal crystallographic axes, two of which intersect obliquely and the third perpendicular to the first two.

monogenetic soil Soil produced

under a single set of continuing conditions, i.e., parent material, climate, relief, biologic activity, etc.

monogeosyncline A primary geosyncline that is long, comparatively narrow, deeply subsided, composed of shallow water sediments and situated within a continent along the inner border of the borderlands.

monomeric Refers to a simple, rather than a multiple molecular structure. In organic chemistry a term used in contrast to *polymeric*.

monomictic lake A lake with only one overturn per year. There are two types: (1) Warm—water never below 4° C., overturn in the winter; and (2) cold—water never above 4° C., overturn in the summer.

monomineralic rock A rock consisting of essentially one mineral; the amounts of other minerals tolerated under the definition vary with the authors, e.g., anorthosite.

Monongahelan Upper Pennsylvanian.

monophyletic Derived through a single line of descent.

monoschematic Applied to a body of rock or mineral deposit whose fabric is identical throughout. Contrasted with macropolyschematic.

monothem A stratigraphic unit; that part of a series of intergrading magnafacies, *q.v.*, that can be bounded by such stratigraphic markers as key beds, bentonite beds, etc., that approach synchroneity.

monotropic forms In certain instances of polymorphism, the vapor-pressure curves of the two forms do not meet below the melting point. They therefore lack a stable inversion point, and the form with the higher vapor pressure is metastable with respect to the other at all temperatures below the melting point. Such higher-vapor-pressure forms are called monotropic and are not interconvertible.

monotype The holotype of a species which was described from a single specimen.

monotypical Applied to a genus that included only one species at the time of publication, or a species that included only one specimen.

monotypy The condition of having been established on a single known form.

monticellite A mineral; $CaMgSiO_4$, usually occurring in contact-metamorphosed limestones. Orthorhombic.

montmorillonite *1.* A group of clay minerals whose formulas may be derived by ion substitution in the general formula $Al_2Si_4O_{10}(OH)_2$ with deficiencies in charge in both the tetrahedral and octahedral positions being balanced by cations located between the three-layer crystal lattice. They are characterized by swelling in water and extreme colloidal behavior. Mineral members of the group are montmorillonite, nontronite, saponite, hectorite, sauconite. Also referred to as the smectite group. *2.* A specific member of the montmorillonite group, $(Na,Ca)_{0.33}(Al,Mg)_2Si_4O_{10}(OH)_2 \cdot nH_2O)$.

monzonite A granular plutonic rock containing approximately equal amounts of orthoclase and plagioclase and thus intermediate between syenite and diorite. Quartz is usually present, but if it exceeds 2% by volume the rock is classified as quartz monzonite or adamellite. Either hornblende or diopside or both are present and biotite is a common constituent. Accessories are apatite, zircon, sphene, and opaque oxides.

moon Any natural satellite of a planet.

moonstone A variety of feldspar, commonly translucent alkali feldspar (sanidine or anorthoclase), or labradorite that exhibits an opalescent play of colors. Used as a gem.

morainal lakes They owe their existence to the blockade of valley or drainage courses by glacial drift. The term drift-barrier lakes would be the more accurate name.

moraine *1.* Drift, deposited chiefly by direct glacial action, and having constructional topography independent of control by the surface on which the drift lies. *2.* An accumulation of drift having initial constructional topography, built within a glaciated region chiefly by the direct action of glacier ice. The term has been used in many different ways and its history is confused. *3.* The floated blocks sometimes carried on lava streams, or the moraines of lava flows as they may suggestively be termed. *4.* The porous, secondary material at the base of the aperture in some species of Endothyra, an important Mississippian foraminifer.

moraine terrace When an alluvial plain or alluvial cone is built against the side or front of a glacier and the glacier is afterward melted away, the alluvial surface becomes a terrace overlooking the valley that contained the ice.

morphogenesis The origin of morphological characteristics. *See also* HAECKEL'S LAW

morphogenetic region A region in which, under a certain set of climatic conditions, the predominant geomorphic processes will give to the landscape characteristics that will set it off from those of other areas developed under different climatic conditions.

morphogeny The production or evolution of morphological characters.

morphologic Pertaining to morphology, *q.v.*

morphologic species Species based solely on the similarity of morphologic characters.

morphologic unit *1.* Rock stratigraphic unit identified by its topographic features, e.g., a Pleistocene glacial deposit. *2.* A surface either depositional or erosional recognized by its topographic character.

morphology *1.* The observation of the form of lands. *2.* The study of the form and structure of organisms.

morphometry Measurement of shape.

morphotype The type specimen of a different form of a dimorphic or polymorphic species.

Morrowan Lowermost Pennsylvanian.

mortar structure A mechanical structure produced by dynamic metamorphism upon granites and gneisses in which small crushed grains of quartz and feldspar occupy the interstices between larger individuals, resembling stones set in mortar.

morvan A region of composite structure, consisting of an older undermass, usually made up of deformed crystalline rocks, that had been long ago worn down to small relief and that was then depressed, submerged, and buried beneath a heavy overmass of stratified deposits, the composite mass then being uplifted and tilted, the tilted mass being truncated across its double structure by renewed erosion, and in this worn-down condition rather evenly uplifted into a new cycle of destructive evolution.

mosaic *1. Petrol:* Applied to the

texture sometimes seen in dynamo-metamorphosed rocks whose crystal fragments are angular and granular and appear, in polarized light, like the pieces of a mosaic. 2. A picture formed by matching together parts of a number of overlapping vertical aerial photographs taken from different camera positions.

mosaic breccia A breccia is one whose fragments have been largely but not wholly disjointed and displaced. Some fragments match along adjacent surfaces.

mosor A monadnock that has survived because of remoteness from the main drainage lines.

mother liquor The residual solution, often impure or complex, which remains after the substances readily and regularly crystallizing have been removed.

mother lode 1. The principal lode or vein passing through a district or particular section of country. 2. The "Great Quartz Vein" in California, traced by its outcrop for 80 miles from Mariposa to Amador.

mottle, n. Irregular small body of material in sedimentary matrix of different texture; difference in color is not essential.

mottled Irregularly marked with spots of different colors. Mottling in soils usually indicates poor aeration and lack of good drainage.

mottled limestone Limestone with narrow branching or anastomosing fucoidlike cylindrical masses of dolomite, often with a central tube or hole. May be organic or inorganic in origin.

moulding or **molding sand** See FOUNDRY SAND

moulin 1. In some places the fields of ice are hollowed out into perpendicular wells, known under the name of moulins (mills), because of the roaring noise of the water which is engulfed in them. 2. An area of a glacier is unbro-

ken, and driblets of water have room to form rills; rills to unite and form streams; streams to combine to form rushing brooks, which sometimes cut deep channels in the ice. 3. A circular depression on the surface of a glacier in the ablation zone into which melt water funnels. *Syn:* GLACIER MILL

mound A low hill of earth, natural or artificial; in general, any prominent, more or less isolated hill.

mount A mountain, or a high hill. Used always instead of "mountain" before a proper name.

mountain 1. A tract of land considerably elevated above the adjacent country. Mountains are usually found connected in long chains or ranges; sometimes they are single, isolated eminences. 2. *Phys. geog:* Any portion of the earth's crust rising considerably above the surrounding surface. The term is usually applied to heights of more than 610 meters, all beneath that amount being regarded as hills, and when of inconsiderable height, as hillocks.

mountain chain A complex, connected series of several more or less parallel mountain ranges and mountain systems grouped together without regard to similarity of form structure, and origin, but having a general longitudinal arrangement or well-defined trend; e.g., the Mediterranean mountain chain of southern Europe. *See also* CORDILLERA

mountain cork A variety of asbestos resembling cork. It is light and floats on water. *Syn:* MOUNTAIN LEATHER

mountain glacier Glacier in which rock projects above the highest levels of the ice and snow.

mountain group A group made up of several or many mountain peaks, or of short mountain ridges. The Catskill Mountains and the Black Hills are examples.

mountain leather *See* MOUNTAIN CORK

mountain pediment A plain of combined erosion and transportation at the foot of a desert mountain range similar in form to the alluvial plains that front the mountains of an arid region, but without alluvial cover and composed of solid rock.

mountain range A single, large mass consisting of a succession of mountains or narrowly spaced mountain ridges, with or without peaks, closely related in position, direction, formation, and age; a component part of a mountain system or of a mountain chain.

mountain slope The sloping surface which forms the side of a mountain; specifically the slope characteristically developed by the processes of erosion in an arid region.

mountain soils Soils, rocky and of variable horizon and thickness, that are typical of mountainous regions.

mountain system Several more or less parallel ranges grouped together. *See* MOUNTAIN CHAIN; MOUNTAIN RANGE; CORDILLERA

mouth *1.* The exit or point of discharge of a stream into another stream or a lake or sea. *2.* An opening or aperture resembling or likened to a mouth and affording entrance or exit, such as the mouth of a cave or canyon.

moveout or **moveout time** Stepout; angularity. The difference in arrival times of a reflection on adjacent traces of seismograph record.

muck *1.* A dark-colored soil, commonly in wet places, which has a high percentage of decomposed or finely comminuted organic matter. *Cf.* PEAT. *2.* Earth,

including dirt, gravel hardpan, and rock, to be, or being, excavated; overburden. *3.* A layer of earth, sand, or sediment lying immediately above the sand or gravel containing, or supposed to contain, gold in placer mining districts. It may itself contain some traces of gold. *4. V:* To excavate or remove muck from.

mucro A short and sharp abrupt spur or spiny tip.

mucronate Terminated by distinct and obvious mucro.

mud *1.* A slimy and sticky mixture of water and finely divided particles of solid or earthy material with a consistency varying from that of a semifluid to that of a soft plastic sediment. *2.* A wet mixture of silt and clay.

mud avalanche Mudflow.

mud crack Desiccation crack, *q.v.*

mud crack polygons *See* POLYGONS, MUD CRACK

mud engineer An engineer who studies and supervises the preparation of various fluids and emulsions, collectively termed mud, used in rotary drilling.

mud flat A muddy, low-lying strip of ground by the shore, or an island, usually submerged more or less completely by the rise of the tide.

mudflow A flowage of heterogeneous debris lubricated with a large amount of water usually following a former stream course. *Syn:* MUD ROCK FLOW; MUDSPATE; MUD AVALANCHE; MUD STREAM

mud geyser A geyser that erupts sulfurous mud. A type of mud volcano.

mud lumps Swellings of bluish-gray clay forming small islands of an acre or more, with a height of 2 to 4 meters above sea level, found at the mouths of the Mississippi; apparently caused by pressure of surface deposits upon buried clays. Formation accompanied by minor flows of marsh gas.

mud-pellet conglomerate Sandstone containing abundant, flattened to rounded, small mudstone masses.

mud polygons Polygonal soil patterns in fine textured material. *Syn:* FISSURE POLYGONS; CELLULAR SOIL

mud pot A type of hot spring consisting of a shallow pit or cavity filled with hot, generally boiling, mud carrying very little water and a large amount of fine-grained mineral matter.

mud rocks A general name applied to sediments composed most commonly of microscopic particles of quartz and clay, sometimes one and sometimes the other predominating.

mudstone *1.* A term originally applied by Sir Roderick Murchison to certain dark-gray, fine-grained shales of the Silurian system in Wales, which, on being exposed to the atmosphere, are rapidly decomposed and converted into their primitive state of mud, but now extended to all similar shales in whatever formation they may occur. *2.* Mudstone includes clay, silt, siltstone, claystone, shale, and argillite. It should only be used when there is doubt as to precise identification of when a deposit consists of an indefinite mixture of clay, silt, and sand particles, the proportions varying from place to place, so that a more precise term is not possible. It may also be used in a general classification of the three common forms of sedimentary rocks into mudstones, sandstones, and limestones. *3.* An indurated clay rock which is not fissile. *4.* A class of carbonate rocks composed of clay size particles. *See* MICRITE

mud volcano Cone-shaped mound

with maximum height of about 250 feet built around a spring by mud brought to the surface by slowly escaping natural gas.

mullion structure Rodding structure, *q.v. 1.* The larger grooves in a fault plane parallel to the direction of displacement. *2.* Series of parallel columns in metamorphic rocks several inches in diameter and several feet long, each column being composed of folded metamorphic rocks.

mullite A mineral $3Al_2O_3.2SiO_2$. Most often observed as a synthetic mineral in ceramic products. Orthorhombic.

multicycle coast A coast with a series of elevated sea cliffs separated from each other in stairlike fashion by narrow wave-cut benches. The sea cliffs at each elevation represent a separate shore line cycle.

multigelation Repeated freezing and thawing.

multigranular particle Sedimentary particle consisting of several adherent crystals.

multipartite map Vertical variability map showing the degrees of concentration of one lithologic type in a stratigraphic unit.

multiple detectors Two, or more, seismic detectors whose combined output energy is fed into a single amplifier-recorder circuit. This technique is used to effect a cancellation of undesirable near-surface waves. *Syn:* MULTIPLE SEISMOMETERS; MULTIPLE GEOPHONES; MULTIPLE RECORDING GROUPS.

multiple dike A dike made up of two or more intrusions of the same kind of igneous rock.

multiple faults System of closely spaced generally parallel faults.

multiple reflection Seismic energy which has been reflected more than once.

multiple seismometers Mutiple

geophones; multiple recording groups; multiple detectors, *q.v.*

multiple sill A sill made up of two or more intrusions composed of the same kind of igneous rock.

multiplex A stereoscopic plotting instrument used in preparing topographic maps by stereophotogrammetry.

multituberculate A member of an order of primitive herbivorous mammals with teeth with more than two cusps; the largest members were about the size of a modern woodchuck.

Muschelkalk A middle series of three series of the Triassic of northern Europe, approximately equivalent to the Anisian and Ladinian.

muscovite A mineral, a member of the mica group, the common white, green, red, or light brown mica of granites, gneisses, and schists. Composition $KAl_2(AlSi_3)O_{10}(OH)_2$. Monoclinic. *Clay mineral:* Illite.

mushroom ice Pillars of ice with roundish, expanded tops formed when a portion of an ice-covered area is protected from the direct effect of sunlight by some surface object, while the ice roundabout is not. *See* GLACIER TABLE

muskeg *1.* Characteristic of northern topography from the International boundary to the Arctic Sea. Alluvial areas with insufficient drainage, over which moss has accumulated to a considerable depth; these swamps are usually covered with tamarack and fir trees. The typical muskeg is traversed by meandering streams, having deep channels but a scarcely perceptible current. Stagnant pools become coated over with a moss of sufficient strength to temporarily sustain the weight of a man. *2.* A term sometimes used in Michigan for a bog lake.

mutant The offspring bearing the mutation.

mutation The effect of a chemical change in the DNA of a chromosome; some are visible, most are not visible; many are deleterious. Mutations are the raw material of evolution.

mutual boundary pattern The pattern made by two adjacent minerals is smooth and regular or forms regular curves with no decided projections of one mineral into another.

mutualism A symbiotic relationship in which both members benefit. Neither can survive without the other.

m.y. Million years.

mylonite A fine-grained, laminated rock formed by extreme microbrecciation and milling of rocks during movement on fault surfaces. Metamorphism is dominantly cataclastic with little or no growth of new crystals. *See* AUGEN SCHIST; CATACLASITE; CRUSH BRECCIA; CRUSH CONGLOMERATE; FLASER GABBRO; FLINT CRUSH ROCK

mylonization; mylonitization The process of forming mylonite, *q.v.*

Myriapoda One of the five main divisions of the phylum Arthropoda; here belong the centipedes and millipedes.

myrmekite A term applied to an intergrowth texture of vermicular quartz in plagioclase.

nacreous Pearly; having the luster of mother-of-pearl.

nacrite An uncommon clay mineral of the kaolin group. It has the same composition, $Al_2Si_2O_5(OH)_4$, as kaolinite and dickite, but is structurally distinct.

nadir The point where the direction of the plumb line extended below the horizon meets the celestial sphere. The nadir is directly opposite the zenith.

nannoplankton Plankton in the size range 5 to 60 micrometers, defined as uncatchable in standard plankton nets.

Nansen bottle A special bottle used for sampling water at ocean depths. It is open at both ends when lowered to the desired depth, but the ends close when the bottle is flipped over, trapping the water sample.

nappe Decke [*Ger.*]. *1.* Faulted overturned folds. *2.* A large body of rock that has moved forward more than one mile from its original position, either by overthrusting or by recumbent folding. *3.* In Belgium the term is a synonym of aquifer. *4.* The term originally meant a covering stratum such as basalt flow and may still be found in foreign literature other than French and English. *5. Hydraul:* A sheet or curtain of water overflowing a structure like a weir or a dam. The nappe has an upper and lower surface.

native An element occurring in nature uncombined.

native metal Any metal found naturally in that state, as, copper, gold, iron, mercury, platinum, silver, etc. *See* NATIVE

native paraffin *See* OZOCERITE

natrolite A mineral of the zeolite group, $Na_2Al_2Si_3O_{10}.2H_2O$. Orthorhombic.

natron A mineral, $Na_2CO_3.10H_2O$. Monoclinic.

natural arch (marine) *See* SEA ARCH

natural bridge *1.* The term "natural bridge" and "natural arch" have been so often used as synonyms, both in common parlance and in scientific literature, that it is necessary to define the terms. In the restricted sense in which the term natural bridge is used, a natural bridge is a natural stone arch that spans a valley of erosion. A natural arch is a similar structure which, however, does not span an erosion valley. *2.* Any bridge spanning a ravine and left in place by erosive agencies. *3.* A synonym for sea arch, *q.v.*

natural gas *1.* A mixture of gaseous hydrocarbons found in nature; in many places connected with deposits of petroleum, to which the gaseous compounds are closely related. *2. Hydrol:* Gases that are entrapped in interstices in the zone of saturation. A body of natural gas may be partly in solution in water transmitting hydrostatic pressure from below or in any petroleum that may intervene between the water and the gas.

natural gas liquids Those liquid hydrocarbon mixtures which are gaseous in the reservoir but are recoverable by condensation or absorption. Natural gasoline, condensate, and liquefied petroleum gases fall in this category.

natural levee *See* LEVEE

natural selection Process by which less vigorous and less well-adapted individuals tend to be eliminated from a population without leaving descendants to perpetuate an inferior stock.

nautilicone Any nautiloid shell coiled in one plane and involute, as in the genus *Nautilus*.

nautiloid *1.* One of the Nautiloidea; shelled cephalopod having external, chambered shell, either straight or variously curved or coiled and with simple septa forming sutures that are simples lines without marked flexures. *2. Adj:* Pertaining to Nautiloidea, *q.v.*

Nautiloidea Class of cephalopods with straight to coiled shells whose septa meet the external shell along a suture that is not thrown into folds. Camb.-Rec.

neanic Describing youthfulness or the stage in which specific characters begin to develop.

neap tide Tide occurring near the time of quadrature of moon. Neap tidal range is usually 10 to 30% less than the mean tidal range.

nearshore circulation The ocean circulation pattern composed of the nearshore currents and coastal currents. *See* CURRENT

nearshore current system Current system caused primarily by wave action in and near breaker zone, which consists of four parts: the shoreward mass transport of water; longshore currents; seaward return flow, including rip currents; and longshore movement of expanding heads of rip currents.

neat line The line which surrounds the map itself. Differs from margin in that the margin is outside the neat line.

Nebraskan The oldest of four classical glacial ages of the Pleis-

tocene of North America; earlier glaciations are now known, but the four classical ages have been retained; also the stage of strata deposited during that age.

neck *1.* A lava-filled conduit of an extinct volcano, exposed by erosion. *Syn:* CHIMNEY; PIPE. *2. Topog:* The narrow strip of land which connects a peninsula with the mainland, or connects two ridges. *See* TOMBOLO. *3.* The narrow band of water flowing seaward through the surf. *Syn:* RIP

neck cutoff The break-through of a river across the narrow neck separating two meanders, where downstream migration of one has been slowed and the next meander upstream has over-taken it. *Cf.* CHUTE CUTOFF

neck, volcanic *See* VOLCANIC NECK

needle *1. Geog.* and *Geol:* A familiar term for pointed, detached masses of rock standing out from cliff or shores to which they geologically belong, and from which they have been severed by erosive action of tides and waves. Applied also to pointed summits of mountains, aiguille or needle-top of the French. *2.* Prominent and sharp rocky pinnacle or spire; aiguille. *3.* Slender needlelike snow crystal usually composed of needlelike components lying parallel, with length of crystal being at least five times greater than diameter.

needle ore *1.* Aikinite. A lead-copper-bismuth sulfide. *2.* Iron ore of very high metallic luster, found in small quantities which may be separated in long slender filaments resembling needles.

negative area Area that subsided conspicuously or repeatedly. Thick stratigraphic sections generally identify such areas.

negative center *Elec. prospecting:* The central region of a closed negative earth potential anomaly, theoretically a point.

negative crystal *1.* An optically

negative crystal. *See* OPTICAL CHARACTER. 2. A pseudomorph consisting of a hollow opening shaped like a crystal, presumably formed by the solution of a previously existing crystal.

negative element 1. Those which have shown a decided tendency to rise are designated positive elements and those which have tended to sink are termed negative elements. 2. A term applied to a large structural feature in the earth's crust, characterized through a long geologic time by a tendency to sink when diastrophism takes place. 3. A portion of the earth's crust which has been submerged again and again during geologic history.

negative elongation Anisotropic crystal elongation parallel to vibration direction of the faster of the two plane-polarized rays.

negative movement Subsidence, actual or relative.

negative shoreline A shoreline resulting from a positive movement of the land or by a negative movement of the sea level; a shoreline of emergence. *Ant:* POSITIVE SHORELINE

nekton 1. Any animal or group of animals that lead or leads an active swimming life. 2. Swimming animals that can direct their own movements against the action of marine currents.

nektoplanktonic Pelagic.

nematoblastic 1. Pertaining to the texture of a recrystallized rock in which the shape of the grains is threadlike. 2. Applied to a fibrous type of schistosity, seen in rocks composed largely of such minerals as glaucophane and actinolite.

Neocene The latter of two epochs into which the Tertiary Period is divided. Also the series of strata deposited during the epoch. *See* NEOGENE

neo-Darwinism Darwinism modified in accordance with modern genetics.

Neogene The later of the two periods into which the Cenozoic Era is divided, it comprises the Miocene, Pliocene, Pleistocene, and Holocene (or Recent) Epochs. Also the system of strata deposited during that period. *See* PALEOGENE

Neolithic New Stone age; a term used to designate those men or times characterized by the use of more sophisticated rock tools associated with early agriculture; approximately the last 8 to 10 thousand years in southern Asia and the Near East, the last 5 thousand years in Europe.

neomagma Newly formed products of metamorphism subjected locally to mass movement of plastic flow, in contrast to hypomagmas of presumably deep-seated sources.

neomineralization Chemical interchange with a rock resulting in alteration of its mineral components and production of new minerals.

neontology *Biol:* the study of existing life.

neotype Specimen selected to replace the holotype if the latter has been lost. *Cf.* LECTOTYPE

nepheline; nephelite A mineral, $(Na,K)AlSiO_4$. Hexagonal. An important rock-forming mineral, especially in nepheline syenites and alkali rich basalts.

nepheline syenite A plutonic rock composed of a granular aggregate of alkalic feldspar, nepheline, and an alkalic ferromagnesian constituent. The ferromagnesian minerals may be amphiboles such as riebeckite, arfvedsonite, barkevikite, etc., or pyroxenes such as aegirite, aegirite-augite, or acmite. Common accessory minerals are cancrinite, sodalite, hauyne, and noselite in addition to apatite, zircon, sphene, and opaque oxides. Other more rare accessories are often present.

nephelinite An extrusive or hypabyssal rock composed primarily of pyroxene (usually titaniferous augite) and nepheline.

nephrite A tough, compact, fine-grained, greenish mineral of the amphibole group. Main constituent of the less rare type of jade. Greenstone.

nepionic Young; referring to the stage or period of the young shell before the appearance of distinctive specific characters.

Neptunian theory A general theory of origin of rocks, proposed by A. G. Werner in the eighteenth century.

Neptunist A follower of Abraham Gottlob Werner (1750–1817) who taught that all rocks were formed in or by water. *Cf.* PLUTONIST

nereite A fossil worm track.

neritic Related to shallow water on the margins of the sea, generally that overlying the continental shelf; combined with oceanic, e.g., neritooceanic, refers to the shallow water above 100 fathoms (200 meters) in the open oceans.

neritic zone 1. That part of the sea floor extending from the low tide line to a depth of 200 meters. 2. A part of the pelagic division of the oceans with the water depths less than 200 meters.

nerito-paralic Water environment of the paralic, *q.v.*

neritopelagic Refers to shallow water swimmers or floaters—those organisms that swim or float in waters of less than 200 meters (100 fathoms) depth.

Nernst's law States that the solubility of a salt is decreased by the presence in solution of another salt that has a common ion.

nesosilicates Silicate structures in which individual SiO_4 tetrahedra are not linked together, i.e., they do not share oxygens. An example is olivine. *Syn:* ORTHOSILICATE. *See* SILICATES, CLASSIFICATION

nested calderas The conditions leading to the formation of a caldera may recur on a smaller scale at the same volcanic center, so that a second depression of this class is developed within the perimeter of the first.

net slip The total slip along a fault; it is the distance measured on the fault surface between two formerly adjacent points situated on opposite walls of the fault.

network Especially in surveying and gravity prospecting, a pattern or configuration of stations, often so arranged as to provide a check on the consistency of the measured values, e.g., a level network, a gravity network based on the integration of torsion-balance gradients.

network structures *See* TECTOSILICATES

neutral stress Pore-water pressure.

neutral surface Surface of no strain.

neutron-gamma log A radioactivity log employing both gamma- and neutron-log curves. The neutron log should respond best to porous fluid-filled rock and the gamma best to shale markers.

neutron log A radioactivity logging method used in boreholes in which a neutron source provides neutrons which enter rock formations encountered and induce additional gamma radiation which is measured by use of an ionization chamber. The gamma radiation so induced is related to the hydrogen content of the rock.

neutron-neutron (n-n) log A technique in which the formation is bombarded by neutrons and the scattered neutrons are measured.

Nevadan orogeny A time of deformation, metamorphism, and plutonism during Jurassic and early Cretaceous time in the western part of the North American cordillera, typified by relations in the Sierra Nevada, California.

névé A mass of snow partly converted into ice; forms the upper part of glaciers. *See* FIRN; SNOW FIELD

New Stone Age *See* NEOLITHIC

Niagaran The middle of three Silurian Epochs of North America; also the rock deposited during that epoch.

niccolite Copper nickel. A mineral, nickel arsenide, NiAs, copper-red. Hexagonal.

nickel bloom *See* ANNABERGITE

nickeliferous Containing nickel.

nicking A flexure which is so sharply exaggerated that breaking begins, although connection still exists. The structure is analogous to the "greenstick" fracture of wood or bone.

nickpoint The point of interruption of a stream profile at the head of a second-cycle valley according to the Treppen concept, *q.v.*

Nicol; Nicol prism A device for producing plane-polarized light, consisting of optically clear calcite so cut and recemented that the ordinary ray produced by double refraction in the calcite is totally reflected, but the extraordinary ray is transmitted. Nicol: Any instrument or device for producing plane-polarized light by absorbing or reflecting one ray produced in a doubly refracting substance such as calcite, polaroid, etc.

Niggli number A series of magma types or categories into which a Niggli molecular norm is fitted.

Niggli's classification A classi-

fication of rocks on the basis of their chemical composition, similar in some respects to the norm system.

Niggli's molecular norm A "norm" calculated from the Niggli numbers in the Niggli classification of rocks.

nigritite A coalified, carbon-rich, fix-bitumen.

niter Saltpeter. A mineral, KNO_3. Orthorhombic.

nitrate A salt of nitric acid, a compound containing the radical $(NO_3)^{-1}$.

nitride A compound of nitrogen with one other more positive element or radical.

nitrite A salt of nitrous acid, a compound containing the radical $NO_2{}^{-1}$.

nitrogen cycle The numerous biogeochemical processes which result in the cycling of nitrogen in the many chemical forms required by various organisms.

nival Characterized by, abounding with, living in or under snow.

nivation Frost action and mass-wasting beneath a snowbank.

noble metal Any metal or alloy of comparatively high value, or relatively superior in certain properties, especially resistance to corrosion or infusibility, as, gold, silver, or platinum; opposed to base metal.

nodal zone An area at which the predominant direction of the littoral drift changes.

node *1.* A point, line, or surface in a vibrating medium at which the amplitude of the vibration is reduced to zero by the interference of oppositely directed wave trains, forming stationary waves; e.g., one of the stationary points on a vibrating string. Such a point is a nodal point and such a cross section a nodal section. *2. Paleobot:* A point on a stem from which a leaf and bud arise.

nodular Having the shape or

composed of nodules. Said of certain ores or strata.

nodule Small more or less rounded body generally somewhat harder than the enclosing sediment or rock matrix.

noise *1. Grav.* and *Mag. prospecting:* Disturbances in observed data due to more or less random inhomogeneities in surface and near-surface material. *2. Seismic prospecting:* All recorded energy not derived from the explosion of the shot. Sometimes loosely used for all recorded energy except events of interest.

nomenclature *1.* The naming of divisions in any scientific taxonomic scheme. *2.* The names used in systematic classification, as distinguished from other technical terms, as, Linnaean nomenclature.

nominal diameter The diameter of a sphere of the same volume as the particle, which is the diameter of a sphere having the same settling velocity as the particle.

nomograph; nomogram A graph or chart reducing a mathematical formula to curves so that its value can be read on the chart coordinates for any value assigned the variables involved.

nonartesian ground water Unconfined ground water.

nonconformity An unconformity in which stratified rocks above the surface rest on unstratified rocks such as igneous or metamorphic.

nonferrous metals Metals other than iron and its alloys in steel. Usually applied to the base metals, such as copper and lead.

nonpiercement salt dome A salt dome in which the salt does not intrude or crosscut the overlying sediments. Contrasts with piercement salt dome, *q.v.*

nonplunging fold A fold with a horizontal axis.

nonrotational or **irrotational strain** Stress and strain axes remain parallel throughout the deformation. Strictly speaking, the stress and strain axes are parallel regardless of the strain.

nonselective diagram *Struct. petrol:* A point or contour diagram made by not consciously selecting representatives of one type.

nonsorted circles Patterned ground, *q.v.*, whose mesh is dominantly circular and has a nonsorted appearance due to the absence of a border of stones such as that characterizing sorted circles.

nonsorted nets Patterned ground with a mesh intermediate between that of a nonsorted circle and a nonsorted polygon and with a nonsorted appearance due to absence of a border of stones such as characterizes a sorted net.

nonsorted polygons Patterned ground whose mesh is dominantly polygonal and has a nonsorted appearance due to the absence of a border of stones such as that characterizing sorted polygons.

nonsorted steps Patterned ground with a steplike form and a nonsorted appearance due to a downslope border of vegetation embanking an area of relatively bare ground upslope.

nonsorted stripes Patterned ground with a striped pattern and a nonsorted appearance due to parallel lines of vegetation-covered ground and intervening strips of relatively bare ground oriented down the steepest available slope.

nontectonite Rock in which the position and orientation of grains have not been influenced by the movement of neighboring grains.

nontronite An iron-rich mineral of the montmorillonite group, $Na_{0.33}Fe_2^{+3}(Al,Si)_4O_{10}(OH)_2 \cdot nH_2O$.

norite A variety of gabbro in which orthopyroxene is dominant over clinopyroxene. Hypersthene-gabbro.

normal In general, a straight line perpendicular to a surface or to another line. Also, a condition of being perpendicular to a surface or line.

normal anticlinorium An anticlinorium, in which the axial planes of the subsidiary folds converge downward.

normal atmospheric pressure Standard pressure, usually taken to be equal to that of a column of mercury 760 mm. in height. Approximately 14.7 pounds per square inch.

normal class Holohedral or holosymmetric class.

normal consolidation Sedimentary compaction in equilibrium with overburden pressure.

normal correction *Mag. prospecting:* The correction made to data to remove the normal field.

normal dip The regional or general inclination of stratified rock over a wide area, as contrasted to local dip due to the presence of local structures.

normal distribution Chance occurrence of one of two possibilities that can be represented by a curve rising from zero to a maximum and then declining symmetrically to zero.

normal or geologic erosion The erosion which takes place on the land surface in its natural environment undisturbed by human activity. It includes (1) rock erosion, or erosion of rocks on which there is little or no developed soil, as in stream channels and rocky mountains, and (2) normal soil erosion, or the erosion of the soil under its natural condition or native vegetative cover undisturbed by human activity.

normal fault A fault at which the hanging wall has been depressed, relative to the footwall.

normal field *Mag. prospecting:* The smoothed value of a magnetic field component as derived from a large-scale survey, worldwide or of continental scope. The normal field of the earth varies slowly with time, and maps of it are as of a certain date.

normal gravity The value of gravity at sea level according to a theoretical formula which assumes the earth to be a spheroid or of some similar regular shape.

normal horizontal separation *See* OFFSET

normal hydrostatic pressure Hydrostatic pressure in porous strata or wells approximately equal to the weight of a column of water whose length is the depth under consideration.

normal limb That limb of an overfold (overturned fold) that is right side up.

normal moveout Reflection arrival time difference due to variable geophone-shot point distances along reflection paths.

normal moveout velocity The velocity of a constant-velocity medium which would give the observed normal moveout. Subsurface P-wave velocity as determined from X^2-T^2 analysis, *q.v.*

normal stress or traction Component of stress or traction perpendicular to a plane.

normal synclinorium A synclinorium in which axial planes of the subsidiary folds diverge downward.

normal travel-time curve In fan shooting, a time-distance curve obtained along a profile in some nearby area which does not contain geologic structures of the type being sought.

normal zoning In plagioclase, the change by which crystals become more sodic in outer parts.

normative *Petrol:* Characteristic

of, pertaining to, agreeing with, or occurring in the norm. Used in the quantitative or norm system of classification of igneous rocks, a normative mode being one which is essentially the same as the norm.

normative mineral Standard mineral, *q.v.*

normative quartz Theoretical quartz calculated according to certain rules from the chemical composition of a rock. *See* NORM

norm system A system of classification and nomenclature for igneous rocks based on the norm, *q.v.,* of each rock. C.I.P.W. system.

nose *1.* A half-developed anticline, i.e., one in which one end is open and without closure. *2.* Place on a map where a bed in a fold shows the maximum curvature. *3.* A projecting buttress of rock, usually overhanging; the projecting end of a hill, spur, ridge, or mountain. *4.* In brittle stars (a type of echinoderm), an articulating projection on arm ossicles.

notch *1.* A short defile through a hill, ridge, or mountain. A deep, close pass; a defile; a gap. Local in New England. *2.* A deep narrow cut at the base of a sea cliff above which the cliff overhangs. *Syn:* NIP

notochord A rod of elastic cells which provides a supporting and stiffening structure in an animal's body; it is replaced by a backbone in the true vertebrates.

novaculite A very dense, even-textured, light-colored crypto-crystalline siliceous rock. Originally, the term was applied to rocks found in the lower Paleozoic rocks of the Ouachita Mountains of Arkansas.

nuclear basin A shallow negative area in an island arc region which is generally conformable to the structural trends of the associated positive elements.

nuclear energy Energy released by reactions involving the nucleus of atoms. The two types of reactions are fission, splitting the nuclei of heavy elements; and fusion, combining the nuclei of light elements.

nuclear engineering The use of nuclear explosions in any applied science, such as fracturing the subsurface for increased hydrocarbon production (as already applied) or for the digging of harbors and canals, as proposed.

nuclear event Any nuclear explosion, premeditated or accidental.

nuclear fuel Any energy-producing substance that involves a change in the number of neutrons or protons in an atom, either by fission or fusion.

nuclear log Radiometric record made in a borehole. *See* GAMMA RAY; SPECTRAL GAMMA RAY

nuclear power plant Any device, machine, or assembly that converts nuclear energy into some form of useful power, such as mechanical or electrical power. In a nuclear electric power plant, heat produced by a reactor is generally used to make steam to drive a turbine that in turn drives an electric generator.

nucleation The beginning of crystal growth at one or more points.

nucleic acid Complex organic substance that is the genetic material in all known organisms. *See* DNA; RNA

nucleus *1.* A kernel; a central mass or point about which other matter is gathered, or to which an accretion is made. *2.* A usually spherical or ovoid protoplasmic body found in most cells and considered as a directive center of many protoplasmic activities, including the transmission of hereditary characteristics. 3. In radiolarians, a rounded mass of protoplasm enclosed in a delicate membrane. 4. Embryonic gastropod shell, commonly consisting of one to four whorls. 5. The central portion of an atom; the chief constituents are protons and neutrons. *Syn:* PROTOCONCH

nuclide A species of atom characterized by the constitution of its nucleus; thus abundances of isotopes are described as abundances of various nuclides.

nuée or **nuée ardente** [*Fr.*] Applied to a highly heated mass of gas-charged lava, more or less horizontally ejected from a vent or pocket at the summit of a volcano, onto an outer slope where it continues on its course as an avalanche, flowing swiftly, however slight the incline, by virtue of its extreme mobility.

nugget A waterworn piece of native gold. The term is restricted to pieces of some size, not mere "colors," or minute particles.

nummulite Lens or coin-shaped shell (test) of rather large foraminifers which were important in the Tertiary, particularly the Eocene. Referring to the type of foraminifer formerly known as the genus Nummulites.

nummulitic limestone An Eocene formation made up chiefly of nummulite shells.

nunatak *1. Glaciol:* An isolated hill or peak which projects through the surface of a glacier. *2. Glacial geol:* A hill or peak which was formerly surrounded but not overridden by glacial ice.

nutation Periodic shifts in the position of the axis of the earth; period is about 19 years.

nutrients Elements or compounds essential as raw materials for organism growth and development; e.g., carbon, oxygen, nitrogen, and phosphorus.

oblique *1.* Neither perpendicular nor horizontal; slanting; inclined. *2. Photog:* An air photograph in which the camera is pointed downward at an angle to the horizontal. A high oblique is one taken at an angle to the vertical sufficiently high so that the horizon is included. A low oblique, with camera pointed more directly downward, does not include the horizon.

oblique extinction Extinction in anisotropic material not parallel to crystal outlines.

oblique fault A fault whose strike is oblique to the strike of the strata.

oblique joint A joint the strike of which is oblique to the strike of the adjacent strata or cleavage.

oblique projection A conformal map projection that is tilted, that is not centered on a pole or on the equator. The equator or a meridian is not used as a center line of orientation; e.g., oblique Mercator projection.

oblique-slip fault A fault in which the net-slip lies between the direction of dip and the direction of strike.

obsequent Flowing in a direction opposite to that of the dip of the strata or the tilt of the surface: said of some streams and contrasted with consequent. Called also reversed stream.

obsequent fault-line scarp A scarp along a fault line, but where the topographically low area is in the block that has been relatively uplifted. *See* RESEQUENT FAULT-LINE SCARP

obsidian An ancient name for volcanic glass. Most obsidians are black, although red, green, and brown ones are known. They are often banded and normally have conchoidal fracture and a glassy luster.

obtuse bisectrix *See* BISECTRIX

ocean The great body of salt water which occupies two-thirds of the surface of the earth, or one of its major subdivisions. The sea as opposed to the land.

ocean basin That part of the floor of the ocean that is more than about 200 m. below sea level.

ocean current *1.* The name current is usually restricted to the faster movements of the ocean, while those in which the movement amounts to only a few miles a day are termed drifts. *2.* A nontidal current constituting a part of the great oceanic circulation. Examples: Gulf Stream, Kuroshio, Equatorial currents.

oceanic Related to the open ocean beyond the edge of the continental shelf, in contrast to shallow, near-shore waters.

oceanic bank Guyot; tablemount, *q.v.* A seamount with a depth less than 200 m.

oceanic crust The type of crust underlying the earth's ocean basins; the sima.

oceanic islands Islands that rise from deep water far from any continent, though they may occur in a close group, like the Hawaiian Islands, the Azores, and the Galápagos.

oceanite A picritic basalt.

oceanography Embraces all studies pertaining to the sea and integrates the knowledge gained in the marine sciences that deal with such subjects as the ocean boundaries and bottom topography, the physics and chemistry of sea water, the types of currents, and the many phases of marine biology. *Syn:* OCEANOLOGY; THALASSOGRAPHY

ocean tides The twice daily rise and fall of sea level caused mainly by the attraction of the moon and earth.

ocellar A rock texture characterized by radiating groups of prismatic or platy minerals such as biotite or pyroxene disposed around the borders of larger euhedral crystals such as analcite or leucite. The structures themselves are called ocelli.

ocher; ochre A pulverulent oxide, usually impure, used as a pigment. Brown and yellow ochers consist of limonite, or goethite, and red ocher of hematite.

Ochoan The youngest of four epochs of the Permian of a large part of North America; also the series of rock deposited during that epoch.

octahedral cleavage In the isometric system, cleavage parallel to the faces of the octahedron.

octahedrite *See* ANATASE

octahedron In the isometric system, a closed form consisting of eight triangular faces each having equal intercepts on all three crystallographic axes.

octane A liquid hydrocarbon of the paraffin series, formula C_8H_{18}.

octaphyllite Trioctahedral phyllosilicate minerals.

Oddo-Harkins rule Elements of even atomic number are more abundant than those of odd atomic number on either side.

oersted The practical, c.g.s. electromagnetic unit of magnetic intensity. A unit magnetic pole, placed in a vacuum in which the magnetic intensity is 1 oersted,

is acted upon by a force of 1 dyne in the direction of the intensity vector. Formerly called gauss.

offlap The reverse of transgressive onlap. Offlap occurs where a shoreline has retreated seaward and progressively younger strata have been deposited in layers offset seaward.

offset *1.* Displacement of formerly contiguous bodies. *2.* In a fault, the horizontal component of displacement, measured parallel to the strike of the fault. *3. Seis:* Shotpoint-to-geophone distance, often measured to the center of the geophone group nearest the shotpoint.

offset well A well drilled at a distance, governed by local field practice, from a productive oil or gas well and on an adjoining leasehold for the purpose of protecting, by recovery through the offset, those reserves that otherwise might be produced by the earlier well. An obligation to drill such offset wells is contained in all oil and gas leases.

offshore bar Barrier beach, *q.v.* An accumulation of sand in the form of a ridge, built at some distance from the shore and under water. It results chiefly from wave action. Not to be confused with barrier island, *q.v.*

offshore beach A long, narrow, low sandy beach with a belt of quiet water separating it from the mainland.

offshore current *1.* Any current in the offshore zone. *2.* Any current flowing away from shore.

Oghurd dune Arabic term used in the Sahara for a massive, mountainous dune.

ogive Any curved dark band, convex downslope, visible on a glacier surface. Specifically, such a band composed of debris-laden ice. Such bands are also called dirt bands, Forbes bands, Alaskan bands.

oil field A district containing a proved subterranean store of petroleum of economic value.

oil pool An accumulation of oil in sedimentary rock that yields petroleum on drilling. The oil occurs in the pores of the rock and is not a pool or pond in the ordinary sense of these words.

oil sand *1.* A general term for any rock containing oil. *2.* Porous sandstone from which petroleum is obtained by drilled wells.

oil shale A fine-grained, laminated sedimentary rock that contains an oil-yielding organic material called kerogen. In the United States, oil shale is found primarily in Colorado; much larger deposits exist in several other countries. Upon heating, oil shales yield from 12 to 60 gallons of oil per ton of rock, although the best U.S. oil shale produces less than 30 gallons per ton.

oil spill The accidental discharge of oil into oceans, bays, or inland waterways. Methods of oil spill control include chemical dispersion, combustion, mechanical containment, and absorption.

oil trap Reservoir rock containing petroleum that is capped or enclosed by an impermeable rock in such manner that migration and escape of the contained hydrocarbons is prevented.

oil-water interface A surface that forms the boundary between a body of ground water and an overlying body of petroleum that saturates the rock.

oil well A dug or bored well, from which petroleum is obtained by pumping or by natural flow.

old age That stage in the development of streams and land forms when the processes of erosion are decreasing in vigor and efficiency or the forms are tending toward simplicity and subdued relief. *Cf.* YOUTH; MATURITY

oleostatic Refers to oil pressure in a natural reservoir.

Oligocene The third of the epochs into which the Tertiary Period is at present ordinarily divided. Also the series of strata deposited during that epoch.

oligoclase A mineral of the plagioclase feldspar series with more sodium than calcium in its composition.

oligomictic lake A lake whose waters are rarely or very slowly mixed. It is thermally stable and usually located in the tropics.

oligomictic rocks [$<Gr.$ oligo-, few; miktos, mixed] Said of detrital sedimentary rocks in which few kinds of detrital minerals or rocks are present.

oligotrophic Refers to lakes with considerable oxygen in the bottom waters and with limited nutrient matter.

olivine Chrysolite. Periodot. A mineral series, solid solutions of forsterite, Mg_2SiO_4, with fayalite, Fe_2SiO_4, the composition often expressed as mol per cent of the constituents (abbreviated Fo and Fa). Orthorhombic. An important rock-forming mineral, especially in the mafic and ultramafic rocks.

olivinite A foliated rock with olivine as the principal constituent. Also used for hornblende picrite with augite and anorthite.

omega structure *1.* Thrust sheet overlying two oppositely directed low-angle thrust planes that steepen abruptly downward. *2.* Structure consisting of overthrusts or overfolds extending outward in opposite directions.

omission solid solution Defective crystal lattice in which atoms or ions are missing from their normal places.

onion weathering Onion-skin weathering; a type of spheroidal weathering in which the successive shells of decayed rock so produced resemble the layers of an onion.

onlap Extension of successive stratigraphic units beyond the marginal limits of their predecessors onto older rocks as in the deposits of a transgressing sea.

onshore A direction landward from the sea.

ontogenetic stage Developmental stage in the growth of an individual organism.

ontogeny The life history, or development, of an individual, as opposed to that of the race (phylogeny).

onyx *1.* A cryptocrystalline variety of quartz, made up of different colored layers, chiefly white, yellow, black, or red. *2.* Translucent layers of calcite from cave deposits, often called Mexican onyx or onyx marble.

oölite *1.* A spherical to ellipsoidal body, 0.25 to 2.00 mm. in diameter, which may or may not have a nucleus, and has concentric or radial structure or both. It is usually calcareous, but may be siliceous, hematitic, or of other composition. *2.* Accretionary oölite usually has a nucleus such as a quartz grain, and radial or concentric structure. It is grown in suspension in an agitating medium. *3.* Replacement oölite usually is without a nucleus, or has a nucleus of quartz. It is less regular and spherical, with concentric and radial structure, less well developed than in accretionary oölites. *4.* A rock composed chiefly of oöliths.

oölith The individual spherite of which an oölite (rock) is composed.

ooze A fine-grained pelagic deposit which contains more than 30% of material of organic origin.

ooze, calcareous An ooze with more than 30% of calcium car-

bonate, which represents the skeletal material of various planktonic animals and plants.

opacite A general term for microscopic, opaque grains in rocks, usually applied to such materials in the groundmass of volcanic rocks. It is generally regarded to consist largely of magnetite dust.

opal A mineral, $SiO_2.nH_2O$. Amorphous. Used as a gem.

opalescence A milky or pearly reflection from the interior of a mineral.

opalized wood Silicified wood, *q.v.*

open bay Bight. A broad indentation between two headlands or points, the bays being sufficiently open so that waves coming directly into the bay are essentially the same in height near the center of the bay as on open portions of the coast.

open burning Uncontrolled burning of wastes in an open dump.

open cut Strip mine.

open flow The rate of flow of a gas well when flowing into the air, unrestricted by any pressure other than that of the atmosphere, usually in units of cubic feet per 24 hours. *See* INITIAL OPEN FLOW

open fold A fold in which the limbs diverge at a large angle.

open form A crystal form whose faces do not enclose space, e.g., a trigonal prism.

open hole A drill hole which, at the depth referred to, contains no casing or pipe, but in which the wall of the hole is formed by the rock penetrated.

open pack Sea ice composed of floes which for the most part do not touch; easily navigable.

ophicalcite A marble containing serpentine. *See* MAGNESIAN MARBLE; MARBLE

ophiolite An assemblage of mafic and ultramafic igneous rocks ranging from spilite and basalt to gabbro and peridotite, including rocks rich in serpentine, chlorite, epidote, and albite derived from them by later metamorphism, whose origin is associated with an early phase of the development of a geosyncline.

ophitic A term applied to a texture characteristic of diabases or dolerite in which euhedral or subhedral crystals of plagioclase are embedded in a mesostasis of pyroxene crystals, usually augite. *See* DIABASIC

optical calcite Calcite crystals so clear that they have value for optical use.

optical character In optical crystallography, the designation, positive or negative, depending on the values of the different indices of refraction of a mineral. For uniaxial crystals with two indices of refraction, if the index of the extraordinary ray exceeds that of the ordinary ray, the mineral has a positive optical character. For biaxial crystals with three indices of refraction, the intermediate index is nearer in value to the smaller index than to the larger one for optically positive crystals.

optical constants *Opt. mineral:* The indices of refraction, axial angle, extinction angle, etc.

optical pyrometer *See* PYROMETER

optic angle The acute angle between the two optic axes of a biaxial crystal; its symbol is 2V.

optic axes Those directions in anisotropic crystals along which there is no double refraction.

optic ellipse Any noncircular section of an index ellipsoid.

optic indicatrix Indicatrix ellipsoid.

optic plane That plane including the two optic axes.

orbicular structure A structure developed in certain phanerocrystalline rocks (e.g., granites, diorites, and corsite) due to the occurrence of concentric shells of different mineral composition around centers that may or may not exhibit a xenolithic nucleus. *Syn:* SPHEROIDAL STRUCTURE; NODULAR STRUCTURE

orbit The path described by a body in its revolution around another body.

orbital current The flow of water accompanying the orbital movement of the water particles in a wave. Not to be confused with wave-generated littoral currents.

orbitoid *1.* One of the foraminiferal family, the Orbitoididae; foraminifera with large, discoidal, saddle-shaped or stellate tests. *2. Adj:* Belonging to the foraminiferal family, the Orbitoididae (the orbitoids).

order *1.* Arrangement, particularly with respect to importance. *2. Tax:* A group of organisms consisting of one or more families and constituting part or all of a class. *3. Geomorph: See* STREAM ORDER; BASIN ORDER

order-disorder inversions An inversion between two polymorphic forms, one of which has a more ordered structure than the other. In general, upon heating the ordered, low-symmetry, low-temperature form a point is reached at which some portion of the lattice becomes disordered (or random), usually with an increase in crystal symmetry, to form the high-temperature form.

order-disorder   polymorphism Two phases, generally related to temperature, in which atoms or ions occur at different positions or at random in a crystal lattice.

order of crystallization The apparent chronological sequence in which crystallization of the various minerals of an assemblage takes place, as evidenced mainly by textural features.

ordinary ray; O-ray In optically

uniaxial crystals, the ray of polarized light that vibrates in the plane of the basal pinacoid perpendicular to the optic axis. Its refraction is not effected by orientation.

Ordovician The second of seven Paleozoic Periods generally used in North America (six in Europe); also the strata of the system of rocks deposited during that period.

ore A mineral, or an aggregate of minerals, more or less mixed with a gangue, which from the standpoint of a miner can be won at a profit or from the standpoint of the metallurgist can be treated at a profit.

ore, developed *See* PROVED ORE

ore blocked out Ore exposed on three sides within a reasonable distance of each other.

ore body Generally a solid and fairly continuous mass of ore, which may include low-grade and waste as well as pay ore, but is individualized by form or character from adjoining country rock.

ore channel The space between the walls or boundaries of a lode which is occupied by ore and veinstone. Also called lode country.

ore cluster Group of ore bodies sometimes differing from each other in structure but interconnected or otherwise closely related genetically. Some ore clusters gather downward into a restricted root.

ore-dressing The cleaning of ore by the removal of certain valueless portions as by jigging, cobbing, vanning, and the like.

ore faces Those ore bodies that are exposed on one side, or show only one face, and of which the values can be determined only in a prospective manner, as deduced from the general condition of the mine or prospect.

ore magma A heavy and highly concentrated solution containing metals and nonmetals.

ore shoot A large and usually rich aggregation of mineral in a vein. It is a more or less vertical zone or chimney of rich vein matter extending from wall to wall, and has a definite width laterally. Sometimes called pay streak, although the latter applies more specifically to placers.

organic Referring to or derived from living organisms. In chemistry, any compound containing carbon.

organic deposits Rocks and other deposits formed by organisms or their remains.

organic reef A sedimentary rock aggregate, composed of the remains of colonial type organisms, mainly marine.

organism Any living individual whether plant or animal.

orient To place, as a map or an instrument, so that some portion of it points in a desired direction.

oriental agate Understood to be all the most beautiful and translucent sorts of agate.

oriental amethyst Strictly speaking, a variety of sapphire, but the term is applied to any amethyst of exceptional beauty.

oriental emerald A green variety of corundum.

oriental garnet Precious garnet.

oriental topaz A yellow variety of corundum, Al_2O_3.

orientation *1.* The assignment or imposition of a definite direction in space. *2. Surv:* The rotation of a map (or instrument) until the line of direction between any two of its points is parallel to the corresponding direction in nature. *3.* The placing of a crystal in the conventional attitude, so as to show its symmetry and the forms to which its faces belong. *4. Struct. petrol:* The arrangement in space of the particles (grains or atoms) of which a rock is composed. 5. *Paleontol:* The determination of the position of various organisms or organic hard parts with reference to such features as the dorsal and ventral sides of the animal, the anterior and posterior of the animal, the axis of coiling, the plane of coiling in shells, etc.

oriented specimen *1. Struct. petrol:* A hand specimen that is so marked that its exact arrangement in space is known. *2. Paleontol:* A fossil whose position is known as regards such features as anterior and dorsal sides, dorsal and ventral sides, the axis of coiling, the plane of coiling, etc.

origin *1.* In a Cartesian coordinate system, the point defined by the intersection of the axes. *2.* An arbitrary zero or starting point on a scale or measuring device.

original Characteristic of, or existing in, a rock at the time of its formation. Said of minerals, textures, etc., of rocks; essentially the same as primary *1,* and contrasted with secondary *1.*

original dip; primary dip The dip of beds immediately after deposition. Because of sinking of the basin these dips may steepen. The dip just prior to folding is initial dip, *q.v.*

original horizontality The state of strata being horizontal or nearly horizontal at the time they were originally deposited. *See also* LAW OF ORIGINAL HORIZONTALITY

ornamentation The pattern of ridges, grooves, nodes, etc., that may interrupt the smooth surface of a shell.

orocline Structural or mountain arc owing its form to differential horizontal displacement after the main features of the structural zone originated.

orocratic Pertaining to a period of time in which there is much diastrophism.

orogen Belt of deformed rocks, in many places accompanied by metamorphic and plutonic rocks. For example, the Appalachian orogen or the Alpine orogen.

orogenic Adjective derived from orogeny, *q.v.*

orogenic belt A linear region subjected to deformation during the orogenic cycle.

orogenic cycle The geotectonic cycle in which a mobile belt (geosyncline) is folded and deformed into a stable orogenic belt, having passed through preorogenic, orogenic, and postorogenic phases.

orogenic facies A term describing the tectonic environment of a geosynclinal facies.

orogenic phase The intermediate phase and the one of shortest duration in the orogenic cycle. The orogenic phase is characterized by a climax of orogenic activity and crustal deformation.

orogeny The process of forming mountains, particularly by folding and thrusting.

orogeosyncline Geosyncline that developed into an orogen.

orometer A sensitive aneroid barometer that is calibrated in feet and/or meters of elevation above sea level. Used for obtaining approximate elevations at various points of observation during reconnaissance mapping.

orthite Allanite.

orthoaxis In the monoclinic system, the axis that is perpendicular to the other two axes.

orthoclase A mineral, a member of the feldspar group. Composition $KAlSi_3O_8$. Monoclinic, dimorphous with microcline. A common mineral of granitic rocks. *Abbr:* Or

orthodolomite Sedimentary dolomite.

orthodome In orthorhombic and monoclinic crystals, the form whose faces are parallel to the ortho axis (b-axis) and intersect the other two axes.

orthogenesis Evolution continuously in a single direction over a considerable length of time; usages vary but the term usually carries the implication that the direction is determined by some factor internal to the organism, or, at least, is not determined by natural selection.

orthogeosyncline A geosyncline between continental and oceanic cratons, containing both volcanic (eugeosynclinal) and nonvolcanic (miogeosynclinal) belts.

orthogneiss A term used to denote a gneiss derived from an igneous rock. See AUGEN GNEISS; GNEISS

orthogonal On a refraction diagram, a line drawn perpendicular to the wave crests.

orthographic projection A perspective azimuthal map projection in which the straight and parallel projecting lines are perpendicular to the plane of the projection. This projection has no prime properties and is used as a visual link between map and globe.

ortholimestone Sedimentary limestone.

orthomagmatic or **orthotectic stage** Applied to the main stage of crystallization of silicates from a typical magma; the stage during which perhaps 90 per cent of the magma crystallizes. *Cf.* PEGMATITIC STAGE

orthopinacoid In orthorhombic and monoclinic crystals, the form whose faces are parallel to the ortho axis (b-axis) and the c-axis.

orthoquartzite A clastic sedimentary rock composed of silica-cemented quartz sand. The cement is commonly deposited in crystallographic continuity with the quartz of the worn grains.

orthorhombic or **rhombic symmetry** *1. Struct. petrol:* Refers to either symmetry of movement or symmetry of fabric. Orthorhombic symmetry of movement is exemplified by the motion that occurs when a sphere is subjected to a single compressive force acting along the vertical axis but is constrained on two opposite sides. Orthorhombic symmetry of fabric is the symmetry of an ellipsoid; there are three planes of symmetry. *2.* The symmetry of a polyhedron with three orthogonal twofold axes of rotatory symmetry. Three crystal classes, 222, 2mm, and 2/m 2/m 2/m have this symmetry.

orthorhombic system *Crystallog:* One of the six crystal systems consisting of crystals whose forms display orthorhombic symmetry and whose faces are referred to three unequal, mutually perpendicular axes. Rhombic system.

orthoschist A schist derived from an igneous rock. See PARASCHIST; SCHIST

orthotectic Designates those processes and products, strictly magmatic in the narrowest sense, exemplified in the normal crystallization of normal igneous rocks.

os; ås Esker.

Osagean The second of five Mississippian epochs in North America; also the series of rocks deposited during that epoch.

Osann's classification A chemical system of classification of igneous rocks.

osar; asar Esker.

oscillation cross-ripple marks Ripple marks consisting of two sets of ripple ridges intersecting at an angle.

oscillation ripple Ripples characterized by symmetry of crests, neither slope being steeper than the other since the ridges are built up by currents which operate from either side with approximately equal force.

oscillatory twinning See POLYSYNTHETIC TWINNING

oscillatory wave A wave in which

each individual particle oscillates about a point with little or no permanent change in position. The term is commonly applied to progressive oscillatory waves in which only the form advances, the individual particles moving in closed or nearly closed orbits. Distinguished from a wave of translation. *See* ORBIT

oscillograph An instrument which renders visible, or automatically traces, a curve representing the time variations of electric phenomena. The recorded trace is an oscillogram.

Osteichthyes Class of vertebrates, the bony fishes.

Ostracoda Subdivision of crustaceans consisting of small bivalved animals inhabiting both salt and fresh water. Shells were molted several times as individuals grew.

Ostwald's rule An unstable phase does not necessarily transform directly to the truly stable phase, but may first pass through successive intermediate phases, presumably due to lower activation energy barriers via that route.

otolith Ear bone of a fish.

outcrop The exposure of bedrock or strata projecting through the overlying cover of detritus and soil.

outer core Outer part of the earth's core between depths of about 2900 and 5000 km. May be liquid.

outer ridge Broad rise, generally more than 160 kilometers wide and from 180 to 1800 meters high, extending parallel to a continental margin; may enclose a marginal basin.

outgassing The dissipation of geothermal energy and matter into the atmosphere through the release of gases through volcanic activity. Most of the atmosphere originated in some such way, as did the hydrosphere.

outlet The opening by or through which any water discharges its content. The lower end of a lake or pond; the point at which a lake or pond discharges into the stream which drains it.

outlet glacier An ice lobe issuing from an ice sheet and occupying a valley.

outlier Portions of any stratified group which lie detached, or out from the main body, the intervening or connecting portion having been removed by denudation.

outwash Drift deposited by meltwater streams beyond active glacier ice.

outwash plain *Obs.* Outwash apron; overwash apron; marginal plain; outwash gravel plain; washed gravel plain. Sandr, sandur [*Icel.*]. A plain composed of material washed out from the ice.

overbank deposit Flood-plain deposit.

overburden Material of any nature, consolidated or unconsolidated, that overlies a deposit of useful materials, ores, or coal, especially those deposits that are mined from the surface by open cuts.

overflow Density currents produced where fresh-water streams enter salt water, or warm-water streams enter bodies of cold water, the lighter water flowing over on the surface of the heavier.

overfold A fold in which the beds on one limb are overturned, i.e., have been rotated through more than 90° so that they are inverted. *Syn:* OVERTURNED FOLD; OVERTHROWN FOLD; INVERTED FOLD. *Obs.*

overgrowth Secondary material deposited in optical continuity with a crystal grain, common in some sedimentary rocks. *Cf.* SECONDARY ENLARGEMENT

overhand stoping The working of

a block of ore from a lower level to a level above.

overhang *1.* The overhanging part of an erosion cliff, where the lower part has been undercut. *2.* A part of the mass of a salt dome that projects out from the top of the dome like the top of a mushroom.

overlap *1. Stratig:* A general term referring to the extension of marine, lacustrine, or terrestrial strata beyond or over older underlying rock whose edges are thereby concealed or "overlapped." The unconformity that commonly accompanies such a relation. *2. Fault:* The horizontal component of separation measured parallel to the strike.

overlay A record or map on a transparent medium which may be superimposed on another record.

overloaded stream An aggrading stream.

overloading The artificial or natural addition to a stable slope of a load of sufficient magnitude to make it unstable.

oversaturated rocks Those rocks which contain an excess of silica, over and above that necessary to form saturated minerals from all bases present. *See* SATURATED; SATURATION; UNDERSATURATED

oversteepening The process of steepening the walls of a valley by the passage of a valley glacier.

overthrust *1.* A thrust fault with low dip and large net slip, generally measured in miles. *2.* A thrust fault in which the hanging wall was the active element; contrasted with underthrust, but it is usually impossible to tell which wall was actively moved. *3.* The process of thrusting the hanging wall (relatively) over the footwall.

overthrust sheet or **block** The block, above a low-angle fault

plane, which has been displaced a matter of miles.

overturn *1.* The exchange of position in fall and spring of bottom and upper waters in a lake, caused by density differences due to temperature changes. *2.* The mixing of hypolimnion and epilimnion zones in lakes.

overturned Having been tilted past the vertical and hence inverted in the outcrop. Said of folded strata and of the folds themselves.

overturned limb That limb of an overfold (overturned fold) that is overturned, i.e., has rotated through more than 90°.

ovoid, *adj.* Ovate or egg-shaped.

oxbow A crescent-shaped lake formed in an abandoned river bend which has become separated from the main stream by a change in the course of the river.

oxidates Sediments formed by the precipitation of the oxidized form of Fe and Mn; ferric oxide and manganese dioxide sediments.

oxidation *1.* Process of combining with oxygen; e.g., the oxidation of Zn gives ZnO. *2.* The removal of one or more electrons from an ion or an atom.

oxide A compound of oxygen with one other more positive element or radical.

oxidize To unite with oxygen. To change an element from lower to higher positive valence.

oxidized zone That portion of an ore deposit which has been subjected to the action of surface waters carrying oxygen, carbon dioxide, etc. That zone in which sulfides have been altered to oxides and carbonates.

oxidizing flame The outer cone of the blowpipe flame, characterized by the excess of oxygen of the air over the carbon of the gas.

oxisol Old, intensely weathered tropical and subtropical lateritic, bauxitic, and silcretic soils; although horizonless in the common sense, analogues of A, B, and C horizons can be picked, but in terms of different components, especially the sesquioxides. Characterized by a high concentration of sesquioxides or a lack of original weatherable minerals.

oxychloride cement; sorel cement A plastic cement formed by mixing finely ground caustic magnesite with a solution of magnesium chloride.

oxyphile elements Elements occurring exclusively, or at least for the most part, combined with oxygen in oxides, silicates, phosphates, carbonates, nitrates, borates, sulfates, etc. Oxygen may be replaced by fluorine or chlorine to a minor extent. Approximately equivalent to Goldschmidt's lithophile elements.

oxysphere Lithosphere. Proposed because 60% of atoms in the earth's crust are oxygen and they occupy more than 90% of the volume of the familiar rocks.

ozocerite One of the solid hydrocarbons. It is light colored, soft, and consists largely of paraffin hydrocarbons of high molecular weight.

P

Pacific series, province, or suite One of two great groups of igneous rocks (along with the Atlantic group) based on their tectonic setting. Originally described as occurring on the margins of the Pacific basin, hence the Circum-Pacific province. Characterized by the tholeiitic magma types, yeilding saturated or oversaturated residues. *See* ATLANTIC SERIES

Pacific-type coastline Trend of folded belts are parallel to the coast. Contrasts with Atlantic type of coast.

pack ice Any large area of floating ice consisting of pieces of ice driven closely together.

packing The spacing or density pattern of the mineral grains in a rock. *Cf.* FABRIC

paha, *n. (sing. and pl.)* Ridges of silt and clay in the area of Iowan glacial drift in northeastern Iowa.

pahoehoe, *n.* A Hawaiian term for basaltic lava flows typified by smooth, billowy, or ropy surface. Varieties include corded, elephant-hide, entrail, festooned, filamented, sharkskin, shelly, and slab pahoehoe. *Cf.* AA

paint pot A type of mud pot containing variegated, highly colored, boiling mud, usually of cream, pink, or reddish tones.

palaeo- *See* PALEO-

palaeoclimatology *See* PALEOCLIMATOLOGY

palagonite, *n.* A yellow or orange, isotropic mineraloid formed by hydration and other alteration (devitrification, oxidation) of sideromelane (basaltic glass), and constituting a characteristic part of palagonite tuffs.

palagonite tuff An indurated deposit of glassy basaltic ash in which the constituent particles are largely altered to palagonite.

paleo-; palaeo- [<*Gr. palaio-, palai-*] A combining form meaning old, ancient, used to denote: (1) Remote in the past; (2) Early, primitive, archaic. Before vowels usually pale-, palae-.

paleobiology That branch of paleontology which treats of fossils as organisms.

paleobotany *See* PALEONTOLOGY

Paleocene Oldest of six epochs of the Cenozoic; also the series of rock strata deposited during that epoch.

paleoclimatology The study of past climates.

paleoecology The science of the relationship between ancient organisms and their environment.

Paleogene The earlier of the two periods comprising the Cenozoic Era, in the classification adopted by the International Geological Congress; it includes the Paleocene, Eocene, and Oligocene Epochs. Also the system of strata deposited during that period. *Cf.* NEOGENE

paleogeography The study of geography throughout geologic time. It may include all, or parts, of the world and of geologic time.

paleogeologic map *1.* Map showing areal geology as it was at some former time. *2.* Areal map of strata below an unconformity.

Paleolithic Old Stone Age; a term

used to designate those men and times characterized by the use of more primitive rock tools. Includes all cultures and times in which such tools were used up to but not including the more advanced tools associated with early agriculture.

paleolithologic map   Map showing lithologic variations at some buried horizon or within some restricted zone.

paleomagnetic north pole   The position of the magnetic north pole determined by the magnetic direction in a rock. This paleomagnetic study enables the scientist to reconstruct the orientation of the earth's magnetic field at the time the rock formed.

paleomagnetic patterns   The striped pattern of alternating strong and weak magnetic attractions possessed by rocks of the mid-ocean ridges. The pattern is nearly symmetrical on opposite sides of the central rift valley of the ridge.

paleomagnetism   Faint magnetic polarization of rocks that may have been preserved since the accumulation of sediment or the solidification of magma whose magnetic particles were oriented with respect to the earth's magnetic field as it existed at that time and place.

paleontologic facies   Facies differentiated on the basis of fossils. *Cf.* BIOFACIES

paleontologic province   Large region characterized by similar fossil faunas.

paleontologic species   A species based on fossil specimens. It is based on morphologic and phylogenetic criteria and cannot be derived from knowledge of soft parts or of breeding criteria.

paleontology; palaeontology   *1.* The science that treats fossil remains, both animal and vegetable. *2.* The science that deals with the life of past geological ages. It is based on the study of the fossil remains of organisms.

paleosol   A buried soil; a soil of the past.

paleotectonic   The crustal deformation at a given time in the geologic past.

paleotectonic map   A map intended primarily to represent the deformation of part of the earth's crust during a certain time interval.

Paleozoic   One of the eras of geologic time—that between the Precambrian and Mesozoic—comprising the Cambrian, Ordovician, Silurian, Devonian, Carboniferous (Mississippian and Pennsylvanian), and Permian systems. Also the erathem of rocks deposited during the Paleozoic Era.

paleozoology   The science of fossil animals; its two subdivisions are invertebrate and vertebrate paleontology.

palimpsest   A structure of metamorphic rocks due to the presence of remnants of the original texture of the rock.

palingenesis   *1.* The process of formation of new magma by the melting or fusion of country rocks with heat from another magma, with or without the addition of granitic material. *2.* The differential melting, in the root parts of folded mountains, to form a pore liquid or ichor. Anatexis. *3. Paleontol.* and *Biol:* The young stages of an organism recapitulate, with only slight modification, some or all of the characters of their ancestors.

palinspastic map   *1.* Map showing restoration of folded and faulted rocks to original relative geographic positions. *2.* A map showing thickness of a sedimentary unit restored to its preerosional or pre-truncation dimensions.

palisade A picturesque, extended rock cliff rising precipitately from the margin of a stream or lake and of columnar structure.

Palisadian disturbance A time of deformation, or orogeny, presumably at the close of the Triassic Period in eastern North America and elsewhere.

palladium gold Porpezite, or gold containing palladium up to 10%.

pallasite *See* SIDEROLITE

palsen Earth mounds, believed to be of periglacial origin and occurring in arctic and alpine regions.

paludal Pertaining to swamps or marshes, and to material deposited in a swamp environment. *See* PALUSTRINE

palustrine Pertaining to material deposited in a swamp environment. *See* PALUDAL

palygorskite A clay mineral characterized by rod- or lath-shaped crystals, $(Mg,Al)_2Si_4O_{10}(OH) \cdot 4H_2O$. Attapulgite.

palynology The study of pollen and other spores and their dispersal, and applications thereof.

pan 1. A natural basin or depression, especially one containing standing water or mud, and, as in South Africa, in the dry season often dried up, leaving a salt deposit. 2. In South Africa, a hollow in the ground where the neck of a volcano formerly existed. 3. Fire clay or underclay of coal seams. 4. A hard, cementlike layer within or just beneath the surface of a soil.

panfan 1. The ultimate surface attained when the last remnants of a range in a region of rising base-level have disappeared and the flanking alluvial fans coalesce at the divide. 2. An end stage in the process of geomorphic development in an arid region in the same sense that the peneplain is an end stage of the general process of degradation in a humid climate.

Pangaea The name Alfred Wegener proposed for the supercontinent comprising all the landmasses or earth which he believed existed about 300 million years ago, prior to continental drift.

panning A technique of prospecting for heavy metals, e.g., gold, by washing placer or crushed vein material in a pan. The lighter fractions are washed away, leaving the heavy metals behind, in the pan.

panplane A nearly flat surface brought about more by the lateral erosion of streams, which pares away the divides and causes a coalescing of all the flood plains of a region to form.

Panthalassa That ocean that surrounded Pangaea before its fragmentation.

pantograph An instrument for copying maps, plans, etc., on any predetermined scale.

paper shale Highly carbonaceous shale that splits in thin, tough, somewhat flexible sheets.

papillate; papillose, *adj.* Bearing minute pimplelike protuberances (papillae).

parabolic dune A dune having, in ground plan, approximately the form of a parabola, with the concave side toward the wind.

paracrystalline deformation Deformation that is contemporaneous with the recrystallization that forms a metamorphic rock.

paraffin A white, tasteless, odorless, and chemically inert waxy substance composed of natural hydrocarbons and obtained from petroleum. Any saturated hydrocarbon of chain structure whose general formula is C_nH_{2n+2}.

paraffin base Term applied to a crude oil containing paraffin wax in solution; such oil is relatively high in hydrogen and low in carbon.

paraffin dirt The paraffin, or "sour dirt" of the Gulf Coast fields; a yellow, waxy substance resembling beeswax, it has often been regarded as indicating the proximity of an oil gas reservoir.

paraffin hydrocarbon One of a series of saturated hydrocarbons with an open-chain structure; general formula C_nH_{2n+2}. A saturated aliphatic hydrocarbon. *Syn:* METHANE HYDROCARBON

paraffin series A homologous series of open-chain saturated hydrocarbons of the general formula C_nH_{2n+2} of which methane (CH_4) is the first member and the type. *Syn:* METHANE SERIES

paragenesis A general term for the order of formation of associated minerals in time succession, one after another.

paragenetic *1.* Refers to the chronological order of the crystallization of minerals as in a vein. *2.* Refers to the genetic relations of sediments in laterally continuous and equivalent facies.

parageosyncline Intracratonal geosyncline. *1.* A geosyncline that lies within a craton, i.e., within a relatively immobile portion of the crust of the earth. *2.* A geosyncline along the margin of a continent.

paragneiss A term used to denote a gneiss derived from a sedimentary rock. *See* GNEISS; ORTHOGNEISS

paragonite A mineral, a member of the mica group, $NaAl_2(AlSi_3)O_{10}(OH)_2$, Monoclinic.

paraliageosyncline A deep geosyncline that passes into coastal plains along the present continental margin.

paralic Pertaining to environments of the marine borders, such as lagoonal, littoral, shallow neritic, etc.

parallax, instrumental A change in the apparent position of an object with respect to the reference mark(s) of an instrument which is due to imperfect adjustment of the instrument or to a change in the position of the observer.

parallel A line extending around the earth parallel to the equator, used to indicate angular distance poleward.

parallel, standard (U.S. public-land surveys) An auxiliary governing line established along the astronomic parallel, initiated at a selected township corner on a principal meridian, usually at intervals of 39 kilometers from the base line, on which standard township, section, and quarter-section corners are established; also known as a correction line. Standard parallels, or correction lines are established for the purpose of limiting the convergence of range lines from the south.

parallel drainage pattern The drainage pattern is called parallel when the streams over a considerable area or in a number of successive cases flow nearly parallel to one another. Parallel drainage implies either a pronounced regional slope, or a slope by parallel topographic features such as glacially remodeled surfaces of the drumloidal or fluted ground moraine type, or control by parallel folded or faulted structures.

parallel evolution The phenomenon whereby related but distinct phyletic stocks develop comparable forms.

parallel extinction Extinction in anisotropic crystals parallel to crystal outlines.

parallel faults A group of faults having essentially the same dip and strike.

parallel or concentric fold A fold in which each bed maintains the same thickness (assuming it was initially of uniform thickness) throughout all parts of the fold.

parallelism The evolution of different lines or families in the same way, with corresponding, successive isomorphs. *Syn:* PARALLEL EVOLUTION; PARALLEL DEVELOPMENT; HOMOEOMORPHY

parallel shot *Seis. prospecting:* A test shot made with all the amplifiers connected in parallel and activated by a single geophone in order to check for lead, lag, polarity, and phasing in the amplifier to oscillograph circuits.

paramagnetic Having a magnetic permeability greater than unity, and susceptibility therefore positive; yet not ferromagnetic.

paramarginal resource The part of subeconomic resources that (1) borders on being economic or (2) is not commercially available because of legal or political circumstances.

parameter Any of the axial lengths or interaxial angles that define a unit cell. The rational multiple of the unit cell dimension measuring the intercept on an axis by a crystal face which determines the orientation of the face relative to the crystallographic axis.

parametric sounding Electromagnetic depth sounding in which the frequency is varied, as opposed to sounding in which the geometry is varied.

paramorph A pseudomorph with the same composition as the original crystal, as calcite after aragonite.

paraschist A term used to denote a schist derived from a sedimentary rock. *See* ORTHOSCHIST; SCHIST

parasitic cone One or more cinder cones which from their position upon the flanks of the larger volcano are referred to as parasitic cones.

paratectonic recrystallization A recrystallization which accompanies deformation.

paratype A specimen other than the holotype, upon which an original specific description is based.

paravane *1.* In seismic water shooting, a planing board used to keep a detector in a vertical position. *2.* In seismic water shooting, a device attached to the end of a towed line and so arranged that the device either travels a path parallel to but offset from the path of the towing vessel, or maintains a fixed depth below the surface, or both.

parental magma That magma from which some other magma was derived.

parent element The radioactive element from which a daughter element is produced by radioactive decay; e.g., radium is the parent element of radon.

parent material (soils) *1.* The horizon of weathered rock or partly weathered soil material from which the soil is formed. Horizon C of the soil profile. *2.* The unconsolidated material from which a soil develops.

parent rock *1.* The original rock from which sediments were derived to form later rocks. *2.* (Soils) The rock from which parent materials of soils are formed.

park *1. Topog:* A grassy, wide, and comparatively level open valley in wooded mountains; also, an open space surrounded by woodland. Local in Rocky Mountains. *2.* Shallow broad solution depression. *Syn:* SINKHOLE; term used locally on the Kaibab Plateau area of Arizona.

paroptesis The changes produced in rocks by dry heat; a baking.

partial or selective diagram

Struct. petrol: A point or contour diagram prepared by deliberately selecting certain grains of one mineral for measurement, such as measuring the large quartz grains in preference to the small ones, or measuring only quartz grains along shear zones.

particle-size histogram A graphic method for presenting the particle-size distribution of sediments as a series of vertical bars whose heights are proportional to the frequency in each class. The term itself is standard statistical usage for such diagrams.

particle velocity For waves, the velocity induced by wave motion with which a specific water particle moves.

particulates *1. Adj:* Of or relating to particles or occurring as minute particles. *2.* Finely divided solid or liquid particles in the air or in an emission. Particulates include dust, smoke, fumes, mist, spray, and fog.

parting *1.* A small joint in coal or rock, or a layer of rock in a coal seam. *2.* The tendency of crystals to separate along certain planes that are not true cleavage planes.

parvafacies The portion of any magnafacies which lies between designated time-stratigraphic planes or key beds traced across the magnafacies.

Pascal's law The principle that the pressure in a fluid not acted upon by external forces is the same at all points or that a fluid transmits pressures equally in all directions.

pass *1.* A gap, defile, or other relatively low break in a mountain range through which a road or trail may pass; an opening in a ridge forming a passageway. *See* COL. *2.* A navigable channel, especially at a river's mouth. *3.* A narrow connecting channel between two bodies of water; an inlet. *4.* An opening through a barrier reef, atoll, or sand bar.

patch reefs Usually small, isolated, organic buildups or areas of limestone deposition supported by a framework of organisms, more or less isolated and surrounded by rocks of different facies.

paternoster lake One of a linear series of small lakes occupying depressions in a glacial stairway.

path *Seis. prospecting* and *Seismol:* The course of travel between two points of a disturbance in an elastic medium.

patina Thin light-colored outer layer produced by weathering.

patterned ground A group term for the more or less symmetrical forms such as circles, polygons, nets, steps, and stripes that are characteristic of, but not necessarily confined to, mantle subject to intensive frost action.

pattern shooting *Seis. prospecting:* The firing of explosive charges arranged in a definite geometric pattern.

pavement A closely packed, smooth, natural bare-rock surface that resembles a paved road surface; e.g., desert pavement.

pay ore Those parts of an ore body which are both rich enough and large enough to work with profit. *See* PAY GRAVEL

pay streak That portion of a vein which carries the profitable or pay ore.

pay zone The vertical interval(s) or thickness of the stratigraphic section of an oil field containing reservoir beds that will yield gas or petroleum in economic quantities.

peacock copper Bornite.

peacock ore Name given to bornite, also less commonly to chalcopyrite, in allusion to the variegated colors on tarnished surfaces.

peak flood The particular flood that covers the flood plain; not to be confused with the flood of flood classification, as some flood plains are covered by the 25-year flood, whereas others are covered only by the 500-year flood. Not to be confused with flood peak, *q.v.*

peak zone *See* EPIBOLE

peat *1.* A dark-brown or black residuum produced by the partial decomposition and disintegration of mosses, sedges, trees, and other plants that grow in marshes and like wet places. *2.* Fibrous, partly decayed fragments of vascular plants which retain enough structure so that the peat can be identified as originating from certain plants (e.g., sphagnum peat or sedge peat).

peat bog A bog containing peat.

peat formation A process of decomposition of vegetable and animal substances intermediate between moldering and putrefaction or rot, during which first the former and then the latter process occurs. *Cf.* DISINTEGRATION; MOLDERING; PUTREFACTION

pebble armor A concentration of pebbles coating a desert area. The pebbles are usually the residual product of wind erosion and are closely fitted together so as to cover the surface in the manner of a mosaic. *Syn:* DESERT PAVEMENT. *See* LAG GRAVEL

pebble dike Vein or dikelike bodies composed of rounded-to-angular pebbles in a finer-grained matrix of pebble material or intrusive igneous rock.

pebble gravel Gravel consisting mainly of rounded rock fragments of pebble size.

pebble phosphate Varieties of natural phosphate that are concretionary or alluvial in origin, hence gravel-like.

pebbles Smooth rounded stones ranging in diameter from 2 to 64 mm.

pectinate, *adj.* Comblike or pinnatifid with very close narrow

divisions or parts; also used to describe spine connections in cacti when small lateral spines radiate like comb teeth from areole.

pectolite A mineral, $NaCa_2Si_3O_8(OH)$, commonly in radiating groups in cavities in diabase. Triclinic.

ped A naturally formed unit of soil structure, e.g., granule, block, crumb, aggregate.

pedalfer An old, general term for a soil characterized by a concentration of sesquioxides. It is the typical soil of a humid region.

pedestal boulder Isolated masses or rock above and resting on a smaller base or pedestal. *See* PERCHED BOULDER

pedestal rock A residual mass of weak rock capped with harder rock. *See* PEDESTAL BOULDER

pediment Gently inclined planate erosion surfaces carved in bedrock and generally veneered with fluvial gravels.

pediment pass Narrow, flat, rock-floored depression connecting pediment slopes on opposite sides of a mountain ridge.

pedion Crystal form with only one face.

pediplain; pediplane Widely extending rock-cut and alluviated surfaces formed by the coalescence of a number of pediments and occasional desert domes.

pedocal Soil of arid or semiarid regions, enriched in lime. Accumulates in regions of low or high temperature and low rainfall; the vegetation is usually prairie, but may be desert scrub.

pedogenesis Soil formation.

pedology The science which treats of soils their origin, character, and utilization.

pedon The smallest unit or volume of soil that represents all the horizons of the soil profile. It is usually about one square meter, but can be larger.

peel-off time *Seis. prospecting:* The time correction to be applied to observed data to adjust them to a depressed reference datum.

peel thrust Overthrust fault block pushed ahead of a resistant mass without affecting a hard under-lying basement and not involving shortening in a folded region.

pegmatite Those igneous rocks of coarse grain found usually as dikes associated with a large mass of plutonic rock of finer grain size. The absolute grain size is of lesser consequence than the relative size. Unless specified otherwise, the name usually means granite pegmatites, al-though pegmatites having gross compositions similar to other rock types are known.

pegmatitic stage or **phase** A stage in the normal sequence of crys-tallization of a magma contain-ing volatiles, at which time the residual fluid is sufficiently en-riched in volatile materials to permit the formation of coarse-grained rocks more or less equiv-alent in composition to the par-ent rock (pegmatites).

pelagic *1.* Pertaining to com-munities of marine organisms which live free from direct de-pendence on bottom or shore; the two types are the free-swim-ming forms (nektonic) and the floating forms (planktonic). *2.* Related to water of the sea as distinct from the sea bottom. *3.* Related to sediment of the deep sea as distinct from that derived directly from the land.

pelean, *adj.* Designating or per-taining to a type of volcanic eruption characterized by explo-sions of extreme violence and the formation of nuées ardentes. The lavas involved in this type of eruption are generally extremely silicic and viscous.

Pelecypoda A division (class) of uncoiled, bivalve mollusks, with straight intestine and sometimes with a foot utilized in locomotion.

Pele's hair Rock material con-sisting of threads of volcanic glass (generally basaltic) drawn out from the lavas by explosion or by bursting of bubbles on lava lakes.

Pele's tears Small drops of vol-canic glass (generally basaltic) with pendant threads, or pairs of drops arranged in dumbbell fashion, thrown out during erup-tions of fluid lava and measur-ing a few millimeters in length.

pelite Mudstone.

pelitic Argillaceous, *q.v.*

pelitomorphic Refers to irregu-lar precipitated grains of calcium carbonate.

pellet Small aggregation of sedi-mentary material. *See* FAECAL PELLETS

pellet structure A feature com-monly shown by clays, formed of small rounded aggregates of clay minerals and fine quartz scattered through a matrix of the same material. The pellets may be separated from the matrix by a shell of organic material. In size, the pellets are 0.1 to 0.3 mm. in diameter, and in a few cases several mm. in length.

pellicular A term applied to wa-ter adhering as films to the sur-faces of openings and occurring as wedge-shaped bodies at junc-tures of interstices in the zone of aeration above the capillary fringe.

Pelmatozoa Subphylum of echi-noderms most of which are per-manently attached by a jointed stem after completion of larval development.

pelyte; pelite Mudstone.

pencil ganister Ganister charac-terized by fine carbonaceous markings and so called from the likeness of these traversing marks to pencil lines.

pencil structure Very pronounced

lineation such as that produced by intersecting bedding and cleavage planes in slate.

pendant A small solutional remnant projecting from the ceiling or an overhanging wall.

pendulum *1.* A body so suspended from a fixed point as to swing freely to and fro under the combined action of gravity and momentum. *2.* A vertical bar so supported from below by a stiff spring as to vibrate to and fro under the combined action of gravity and the restoring force of the spring. Also called an inverted pendulum.

penecontemporaneous A term used in connection with the formation of a sedimentary rock such as a cherty limestone or a concretionary shale.

peneplain *1.* A land surface worn down by erosion to a nearly flat or broadly undulating plain; the penultimate stage of old age of the land produced by the forces of erosion. *2.* By extension, such a surface uplifted to form a plateau and subjected to renewed degradation and dissection.

peneplanation The subacrial degradation of a region approximately to base level, forming a peneplain.

penetration twin *See* TWIN

penetrometer A weight-driven rod or drill for measuring the vertical resistance of snow to penetration.

peninsula A body of land nearly surrounded by water, and connected with a larger body by a neck or isthmus; also, any piece of land jutting out into the water.

Pennsylvanian In the United States the sixth of seven periods of the Paleozoic. Equivalent, approximately, to the Upper Carboniferous outside of the United States. Also the system of rocks deposited during that period.

Penokean orogeny A time of de-

formation and granite emplacement during the Precambrian in Minnesota and Michigan dated radiometrically at about 1700 m.y. ago.

pentagonal dodecahedron See PYRITOHEDRON. *Obs.*

pentane A liquid hydrocarbon of the paraffin series, formula C_5H_{12}.

pentlandite A mineral, $(Fe,Ni)_9S_8$. Isometric. An important ore of nickel.

pentremite A type of blastoid that was particularly common in Mississippian time and illustrates in the extreme the almost perfect pentameral symmetry of this group of animals.

pepino Rounded, conical-shaped hills resulting from tropical karst action. *Syn:* HUM; HAYSTACK; MOGOTES; COCKPIT

peptize To bring into stable colloidal suspension.

peralkaline In the Shand classification of igneous rocks, a division embracing those rocks in which the molecular proportion of alumina is less than that of soda and potash combined.

peraluminous In the Shand classification of igneous rocks, a division embracing those rocks in which the molecular proportion of alumina exceeds that of soda, potash, and lime combined.

percentage log A well record made from an examination of cuttings in which the percentage of each type of rock present in each sample of cuttings is estimated and plotted.

perched boulder A large erratic lying in an unstable position on top of a hill or boss.

perched ground water Ground water separated from an underlying body of ground water by unsaturated rock. Its water table is a perched water table.

perched rock or **block** A large mass of rock which, after glacial transportation, has been lodged in some conspicuous isolated position.

perched water table Water table above an impermeable bed underlain by unsaturated rocks of sufficient permeability to allow movement of ground water.

percolate To pass through fine interstices; to filter; as water percolates through porous stones.

percussion mark Crescentic chatter or percussion marks on the finer grained and well-rounded pebbles, especially porphyries.

perennially frozen ground *See* PERMAFROST

perennial stream Streams that flow throughout the year and from source to mouth.

perfect gas *See* IDEAL GAS

perforation Puncturing of well casing opposite an oil-bearing zone to permit oil to flow into a cased borehole.

pergelation The act or process of forming permanently frozen ground in the present or in the past.

pergelisol *See* PERMAFROST

periclase Magnesia. A mineral, MgO. Isometric.

periclinal Dipping on all sides from a central point or apex. Applied to strata which dip in this manner from some common center of elevation. *Syn:* QUAQUA-VERSAL

pericline Dome. *See* QUAQUA-VERSAL (*Brit.*)

pericline twin *See* TWIN LAW

peridot The gem variety of olivine.

peridotite A coarse-grained, ultramafic rock consisting of olivine and pyroxene with accessory constituents.

perigee Shortest distance from an earth satellite orbit to the earth's center.

periglacial Refers to areas, conditions, processes, and deposits adjacent to the margin of a glacier.

perimagmatic Close to the magma.

period *1.* A major, worldwide, standard geologic time unit corresponding to a system. *2.* An interval of time characterized in some particular way. *3.* Time required for a recurrent motion or phenomenon to complete a cycle and begin to repeat itself.

periodic or **tidal current** A current, caused by the tide-producing forces of the moon and the sun, which is a part of the same general movement of the sea manifested in the vertical rise and fall of the tides.

peripheral faults Arcuate border faults surrounding an area of subsidence, or an area of elevation such as a diapir.

perlite A volcanic glass having numerous concentric cracks which give rise to perlitic structure. Most perlites have a higher water content than obsidians. A high proportion of all perlites are rhyolitic in composition.

perlitic structure A structure produced in homogeneous material by contraction during cooling, and consisting of a system of irregular, convolute, and spheroidal cracks; generally confined to natural glass, but occasionally found in quartz and other noncleavable minerals and as a relict structure in devitrified rocks.

permafrost Permanently frozen ground (subsoil). Permafrost areas are divided into more northern areas in which permafrost is continuous, and those more southern areas in which patches of permafrost alternate with unfrozen ground.

permafrost table A more or less irregular surface which represents the upper limit of permafrost.

permanent current A current that runs continuously independent of the tides and temporary causes. Permanent currents include the fresh water discharge of a river and the currents that form the general circulatory systems of the oceans.

permanently frozen ground See PERMAFROST

permeability *1.* The permeability (or perviousness) of rock is its capacity for transmitting a fluid. Degree of permeability depends upon the size and shape of the pores, the size and shape of their interconnections, and the extent of the latter. It is measured by the rate at which a fluid of standard viscosity can move a given distance through a given interval of time. The unit of permeability is the darcy, *q.v.* See also MILLIDARCY. *2. Geophys:* The ratio of the magnetic induction to the magnetic intensity in the same region. In paramagnetic matter the permeability is nearly independent of the magnetic intensity; in a vacuum it is strictly so. But in ferromagnetic matter the relationship is definite only under fully specified conditions.

permeability, relative The ratio of the permeability of a porous medium under any given conditions to the absolute permeability. This term usually signifies the permeability to one fluid phase when two or more phases are present in the porous medium.

permeability coefficient Coefficient of permeability. The rate of flow of water in gallons a day through a cross section of 1 square foot under a unit hydraulic gradient. The standard coefficient is defined for water at a temperature of 60° F. The field coefficient requires no temperature adjustment and the units are stated in terms of the prevailing water temperature.

permeability trap A condition in which a permeable part of a bed or group of beds is bounded, particularly on the updip side, by relatively impermeable rock.

permeable Pervious. *Hydrol:* Having a texture that permits water to move through it perceptibly under the head differences ordinarily found in subsurface water. A permeable rock has communicating interstices of capillary or supercapillary size.

Permian The last of seven periods (six outside of North America) of the Paleozoic Era; also the system of rocks deposited during that period.

permineralization The process of fossilization wherein the original hard parts of an animal have additional mineral material deposited in their pore spaces.

Permo-Carboniferous Strata not differentiated between the Permian and Carboniferous Systems, particularly in regions where there is no conspicuous stratigraphic break and fossils are transitional.

Permo-Triassic Strata not differentiated between the Permian and Triassic Systems, particularly in regions where the boundary occurs within a nonmarine red beds succession.

perovskite A mineral, $CaTiO_3$. Pseudo isometric, usually in yellow, brown, or black cubes.

perpendicular slip The component of the net slip measured perpendicularly to the trace on the fault of the disrupted index plane (bed, dike, vein, etc.) in the fault plane.

perpendicular throw The distance between the two parts of a disrupted bed, dike, vein, or of any recognizable surface measured perpendicular to the bedding plane or to the surface in question.

Perret phase (of eruption) Emission of much high energy gas which may greatly enlarge a volcanic conduit.

perthite A variety of feldspar consisting of intergrown orthoclase or microcline with albite.

pervious *See* PERMEABLE

petrifaction The process of petrifying, or in the older literature an object that has been petrified.

petrified rose; desert rose Aggregates of clusters of tabular crystals of barite or gypsum that form chiefly in sandstone, enclosing the sandy matrix within the crystal and resembling a rose.

petrified wood Silicified wood, *q.v.*, often with accessory minerals.

petrify To become stone. Organic substances, such as shells, bones, wood, etc., embedded in sediments, become converted into stone by the gradual replacement of their tissues, particle by particle, with corresponding amounts of infiltrated mineral matter.

petro-; petr- [*Gr.*] A combining form meaning rock or stone.

petrochemistry *1*. The chemistry of rocks. *2*. The chemistry of petroleum; disapproved by some geochemists.

petrofabric analysis Petrofabrics, *q.v.*

petrofabric diagram A diagram used in petrofabric analysis. It may be a point diagram or contour diagram, *q.v.*

petrofabrics Petrofabric analysis. The study of spatial relations, especially on a microscale, of the units that comprise a rock, including a study of the movements that produced these elements. The units may be rock fragments, mineral grains, or atomic lattices.

petrofacies Facies distinguished by petrographic characters.

petrogenesis A branch of petrology which deals with the origins of rocks, and more particularly with the origins of igneous rocks.

petrogenic grid Pressure-temperature diagram with equilibrium curves showing the stability fields of specific minerals and mineral associations.

petrogeny's residual system The system $NaAlSiO_4$-$KAlSiO_4$-SiO_2, which represents a close approximation to the composition of many residual liquids from magmatic differentiation.

petrographer One versed in the science of petrography or the systematic description and classification of rocks.

petrographic microscope Polarizing microscope.

petrographic period Time represented by a rock kindred.

petrographic province A region or district in which some or all of the igneous rocks are regarded as consanguineous, or as derived from a common parent magma; a comagmatic district or province.

petrography That branch of geology dealing with the description and systematic classification of rocks, esp. igneous and metamorphic rocks and esp. by means of microscopic examination of thin sections. Petrography is more restricted in scope than petrology, which is concerned with the origin, occurrence, structure, and history of rocks. *Adj:* petrographic. *Cf.* LITHOLOGY. *See also* PETROGRAPHER

petroleum A naturally occurring complex liquid hydrocarbon that may contain varying degrees of impurities (sulfur, nitrogen) which after distillation yields a range of combustible fuels, petrochemicals, and lubricants. *Syn:* CRUDE OIL

petroleum geologist A geologist engaged in the exploration or production processes of hydrocarbon fuels. *See also* PETROLEUM GEOLOGY

petroleum geology That branch of economic geology which relates to the origin, occurrence, migration, accumulation, and exploration for hydrocarbon fuels. Its practice involves the application of geochemistry, geophysics, paleontology, structural geology, and stratigraphy to the problems of finding hydrocarbons. *See also* PETROLEUM GEOLOGIST

petroliferous Containing or yielding petroleum.

petroliferous province An area containing known commercial accumulations of petroleum in a tectonic unit, such as a sedimentary basin or a geosyncline.

petrologist One who does petrology.

petrology That branch of geology dealing with the origin, occurrence, structure, and history of rocks, esp. igneous and metamorphic rocks. Petrology is broader in scope than petrography, which is concerned with the description and classification of rocks. *Adj:* petrologic *Cf.* LITHOLOGY. *See also* PETROLOGIST

phacolith A concordant intrusive in the crest of an anticline and trough of a syncline, hence in cross section it has the shape of a doubly convex lens.

phanerite A general term applied to wholly crystalline rocks, the constituents of which may be distinguished with the unaided eye.

phaneritic A textural term applied to igneous rocks in which all the crystals of the essential minerals can be distinguished with the unaided eye; i.e., megascopically crystalline. Contrasted with cryptocrystalline; microcrystalline.

Phanerozoic Post-Precambrian time or rocks; originally thought to be the time of evident life.

phantom crystal A crystal in which an earlier stage of crystallization is outlined in the interior.

phantom horizon In seismic reflection prospecting, a line so constructed that it is parallel to the nearest actual dip segment everywhere along a profile.

phase *1.* A variety differing in some minor respect from the dominant or normal type; a facies; ordinarily used in the detailed description of igneous rock masses. *2. Phys. chem:* A homogeneous, physically distinct portion of matter in a nonhomogeneous (i.e., heterogeneous) system, as the three phases—ice, water, and aqueous vapor. *3.* The point or stage in the period to which the rotation, oscillation, or variation has advanced, considered in its relation to a standard position or assumed instant of starting. This relation is commonly expressed in angular measure. *4. Earthquake seismol:* An event on a seismogram marking the arrival of an impulse or a group of waves at a detecting instrument and indicated by a change of period or amplitude, or both. *5.* Facies, *q.v.,* as it was used prior to 1849. *6.* Geologic time unit smaller than an age. (Uncommon.) *7.* Rock facies identified by both original and secondary characters of the strata. (Uncommon.)

phase boundary Boundary line. The line separating any two phase areas (in binary systems) or any two liquidus surfaces (in ternary systems).

phase diagram A graph in which two or more of the variables temperature, pressure, and concentrations are plotted against one another, designed to show the boundaries of the fields of stability of the various phases of a heterogeneous system.

phase equilibria Heterogeneous equilibria, *q.v.*

phase rule The statement that for any system in equilibrium, the number of degrees of freedom is two greater than the difference between the number of components and the number of phases; in symbols, $F=C-P+2$.

phase velocity The velocity with which a seismic disturbance of a given frequency is propagated. When the phase velocity of a material varies with frequency, the material is said to exhibit DISPERSION, *q.v.*

phassachate A lead-colored agate.

phenacite A beryllium orthosilicate, Be_2SiO_4. Sometimes used as a gem.

phenocryst One of the relatively large and ordinarily conspicuous crystals of the earliest generation in a porphyritic igneous rock.

Phi grade scale *See* GRADE SCALE, PHI

Phi mean particle diameter A logarithmic mean particle diameter obtained by using the negative logs of the class midpoints to the base 2.

Phi Sigma Sigma Phi. The standard deviation of a particle-size distribution computed in terms of Phi grades.

phlogopite A mineral, a member of the mica group, $K(Mg,Fe'')_3(AlSi_3)O_{10}(F,OH)_2$.

phonolite Extrusive equivalent of nepheline syenite. The principal mineral is soda orthoclase or sanidine.

phorogenesis Slipping of the earth's crust over the mantle.

phosphate A salt or ester of phosphoric acid; a compound containing the radical PO_4^{-3}.

phosphate rock A sedimentary rock containing calcium phosphate. The form in which the phosphate occurs is obscure. The chief mineral commonly is apatite.

phosphorescence Luminescence caused by exposure to light or other forms of radiation, and lasting after exposure has ceased. *Cf.* FLUORESCENCE

phosphorus cycle One of the mineral cycles, *q.v.*

photogeology The geologic interpretation of aerial photographs.

photogeomorphology Study of earth forms as revealed by aerial photographs.

photogrammetry The science and art of obtaining reliable measurements from photographs.

photomap The reproduction of a single photograph, composite, or mosaic, complete with grid lines and marginal data.

photomicrograph An enlarged or macroscopic photograph of a microscopic object, taken by attaching a camera to a microscope.

photosynthesis Synthesis of chemical compounds effected with the aid of radiant energy, especially light.

phreatic cycle The period of time during which the water table rises and then falls. It may be a daily, annual, or other cycle.

phreatic explosion A volcanic explosion, ordinarily of extreme violence, caused by the conversion of ground water to steam. Such steam explosions have a low temperature and do not expel essential ejecta.

phreatic water A term that originally was applied only to water that occurs in the upper part of the zone of saturation under water-table conditions (*syn.* of unconfined ground water, or well water) but has come to be applied to all water in the zone of saturation, thus making it an exact synonym of ground water.

phreatic zone Zone of saturation.

phreatophyte A plant that consumes and then transpires inordinate amounts of water.

phyla Plural of phylum, *q.v.*

phyletic Pertaining to a line of organic descent.

phyletic evolution Evolution involving changes in lineages but little or no increase in the number of taxonomic groups.

phyllite An argillaceous rock intermediate in metamorphic grade between slate and schist. The mica crystals impart a silky sheen to the surface of cleavage (or schistosity). *See* SCHIST; SLATE

phyllitization Development of phyllitic rocks.

phyllosilicates A group of silicate minerals with similar structures characterized by silicon-oxygen tetrahedra linked in a planar arrangement. Micas.

phylogenetic species A species that is a segment of an evolved lineage—true of all true species—but *cf.* HORIZONTAL EVOLUTION

phylogeny The line, or lines, of direct descent in a given group of organisms. Also the study or the history of such relationships.

phylogerontism The condition of racial deterioration and approaching extinction. Now doubted as a valid concept.

phylum One of the primary divisions of the animal and plant kingdom; a group of closely related classes of animals or plants.

physical processes maps Derivative maps for planning that combine for any area the reaction of those physical properties of soil and substrate that may affect man's activities.

physiographic cycle The sequence of changes from the beginning of youth to old age.

physiographic province Region of similar structure and climate that has had a unified geomorphic history.

physiography [<*Gr.* physis, nature; and graphe, description] The study of the genesis and evolution of land forms.

phytolith A stony or mineral structure secreted by a living plant. Often composed of opal or calcite.

phytoplankton All the floating plants such as diatoms, dinoflagellates, coccolithophores, and sargassum weed.

picacho A peak or sharply pointed hill or mountain. Because of the steep slopes of mountains in the desert region, picacho appears as the name of numerous mountains in southwestern Arizona.

pick *Seis. prospecting:* 1. The selection of an event on a seismic record. Also used as a verb, as, to pick reflections. 2. Any selected event on a seismic record.

picrite Olivine-rich basalt as is often formed by the settling of olivine in thick flows and sills, etc. These often contain 50 or more per cent of olivine.

pictograph 1. Any conventionalized representation of an object. 2. Diagram showing range of variability, commonly a scatter diagram. 3. More often refers to rock pictographs of primitive peoples.

piecemeal stoping A process whereby magma eats into its roof by engulfing relatively small isolated blocks, which presumably sink to depth where they are assimilated. *See* MAGMATIC STOPING

piedmont Lying or formed at the base of mountains, as, a piedmont glacier.

piedmont alluvial plain A plain formed by the coalescence of alluvial fans.

piedmont glacier A glacier formed by coalescence of two or more valley glaciers beyond the base of a steep slope.

piedmont steps, benchlands, or treppen Regional terraces sloping outward (down valley) to correspond with the several graded reaches of the streams are postulated to develop as the response to a continually accelerated upheaval of an expanding dome.

piercement dome A salt dome in which the salt core has broken through the overlying strata until it reaches or approaches the surface.

piercement fold; piercing fold Diapir fold, q.v.

piezocrescence The growth of one crystallographic orientation out of another under the influence of stress produced by either mechanical or thermal means.

piezoelectric 1. Having the ability to develop surface electric charges when subjected to elastic deformation, and conversely. 2. Oscillates in alternating current circuits with frequencies harmonic with the stimulating frequency.

piezometric surface Hydrol: 1. An imaginary surface that everywhere coincides with the static level of the water in the aquifer. 2. The surface to which the water from a given aquifer will rise under its full head.

pig A container usually made of lead used to ship or store radioactive materials.

pigeonite See PYROXENE

Piggot corer Device for sampling bottom sediments. A core barrel is driven into unconsolidated material by an explosive charge.

pile A nuclear reactor.

pillar A column of rock remaining after solution of the surrounding rock. See HOODOOS

pillow lavas A general term for lavas that exhibit pillow structure, occurring mostly in basic lavas (basalts and andesites) and especially in the sodium-rich basalts known as spilites.

pillow structure The peculiar structure exhibited by some basic lavas (especially spilites) which consist of an agglomeration of rounded masses that resemble pillows, bolsters, or filled sacks.

pilotaxitic A textural term applied to the groundmasses of holocrystalline, glass-free (volcanic) rocks consisting of a feltlike interweaving of lath-shaped microlites (ordinarily plagioclase), commonly in flow lines. See FELTY

pimple mound Mima mound; pimple plain.

pimple plain Characterized by numerous conspicuous, small, rounded, circular elevations 4½ meters to 9 meters in diameter and one-half meter to about 2 meters in height.

pinacoid An open crystal form consisting of two parallel faces.

pinch A compression of the walls of a vein, or the roof and floor of a coal bed, which more or less completely displaces the ore or coal. Called also pinch out.

pinch out Thin out, q.v.

pingo Relatively large mound raised by frost action above the permafrost and generally persisting for more than a single season.

pingok Pingo.

pingo remnant Kettlelike depression resulting from the melting of a mass of ground ice.

pinnacle 1. Topog: Any high tower or spire-shaped pillar of rock, alone or cresting a summit. A tall, slender, pointed mass; especially a lofty peak. 2. A sharp pyramid or cone-shaped rock under water or showing above it.

pinnate joints See FEATHER JOINTS

pipe A tabular opening or cylindrical rock body filling a tabular opening. It is usually more or less vertical.

pipe, volcanic See VOLCANIC PIPE

pipe clay Potter's clay, q.v.

pipe ore Iron ore (limonite) in vertical pillars, sometimes of conical, sometimes of hourglass form, embedded in clay. Probably formed by the union of stalactites and stalagmites in caverns.

piperno, *n.* A local Italian name applied to the trachytic tuffs of the Phlegrean Fields in the vicinity of Naples. The rock is characterized by a eutaxitic structure and the presence of numerous stringers and lenticles of dark glass (fiamme) in a light-colored, porous, glassy matrix and is generally considered to be a type of welded tuff.

pipestone Catlinite, *q.v.*

piracy The diversion of the upper part of a stream by the headward growth of another stream. Also called beheading, stream capture, and stream robbery.

pirate stream See PIRACY

Pisces Subphylum of vertebrates; fish.

pisolite A spherical or subspherical accretionary body over 2 mm. in diameter.

pisolith A small spheroidal particle with concentrically laminated internal structure, ranging from 1 to 10 mm. in diameter. The unit particle in the rock, "pisolite."

pisolitic Consisting of rounded grains like peas or beans.

pisolitic tuff An indurated pyroclastic deposit made up chiefly of accretionary lapilli or pisolites.

pit *1. Topog:* A cavity or hole in the ground, natural or artificial, such as the La Brea Pits of tar in California. *Speleol:* 2. A deep hole, generally circular in outline, with vertical or nearly vertical walls; 3. Small holes made in cave fills by cave beetles. 4. *Paleobot:* A thin place in a cell wall. A simple pit has no overarching wall, a bordered pit has such a wall.

pitch *1.* Often used as synonymous with plunge. 2. Of an ore shoot in a vein, the angle between

the axis of the ore shoot and the strike of the vein. The pitch is measured in the plane of the vein. *Petroleum geol:* 3. A solid hydrocarbon belonging to the group of asphaltites; 4. One of the residues formed in the distillation of wood or coal tar. It is also obtained from petroleum. The term pitch is sometimes employed indiscriminately to mean bitumen or asphalt. 5. *Speleol:* A vertical shaft in a pothole.

pitchblende Uraninite.

pitch length The length of an ore shoot in its greatest dimension.

pitchstone A volcanic glass characterized by a pitchy rather than glassy luster. They may be almost any color and have compositions equivalent to a wide range of volcanic rocks. They contain a rather high percentage of water compared to other glassy rocks.

Pitot tube A small tube, bent at one extremity to form a right angle. The bent end of the tube is inserted in the flowing stream so that the plane of the opening is perpendicular to the direction of flow. This determines the impact pressure of the flowing stream, indicated by the height to which it will force a fluid column, usually water or mercury. Commonly used in the form of a "U" tube to measure the flow of gas wells.

pitted plain *Topog:* 1. An outwash plain of gravel or sand with kettle holes; 2. Plain with numerous, small, closely spaced sinkholes.

pivotal fault Hinge fault, *q.v.*

placental *1.* Member of the Mondelphia. 2. Bearing a placenta.

placer [*Sp.*] A place where gold is obtained by washing; an alluvial or glacial deposit, as of sand or gravel, containing particles of gold or other valuable mineral.

placer claim *1.* A mining claim located upon gravel or ground whose mineral contents are extracted by the use of water, by sluicing, hydraulicking, etc. *See* PLACER. *2.* Ground with defined boundaries which contains mineral in the earth, sand, or gravel; ground that includes valuable deposits not fixed in the rock.

placer deposit A surficial mineral deposit formed by mechanical concentration of mineral particles from weathered debris. The mechanical agent is usually alluvial but can also be marine, eolian, lacustrine, or glacial, and the mineral is usually a heavy metal such as gold.

placer mining That form of mining in which the surficial detritus is washed for gold or other valuable minerals. When water under pressure is employed to break down the gravel, the term hydraulic mining is generally employed.

Placodermi Class of vertebrates consisting of primitive jawed fish of varied characters, some armored, some sharklike.

plagioclase A mineral series ranging from the composition NaAlSi$_3$O$_8$ (albite) to CaAl$_2$Si$_2$O$_8$ (anorthite). Triclinic. One of the most common rock-forming minerals.

plain A region of general uniform slope, comparatively level, of considerable extent, and not broken by marked elevations and depressions; it may be an extensive valley floor or a plateau summit. Any extent of level or nearly level land.

plain, coastal A plain fronting the coast and generally representing a strip of recently emerged sea bottom.

plain of denudation A nearly plane surface, produced by erosion.

plain tract Lower portion of a stream course characterized by low gradient and a wide flood plain.

planar cross-stratification Compound stratification. A type of cross-stratification in which the lower bounding surfaces of the sets are planar surfaces of erosion.

planar flow structure Any planar structure that develops during the intrusion of magma. May be expressed by the parallel arrangement of platy minerals (giving a foliation), by slablike inclusions, by schlieren, or by bands of different mineralogy or texture. Synonymous with planar structure and platy flow structure.

planar gliding Uniform slippage along plane surfaces.

planation *1.* The widening of valleys through lateral corrasion by streams after they reach grade and begin to swing, and the concurrent formation of flood plains. Also, by the extension of the above processes, the reduction of divides and the merging of valley plains to form a peneplain; peneplanation. *2.* The grading of an area or district by any erosive process, either subaerial or marine.

plane correction A correction applied to observed data to reduce them to a common reference plane.

plane group A set of symmetry operations in a plane; there are 17 of them.

plane of flattening *See* FLATTENING, PLANE OF

plane of symmetry *1.* A plane to which a crystal is symmetrical; i.e., each face, corner, and edge of an ideally developed crystal is the mirror image, with respect to this plane, of another face, corner, or edge. *2.* In bilateral animals the plane that divides the animal into two halves of which each is the mirror image of the other.

plane-polarized light Light constrained to vibrate in a plane; a single Nicol prism produces plane-polarized light. *Cf.* CROSSED NICOLS

plane strain State of strain in which intermediate strain axis is unity or can be ignored.

plane stress State of stress in which tractions (stresses) involving the intermediate principal stress vanish.

plane surveying Surveying in which the curvature of the earth is disregarded, as in ordinary field and topographic surveying.

plane table *1.* A simple surveying instrument by means of which one can plot the lines of a survey directly from the observations. It consists of a drawing board on a tripod, with a ruler, the ruler being pointed at the object observed. *2.* An inclined ore-dressing table.

planetesimal One of numerous small solid planetary bodies which, according to the planetesimal hypothesis, had individual orbits about the Sun and of which the planets were formed by aggregation.

planetesimal hypothesis or theory The hypothesis that the earth and other planets were formed by the collision and coalescence of planetesimals and have never been wholly molten.

planimeter An instrument for measuring the area of any plane figure by passing a tracer around the bounding plane.

planimetric analysis Analysis of patterns in a fabric diagram based on distribution of points and areal comparisons.

planimetric map A map which presents the horizontal positions only for the features represented; distinguished from a topographic map by the omission of relief in measurable form.

planimetry The determination of horizontal distances, angles, and areas by measurements on a map.

plankton Holoplankton, *q.v.* Floating organisms. All of the floating or drifting marine life.

plankton bloom A sudden rapid increase (usually geometric) to an enormous number of individual plankters under certain conditions. *See also* BLOOM

planktonic Floating.

planosol An intrazonal soil having a leached surface layer underlain by a definite clay pan, or hard pan, developed on a nearly flat, upland surface under grass or forest vegetation in humid or subhumid climate.

Plantae The vegetable kingdom.

plastering-on Process of addition of material to a ground moraine by melting at the base of a glacier.

plastic Capable of being molded into any form, which is retained.

plastic deformation A permanent change in shape of a solid that does not involve failure by rupture.

plastic flow In structural geology, synonym of plastic deformation.

plastic index The difference between liquid and plastic limits, indicating the range of moisture content within which a soil-water mixture is plastic.

plasticity The property of a material that enables it to undergo permanent deformation without appreciable volume change or elastic rebound, and without rupture.

plasticity index The range of water content, expressed as a percentage of the weight of the oven-dried soil, through which the soil is plastic. It is defined as the liquid limit minus the plastic limit.

plastic limit The lowest moisture content, expressed as a percentage of the weight of the oven-dried soil, at which the soil can be rolled into threads one-eighth inch in diameter without the thread breaking into pieces. One of the Atterberg limits, *q.v.*

plastic strain *1.* In the case of a single mineral the term connotes permanent deformation accomplished by gliding within the crystal lattice without loss of cohesion. *2.* In rocks, which are composed of many crystals often belonging to several mineral species, the term is conveniently applicable to any permanent deformation throughout which the rock maintains essential cohesion, and strength, regardless of extent to which local microfracturing and displacement of individual grains may have entered into the process.

plat A diagram drawn to scale showing all essential data pertaining to the boundaries and subdivisions of a tract of land, as determined by survey or protraction.

plate *Geol:* One of many large segments of the earth's crust (lithosphere) varying in thickness from 50 to 250 km. and including a portion of the upper mantle above the asthenosphere.

plateau A relatively elevated area of comparatively flat land which is commonly limited on at least one side by an abrupt descent to lower land.

plateau basalt A term applied to those basaltic lavas that occur as vast composite accumulations of horizontal or subhorizontal flows and which, erupted in rapid succession over great areas, have at times flooded sectors of the earth's surface on a regional scale. They are generally believed to be the product of fissure eruptions.

plateau glacier An ice sheet that

occupies a more or less flat, plateaulike area.

plateau mountain The folds of a mountain chain frequently pass abruptly into the horizontal strata of a basal plateau which, when largely denuded and eroded, may remove itself into a series of plateau mountains.

plate boundary Zone of seismic and tectonic activity along the edges of lithosphere plates, presumed to indicate relative motion between plates.

plate tectonics A theory of global-scale dynamics involving the movement of many rigid plates of the earth's crust. Considerable tectonic activity occurs along the margins of the plates where buckling and grinding occurs as the plates are propelled by the forces of deep-seated mantle convection currents. This has resulted in continental drift and changes in shape and size of oceanic basins and continents.

platform *1. Stratig:* The area of thinner sediments adjoining a geosynclinal wedge of thicker equivalent beds or a basin of thicker equivalent sediments. *2. Paleon:* In a coral the flat bottom or floor of the calyx.

platy flow structure *See* PLANAR FLOW STRUCTURE

playa [*Sp.*] *1.* A shore, strand, beach, or bank of a river. Generally sandy, and sometimes auriferous. *2.* The shallow central basin of a desert plain, in which water gathers after a rain and is evaporated.

playa lake Broad, shallow sheets of water which quickly gather and almost as quickly evaporate, leaving mud flats or playas to mark their sites.

Playfair's law Every river appears to consist of a main trunk, fed from a variety of branches, each running in a valley proportioned to its size, and all of them together forming a system of valleys communicating with one another, and having such a nice adjustment of their declivities that none of them join the principal valley either on too high or too low a level; a circumstance which would be infinitely improbable if each of these valleys were not the work of the stream which flows in it.

Pleistocene The earlier of the two epochs comprising the Quaternary Period. Also the Post-Pliocene (post-Tertiary) glacial age, which in the above terminology implies the glacial age is over. Also the series of sediments deposited during this epoch. Some geologists use Quaternary and Pleistocene synonymously, implying that the glacial age is still with us.

pleochroic, *adj. See* PLEOCHROISM

pleochroism; dichroism The property of differentially absorbing light that vibrates in different directions in passing through a crystal.

pleomorphous; pleomorphic Polymorphous, *q.v.*

Pleospongia Group of Lower and Middle Cambrian fossils generally having the form of corals, but possessing structures suggesting relationships to sponges.

plexus In evolution, a network of anastomosing lineages, which never completely isolate and are able to recombine through parts of geologic time.

plicated *1.* Folded together as highly inclined and contorted strata. *2. Paleobot:* Folded as in a fan, or approaching this condition. *Paleozool:* Ribbed as in plicate brachiopods or pelecypods.

Plinian, *adj.* Designating or pertaining to a type of volcanic eruption of extreme violence like that described by Pliny the

Younger, in A.D. 79, which wrecked the ancient mountain centered approximately on the site of the present-day Vesuvius and buried the cities of Herculaneum and Pompeii under thick deposits of volcanic debris. The characteristics of Plinian eruptions have been only rather vaguely defined, but it is generally believed that they are the result of hydro-explosions, i.e., steam explosions of colossal power.

plinthite The hard, illuviated horizon, formed at depth in a lateritic latosol.

Pliocene The last epoch of the five epochs of the Tertiary Period (middle of three Neogene epochs or fifth of seven epochs of the Cenozoic); also the series of rocks deposited during that epoch.

plot To place survey data upon a map or plat. In past use, no clearly defined difference existed between plat and plot.

pluck To tear away projecting pieces of rock; said of the action of glaciers on contiguous rock.

plucking *1.* Monoliths up to many feet in diameter, bounded by structural surfaces, are lifted from the rock by the flowing ice and removed. This process has been termed plucking, and also quarrying. *2.* The process of erosion, by glaciers and streams, whereby blocks are removed from bedrock along joints and stratification surfaces. Quarrying.

plug, umbilical The filling of secondary shell material which is found in the umbilical region of certain Foraminifera.

plug, volcanic *See* VOLCANIC PLUG

plug dome *See* VOLCANIC DOME

plumb *1.* Vertical. *2.* A plumb bob; a plummet. *3.* To carry a survey into a mine through a shaft by means of heavily weighted fine wires hung vertically in the shaft. The line of sight passing through the wires at the surface is thus transferred to the mine workings.

plumbago Graphite.

plumb point The point on the ground vertically beneath the perspective center of the camera lens.

plunge 1. *Surv:* To set the horizontal cross wire of a theodolite in the direction of a grade. 2. To turn over the telescope of a transit on its horizontal transverse axis. 3. *Struc. geol:* The inclination of a fold axis or other geologic structure, measured by its departure from horizontal, it is mainly used for the geometry of folds. The attitude of the axial line of a fold defined by the bearing and inclination of that line. The plunge inclination is the angle between the sloping axial line and the horizontal as measured in a vertical plane.

plunge pool Potholes, in general, of large size, occurring at the foot of a vertical or nearly vertical waterfall.

plunging fold A fold in which the axis of folding lies at a relatively steep angle to the horizontal.

pluton In the strictest sense, a body of igneous rock that has formed beneath the surface of the earth by consolidation from magma. In a broader sense, it may include bodies composed of pseudoigneous rock that formed beneath the surface of the earth by the metasomatic replacement of an older rock.

plutonic Rocks, usually igneous, formed at great depth.

plutonic cognate ejecta Coarsely crystalline fragments consanguineous with the lavas of a given volcano, which solidified at depth, generally as dikes and sills, but were brought to the surface by pyroclastic eruption.

plutonic emanations The volatile material given off by a deep-seated magma.

plutonic rock An igneous rock formed at great depth by magmatic crystallization or chemical alteration.

plutonic series Series of different rocks that evolved from the same original material through various metamorphic stages until final crystallization ceased.

plutonic water Water in or derived from magma at considerable depth; the minimum depth is not known but it is that of plutonic rocks, probably in the order of several miles.

Plutonism The obsolete belief that all of the rocks of the earth solidified from an original molten mass. *Cf.* NEPTUNISM

Plutonist A follower of James Hutton (1726–1797) who recognized that internal heat of the earth has been important in geologic development and that some rocks are igneous in origin. *Cf.* NEPTUNIST; MAGMATIST

plutonite Coarse-grained intrusive igneous rock.

pluvial 1. *Geol:* Due to the action of rain. 2. Pertaining to deposits by rain water or ephemeral streams.

pneumatolitic metamorphism Contact metamorphism in which the composition of rock has been altered by introduced magmatic material.

pneumatolysis The alteration of rocks and the formation of minerals during or as a result of the emanation of gases and vapors from solidifying igneous rocks. *See* PNEUMATOLYTIC

pneumatolytic A term used in various connotations by various authors and perhaps best abandoned. It has been used to describe processes such as (1) surface effects of gases near volcanoes, (2) contact metamorphic effects surrounding deep-seated intrusives without any knowledge of gas vs. liquid state, (3) that stage in igneous differentiation between pegmatitic and hydrothermal, which is supposed to be characterized by gas-crystal equilibria, and (4) (very loosely) any deposit containing "pneumatolytic" minerals or elements, such as tourmaline, topaz, fluorite, lithis, and tin, and hence presumed to have formed from a "gas" phase.

pneumotectic Processes and products of magma consolidation in which fundamental influences of a sort that was magmatic in the strictest sense were recognizably modified and to some extent controlled by gaseous constituents or so-called mineralizers.

pocket *1*. A small body of ore; an enlargement of a lode or vein; an irregular cavity containing ore. *2*. A natural underground reservoir of water. *3*. A receptacle, from which coal, ore, or waste is loaded into wagons or cars. *4*. A ganister quarryman's local term for masses of rock 9 to 15 meters in width that are worked out and loaded, buttresses of untouched rock being left between them to support the upper masses. *5*. A hole or depression in the wearing course of a roadway. *6*. A glen or hollow among mountains. *7*. Solutional concavity in a cave ceiling, wall, or floor whose location is not determined by a joint but is localized by stream action.

pod *1*. A rudely cylindrical ore body that decreases at the ends like a cigar. *See* LENS. *2*. A very shallow depression up to more than 10 km. in diameter of the south Russian steppes containing temporary lakes; may reflect uneven loess deposition, preloess topography, deflation, solution, etc.

podsol; podzol A zonal soil hav-

ing a surface layer of mats of organic material overlying gray, leached horizons and dark brown, illuvial horizons. Develops under coniferous or mixed forests in cool, moist climates.

poeciloblastic Poikiloblastic, *q.v.*

poikilitic; poecilitic A textural term denoting a condition in which small granular crystals are irregularly scattered without common orientation in a larger crystal of another mineral.

poikiloblastic; poeciloblastic A texture due to the development, during recrystallization, of a new mineral around numerous relics of the original minerals.

point *1*. A small cape; a tapering projection from the shore of a lake, river, or sea. *See* CAPE; TONGUE. *2*. A position on a reference system determined by a survey. A fix.

point-bar deposit Sediment deposited on the inside of a growing meander loop.

point diagram *Struct. petrol:* A petrofabric diagram on which each item measured is represented by a point. Each point may represent an optical direction in some mineral, such as the *c* crystallographic axes of quartz grains, or the perpendicular to the cleavage of micas, etc.

point group One of the 32 crystal classes based on possible collections of symmetry elements about axes intersecting in a point.

point maximum Concentration of poles around a point in a fabric diagram; a polar cap generally indicating slipping movement.

poised stream *1*. Condition of a river that is neither eroding nor depositing sediment. *2*. *Engin:* Grade.

Poisson's ratio The ratio of the fractional transverse contraction to the fractional longitudinal extension of a body under tensile stress.

polar cap Concentration of points in a fabric diagram.

polar diagram Polar projection.

polar front The line of discontinuity, which is developed in suitable conditions between air originating in polar regions and air from low latitudes, on which the majority of the depressions of temperate latitudes develop.

polar glaciers *1.* Glaciers formed at high latitudes and developed on plateaus and from which ice tongues extend down valleys trenched into the plateau. *2.* In Ahlmann's classification, a glacier with the accumulation area covered by firn formed by slow recrystallization of solid precipitation and with subsurface temperatures below freezing to considerable depth throughout the year. Two subtypes are highpolar and subpolar glaciers.

polariscope An instrument for studying the properties of and examining substances in polarized light.

polarity *1.* The electrically positive or negative condition of a battery or generator terminal. *2.* The magnetically positive (north) or negative (south) character of a magnetic pole. *3.* Capability of dissociation as ions.

polarization The modification of lights so that its vibrations are restricted to a single plane.

polarize *1.* To endow with poles, as a magnet. *2.* To produce an electrical separation or orientation, especially in the molecules of a dielectric. *3.* To impress some spatial characteristic, as upon the vibrations identified with radiation, e.g., in elliptically polarized light.

polarized light Light that has been changed from the ordinary state in which the transverse vibrations occur in all planes passing through the line of propagation to a state in which they are in a single plane (plane polarized).

polarizer An apparatus for polarizing light. In a polarizing microscope, it may be the lower Nicol prism or polaroid.

polarizing microscope A microscope that uses polarized light and a revolving stage for analysis of petrographic thin sections. Two prisms, one above and the other below the stage, polarize and analyze the light; the stage rotates about the line-of-sight axis. *Syn:* PETROGRAPHIC MICROSCOPE

polar projection Projection of points on the surface of a sphere to a plane tangent at its pole.

polar symmetry A type of crystal symmetry in which the two ends of the central crystallographic axis are not symmetrical. Such a crystal is said to display hemimorphism.

polar wandering *1.* Short-period movement of earth's poles resulting from wobbling of its axis. *2.* Long-period, more or less systematic displacement of the earth's poles along curved paths which may have occurred during the passage of geologic time.

polar wandering curve The path of movement of a continent over the earth's surface during the geologic past. This path is located by identifying the paleomagnetic north pole position of a series of rock samples of different ages obtained from that continent.

polders Flat tracts in Holland below the level of the sea or the nearest river, such as a lake or morass which has been drained and brought under cultivation. They are protected from inundation by embankments called dikes. Similar to these are the fens of England.

pole *1. Struct. petrol:* A point on the reference sphere that represents the intersection of the sphere with a line passing through the center of the sphere. The line may be some optical direction in a mineral or it may be a perpendicular to some plane. 2. The ends of the axis of coiling in planispirally coiled shells or tests, as in the fusulines. 3. One of the extremities of the axis of symmetry of radio symmetrical pollen grains.

pole-fleeing force Force supposedly resulting from the oblate shape of the earth causing landmasses to move toward the equator.

polish, *n.* An attribute of surface texture related to the regularity of reflections; a surface presenting a high luster or characterized by highlights as distinct from a surface which is "dull."

polje; polye An isolated depression, generally several miles long or wide, floored with flat alluvium; walls generally steep. *Syn:* INTERIOR VALLEY. *See* UVALA

pollutant Any natural or artificial substance that enters the ecosystem in such quantities that it does harm to the ecosystem in any way.

pollution The natural or unnatural addition of anything to an ecosystem to the extent that it harms all or any part of that ecosystem.

polyconic map projection A map projection having the central geographic meridian represented by a straight line, along which the spacing for lines representing the geographic parallels is proportional to the distances apart of the parallels; the parallels are represented by arcs of circles which are not concentric, but whose centers lie on the line representing the central meridian, and whose radii are determined by the lengths of the elements of cones which are tangent along the parallels. All meridians except the central one are curved.

polycrystal Assemblage of crystal grains of unspecified number, shape, size, orientation, or bonding that together form a solid body.

polye; polje *See* POLJE; UVALA

polygenetic Originating in various ways or from various sources; formed at different places or times or from different parts; said specifically, in geology, of mountain ranges. Opposed to monogenetic.

polygenetic soil Soil produced under conditions that have changed importantly with time.

polygeosyncline A wide, long-enduring, primary geosyncline in which shallow water sediments accumulated but in which one or more parallel geanticlines arise to separate the primary geosyncline into two or more sequent geosynclines.

polygonal ground A form of patterned ground marked by polygonal or polygon-like arrangements of rock, soil, and vegetation, produced on a level or gently sloping surface by frost action; esp. a ground surface consisting of a large-scale network of ice-wedge polygons. *Syn:* POLYGON GROUND; POLYGONAL SOIL.; POLYGONBODEN; POLYGONAL MARKINGS; CELLULAR SOIL

polygons, mud crack Mud cracks (sun cracks, shrinkage cracks) form as sediments lose contained water. The cracks bound polygons, which vary in number of sides and dimensions of angles between the sides. Cracks are rarely straight, and polygons may be bounded by as few as three and as many as eight cracks.

polyhalite A mineral, $K_2Ca_2Mg(SO_4)_4.2H_2O$, commonly in pink, red, or gray masses in potassium salt deposits. Triclinic.

polymerization *1.* The joining of identical molecules to form larger ones without altering the total chemical composition. 2. The joining of similar molecules to form larger and more complex ones. (Loose usage)

polymerize To change (by union of two or more molecules of the same kind) into another compound having the same elements in the same proportions, but a higher molecular weight and different physical properties.

polymictic lake A lake which is continually mixing or with very short stagnation periods.

polymictic rocks Rocks characteristic of geosynclinal regions and including arkoses and graywackes.

polymorph One of several different morphologic kinds occurring in a species, a mineral, etc.

polymorphism Pleomorphism. *1.* The existence of a species in several forms independent of the variations of sex. *2.* A substance which can exist in more than one solid form is said to have polymorphic forms, e.g., rhombic and monoclinic sulfur, α and β quartz. *3.* The property possessed by certain chemical compounds of crystallizing in several distinct forms.

polymorphous *1.* Having the same chemical composition but crystallizing in different crystal systems or classes. *See* ALLOTROPIC; DIMORPHOUS; PLEOMORPHOUS; TRIMORPHOUS. *2.* Existing in several forms, independent of variations of sex.

polyp A coelenterate living singly or in colonies; a scyphozoan (jellyfish) or the asexual generation of a hydrozoan.

polypheletic An artificial construction of a classification of animals, usually the result of lack of knowledge, in which a taxon contains lineages of ancestral lines that are not closely related.

polysynthetic twinning Successive twinning of three or more individuals according to the same twinning law and with the composition planes parallel. Often revealed by striated surfaces.

polytypic *1. Tax:* Including several units of the next lower category, as, for example, a species with several subspecies. *2.* Referring to a species which has a group of subspecies which replace each other geographically. Proposed as "polytypic species."

polytypic species A species which consists of a group of subspecies which replace each other geographically.

ponding The natural formation of a pond or lake in a watercourse; chiefly: (a) by a transverse mountain uplift whose rate of elevation exceeds that of the stream's erosion, or (b) by a dam caused by glaciers, volcanic ejecta, landslips, or alluvial cones.

pontic Euxinic. Applied to an association of black shale and dark limestone, deposited in stagnant waters.

population *1. Biol:* All individuals of species living in more or less intimate association with each other. *2. Stat:* Those individuals or that group from which a particular sample is taken or derived.

porcelaneous Having the appearance of porcelain.

porcelaneous chert A type of smooth chert which has a smooth fracture surface, hard, opaque to subtranslucent, typically chinawhite resembling chinaware or glazed procelain, grades to chalky.

porcelanite *1.* A light-colored, porcelaneous rock resulting from the contact-metamorphism of marls. *See* HORNFELS. *2.* Fused shales and clay, that occur in roof and floor of burned coal seams.

pore *1.* Interstice; void, *q.v.* A space in rock or soil not occupied by solid mineral matter. *2.* In blastoids, an opening at margin of an ambulacrum leading to one of the hydrospires. *3.* In cystoids: horizontal tubes or slits occupying parts of two adjoining plates. *4.* In echinoids: pit for attachment of a ligament which fastens spine to tubercle. *5.* A minute opening or foramen, or orifice.

pore diameter Diameter of the largest sphere that might be contained within a pore.

pore fluid *See* INTERGRANULAR FILM

pore space The open space between particles in a rock or soil.

Porifera Phylum of simply organized metazoans without specialized tissues or organs; sponges.

porosimeter An instrument used to determine the porosity of a rock sample by comparing the bulk volume of the sample with the aggregate volume of the pore spaces between the grains.

porosity The ratio of the aggregate volume of interstices in a rock or soil to its total volume. It is usually stated as a percentage.

porous Containing voids, pores, interstices, or other openings which may or may not interconnect.

porphyrin A complex organic substance found in green plants and forming the basic feature of unit of the chlorophyll.

porphyritic; porpyrite A textural term for those igneous rocks in which larger crystals (phenocrysts or insets) are set in a finer groundmass which may be crystalline or glassy, or both.

porphyroblast *1.* A term given to the pseudo-phenocrysts of rocks produced by thermodynamic metamorphism. The corresponding texture is called porphyroblastic. *2.* Large grains or crystals, commonly perfect, developed in schists resulting from deformation of rocks originally containing phenocrysts. *Syn:* METACRYSTS

porphyry A term first given to an altered variety of porphyrite (porphyrites lapis) on account of its purple color, and afterward extended by common association to all rocks containing conspicuous phenocrysts in a fine-grained or aphanitic groundmass. The resulting texture is described as porphyritic.

porphyry copper Disseminated copper minerals in a large body of porphyry. In the commercial sense the term is not restricted to ore in porphyry but is applied to deposits characterized by huge size (particularly with respect to horizontal dimension), uniform dissemination, and low average per-ton copper content.

portal *1.* Gap in a borderland by which an epicontinental sea communicated with the permanent ocean. *2.* Surface entrance to a mine, particularly to a drift, tunnel, or adit. *Syn:* ENTRY

Portland cement A hydraulic cement consisting of compounds of silica, lime, and alumina. It is obtained by burning to semi-fusion an intimate mixture of pulverized materials containing lime, silica, and alumina in varying proportions within certain narrow limits, and by pulverizing finely the clinker that results.

positive An arch of the craton which persistently tends to stand higher than the surrounding shelves.

positive area Area that has been uplifted conspicuously or repeatedly.

positive crystal An optically positive crystal. *See* OPTICAL CHARACTER

positive element Large structural feature or area that has had a long history of progressive uplift; also in a relative sense one that has been stable or has subsided much less than neighboring negative elements.

positive elongation Elongation of anisotropic crystals parallel to vibration direction of the slower of the two plane-polarized rays.

positive movement Uplift, actual or relative.

positive ore Ore exposed on four sides in blocks of a size variously prescribed. *See* PROVED ORE. Ore which is exposed and properly sampled on four sides, in blocks of reasonable size, having in view the nature of the deposit as regards uniformity of value per ton and of the third dimension, or thickness.

positive shoreline A shoreline resulting from a negative movement of the land or by a positive movement of the sea level; a shoreline of submergence. *Ant:* NEGATIVE SHORELINE

possible ore Ore which may exist below the lowest workings, or beyond the range of actual vision.

postmagmatic reactions A general term covering reactions occurring after the bulk of the magma has crystallized. The exact range covered by the term varies with different authors, but generally includes the hydrothermal stage.

postorogenic Apotectonic; postkinematic. An event that takes place after an orogeny.

posttectonic crystallization Recrystallization in a tectonite that continued after deformation ceased.

potable Drinkable. Said of water and beverages.

potash Potassium carbonate, K_2CO_3. Loosely used for potassium oxide, potassium hydroxide, or even for potassium in deplorable expressions such as potash feldspar.

potash feldspar A mineral member of the feldspar group containing a high potassium content. Orthoclase, microcline, sanidine.

potash fixation Retention of potassium in clays either by chemical combination in clay minerals or by adsorption.

potassic Of, pertaining to, or containing potassium.

potassium bentonite Metabentonite. *See* K-bentonite.

potassium 40 Radioactive potassium; half-life about 1.42×10^8.

potato stone Geode.

pot clay A highly refractory fire clay used in the manufacture of pottery.

potential A term applied to several different scalar quantities, the measure of each of which involves energy as a function of position or of condition.

potential barrier *See* ACTIVATION, ENERGY OF

potential electrode *Elec. prospecting:* One of two electrodes between which is measured the difference of potential due to natural currents or those artificially introduced into the ground.

potential energy of waves In a progressive oscillatory wave, the energy resulting from the elevation or depression of the water surface from the undisturbed level.

potentiometer *1.* An instrument for measuring or comparing electromotive forces. It consists essentially of a resistance with a sliding contact, or its equivalent, used in connection with a galvanometer and a standard cell. *2.* A source of adjustable voltage, consisting of a resistor through which current is flowing, provision being made to connect to any desired point along the resistor.

potentiometric map A subsurface contour map showing the elevation of a potentiometric surface of an aquifer (top of the water table).

potentiometric surface Surface to which water in an aquifer would rise by hydrostatic pressure.

pothole *1.* A hole generally deeper than wide, worn into the solid rock at falls and strong rapids by sand, gravel, and stones being spun around by the force of the current. *2.* In Death Valley, a circular opening, two to four feet in diameter filled with brine and lined with salty crystals. *3.* A rounded, steep-sided depression resulting from downward surface solution. *4.* An underground system of pitches and slopes. Applied in some cases to single pitches reaching the surface. *5.* A rounded cavity in the roof of a mine caused by a fall of rock, coal, ore, etc. *6.* A hole in the ground from which clay for pottery has been taken. *7.* Depression between dunes that contains water.

potter's or **pipe clay** *1.* Pure plastic clay, free from iron, and consequently white after burning. *2.* A clay adapted for use on a potter's wheel, for manufacture of pottery.

Pottsvillian Lower Pennsylvanian.

powder method; powder diffraction X-ray diffraction from a powdered, crystalline sample, commonly observed by the Debye-Scherrer camera method or by a recording diffractometer.

powder snow Dry fallen snow composed of crystals or grains which lie loosely.

Poynting's law A special case of the Clapeyron equation in which the fluid is removed as fast as it forms (as under metamorphic stress) so that its volume may be ignored.

pozzuolana; pozzolan; pozzolana; pozzuolane A leucitic tuff quarried near Pozzuoli, Italy, and used in the manufacture of hydraulic cement. The term is now applied more generally to a number of natural and manufactured materials (ash, slag, etc.) which impart specific properties to cement. Pozzuolanic cements have superior strength at a late age and are resistant to saline and acidic solutions.

prairie soil A dark-brown soil with some leaching in the A horizon; calcareous B horizon.

prase A translucent green variety of chalcedony. Used as a gemstone.

Pratt isostasy A suggested type of hydrostatic support for the earth's solid outer crust in which the crustal density is supposed to be greater under mountains than under oceans.

Precambrian All rocks formed before the Cambrian. It has been recommended that Precambrian be divided into Early Precambrian and Late Precambrian.

precession camera An X-ray camera used to register the diffraction from a single crystal showing individual layers of the reciprocal lattice without distortion.

precious metal Gold, silver, or any of the minerals of the platinum group.

precious stone A relatively rare, durable gemstone of unusual beauty including diamond, ruby, emerald, and sapphire.

precipitation *1. Hydrol:* The discharge of water, in liquid or solid state, out of the atmosphere, generally upon a land or water surface. The quantity of water that has been precipitated (as rain, snow, hail, sleet) measured as a liquid. *2.* The process of separating mineral constituents from a solution by evaporation (halite, anhydrite) or from magma to form igneous rocks.

precision depth recorder An echo sounder having an accuracy better than 1 in 3000.

preconsolidation pressure Pressure exerted on unconsolidated sediment by overlying material that resulted in compaction; the overburden may have been removed later by erosion.

precrystalline deformation Nonruptural deformation in a tectonite where recrystallization continued after deformation ceased.

predazzite A marble containing calcite and brucite, the molecular proportion of MgO to CaO, being less than in pencatite; i.e., less than 1:1. *See* MAGNESIAN MARBLE; MARBLE

preferred orientation *Struct. petrol:* A rock in which the grains are more or less systematically oriented by shape or in which the atomic structure shows a more or less systematic arrangement.

pregeologic Before the time when the surface of the earth became generally similar to what it is today, certainly 3 and perhaps 4.5 billion years ago.

preglacial Of, pertaining to, or occurring in geologic time before a glacial epoch, usually referring to the last or Pleistocene, but not always.

prehnite A mineral, $Ca_2Al_2Si_3O_{10}$-$(OH)_2$, commonly in green botryoidal masses. Orthorhombic, hemimorphic.

preliminary waves The body waves of an earthquake. They record on the seismograph before the surface waves by virtue of their high speeds in the interior of the earth which they penetrate. A collective term including both P-waves (first preliminary waves) and S-waves (second preliminary waves), *q.v.*

preoccupied Previously used; said of a taxonomic name. The law of priority does not permit the reuse of a preoccupied name.

preorogenic *1.* Dating from a time preceding the formation of mountains. *2.* Refers to beginning of orogenic disturbance; early orogenic.

pressure Force per unit area applied to outside of a body.

pressure, geostatic The pressure exerted by a column of rock. It averages about 1 pound per square inch per foot of column height, but, until it is sufficiently great that the rocks collapse, it is not transmitted to any fluid contained within them.

pressure arches Wavelike formations on a glacier surface; a stage in the formation of Forbes bands, *q.v.*, or ogives.

pressure figure A figure produced by intersecting lines of parting, due to gliding when certain minerals, like mica, are compressed by a blunt point. They are similar in character, but not necessarily in position, to the so-called percussion figures produced by a blow with a sharp point.

pressure head Hydrostatic pressure expressed as the height of a column of water that can be supported by the pressure. It is the height that a column of water rises in a tightly cased well that has no discharge. The pressure head is commonly expressed with reference to the land surface at the well or to some other convenient level.

pressure release An outward-expanding force resulting from unloading of confining pressures of deep burial. Massive plutonic rocks split into great shells or spalls as a result of confining pressure release.

pressure-release jointing Exfoliation resulting from confining pressure release.

pressure shadow *Struct. petrol:* An area adjoining a porphyroblast characterized by a growth rather than a deformation fabric, as seen in a section normal to the b fabric axis. Its sigmoid form indicates the direction sense of the movement.

pressure solution Solution occurring preferentially at grain contacts where static pressure exceeds hydraulic pressure of interstitial fluid. See RIECKE'S PRINCIPLE

pressure tube When an isolated rock or stone rests upon the surface of the solid ice of a glacier, it may be near the center of one of the crevice-surrounded masses, where some emboulement has deposited it, by its greater conducting power and heat-absorbing surface, it becomes warmed by the Sun's rays, much more than the ice on which it rests, and sinks down through the ice, forming a tube of its own diameter, often of an enormous depth.

pressure wave P-wave, *q.v.* Compressional wave.

pretectonic pluton Intrusion older than a particular period of folding; may be genetically related to orogeny or much older.

pretectonic recrystallization Recrystallization in a tectonite that ceased before deformation was completed.

primacord *Seis. prospecting:* A detonating fuse which consists of an explosive core contained within a waterproof textile covering.

primärrumpf An upwarped, progressively expanding dome, with a rise so slow that degradation keeps pace with uplift.

primary *1.* Characteristic of or existing in a rock at the time of its formation. Said of minerals, textures, etc., of rocks. Essentially the same as original, and contrasted with derived or secondary, *1. 2.* Formed directly by solidification from fusion or deposition from solution. Said of igneous rocks and chemical sediments and contrasted with derivative (little used). *3.* Originally the same as the present Precambrian, then extended to include the present Paleozoic, and later restricted to Paleozoic; finally abandoned and now obsolete.

primary arc Mountain or island arc convex outward from a continent and overlying a deep-seated structure such as a great shear zone or tectogene; consists of an inner volcanic arc and an outer arc that may be represented either by distorted sediments or an oceanic trench.

primary basalt A presumed original magma, from which all other rock types are obtained by various processes.

primary dip See ORIGINAL DIP

primary dolomite A dense finely textured well-stratified unfossiliferous dolomite rock formed in place by direct precipitation from sea water or lake water by direct chemical or biochemical processes.

primary flat joints See L-JOINTS

primary flowage Movement within an igneous rock which is partly fluid.

primary geosyncline Major undivided geosyncline.

primary gneiss A term applied to a rock that exhibits foliation, lineation, or other planar or linear structures such as are generally characteristic of metamorphic rocks, but which because of the absence of observable granulation or recrystallization is considered to be igneous.

primary magma A magma directly erupted from the earth's simatic substratum. Usually considered also as synonymous with parental magma.

primary minerals Those minerals that were deposited in the original ore-forming or rock-forming episode. *Cf.* SECONDARY MINERALS

primary openings Openings or voids existing when the rock was formed.

primary phase In an isoplethal study, the first crystalline phase to appear on cooling a composition from the liquid state, i.e., at the intersection of the isopleth and liquidus.

primary structure *1.* Structure of a sedimentary rock which is dependent on the conditions of deposition, mainly current velocity and rate of sedimentation. *2.* Those structural features that are contemporaneous with the first stages in the formation of a rock.

primary tectonite Tectonite whose fabric was determined by movement of the medium in which the rock developed.

primary wave *See* P-WAVE.

prime meridian The meridian of Greenwich; arbitrarily selected meridian having a longitude of zero, and used to reckon all other longitudes east and west to 180°.

primitive circle *Crystallog:* The great circle in the plane of a stereographic projection; the circle inscribed in the plane of a gnomonic projection to define the scale.

principal axes of strain In elastic theory, the principal axes of the reciprocal strain ellipsoid. The extensions of lines drawn in these directions, in the unstrained state, are stationary for small variations of direction. One of them is the greatest extension, the other is the smallest.

principal axes of stress The coordinate axes along which no shearing stresses exist.

principal axis In the tetragonal and hexagonal systems, the vertical crystallographic axis; hence

in uniaxial crystals, the optic axis.

In orthorhombic and triclinic crystals: (1) the axis of the principal zone; (2) the axis with the shortest period, often the axis of the principal zone.

In monoclinic crystals: (1) the axis c, usually the axis of the principal zone excluding the symmetry axis; (2) the symmetry axis b.

principal meridian A meridian line accurately located and used as a basis from which to construct interior lines of monuments, called guide meridians, for the use of surveyors.

principal stresses Intensities of stress (maximum, minimum, and intermediate) along each of three mutually perpendicular axes in terms of which any state of stress can be described.

prism *1. Crystallog:* An open form of three or more similar faces parallel to a single axis; the shape of its cross section is generally used as a modifier, as trigonal prism, rhombic prism, dihexagonal prism; *2. Obs.* Any prism that is parallel to the vertical axis c. *3.* Long relatively narrow, wedge-shaped sedimentary deposit, particularly one of great thickness as in a geosyncline.

prism, geosynclinal *See* GEOSYN-CLINAL PRISM

prism level A kind of dumpy level with a mirror over the level tube, and a pair of prisms so placed that the position of the level bubble can be determined at any time by the levelman without the necessity of moving his head from the eyepiece.

probable ore A class of ore whose occurrence is, to all essential purposes, reasonably assured but not absolutely certain.

probe Any instrument for meas-

as orthoclase. A monoclinic disordered orthoclase.

sanidinite facies Metamorphic rocks that crystallized under maximum temperature and minimum pressure conditions.

saponite Clay mineral of the montmorillonite group, $(Ca,Na)_{0.33}$-$(Mg,Fe)_3(SiAl)_4O_{10}(OH)_2$.

sapphire Blue gem-quality corundum, or, with the proper adjective, any other color except red, e.g., yellow sapphire.

saprolite A soft, earthy, clay-rich, thoroughly decomposed rock formed in place by chemical weathering of igneous or metamorphic rocks. Forms in humid or tropical or subtropical climates. The color is commonly red or brown.

sapropel 1. An aquatic ooze or sludge that is rich in organic (carbonaceous or bituminous) matter. 2. A fluid organic slime originating in swamps as a product of putrefaction. In its chemical composition it contains more hydrocarbon than peat. When dry, it is a lusterless, dull, dark, and extremely tough mass which is hard to break up.

sapropelite series Series of organic and coaly materials in order of increasing rank: sapropel, saprocol, saprodil, saprodite, sapanthracone, sapanthracite.

Sarcodina Class of Protozoa of changeable body form which extrude mobile parts of pseudopodia; includes Foraminifera and Radiolaria.

sard A translucent brown or reddish-brown variety of chalcedony. Used as a gemstone. Similar to carnelian.

sardonyx A gem variety of chalcedony that is similar to onyx in having a banded structure but is colored red and white or black.

Sargasso Sea A region of clear, warm water in the North Atlantic Ocean that contains floating seaweed called sargassum.

satelite The gem-trade name for a fibrous serpentine having a chatoyant effect.

sathrolith; saprolith Regolith, q.v.

satin spar Fibrous gypsum.

saturated 1. Hydrol: A rock or soil is saturated with respect to water if all its interstices are filled with water. 2. Petrol: Applied to minerals capable of crystallizing from rock magmas in the presence of an excess of silica. Such minerals are said to be saturated with regard to silica and include the feldspars, pyroxenes, amphiboles, micas, tourmaline, fayalite, spessartite, almandine, and accessory minerals such as sphene, zircon, topaz, apatite, magnetite, and ilmenite. The term is also applied to igneous rocks composed wholly of saturated minerals.

saturated zone The zone of saturation; a subsurface zone below which all rock pore space is filled with water.

saturation 1. The extent or degree to which the voids in rock contain oil, gas, or water. Usually expressed in per cent related to total void or pore space. 2. Petrol: A principle developed by S. J. Shand for the classification of igneous rocks, based on the presence or absence of saturated or unsaturated minerals.

saturation line The line, on a variation diagram of an igneous rock series, representing saturation with respect to silica; rocks to the right of it are oversaturated, those to the left are undersaturated.

sauconite A clay mineral. A zinc-bearing member of the montmorillonite group. See MONTMORILLONITE

sausage structure See BOUDINAGE

saussurite A tough, compact,

white, greenish, or grayish mineral aggregate, produced in part by the alteration of feldspar, and consisting chiefly of zoisite or epidote.

savanna; savannah A tract of level land having a wet soil except during periods of dry weather, and supporting grass and other low vegetation, with but a scattered growth of pine or other trees and bushes. Sometimes applied to tracts of open prairie land.

scabland; scabrock Used in the Pacific Northwest to describe areas where denudation has removed or prevented the accumulation of a mantle of soil and the underlying rock is exposed or covered largely with its own coarse, angular debris.

scalar *Struct. petrol:* Applied to the physical features of a fabric which are nondirectional; e.g., grain shape, porosity, crystal habit, etc., are scalar quantities.

scale Crude paraffin obtained in petroleum refining by filtering from the heavier oils. Loose, thin fragments of rock, threatening to break or fall from either roof or wall.

scalped anticline Anticline whose upper part was eroded before the deposition of overlying unconformable strata.

Scaphopoda Class of mollusks whose noncoiled, elongate body is covered by a gently curved, tapering, cylindrical shell open at both ends; tusk shells.

scapolite A group of minerals of general formula $(Na,Ca)_4Al_3(Al,Si)_3Si_6O_{24}(Cl,CO_3SO_4)$. Tetragonal.

scar [*Sax.*] *1.* Any bluff precipice of rock. *2.* An isolated or protruding rock; a steep rocky eminence; a bare place on the side of a mountain or other steep slope. *3. See* SHORE PLATFORM

scarp An escarpment, cliff, or steep slope of some extent along the margin of a plateau, mesa, terrace, or bench. *See* FAULT SCARP; FAULT-LINE SCARP

scatter diagram *1.* Coordinate diagram showing by points the relations of specimens observations with respect to two or three variables. *2. Struct. petrol:* An orientation diagram which has not been contoured; lineations, axes, or poles of planes are represented by points.

scheelite A mineral, $CaWO_4$, commonly containing molybdenum. Fluoresces blue to creamy white, with color depending on molybdenum content.

schiller A bronzelike luster or iridescence due to internal reflection in minerals that have undergone schillerization.

schillerization The development of poikilitic texture by the formation of inclusions and cavities along particular crystal planes, largely by solution somewhat as are etch figures.

schist A medium or coarse-grained metamorphic rock with subparallel orientation of the micaceous minerals which dominate its composition. *See* AUGEN SCHIST; ORTHOSCHIST; PARASCHIST; PHYLLITE

schistose *See* SCHIST

schistosity That variety of foliation that occurs in the coarser-grained metamorphic rocks. Generally the result of the parallel arrangement of platy and ellipsoidal mineral grains.

schlieren Tabular bodies generally a few inches to tens of feet long that occur in plutonic rocks; they have the same general mineralogy as the plutonic rocks, but because of some differences in the ratios of the minerals they are darker or lighter; the boundaries with the plutonic rock tend to be transitional. Some schlieren are modified inclusions, others may be segregations of minerals.

Schmidt net Lambert projection. An equal-area azimuthal projection of the lower hemisphere of a sphere onto a plane.

schorl Black tourmaline.

schuppen structure Imbricate structure, *q.v.*

scintillation A small flash of light produced by an ionizing radiation in a phosphor or scintillator.

scintillation counter An instrument that measures ionizing radiation by counting individual scintillations of a substance. It consists of a phosphor and a photomultiplier tube that registers the phosphor's flashes.

scintillometer A scintillation counter.

scissor fault A fault on which there is increasing offset or separation along the strike from an initial point of no offset, with reverse offset in the opposite direction. The separation is commonly attributed to a scissorlike or pivotal movement on the fault, whereas it is actually the result of uniform strike-slip movement along a fault across a synclinal or anticlinal fold.

scolecite A mineral, $CaAl_2Si_3O_{10}.3H_2O$, a zeolite. Monoclinic.

scolecodont Tiny toothed or jawlike fossils composed of about equal amounts of silica and organic material; probably remains of annelid worms.

scoria Volcanic slag. Pyroclastic ejecta, usually of basic composition, characterized by marked vesicularity, dark color, heaviness, and a texture that is partly glassy and partly crystalline. Fragments of scoria between 4 mm. and 32 mm. are essentially equivalent to volcanic cinders. *Adj:* Scoriaceous.

scoriaceous Cellular, *q.v.*

scorodite A mineral, $FeAsO_4.2H_2O$. Orthorhombic.

scour Erosion, especially by moving water. *See* EROSION

scour and fill The process of cutting and refilling channels in sediments.

scour depression Where the channel of a stream is curved, the swiftest thread of the current is near the outside of the bend. The maximum erosive force of the current is exerted over a crescentic area in the bend.

scout 1. *Petroleum geol:* A person who gathers information from drilling wells of others, or operations, for the benefit of his own company. 2. Frequently used for an engineer who makes preliminary examinations of promising mining claims and prospects.

scree A heap of rock waste at the base of a cliff or a sheet of coarse debris mantling a mountain slope. By most writers "scree" is considered to be a synonym of "talus," but it is a more inclusive term. Whereas talus is an accumulation of material at the base of a cliff, scree also includes loose material lying on slopes without cliffs.

screen analysis The determination of weights of crushed material which passes through or is held on a series of screens of varying mesh.

screened sand Sand freed of finer material by the winnowing action of waves and currents.

screw axis Axis around which spiral movement of the components of a space lattice may occur involving both rotation and translation along the axis.

scroll A long closely fitting ridge within the large meander loop that has been built, during bankfull stages, at the inner edge of the low-water channel.

scroll meander Long, curving, parallel ridges (scrolls) that during stages of high water have been aggraded against the inner bank of the meandering channel, while the opposite bank experienced erosion.

Scyphozoa Class of coelenterates represented by medusae without hard parts; true jellyfish.

S-dolostone Stratigraphically controlled dolostone occurring in extensive beds generally intertongued with limestone.

se Foliation of groundmass around metamorphic crystals which indicates external structure.

sea 1. An ocean or a large body of (usually) salt water less than an ocean. 2. Waves caused by wind at the place and time of observation. 3. State of the ocean or lake surface in regard to waves.

sea arch At places where two sides of a headland are attacked by waves, a weaker or narrower section may be cut through completely by sea-cave enlargement. The opening so made is called a sea arch. *Syn:* NATURAL ARCH; NATURAL BRIDGE; MARINE ARCH

sea breeze The breeze that blows from the sea to the land on many coasts from about 10 or 11 A.M. to sunset on sunny days in summer. *See* LAND BREEZE

sea cave A cleft in a sea cliff excavated in easily weathered rocks by waves and currents.

sea cliff A cliff formed by wave action.

seacoast The coast adjacent to the sea or ocean.

sea-floor spreading The concept of continental drift by lateral movement of the lithosphere plates (crustal blocks) associated with magma upwellings in the central rifts of the mid-ocean ridges. The plates are believed to move

because of circulation of material within the earth which causes flow at depth (in the asthenosphere) dragging the plates along.

sea level *See* MEAN SEA LEVEL

seam 1. A stratum or bed of coal or other mineral. 2. A plane in a coal bed at which the different layers of coal are easily separated.

seamount A submarine mountain rising more than 500 fathoms (915 meters) above the ocean floor. Generally a volcanic cone.

sea or seamount range An elongated series of seamounts, sea peaks, or table mounts, the bases of which may be confluent, rising from a prominent elevation of the sea floor (ridge).

sea slide Submarine slide similar to a landslide.

sea stack Small steep-sided rocky projection above sea level near a coast.

sea valley A submarine depression of broad valley form without the steep side slopes which characterize a submarine canyon.

sea wall 1. A long embankment of smooth boulders, without gravel, built by powerful storm waves at the high-water mark. 2. A structure built by man along a portion of a coast primarily to prevent erosion and other damage by wave action. It retains earth against its shoreward face.

secondary 1. A general term applied to rocks and minerals formed as a consequence of the alteration of pre-existing minerals. Secondary minerals may thus be formed *in situ* as pseudomorphs or paramorphs, or they may be deposited from solution in the interstices of a rock through which the solution is percolating. 2. Formed of material derived from the erosion or disintegration of other rocks; derivative: said of clastic sedimentary rocks. *3.* The output coil of a transformer.

secondary arc Mountain arc raised behind the junction point of two primary arcs and convex in the opposite direction.

secondary consolidation Compaction of sediment occurring at essentially constant pressure resulting from internal processes such as recrystallization.

secondary enlargement The deposition around a nucleus, in optical continuity with it, of material of the same composition as the nucleus. Under proper conditions good crystal faces may be developed in this way.

secondary enrichment An enrichment, generally in sulfide deposits, by solutions of supergene origin. Oxidation of sulfide minerals produces acidic solutions that leach metals, carrying them downward to further enrich other sulfide minerals. A common process of upgrading copper porphyry deposits.

secondary recovery The recovery obtained by any method whereby oil or gas is produced by augmenting the natural reservoir energy, as by fluid injection. It usually implies substantial depletion of the reservoir before the injection of fluids, followed by a secondary development period.

secondary reflections Multiple reflections, *q.v.*

secondary structure Structure in a sedimentary rock which developed penecontemporaneously with sedimentation or shortly thereafter.

secondary waves Distortional, equivolumnar, shear, transverse, or shake waves. S-waves, *q.v.*

second boiling point The development of a gas phase from a liquid upon cooling. During the cooling crystallization of large quantities of compounds low in or lacking volatile materials (such as feldspar) results in a sufficient increase in the concentration of volatile materials (such as water) in the residual liquid that finally the vapor pressure of this liquid becomes greater than the confining pressure and a gas phase develops (i.e., the liquid boils).

second law of thermodynamics For all reversible processes the change in entropy is equal to the heat which the system exchanges with the outside world divided by the absolute temperature. In irreversible processes the change in entropy is greater than the quotient of heat and temperature. The second law introduces entropy as a function of the state of the system.

secretion *1.* The process by which animals and plants transform mineral material from solution into skeletal forms. *2.* Material which has been deposited from solution by infiltration into the cavity of a rock. *Cf.* CONCRETION. *3.* The term is the antonym of concretion in which growth is outward from a nucleus.

sectile Capable of being cut with a knife without breaking off in pieces.

section *1. Geol:* Either a natural or an artificial rock cut, or the representation of such on paper. *2.* A vertical exposure of strata. *3.* A drawing or diagram of the strata sunk through in a shaft or inclined plane, or proved by boring. *4.* In Scotland, a division of the mine workings. *5.* One of the portions, of 1 mile square, into which the public lands of the United States are divided; contains 640 acres (259 hectares). One thirty-sixth of a township. *6.* A very thin slice of anything, especially for microscopic examination. *7.* The local series of beds constituting a group or formation, as, the Cambrian section of Wales. *8.* An important division of a genus.

secular movements Movements of the earth's crust which take place slowly and imperceptibly.

secular variation A relatively large, slow change in part of the earth's magnetic field caused by the internal state of the planet and having a form roughly to be expected from a simple but not quite uniformly polarized sphere.

sedentary *1. Paleontol:* Attached, as, an oyster, barnacle, or similar shelled invertebrate. *2.* In sedimentation, formed in place without transportation by the underlying rock or by the accumulation of organic material; said of some soils, etc.

sedifluction The subaquatic or subaerial movement of material in unconsolidated sediments which takes place in the primary stages of diagenesis.

sediment *1.* Solid material settled from suspension in a liquid. *2.* Solid material, both mineral and organic, that is in suspension, is being transported, or has been moved from its site of origin by air, water, or ice, and has come to rest on the earth's surface either above or below sea level.

sedimentary Descriptive term for rock formed of sediment, especially: (1) Clastic rocks, as, conglomerate, sandstone, and shales, formed of fragments of other rock transported from their sources and deposited in water. (2) Rocks formed by precipitation from solution, as, rock salt and gypsum, or from secretions of organisms, as, most limestone.

sedimentary basin Geologically depressed area with thick sediments in the interior and thinner sediments at the edges.

sedimentary cycle The major

sedimentary rhythm which is the complement of the geographic cycle of W. M. Davis. This cycle determines the ordered sequence of orthoquartzite, graywacke, and arkose.

sedimentary facies Any areally restricted part of a designated stratigraphic unit which exhibits characters significantly different from those of other parts of the unit.

sedimentary mantle Sedimentary rocks overlying the crystalline basement.

sedimentary rocks Rocks formed by the accumulation of sediment in water (aqueous deposits) or from air (eolian deposits). The sediment may consist of rock fragments or particles of various sizes (conglomerate, sandstone, shale); of the remains or products of animals or plants (certain limestones and coal); of the product of chemical action or of evaporation (salt, gypsum, etc.); or of mixtures of these materials. Some sedimentary deposits (tuffs) are composed of fragments blown from volcanoes and deposited on land or in water. A characteristic feature of sedimentary deposits is a layered structure known as bedding or stratification. Each layer is a bed or stratum. Sedimentary beds as deposited lie flat or nearly flat.

sedimentary structure Any structure in a sedimentary rock. Primary structures such as cross beds are formed at the time of deposition, while secondary structures such as nodules and concretions are epigenetic.

sedimentary tectonics Buckling and folding of strata in geosynclinal basins produced by subsidence of the geosyncline.

sedimentary trap An area between a high-energy and a low-energy environment in which sedimentary materials accumulate.

sedimentation *1.* That portion of

the metamorphic cycle from the separation of the particles from the parent rock, no matter what its origin or constitution, to and including their consolidation into another rock. Sedimentation, thus, includes a consideration of the sources from which the sediments are derived; the methods of transportation from the places of origin to those of deposition; the methods, agents, and environments of deposition; the chemical and other changes taking place in the sediments from the times of their production to their ultimate consolidation; the climatic and other environmental conditions prevailing at the place of origin, over the regions through which transportation takes place, and in the places of deposition; the structures developed in connection with deposition and consolidation; and the horizontal and vertical variations of sediments. 2. In waste water treatment, the settling out of solids by gravity.

sedimentation curve An experimental curve showing cumulatively the quantity of sediment deposited or removed from suspension in successive units of time from an originally uniform suspension.

sedimentation unit That thickness of sediment which was deposited under essentially constant physical conditions.

sediment concentration Ratio of the weight of the sediment in a water-sediment mixture to the total weight of the mixture. It is ordinarily expressed in per cent for high values of concentration and in parts per million (ppm) for the low values.

sediment load The solid material transported by a stream, including bed-material load and wash load.

sedimentology The study of sedimentary rocks and the processes by which they were formed.

sediment station A river section

where samples of suspended load are taken each day, or periodically.

seep A spot where water or petroleum oozes from the earth, often forming the source of a small trickling stream.

segregate 1. To separate the undivided joint ownership of a mining claim into smaller individual "segregated" claims. 2. Geol: To separate from the general mass, and collect or become concentrated at a particular place or in a certain region, as in the process of crystallization and solidification.

segregation, magmatic See MAGMATIC DIFFERENTIATION

segregation banding A compositional banding in gneisses that is not primary in origin, but rather is the result of segregation of material from an originally homogeneous rock.

seiche A periodic oscillation of a body of water whose period is determined by the resonant characteristics of the containing basin as controlled by its physical dimensions. These periods generally range from a few minutes to an hour or more.

seif dune A very large, sharp-crested, tapering longitudinal dune or chain of sand dunes, commonly found in the Sahara Desert; its crest in profile consists of a succession of peaks and cols, and it bears on one side a succession of curved slip faces produced by strong but infrequent cross winds that tend to increase its height and width. A seif dune may be as high as 100–200 m., it may range in length from 400 m. to more than 100 km. (300 km. in Egypt). Etymol: Arabic *saif*, "sword"; the term originated in North Africa but is applied elsewhere to similar dunes of appreciably smaller size. Pron: *safe*. Syn: SEIF; SIF; SAIF; SWORD DUNE

seism Earthquake.

seismic Pertaining to an earthquake or earth vibration, including those that are artificially induced.

seismic activity Seismicity.

seismic area *1.* An earthquake zone. *2.* The region affected by a particular earthquake.

seismic belt An elongate earthquake zone, esp. the belts of the circum-Pacific, the Mediterranean and Trans-Atlantic, the Mid-Atlantic, and the mid-Indian belt.

seismic detector Any instrument, e.g., seismometer, seismograph, geophone, that receives seismic impulses. *Colloquial syn:* POT; SEIS

seismic discontinuity Physical discontinuity within the earth separating materials in which seismic waves travel at significantly different velocities.

seismic-electric effect A phenomenon in which a periodic change in current is caused to flow between two electrodes inserted in the ground when a seismic wave passes through the region between the electrodes.

seismic event An earthquake or a somewhat similar transient earth motion caused by an explosion. *Syn:* EVENT; QUAKE

seismic exploration The use of seismic techniques, usually involving explosions, to map subsurface geologic structures with the aim of locating economic deposits. *Syn:* PROSPECTING SEISMOLOGY; APPLIED SEISMOLOGY

seismic intensity The average rate of flow of seismic wave energy through a unit section perpendicular to the direction of propagation. *See also* SOUND INTENSITY

seismicity The likelihood of an area being subject to earthquakes. The phenomenon of earth movements; seismic activity.

seismic map A contour map constructed from seismic data. The z

coordinate could be either time or depth: when it is time, the map is called a raw map; when depth, a migrated map. A depth map may be tied by well data.

seismic method A method of geophysical prospecting using the generation, reflection, refraction, detection, and analysis of elastic waves in the earth.

seismic prospecting A type of geophysical prospecting that is based on analysis of elastic waves artificially generated in the earth. *See also* SEISMIC SURVEY

seismic record In geophysical prospecting, a photographic or magnetic record of reflected or refracted seismic waves; in earthquake seismology, a record of all seismic activity, including background noise and body and surface waves from both natural and artificial events.

seismic risk The probability that any particular area will undergo a certain number of earthquakes per unit time; the danger of a devastating earthquake to any area.

seismic shooting A method of geophysical prospecting in which elastic waves are produced in the earth by the firing of explosives.

seismic survey The gathering of seismic data from an area; the initial phase of seismic prospecting.

seismic swarm The occurrence of many, usually small or very small (less than magnitude 2) earthquakes in a very short time.

seismic velocity The rate of propagation of an elastic wave, usually measured in km./sec. The wave velocity depends upon the type of wave, as well as the elastic properties and density of the earth material through which it travels.

seismic wave *1.* A general term for all elastic sea waves produced by earthquakes or generated artificially by explosions. It includes both body waves and surface waves. *Obs. syn:* EARTH

WAVE. 2. A seismic sea wave, or tsunami. *Syn:* EARTHQUAKE WAVE

seismogram The record made by a seismograph.

seismograph Instrument which records seismic waves. *Syn:* DETECTOR

seismologist One who applies the principles of seismology to his work, e.g., oil exploration, earthquake detection, and analysis.

seismology 1. The science of earthquakes: all that relates to their forces, duration, lines of direction, periodicity, and other characteristics. 2. A geophysical science which is concerned with the study of earthquakes and measurement of the elastic properties of the earth.

seismometer Detecting device which receives seismic impulses. *Syn:* GEOPHONE; DETECTOR; PICKUP; JUG

selective fusion The fusion of only a portion of a mixture or rock; the liquid portion will contain, in the general case, a greater amount of the more readily fused materials than the solid portion contains.

selective or **preferential replacement** Replacement of one mineral in preference to, or more rapidly than, another.

selenate A salt of selenic acid; a compound containing the radical $(SeO_4)^{-2}$.

selenide A compound of selenium with one other more positive element or radical.

selenite A clear, transparent variety of gypsum, $CaSO_4 \cdot 2H_2O$.

selenology A branch of astronomy that deals with the moon; the science of the moon, including lunar geology.

selenomorphology The geomorphology of the moon.

self-potential method An electrical exploration method in which one determines the spontaneous electrical potentials (spontaneous polarization) that are caused by electrochemical reactions associated with metallic mineral deposits. *Syn:* SPONTANEOUS-POTENTIAL METHOD

selvage 1. A zone of altered material along a fault, joint, vein, or fissure showing effects of circulating solutions or vapors. It is usually a layer of soft clayey material separating ore from country rock in a vein. 2. The chilled glassy border of a dike or lava flow. 3. A marginal zone, as in a dike or vein, having some distinctive feature of fabric or composition.

semianthracite Nonagglomerating anthracitic coal having 86% or more, and less than 92% of fixed carbon (dry, mineral-matter-free) and 14% or less and more than 8% of volatile matter (dry, mineral-matter-free).

semiarid Pertaining to a subdivision of climate in which the associated ecological conditions are distinguished by short grass (whereas a subhumid climate is characterized by tall grass), and best exemplified by the climate of the steppes.

semibituminous Half or somewhat bituminous; applies to a variety of coal intermediate between bituminous coal and anthracite, averaging 15 to 20% of volatile matter.

semidiurnal tides A tide with two high and two low waters in a tidal day, with comparatively little diurnal inequality.

senescence The process of growing old. Sometimes used to refer to lakes nearing extinction.

senility Condition of organisms that in old age revert to development resembling younger stages.

separation Indicates the distance between any two parts of an index plane (bed, vein, etc.) disrupted by a fault. Horizontal separation is separation measured in any indicated horizontal direction; vertical separation is measured along a vertical line; stratigraphic separation is measured perpendicular to the bedding planes.

sepiolite A chain-lattice clay mineral, $Mg_4Si_6O_{15}(OH)_2 \cdot 6H_2O$, occurring in fibrous crystals matted into an extremely lightweight absorbent white mass. Meerschaum.

septarian A structure developed in certain concretions known as septarian nodules, consisting of an irregular polygonal system of internal cracks, which are almost always occupied by calcite or other minerals.

septarium A roughly spheroidal concretion, generally of limestone or clay-ironstone, cut into polyhedral blocks by radiating and intersecting cracks which have been filled (and the blocks cemented together) by veins of some material, generally calcite. *Pl.* septaria. *Syn:* SEPTARIAN BOULDER; SEPTARIAN NODULE; TURTLE STONE

septum *1.* The internal partition or division between the chambers of a cephalopod; also in the pterobranches. *2.* Radially arranged vertical plates of stony substance which project inward and upward from the wall and base of individual coral skeletons. *3.* In echinoid spines, the platelike structures which radiate from the axial zone toward the anterior of the spine and are seen in cross sections of the spine.

seracs When two or more sets of crevasses intersect, the surface of the glacier is torn into a broken mass of jagged ice pinnacles known as seracs.

sere A sequence of ecologic communities from pioneer stage to climax community.

serial samples Samples collected according to some predetermined plan, such as along the intersections of grid lines, or at stated distances or times. The method is used to insure random sampling.

sericite A fine-grained variety of mica occurring in small scales, especially in schists. Usually muscovite, but may consist of paragonite or hydrous micas.

series A time-stratigraphic unit ranked next below a system. Loosely used in petrology for related igneous rocks.

series circuit An electrical circuit so connected that there is a single continuous path and all the current flows through each component.

serpentine A mineral group with the general formula, $(Mg,Fe)_3Si_2O_5(OH)_4$. Characterized by long fibrous crystals; the group includes the minerals antigorite and chrysotile.

serpentinite A rock consisting almost wholly of serpentine minerals derived from the alteration of previously existing olivine and pyroxene.

serrate, *adj.* [<*Lat.* serra, saw] *1.* Pertaining to the rocky summit of a mountain having a sawtooth profile; a small sierra-shaped ridge. Local in Southwest. *2. Paleobot:* Said of a margin when saw-toothed with the teeth pointing forward.

sessile Applied to organisms that are closely attached to other objects or to the substrate.

sessile benthos Sea-bottom-dwelling forms which are anchored in position by attachment to the bottom.

set A group of essentially parallel planar features, especially joints, dikes, faults, veins, etc.

settling The sag in outcrops of laminated sedimentary rocks caused by rock creep. *Syn:* OUTCROP CURVATURE; TERMINAL CREEP

sexual dimorphism Differences in

form exhibited by males and females of the same species.

shaft *1.* An excavation of limited area compared with its depth, made for finding or mining ore or coal, raising water, ore, rock, or coal, hoisting and lowering men and material, or ventilating underground workings. Often specifically applied to approximately vertical shafts as distinguished from an incline or inclined shaft. *2. Speleol:* A vertical passage.

shale *1.* A laminated sediment in which the constituent particles are predominantly of the clay grade. *2.* Shale includes the indurated, laminated, or fissile claystones and siltstones. The cleavage is that of bedding and such other secondary cleavage or fissility that is approximately parallel to bedding. The secondary cleavage has been produced by the pressure of overlying sediments and plastic flow.

shale break Thin layer or parting of shale between harder strata, primarily a drillers' term.

shale oil A crude oil obtained from bituminous shales, especially in Scotland, by submitting them to destructive distillation in special retorts.

shallow-focus earthquake Earthquake whose focus occurs at a depth of less than 65 km.

shaly bedded Term applied to a sedimentary deposit whose stratification is in the form of laminae 2 to 10 mm. in thickness.

Shand's classification A classification of igneous rocks based on crystallinity, degree of saturation with silica, degree of saturation with alumina, and color index.

shard A curved, spiculelike fragment of volcanic glass.

sharpstone A sedimentary rock made up of angular particles more than 2 mm. in greatest dimension.

shatter cone A distinctively striated conical fragment of rock along which fracturing has occurred, ranging in length from less than a centimeter to several meters, generally found in nested or composite groups in the rocks of cryptoexplosion structures, and generally believed to have been formed by shock waves generated by meteorite impact. Shatter cones superficially resemble cone-in-cone structure in sedimentary rocks; they are most common in fine-grained homogeneous rocks such as carbonate rocks (limestones, dolomites), but are also known from shales, sandstones, quartzites, and granites. The striated surfaces radiate outward from the apex in horsetail fashion; the apical angle varies but is close to 90 degrees. *Syn:* SHEAR CONE; PRESSURE CONE

shear *1. V:* To subject a body to shear, similar to the displacement of the cards in a pack relative to one another. *2. N:* The effect produced by action of a shearing stress.

shear cleavage Slip cleavage; strain-slip cleavage.

shear or slip fold A fold formed as a result of the minute displacement of beds along closely spaced fractures or cleavage planes.

shear fracture *1.* A fracture that results from stresses which tend to shear one part of a specimen past the adjacent part. Contrasts with tension fractures. *2.* Fracture at a more or less acute angle to applied force generally with some pulverized material along its surface; splitting fracture.

shearing off Extensive lateral shearing movement within, between, or below layers of the earth's crust or at top of basement.

shear joints Joints that formed as shear fractures.

shear modulus Rigidity or shear modulus is a measure of the stress-strain ratio for a simple shear.

shear strength The internal resistance offered to shear stress. It is measured by the maximum shear stress, based on original area of cross section, that can be sustained without failure.

shear stress or **tangential stress** A stress causing or tending to cause two adjacent parts of a solid to slide past one another parallel to the plane of contact.

shear wave S-wave. Distortional, equivolumnar, secondary, or transverse wave.

shear zone *Geol:* A zone in which shearing has occurred on a large scale so that the rock is crushed and brecciated.

sheepback Roche moutonnée, *q.v.*

sheet *1.* A tabular mass of igneous rock, either a flow, sill, or dike. *2.* A widespread tabular body of sedimentary rock. *See* SHEET SAND; BLANKET SAND. *3. Speleol:* Thin coating of calcium carbonate formed on walls, shelves, benches, and ledges by trickling water. *4.* In Australia, a solid body of pure ore filling a crevice. *5.* In Upper Mississippi lead region, galena in thin and continuous masses. The ore itself is called sheet mineral.

sheeted vein A group of closely spaced, distinct, parallel fractures filled with mineral matter and separated by layers of barren rock.

sheeted zone A zone of closely spaced fractures whether mineralized or not.

sheet or **sheetflood erosion** Erosion accomplished by sheets of running water, as distinct from streams.

sheetflood Movement of near uniform sheets of flood water down the surface of a slope.

sheet flow Laminar flow.

sheeting. In a restricted sense, the gently dipping joints that are essentially parallel to the ground surface; they are more closely spaced near the surface and become progressively farther apart with depth. Especially well developed in granitic rocks. *2.* In a general sense, a set of closely spaced joints.

sheet minerals Those minerals belonging to the phyllosilicate group. Includes the micas, the chlorites, and most of the clay minerals.

sheet sand A sandstone of great areal extent, presumably deposited by a transgressing sea advancing over a wide front and for a considerable distance. *See* BLANKET SAND

sheet structure The type of fracture or jointing formed by pressure-release jointing or foliation.

shelf *1.* In the ocean, the zone extending from the line of permanent immersion to the depth (usually about 65 fathoms) where there is a marked or rather steep descent toward the great depths. *See* CONTINENTAL SHELF. *2.* In Cornwall, the solid rock or bedrock, especially under alluvial tin deposits. *3.* A rock, ledge of rocks, reef, or sandbank in the sea. *4.* A projecting layer or ledge of rock on land. *5.* In gastropod shells, the subhorizontal part of whorl surface next to a suture, bordered on side toward periphery of whorl by a sharp angulation or by a carina.

shelf ice Ice shelf, *q.v.* The extension of glacial ice from land into coastal waters. Shelf ice, which may be several hundred feet thick, is in contact with the bottom nearshore, but not at its seaward terminus.

shelf sea The water that rests upon a sea shelf (continental shelf).

shell *1.* The generally hard rigid covering of an animal, commonly calcareous, in other cases chiefly or partly chitinous, horny, or even siliceous. 2. *Petroleum geol:* A torpedo used in oil wells. 3. A cylinder or tube of light metal which is filled with nitroglycerin or other explosive, lowered into a drill hole, and detonated when a well is shot. 4. A thin and usually hard layer of rock found in drilling a well.

shellfish Any aquatic invertebrate with a hard external covering; more commonly mollusks and crustaceans.

shelves The most stable areas of the craton that are periodically flooded by marine water.

shield 1. A continental block of the earth's crust that has been relatively stable over a long period of time and has undergone only gentle warping in contrast to the strong folding of bordering geosynclinal belts. Mostly composed of Precambrian rocks. 2. A disk-shaped formation standing edgewise at a high angle in a cave. 3. In animals, a protective structure likened to a shield, as, a large scale, carapace, or lorica. 4. A wall that protects workers from harmful radiation released by radioactive materials.

shield basalts Multiple vent basalts. Basaltic accumulations of smaller size than the plateau or flood basalts, arising from the confluence of lava flows from a large number of small and closely spaced volcanoes.

shield volcano A broad, gently sloping volcanic cone of flat domical shape, usually several tens or hundreds of square miles in extent, built chiefly of overlapping and interfingering basaltic lava flows. Typical examples are the volcanoes Mauna Loa and Kilauea on the island of Hawaii, and the great basaltic volcanoes of Iceland.

shift The relative displacement of the units affected by a fault but outside the fault zone itself.

shingle Loosely and commonly, any beach gravel which is coarser than ordinary gravel, especially if consisting of flat or flattish pebbles and cobbles.

shingle beach A beach whose surface is covered with shingle rock. *See* SHINGLE

shingle rampart A ridge of shingle, 1 or 2 meters high, built up by waves on the seaward edge of a reef flat.

shingle structure Arrangement of veins en echelon in the manner of shingles on a roof.

shingling Imbricate structure. The overlapping upstream of platy or tabular pebbles in stream deposits and seaward overlapping in beach deposits.

shipping ore Any ore of greater value when broken than the cost of freight and treatment.

shoal 1. A part of the area covered by water, of the sea or lake or river, when the depth is little; a bank always covered, though not deeply. 2. *V:* To become shallow gradually. 3. A detached elevation of the sea bottom comprised of any material except rock or coral, and which may endanger surface navigation.

shoal reefs 1. Bank reefs. 2. Reef growths developed in irregular patches amidst submerged shoals of calcareous debris.

shock Earthquake.

shock breccia Fragmental rock shattered by the action of shock waves.

shock metamorphism The permanent alteration of rocks and minerals as a result of the passage of high-pressure shock waves produced by the hypervelocity impact of large meteorites.

shock wave A compressional wave of supersonic velocity formed whenever the speed of a body exceeds the ability of a medium to transmit sound. Shock waves have a pattern of flow that changes abruptly with corresponding changes in temperature, pressure, and density.

shock zone A volume of rock surrounding an explosion or impact crater in which the effects of shock-metamorphic deformation are present.

shoestring sands A narrow and relatively long body of sandstone. Examples have been described from the Pennsylvanian system of east-central Kansas. They originated as channel fillings, bars, beaches, etc.

shonkinite A melanocratic syenite often containing a small amount of nepheline. The principal minerals are augite and orthoclase.

shoot *Seis. explor:* The firing of the explosive by an electrical impulse; also the process of carrying out a seismic survey, to "shoot" an area or prospect.

shooting a well Exploding a charge of nitroglycerin in a drill hole, at or near an oil-bearing stratum, for the purpose of increasing the flow of oil.

shoot of ore A body of ore with relatively small horizontal dimensions and steep inclination in a lode; in contradistinction to a course of ore, which is flatter. *See* CHUTE

shoran A high-frequency radio wave location system using microwave pulses used for offshore and airborne prospecting operations. Two stations are located at fixed points, the third is on the mobile station whose location is desired. The fixed stations broadcast pulses, the mobile station rebroadcasts them, and the round-trip time is measured by means of cathode-ray screens to an accuracy of ±7.6 meters.

shore The common margin of dry land and a body of water.

shoreface The narrow zone seaward from the low-tide shoreline permanently covered by water, over which the beach sands and gravels actively oscillate with changing wave conditions.

shoreline The line of intersection of the sea with the land. The region immediately to the landward of the shoreline is called the coast.

shoreline cycle Coasts are sequential forms developed by marine erosion, and in part by accumulation, from varied initial forms. A succession of stages, or shoreline cycle, through which the coastal features normally pass, can be developed for each kind of initial coast. *See* CYCLE OF SHORE DEVELOPMENT

shoreline of depression *See* SHORELINE OF SUBMERGENCE

shoreline of elevation *See* SHORELINE OF EMERGENCE

shoreline of emergence or **elevation** Results when the water surface comes to rest against a partially emerged sea or lake floor.

shoreline of submergence or **depression** Produced when the water surface comes to rest against a partially submerged land area.

shore platform Plane of marine abrasion; wave-cut terrace.

shortening *Seismol:* The decrease in distance between two accurately measured points as a result of earth stresses prior to earthquakes.

short limb That side of an asymmetrical fold in which the direction is sooner reversed; generally the steeper side.

short range order Lack of orderliness in arrangement in a solid or liquid over distances a little beyond immediately neighboring atoms or molecules; typical of glass and liquids.

shot The explosion in seismic operations.

shot break *Seis. explor:* The electrically recorded instant of explosion.

shot copper Small rounded particles of native copper, somewhat resembling small shot in size and shape.

shot datum Seismic calculations are usually reduced to a convenient reference surface or plane. These calculations simulate a condition where the charge is shot on the reference surface and the arrival of seismic waves is also recorded on this same reference surface. At this reference surface, the time-depth charts have their origin.

shot depth The distance from the surface to the charge. In the case of small charges, the shot depth is measured to the center of the charge or to the bottom of the hole. In the case of large charges, the distances to the top and to the bottom of the column of explosives are frequently given, and may be reduced to effective shot depth to give the equivalent of a concentrated charge.

shot drill An earth-boring drill using steel shot as an abrasive.

shot elevation Elevation of the dynamite charge in the shot hole.

shothole *Seis. prospecting:* The borehole in which an explosive is placed for blasting.

shothole fatigue Phenomenon causing observed travel times to another point to increase with successive shots in the same hole.

shot instant *Seis. explor:* The instant of detonation of the dynamite charge.

shot point That point at which a charge of dynamite is exploded for the generation of seismic energy. In field practice, the shot point includes the hole and its immediately surrounding area.

show (of oil or gas) A noncommercial quantity of oil or gas, encountered in drilling.

shrinkage crack Mud crack; sun crack.

shrink-swell Refers to clays or soils that alternately expand and contract in a semiarid climate where drying out is possible.

shrub-coppice dune A mound of wind-blown sand built up by bush or clump vegetation.

si Orderly arrangement of inclusions in crystals that have grown during metamorphism and indicate internal structure.

sial A layer of rocks underlying all continents, which ranges from granitic at the top to gabbroic at the base. The thickness is variously placed at 30–35 km. The name derives from the principal ingredients, silica and alumina. Specific gravity is considered to be about 2.7.

sialma A mnemonic term derived from "si" for silica, "al" for alumina, and "ma" for magnesia and applied as a compositional term to a layer within the earth which occupies a position intermediate between sial and sima, *q.v.*

sibling species One of two or more species that are closely related, very similar morphologically but reproductively isolated.

siderite Chalybite. *1.* A mineral, $FeCO_3$, commonly containing also Mg and Mn *2.* An iron meteorite.

siderolite A type of meteorite consisting of approximately equal parts of metal and silicate phases. *Syn:* PALLASITES; STONY IRONS; SYSSIDERITES

sideromelane Basaltic glass. Characteristic of palagonite tuffs.

siderophile elements Elements with a relatively weak affinity for oxygen and sulfur, characterized by ready solubility in molten iron, hence concentrated in iron meteorites and probably in the earth's iron core.

siderosphere Central iron core of the earth.

side shot A reading or measurement from a survey station to locate a point which is not intended to be used as a base for the extension of the survey.

sidewall core A core or rock sample extracted from the wall of a drill hole, either by shooting a retractable hollow projectile, or by mechanically removing a sample.

sidewall sampling The process of securing samples of formations from the sides of the borehole anywhere in the hole that has not been cased.

sienna Mineral paint.

sierra [*Sp.*, <*Lat.* serra, saw] A chain of mountains whose successive peaks present the resemblance of a saw.

sieve analysis Determination of the percentage distribution of particle size by passing a measured sample of soil or sediment through standard sieves of various sizes.

sieve texture *See* POIKILOBLASTIC

sif Seif.

sight *1.* A bearing or angle taken with a compass or transit when making a survey. *2.* Any established point of a survey.

sigmoidal fold Fold with steeply inclined axis whose outcrop pattern shows S-shaped strike curves.

signal correction A correction to eliminate the time differences between reflection times, resulting from changes in the outgoing signal from shot to shot.

signal effect Variations in arrival times of reflections as a result of changes in the outgoing signal, the reflection being recorded with identical filter settings.

silcrete The harder, illuviated horizon in a siliceous latosol, usually formed at depth.

silica Silicon dioxide, SiO_2.

silica sand Sand very high in SiO_2, hence a source of silicon; also has industrial uses.

silicate A salt of any of the silicic acids, real or hypothetical. A compound whose crystal structure contains Si-O tetrahedra either isolated or joined together through one or more of the oxygen atoms.

silicates, classification Based on types of linkages of Si-O tetrahedra, in which Si may be partly replaced by Al. The types of silicates are: nesosilicates, sorosilicates, cyclosilicates, inosilicates, phyllosilicates, and tectosilicates.

siliceous; silicious Of or pertaining to silica; containing silica, or partaking of its nature. Containing abundant quartz.

siliceous latosols Gray and reddish soils leached of soluble minerals and of aluminum and iron, but retaining or accreting silica.

siliceous ooze A fine-grained pelagic deposit with more than 30% material of organic origin, a large percentage of which is siliceous skeletal material produced by planktonic plants and animals.

siliceous residue Insoluble residue, *q.v.*

siliceous shale Hard fine-grained rock of shaly structure generally believed to be shale altered by silicification.

siliceous sinter Siliceous sinter, geyserite, and fiorite are names given to the nearly white, often soft, and friable, hydrated varieties formed on the evaporation of the siliceous waters of hot springs and geysers, or through the eliminating action of algous vegetation. *See* SINTER

silicification The introduction of or replacement by, silica. Generally the silica formed is fine-grained quartz, chalcedony, or opal, and may both fill up pores and replace existing minerals. The term covers all varieties of such processes, whether late magmatic, hydrothermal or diagenetic.

silicified wood A material formed by replacement of wood by silica in such manner that the original form and structure of the wood is preserved. The silica is generally in the form of opal or chalcedony.

silky Having the luster of silk, like fibrous calcite, fibrous gypsum.

sill 1. An intrusive body of igneous rock of approximately uniform thickness and relatively thin compared with its lateral extent, which has been emplaced parallel to the bedding or schistosity of the intruded rocks. 2. A submarine ridge or rise separating partially closed basins from one another or from the adjacent ocean.

silled basin Submarine basin of deposition separated from the main water body by a relatively narrow submerged ridge; deeper water in the basin is likely to be more or less stagnant. *Syn*: BARRED BASIN

sillimanite Fibrolite. A mineral, Al_2SiO_5, trimorphous with kyanite and andalusite. Orthorhombic.

silt 1. A clastic sediment, most of the particles of which are between 1/16 and 1/256 mm. in diameter. 2. Soil consisting of 80% or more silt (.05–.002 mm.) and less than 12% clay.

siltstone A very fine-grained consolidated clastic rock composed predominantly of particles of silt grade.

Silurian The third of seven periods (before Devonian and after Ordovician) of the Paleozoic; also the system of rocks deposited during that period.

silver The native element, commonly containing Au and Hg. Isometric.

silver glance Argentite.

sima The basic outer shell of the earth; under the continents it underlies the sial, but under the Pacific Ocean it directly underlies the oceanic water. Originally, the sima was considered basaltic in composition with a specific gravity of about 3.0. In recent years it has been suggested that the sima is peridotitic in composition with a specific gravity of about 3.3.

similar folding That type of folding in which each successively lower bed shows the same geometrical form as the bed above.

simple harmonic motion Oscillatory motion of a particle or mass, with constant amplitude, and which is sinusoidal with time. Such motion results when the restoring force is proportional to the displacement (as in a simple pendulum of small amplitude).

simple shear A homogeneous strain that consists of a movement in one direction of all straight lines initially parallel to that direction.

Sinian A system of rocks in China and Siberia variously referred to the Cambrian or the Precambrian; typical Cambrian fossils have not been recovered from these rocks and the trend is to consider them Precambrian although they are conformable with overlying Cambrian rocks.

sinistral fault Left-lateral fault.

sinistral fold Asymmetric fold in which the long limb is apparently offset to the left as one looks along the long limb (i.e., offset in the same manner as in a left-hand fault).

sinkhole A funnel-shaped depression in the land surface generally in a limestone region communicating with a subterranean passage developed by solution.

sinking A method of controlling oil spills that employs an agent to entrap oil droplets and sink them to the bottom of the body of water. The oil and sinking agent are eventually biologically degraded.

sinter A chemical sediment deposited by a mineral spring, either hot or cold. Siliceous sinter, consisting of silica, is also called geyserite and fiorite; calcareous sinter, consisting of calcium carbonate, is also called tufa, travertine, and onyx marble.

sinus *1.* In gastropod shells, a reentrant in the outer edge of the aperture with nonparallel sides. *2.* In a brachiopod shell, the major rounded depression along the longitudinal midline, generally found on the pedicle valve. *Syn:* SULCUS (usually called sulcus, since 1932). *3.* The space or recess between two lobes or divisions of a leaf or other expanded organ.

siphon; syphon Small upright or inverted U-shaped channel with water in hydrostatic equilibrium. *Biol:* Various, not always homologous, organs for bringing water into the mantle cavity of mollusks, particularly pelecypods, gastropods, and cephalopods.

size *1.* In brickmaking, plasticity, as of tempered clay. *2.* To separate minerals according to various screen meshes.

s-joints *See* LONGITUDINAL JOINTS

skarn The term is generally reserved for rocks composed nearly entirely of lime-bearing silicates and derived from nearly pure limestones and dolomites into which large amounts of Si, Al, Fe, and Mg have been introduced.

skeleton crystals Hollow or imperfectly developed crystals.

skin effect *1.* The phenomenon in which alterations in permeability in the vicinity of a drill hole are caused by drilling and completion operations. *2.* The concentration of alternating current in a conductor toward its exterior boundary.

slab or slabstone Cleaved or finely parallel jointed rocks, which split into tabular plates from 1 to 4 inches thick.

slack water Essentially currentless water such as that occurring in flooded areas beyond a stream channel or in an estuary at high or low tide.

slag, volcanic *See* SCORIA

slaking Loosely, the crumbling and disintegration of earth materials when exposed to air or moisture. More specifically, the breaking up of dried clay when saturated with water, due either to compression of entrapped air by inwardly migrating capillary water or to the progressive swelling and sloughing off of the outer layers.

slate *1.* A fine-grained metamorphic rock possessing a well-developed fissility (slaty cleavage). *See* CLAY SLATE; PHYLLITE; SPOTTED SLATE. *2.* A coal miner's term for any shale accompanying coal; also sometimes applied to bony coal.

slate coal *1.* In England, a hard dull variety of coal. *2.* Coal that has pieces of slate of greater or less size attached to it, which can be separated by breaking the coal into smaller pieces and subjecting the coal to a washing process.

slaty cleavage That variety of foliation typical of slates but found in many other kinds of rocks. Generally the result of parallel arrangement of platy or ellipsoidal minerals. *See* FOLIATION

slice *1.* A large block caught along a thrust. *2.* Arbitrary informal division, either of uniform thickness or constituting some uniform fraction, of an otherwise indivisible stratigraphic unit that is distinguished for individual facies mapping or analysis.

slickenside Polished and striated (scratched) surface that results from friction along a fault plane.

slide *1.* The descent of a mass of earth or rock down a hill or mountain side. *2.* The track of bare rock left by a landslide. *3.* Material moved in a landslide.

slim hole Drill hole of the smallest practicable size, often bored as a stratigraphic test.

slip *1.* A fault. *2.* A smooth joint or crack where the strata have moved upon each other. *3.* The relative displacement of formerly adjacent points on opposite sides of the fault, measured in the fault surface. *See* DIP SLIP; STRIKE SLIP. *4.* The downhill movement of a mass of soil under wet or saturated conditions. The movement is only a short distance and the soil mass stays relatively intact. A form of landslide.

slipband Luder's line, *q.v.*

slip bedding The contortion of stratification planes into complex folds caused by gliding.

slip or shear cleavage That variety of foliation along which there has been visible displacement, usually shown by bedding that is cut by the cleavage.

slip fold Shear fold, *q.v.*

slip-off slope A streamward sloping erosion surface developed along the inner bends of rivers. The surface is the result of the interaction of lateral and downward erosion by the river.

slip plane Closely spaced surfaces along which differential movement takes place in rock. Analogous to surfaces between playing cards. *Syn:* GLIDE PLANE; GLIDING PLANE.

slip sheet A gravity-collapse structure. A bed that has slid down the flanks of an anticline, fractured at its base, and slid out over the adjacent strata.

slip surface Flow surface.

slip tectonite Rock whose particles have been oriented by movement along s-planes.

slope *1.* The inclined surface of a hill, mountain, plateau, plain, or any part of the surface of the earth; the angle at which such surfaces deviate from the horizontal. *2. Min:* An inclined passage driven from the dip of a coal vein. *Cf.* SLANT

slope stability The resistance of a natural or artificial slope to landslide failure.

slope stability maps Derivative maps that combine those areas that have similar stabilities of slope for different human activities.

slope wash Soil and rock material that is being or has been moved down a slope predominantly by the action of gravity assisted by running water that is not concentrated into channels. The term may also designate the process by which such material is moved.

slough A place of deep mud or mire. *Syn:* BOG; QUAGMIRE; SWALE

sluicing Concentrating heavy minerals, e.g., gold, cassiterite, diamonds, by washing unconsolidated material through boxes (sluices) equipped with riffles that trap the heavier minerals on the floor of the box.

slump *1. N:* Material that has slid down from high rock slopes. *V:* To slip down en masse. *2.* The downward slipping of a mass of rock or unconsolidated material of any size, moving as a unit or as several subsidiary units, usually with backward rotation on a more or less horizontal axis parallel to the cliff or slope from which it descends.

slump bedding Deformation in an unconsolidated or plastic sediment due to subaqueous slump or gliding. The disturbance may be restricted to layers only an inch or two thick, and are confined to a single bed or zone between undisturbed beds. *Syn:* CURLY BEDDING; GLIDE BEDDING; HASSOCK STRUCTURE

slump block A single coherent mass of material torn away during a block slump or block glide in which the slide mass remains virtually intact and moves outward and downward.

slump fault A gravity fault, or normal fault.

slump fold An interformational fold produced by the slumping of soft sediments on an incline, such as the continental slope.

slush pit A pit used in rotary drilling where water can be stored for circulation through the hole while drilling. Mud can be mixed in the slush pit if necessary in drilling the hole.

smaltite A mineral $(Co,Ni)As_{3-x}$. Isometric. An ore of cobalt.

smectite A clay mineral group synonymous with montmorillonite group.

smithsonite Dry bone ore. A mineral of the calcite group, $ZnCO_3$. Hexagonal rhombohedral. An ore of zinc.

smoky quartz Cairngorm. A smoky, brown-colored crystalline variety of quartz.

Snell's law Law of refraction, q.v.

snow avalanche The rapid downslope movement of large quantities of snow, usually in a mountain region since steepness of slope is an important factor.

snowball garnet Pinwheel garnet.

snow field Where snow endures from year to year over any considerable area.

snowline The altitude above which there is permanent snow.

soapstone A massive impure variety of talc.

soda Sodium carbonate, Na_2CO_3; especially the decahydrate, $Na_2CO_3.10H_2O$. Loosely used for sodium oxide, sodium hydroxide, sodium bicarbonate, and even for sodium in deplorable expressions such as soda feldspar.

soda feldspar Misnomer for sodium feldspar. Albite. *See* PLAGIOCLASE

soda lake Undrained depressions containing large amounts of salts of sodium and magnesium, chiefly sulfates.

sodalite A mineral, $Na_4Al_3(SiO_4)_3$-Cl. Isometric.

soda niter Chile saltpeter. A mineral, $NaNO_3$. Hexagonal rhombohedral.

soft coal Bituminous coal as opposed to anthracite.

soft ground That part of a mineral deposit that can be mined without drilling and shooting hard rock; commonly occurs in the upper weathered portion of a vein.

soft rock Rock that can be removed by air-operated hammers, but cannot be handled economically by pick. Loosely used to distinguish sedimentary from igneous and metamorphic rock.

soft water Water with practically no dissolved calcium or magnesium salts. *Cf.* HARD WATER

soil That earth material which has been so modified and acted upon by physical, chemical, and biological agents that it will support rooted plants. The term as used by engineers includes, in addition to the above, all regolith.

soil analysis *Geochem. prospecting:* Method which consists of taking soil samples and analyzing them for the various hydrocarbons and other gases and waxes, minerals, or other rare components which they may contain.

soil conditioner A biologically stable organic or inorganic material such as humus compost, vermiculite, or perlite, that makes soil more amenable to the passage of water and to the distribution of fertilizing material, providing a better medium for necessary soil bacteria growth.

soil creep Slow movement of rock fragments down even, gentle slopes.

soil horizon A layer of soil that is distinguished from adjacent layers by characteristic physical properties such as structure, color, or texture. The letters A, B, and C are used to designate soil horizons. The A horizon is the upper part. It consists of mineral layers of maximum orgqnic accumulation; or layers from which clay materials, iron, and aluminum have been lost; or both. The B horizon lies under the A. It consists of weathered material with an accumulation of clay, iron, or aluminum; or with more or less blocky or prismatic structure; or both. The C horizon under the B is the layer of unconsolidated, weathered parent material. Not all these horizons are present in all soils. Sometimes O or H is used for the unaltered organic debris at the surface.

soil map A map showing the distribution of various soil types of a particular area or region.

soil mechanics The science of the mechanical properties of a mass of loose or unbonded particles, particularly of their composition, shear resistance, and effects of water.

soil profile Succession of zones or horizons, q.v., beginning at the surface that have been altered by normal soil-forming processes of which leaching, oxidation, and accretion are particularly important.

soil stratigraphic unit Rock stratigraphic unit consisting of the upper weathered part of an older sedimentary deposit.

soil stripes See STRIPED GROUND

soil surveys Dynamic methods used in engineering to test the site of buildings, dams, bridges and similar structures for possibilities of compaction or of earthquake or other vibration damage.

soil zone Soil horizon.

sol A colloidal dispersion of a solid in a liquid.

solar aberrations Those movements of the sun in relation to the earth that are usually cyclical but not constant, such as the precession of the equinoxes.

solar constant The rate at which solar radiant energy is received outside the atmosphere on a surface normal to the incident radiation at the earth's mean distance from the sun. The value of the mean solar constant is 1.94 gram calories per minute per square centimeter.

solar energy Energy emitted by the sun.

sole *1.* The lowest thrust plane in an area of overthrusting. Commonly rocks above are imbricated. *2.* Lower surface of a sedimentary stratum.

sole fault A low-angle thrust fault forming the sole of the thrust nappe; also, the basal main fault of an imbrication.

sole injection Discordant pluton, generally mafic, formed by intrusion of magma along a relatively flat thrust plane.

sole mark A cast of primary sedimentary structures (such as cracks, tracks, and grooves) formed on the lower surface or underside of a siltstone or sandstone bed, and revealed as sole casts after the original underlying sedimentary layer has weathered away.

solfatara A semiextinct volcano, emitting only gaseous sulfurous exhalations, and aqueous vapors, so called from the Solfatara, near Naples.

solid Matter with a definite shape and volume and some fundamental strength. Crystalline solids are composed of grains of one or more minerals in which the atoms have an orderly arrangement.

solid flow Flowage within a solid body accomplished by rearrangement between or within its particles.

solid solution A single crystalline phase which may be varied in composition within finite limits without the appearance of an additional phase.

solid stage That stage in the cooling of a magma when it has become completely solid, but while the magma is still present below. The structural features that develop are controlled by the movement of the still-liquid magma elsewhere in the body.

solidus The locus of points in a temperature-composition diagram in a system at temperatures above which solid and liquid are in equilibrium and below which the system is completely solid. In binary systems without solid solutions, it is a straight line, and with solid solutions, it is a curved line or a combination of curved and straight lines. Likewise, in ternary systems, the solidus is a flat plane or a curved surface, respectively.

solid waste Useless, unwanted or discarded material with insufficient liquid content to be free flowing. *Also see* WASTE. (1) agricultural—solid waste that results from the raising and slaughtering of animals, and the processing of animal products and orchard and field crops. (2) commercial—waste generated by stores, offices, and other activities that do not actually turn out a product. (3) industrial—waste that results from industrial processes and manufacturing. (4) institutional—waste originating from educational, health care, and research facilities. (5) municipal—residential and commercial solid waste generated within a community. (6) pesticide—the residue from the manufacturing, handling, or use of chemicals intended for killing plant and animal pests. (7) residential—waste that normally originates in a residential environment. Sometimes called domestic solid waste.

solifluction The process of slow flowage from higher to lower ground of masses of waste saturated with water. Also applied to similar subaqueous flowage.

solifluction lobe Tonguelike mass of solifluction debris commonly with steep front and relatively gentle upper surface.

solifluction sheet Broad solifluction mantle.

solifluction slope Smooth slope produced by solifluction.

solifluction stream Narrow, laterally confined, streamlike solifluction mantle.

solitary coral Cup coral; fossil cow's horn; horn coral, *q.v.* An individual corallite that exists unattached to other corallites.

solubility The equilibrium concentration of solute when undissolved solute is in contact with the solution.

solubility product The equilibrium constant for the process of solution of a substance (generally in water).

solum The upper part of the soil profile, *q.v.*, above the parent material, in which the processes of soil formation are taking place. In mature soils this includes the A and B horizons.

solution *1.* The change of matter from the solid or gaseous into the liquid state by its combination with a liquid. When unaccompanied by chemical change, it is called physical solution; otherwise, chemical solution. *2.* The result of such change; a liquid combination of a liquid and a nonliquid substance.

solution potholes Includes all the holes that are formed primarily by solution action. Such holes are more numerous in soluble rocks, notably limestones.

solution transfer Solution of detrital grains at points of contact followed by deposition of dissolved material on other parts of the grain surfaces.

solution valley Broadly U-shaped valley in carbonate rocks formed by solution.

solvate Chemical compound consisting of a dissolved substance and its solvent, e.g., hydrated calcium sulfate.

solvent The dissolving medium; the substance in which solution takes place.

solvus The curved line (binary systems) or surface (ternary systems) separating a field of homogeneous solid solution from a field of two or more solid phases which may form from the homogeneous one by unmixing.

sonic log An acoustic log continuously recording travel time of sound from surface to an instrument lowered down a borehole.

sonobuoy An anchored floating buoy which transmits a radio impulse when actuated by a sound wave in the water. Used in radioacoustic ranging as a method of location in marine operations, especially seismic prospecting.

sonoprobe Type of echo sounder that generates sound waves and records their reflections from inequalities beneath a sedimentary surface.

sooty chalcocite A black, pulverant variety of chalcocite of supergene origin.

Soret's principle That principle by which, if differences of temperature are induced in a solution of common salt or other substance in water, the dissolved

material will become relatively concentrated in those portions in which the temperature is lowest.

sorosilicate A structural class of silicate minerals characterized by the linkage of two Si-O tetrahedra by the sharing of one oxygen. An example is hemimorphite, $Zn_4(Si_2O_7)(OH) \cdot H_2O$.

sorption A term including both adsorption and absorption. Sorption is basic to many processes used to remove gaseous and particulate pollutants from an emission and to clean up oil spills.

sorted circles Patterned ground whose mesh is dominantly circular and has a sorted appearance commonly due to a border of stones surrounding finer material.

sorted nets Patterned ground with a mesh intermediate between that of a sorted circle and a sorted polygon and with a sorted appearance commonly due to a border of stones surrounding finer material.

sorted polygons Patterned ground whose mesh is dominantly polygonal and has a sorted appearance commonly due to a border of stones surrounding finer material.

sorting 1. In a genetic sense the term may be applied to the dynamic process by which material having some particular characteristic, such as similar size, shape, specific gravity, or hydraulic value, is selected from a larger heterogeneous mass. 2. In a descriptive sense the term may be used to indicate the degree of similarity, in respect to some particular characteristic, of the component parts in a mass of material. 3. *Stat:* A measure of the spread of a distribution on either side of an average.

sorting coefficient A mathematical measure of the degree of sorting of a sediment. *See* SORTING INDEX

sorting index A measure of the uniformity of particle size in a sediment, usually based on the statistical spread of the particle-size frequency curve.

sound *1.* A relatively long arm of the sea or ocean forming a channel between an island and a mainland or connecting two larger bodies, as, a sea and the ocean, or two parts of the same body; usually wider and more extensive than a strait. *2. V:* To measure or ascertain the depth of water, as with sounding lines. *3. Geophys:* Elastic waves in which the direction of particle motion is longitudinal, i.e., is parallel with the direction of propagation in the air, but also applies to wave motion in liquids and solids. It is the type of wave motion applied in the reflection seismograph method of geophysical prospecting.

sounding A measured depth of water. On hydrographic charts the soundings are adjusted to a specific plane of reference (sounding datum).

sounding datum The plane to which soundings are referred.

sounding line A line, wire, or cord used in sounding. It is weighted at one end with a plummet (sounding lead). *Syn:* LEAD LINE

sound intensity Average rate of flow of sound energy through a unit section normal to the direction of propagation. Average power transmission per unit area. *Syn:* ACOUSTIC INTENSITY; SEISMIC ITENSITY

sound ranging *Geophys:* Method of locating a source of unknown position by acoustic triangulation, i.e., by recording sound impulses on receivers at known positions; has been used in military application to locate gun positions.

source *1. Seis. prospecting:* Either the point of origin or shot from which elastic waves are propagated, or the formation, horizon, interface or boundary at which the seismic wave is refracted and/or reflected and returned to the surface. *2. Earthquake seismol:* The point of origin of an earthquake. *3.* In neutron logging, the source of neutrons at one end of the logging tool.

source bed concept Hypothesis that many sulfide ore bodies were derived from sulfides deposited syngenetically in a particular stratigraphic zone of a sedimentary basin.

source beds Rocks in which oil or gas has been generated.

source rock The geological formation in which oil, gas, and/or other minerals originate.

sour oil or **gas** Crude oil containing an abnormally large amount of sulfur and sulfur compounds; or natural gas which contains objectionable amounts of hydrogen sulfide and other sulfur compounds.

space group A set of symmetry operations of the indefinite repetition of motif in space; there are 230 of them.

space lattice *See* SPACE GROUP

spall *1.* Relatively thin, commonly curved and sharp-edged pieces of rock produced by exfoliation. *2. V:* To break off in layers parallel to a surface.

span *1.* Length of a time interval. *2.* Informal designation for a local geologic time unit.

Spanish ocher A variety of red ocher.

spar *1.* As used, loosely, almost any transparent or translucent, readily cleavable, crystalline mineral having a vitreous luster, as, calcspar (calcite), fluorspar (fluorite) heavy spar (barite). Most commonly used for feldspar. 2. *Min:* Small clay veins found in coal seams.

sparite A limestone rock composed largely of allochems plus sparry calcite cement; sparry allochem limestones.

sparker A marine seismic energy source employing a high-voltage electrical discharge underwater.

spathic Having good cleavage.

spathization The widely distributed crystallization of calcite or dolomite, forming either "open" or "closed" fabrics.

spatter cone A low, steep-sided hill or mound of spatter built by lava fountains along a fissure or around a central vent. The great bulk of the structure consists of spatter, the glassy skins of which adhere to form agglutinate.

special creation Theory antedating an understanding of evolution that each species of organisms inhabiting the earth was created fully formed and perfect by some divine process.

speciation 1. Any process of evolution that produces new species. 2. Origin of genetic isolation separating two or more parts of a population.

species No entirely satisfactory definition can be formulated because theoretical and practical species are not necessarily the same. Ideally, the species concept embraces (a) interbreeding, (b) morphologic similarity, (c) physiologic compatibility, (d) ecologic association, (e) geographic distribution, and (f) continuity in time. Practically, a species is the type specimen (holotype) and other individuals considered to be so closely related and similar that they should be referred to by a single species name.

species diversity index The ratio between the number of species in an area and the importance values (e.g., numbers, biomass, productivity) for each species. There are several formulas for species diversity indices varying primarily in the determination of the importance values. Each index has its own advantages and limitations. *See also* DIVERSITY

species group A group of species that may be regarded as all descended from a common stock, which sets that group apart from other species of the same genus.

specific capacity (of a well) The discharge expressed as rate of yield per unit of drawdown.

specific conductance A measure of conductivity of liquids. In water this is a measure of the ions in solution, therefore a measure of the dissolved solids, and therefore, also, a measure of the pollution load.

specific gravity Ratio of the mass of a body to the mass of an equal volume of water at a specified temperature.

specific heat Quantity of heat necessary to raise the temperature of 1 gram of a given substance 1 degree centigrade.

specific name 1. The second or trivial name of a species. 2. Less properly, the name of a species consisting of two words.

specific retention *Hydrol:* As applied to a rock or soil it is the ratio of (1) the volume of water which, after being saturated, it will retain against the pull of gravity to (?) its own volume. It is stated as a percentage.

specific rotation Rotatory power (rotation of plane of polarization of polarized light) given in degrees per decimeter for liquids and solutions, and in degrees per millimeter for solids.

specific volume Reciprocal of density or volume per unit mass expressed in cubic centimeters per gram, or cubic feet per pound.

specific yield As applied to a rock or soil it is the ratio of (1) the volume of water which, after being saturated, it will yield by gravity to (2) its own volume. This ratio is stated as a percentage.

specimen Properly speaking, a sample of anything; but among miners it is often restricted to selected or handsome minerals, as, fine pieces of ore, crystals, or pieces of quartz containing visible gold.

spectral gamma-ray log Record of the radiation spectrum and relative intensities of gamma rays emitted by strata penetrated in drilling. Because of their different energies the relative amounts of radioactivity contributed by different elements can be determined.

spectrographic analysis Analysis by obtaining the spectrum of a substance and matching lines in the spectrum with known wavelengths of lines in the spectra of the elements. The analysis can be made quantitative by comparing intensities of the spectral lines.

spectrometer Instrument used in determining the index of refraction. Spectroscope fitted for measurements of the spectra observed with it.

spectrophotometer Optical instrument for comparing the intensities of the corresponding colors of two spectra.

spectroscope Optical instrument for forming and examining spectra.

spectrum An array of visible light ordered according to its constituent wavelengths or color.

specular hematite Specularite, *q.v.*

specularite Hematite, Fe_2O_3, occurring in tabular or disklike crystals of gray color and splendent metallic luster. *Syn:* GRAY HEMATITE; SPECULAR HEMATITE; SPECULAR IRON

speculative resources Undiscovered resources that may occur either in known types of deposits in a favorable geologic setting where no discoveries have been made, or in as yet unknown types of deposits that remain to be recognized.

speed Time rate of motion measured by the distance moved per unit time. *Seis. prospecting:* The rate of propagation of elastic waves.

speleology The scientific study or exploration of caverns and related features.

spergenite A calcarenite containing less than 10% of quartz and composed primarily of oölites and fossil detritus.

spessartite A mineral of the garnet group, $Mn_3Al_2(SiO_4)_3$.

sphalerite A mineral, $(Zn,Fe)S$, dimorphous with wurtzite. Isometric. The principal ore of zinc. *Syn:* BLENDE; BLACKJACK; JACK; ROSIN JACK; ZINC BLENDE

sphene Titanite. A mineral, $CaTiSiO_5$. Monoclinic.

sphenoid *1. Crystallog:* An open form in the monoclinic system, consisting of two intersecting similar faces related to one another only by a twofold symmetry axis perpendicular to their line of intersection and bisecting the dihedral angle between them. *Cf.* DOME. *2.* A disphenoid.

spherical coordinates A system of three-dimensional coordinates defined by a radius and two angles (like latitude and longitude). *Seis. prospecting:* The radial distance and angular measures which give the orientation of pulses originating at a point source, such as a shothole.

spherical wave front Spherical surface which a given phase of a seismic impulse (in an isotropic medium) occupies at any particular time.

spherical weathering Spheroidal weathering, *q.v.*

sphericity The degree in which the shape of a fragment approaches the form of a sphere.

spheroid In general, any figure differing but little from a sphere. *Geodesy:* A mathematical figure closely approaching the geoid in form and size, and used as a surface of reference for geodetic surveys.

spheroidal structure Orbicular structure, *q.v.*

spheroidal symmetry Axial symmetry, *q.v.*

spheroidal weathering Boulders produced chiefly by chemical weathering of rock along fractures. Such boulders are called boulders of decomposition.

spherulite A small, radiating, and usually concentrically arranged aggregation of one or more minerals generally of spherical or spheroidal shape, formed by the radial growth of acicular crystals in a rigid glass about a common center or inclusion.

spicule Tiny siliceous or calcareous object, commonly needle-shaped or branched, contained in the tissues of certain invertebrate animals such as sponges.

spike, isotopic Known amount of an isotopic mixture of known and abnormal composition.

spilite A basaltic rock with albitic feldspar. The albitic feldspar is usually accompanied by autometamorphic minerals or minerals characteristic of low-grade greenstones such as chlorite, calcite, epidote, chalcedonic silica or quartz, actinolite, and others.

spilosite A rock representing an early stage of the formation of adinole or spotted slate.

spinel *1.* A mineral, $(Mg,Fe)Al_2O_4$. Isometric. Used as a gem.
2. A mineral group of general formula AB_2O_4, where $A=Mg,Fe'',Zn,Mn'',Ni,$ $B=Al,Fe''',Mn''',Cr$. Isometric.

spinning fiber Asbestos suitable for the spinning of asbestos fabrics.

spinoff The concept that, when the rocks of the earth melted as a result of gravitational and impact heat, the heavier constituents (iron-related compounds) sank toward the center and the resulting increase in angular momentum threw the original atmosphere off into space.

S-P interval *Earthquake seismol:* The time interval between the first arrivals of longitudinal and transverse waves, which is a measure of the distance from the earthquake source.

spit A small point of land or narrow shoal projecting into a body of water from the shore.

s-plane An upwarped s-surface.

splash zone The zone immediately landward of the mean higher high water level affected by the wave spray. *See also* SUPRALITTORAL

splays *1.* Divergent small faults at the extremities of large normal faults, especially rifts. *2.* Sloping and spreading or fanning out units of deposition, such as the deposits from a crevasse in a levee or the delta front sands or siltstones spreading out from the mouth of a distributary.

splendent Applied to the luster of a mineral that reflects with brilliancy and gives well-defined images, as, hematite, cassiterite.

spliced Applied to veins when they pinch out and are overlapped at that point by another parallel one.

split spread *Seis. prospecting:* A line of detectors symmetrically disposed on two sides of the shot point.

spodosol Humid forest soils, mostly under conifers, with an ashy gray, leached A horizon and an iron and organic rich B horizon; comparable to gray forest podsols.

spodumene A mineral, $LiAlSi_2O_6$. Monoclinic. An ore of lithium. The clear green and pink varieties are used as gems.

spoil *1.* Debris or waste material from a coal mine. *2.* In England, a stratum of coal and dirt mixed. *3.* Dirt or rock that has been removed from its original location, specifically materials that have been dredged from the bottoms of waterways.

spoil banks Submerged embankments of dumped material dredged from a channel which lie alongside the channel.

sponge An organism belonging to the simplest and least advanced phyla of mutiple-celled animals or Porifera; generally possesses a spicular skeleton.

spontaneous polarization Self-potential method; spontaneous potential method, *q.v.*

spontaneous potential method An electrical method in which a potential field caused by spontaneous electrochemical phenomena is measured. *Syn:* SELF-POTENTIAL METHOD; SPONTANEOUS POLARIZATION

sporadic permafrost zone Regional zone predominantly free of permafrost, with only scattered areas underlain by permafrost.

spore An asexual reproductive structure, commonly unicellular and usually produced in sporangia.

spore coal Coal in which the attritus may contain a large amount of spore matter together with a certain amount of transparent attritus.

spot correlation Correlation of reflections on isolated seismograms by "spotting" similarities in character and interval. *Syn:* CORRELATION METHOD

spotted slate An argillaceous rock in which low-grade metamorphism has caused the growth of incipient porphyroblasts.

spot test Delicate chemical test for ores or other substances in which a drop of solution is applied to filter paper impregnated with a sensitive reagent and a colored spot is formed.

spread, seismic Arrangement of the geophones in relation to the shot point. Several patterns are used in field practice such as correlation, continuous, interlocking, reversed, removed, leap frog, in-line, end-to-end, parallel, cross, "L," perpendicular, star, arc, fan, circular, split, and straddle spreads.

spread correction Reduction of the seismic reflection time to vertical incidence by applying the ratio of the depth of the reflection horizon to the distance between the shot point and the observed geophone.

spring A place where, without the agency of man, water flows from a rock or soil upon the land or into a body of surface water.

spring tide A tide that occurs at or near the time of the new and full moon and which rises highest and falls lowest from the mean level.

spud in To commence the actual drilling of a well.

spurs The subordinate ridges which extend themselves from the crest of a mountain like ribs from the vertebral column.

squeeze job Usually a secondary cementing job where the cement is pumped into the formation through the bottom of the casing or through perforations in casing to obtain a shutoff of undesirable fluids.

s-surface; s-plane *Struct. petrol:* Any planar surface. May be a plane of stratification, schistosity, shear, or a statistical surface based on petrofabric analysis.

stability *Thermodyn:* A phase is

said to have stability if a slight perturbation in the variables defining the system, temperature, pressure, or composition does not result in the appearance of a new phase.

stability field (critical level) The temperature and pressure within which a mineral is stable.

stability series Order of persistence. The order of resistance of minerals to alteration or destruction by weathering of the parent rock, abrasion during transportation, and postdepositional solution.

stabilization 1. The process of converting active organic matter into inert, harmless material. 2. Any process or method of increasing soil or rock stability.

stabilized dune A dune protected from further wind action by a cover of vegetation or by cementation of the sand; known also as a "fixed" or "anchored" dune.

stable gravimeter A gravimeter having a simple weight on a spring such that the sensitivity is proportional to the square of the period.

stack 1. Under favorable conditions waves are able to cut back on the two sides of a tiny promontory, and then, aided by weathering, to cut behind the end of this, leaving it as an island, or stack, entirely removed from the mainland. *Syn:* CHIMNEY; SKERRIES. 2. *Seismol:* A composite of several seismic traces usually corrected for moveout and statics before being summed. *See* CDP-STACKING

stade Time represented by glacial deposits formerly termed a substage; differs from a stratigraphic substage because it is not a rock unit and has somewhat variable time value from place to place.

stadia *Surv:* 1. A temporary station. 2. A stadia rod. 3. An instrument for measuring distances, consisting of a telescope with

special horizontal parallel lines or wires, used in connection with a vertical graduated rod; also, the rod alone, or the method of using it.

stadial 1. *N:* Stade. 2. *Adj:* Pertaining to a stade.

stadial moraine Recessional moraine.

stadia rod A graduated rod used with an instrument of the stadia class to measure the distance from the observation point to the place where the rod is positioned.

stadia tables Mathematical tables from which may be found, without computation, the horizontal and vertical components of a reading made with a transit and stadia rod.

stage 1. *Hydraul:* Elevation of a water surface above any chosen datum plane, often above an established low-water plane; gauge height. 2. *Stratig:* Those time-stratigraphic units next in rank below a series. Their boundaries can be (a) by correlation to bounding unconformities, (b) by correlation to a type section (stratotype), or preferably (c) those biological changes that represent true time, e.g., evolutionary changes.

stagnation 1. Condition of water not stirred by currents or waves; commonly implies oxygen deficiency and accumulation of noxious substances. 2. Condition of a glacier that has ceased to move.

stalactite A cylindrical or conical deposit of minerals, generally calcite or aragonite, hanging from the roof of a cavern.

stalagmite Columns or ridges of carbonate of lime rising from a limestone cave floor, and formed by water charged with carbonate of lime dripping from the stalactites above. Stalactites and stalagmites often meet, and then form a column from floor to roof.

standard atmosphere The International Standard Atmosphere which is used as the basis of graduation of altimeters assumes at mean sea level a temperature of 15° C., a pressure of 760 mm. of Hg. (1013.2 millibars) and a lapse rate of 6.5° C./km. from sea level up to 11 km.

standard cell A method of studying the chemical relationships between rocks by calculating the number of various cations in the rocks per 160 oxygen ions; the results of such calculations.

standard deviation map Vertical variability map showing the degree of dispersion of one lithologic type about its center of gravity in a stratigraphic unit.

standard mineral A hypothetical mineral composition, as used in the calculation of the norm, *q.v. Syn:* NORMATIVE MINERAL

standard state The condition within the crust of the earth such that the pressures are uniform in all directions. That is, the stresses are essentially hydrostatic.

standing wave A type of wave in which there are nodes, or points of no vertical motion and maximum horizontal motion, between which the water oscillates vertically.

stand of tide An interval at high or low water when there is no sensible change in the height of the tide. The water level is stationary at high and low water for only an instant, but the change in level near these times is so slow that it is not usually perceptible.

standstill The condition or time of static sea level.

stanniferous Yielding or containing tin, as, stanniferous ores.

stannite A mineral, Cu_2FeSnS_4. Tetragonal.

star, *adj.* Asteriated, as a star ruby.

starved basin *See* BASIN, STARVED

state-line fault A tongue-in-cheek term for the discontinuity of geologic structures appearing at the borders of geologic maps of adjacent geographic areas, such as state boundaries, due to differences in interpretation.

state of matter Matter is generally considered to exist in three states: gaseous, liquid, and solid. The relations between liquid and solid states are not entirely clear and it has been suggested that a better division is: gaseous, vitreous (amorphous), and crystalline.

static *1.* Noises which interfere with radio waves. The term is also applied to interferences with acoustic and seismic waves. *2. Seismol:* Time corrections applied to reflection seismic traces to eliminate delays caused by near surface variations in elevation and weathering layer thicknesses.

static metamorphism Diagenism. *Geol:* Metamorphism produced by the internal heat of the earth and the weight of the superincumbent rocks and not accompanied by appreciable deformation. A term used in contradistinction to dynamic metamorphism which involves stresses principally due to thrust.

static pressure Pressure that is "standing" or stabilized due to the fact that it has attained the maximum possible from its source and is not being diminished by loss.

static zone The zone which extends below the level of the lowest point of discharge, and in which the water is stagnant or moves with infinitesimal velocity.

station Ground position at which the geophysical instrument is set up for observation in the field.

stationary field Natural field of force as a gravimetric or a magnetic field. *Elect. prospecting:* A field which does not change with time and is produced by direct current after equilibrium has been reached.

stationary wave A wave of essentially stable form which does not move with respect to a selected reference point; a fixed swelling. Sometimes called standing wave.

staurolite A mineral, $(Fe,Mg)_2\cdot Al_9Si_4O_{23}(OH)$, occurring in metamorphic rocks. Orthorhombic.

steady-state stream A graded stream.

steady-turbidity current Persistent turbidity current, e.g., one produced by a heavily sediment-laden river flowing into a body of deep standing water.

steatite Massive, in many cases impure, talc-rich rock.

steatite talc High-grade variety of talc suitable for use in electronic insulators.

s-tectonite Rock whose predominant structures are s-planes which may have been produced by either slipping or flattening.

steephead A nearly vertical bluff or streamhead, at the base of which springs emerge, supplying small streams, assumes in time a semicircular form, which is the steephead.

stellate Starlike; stellate hairs having radiating branches or, when falsely stellate, are separate hairs aggregated into starlike clusters; hairs once or twice forked are often treated as stellate.

Stelleroidea Class of echinoderms of starlike form; includes asteroids, auluroids, and ophiuroids.

stenohaline Refers to relatively low tolerance for variation in salinity.

stepout Angularity; moveout or moveout time, *q.v.*

stepout correction The correction, determined from the geometry of the detector spread, which eliminates the effect on the reflection time of the horizontal distance between seismometers and shot point.

stepout time *Seis. prospecting:* The time differential in arrivals of a given peak or trough of a reflected or refracted event for successive detector positions on the earth's surface. This difference gives information on the dip of the reflecting or refracting horizon in the earth.

steppe An extensive, treeless grassland area in southeastern Europe and Asia developing in the semiarid mid-latitudes of that region. They are generally considered drier than the prairie which develops in the subhumid mid-latitudes of the United States.

steptoe An islandlike area in a lava flow. *Cf.* NUNATAK

stereogram A stereographic projection.

stereographic projection A two-dimensional array of points each representing the orientation of a single plane in three dimensions.

stereo net A term used in structural geology and crystallography for a Wulff net, *q.v.*

stereoscope An optical instrument for assisting the observer in obtaining stereoscopic vision from two properly prepared photographs.

stereoscopic image or **model** That mental impression of a three-dimensional model which results from stereoscopic fusion of a stereoscopic pair.

stereoscopic pair Two photographs of the same area taken from different camera stations in such a manner that a portion of the area appears on both photographs. *Syn:* STEREOGRAM

stereoscopic vision Binocular vision in which an image from each eye is superimposed on the image from the other eye so as to give the effect of depth. *Cf.* BINOCULAR VISION

Sternberg's law Wearing away of transported pebbles is proportional to their weight in water and distance traveled.

stibnite Antimony glance; gray antimony. A mineral, Sb_2S_3. Orthorhombic. The principal ore of antimony.

stiff clay Clay of low plasticity.

stillbite A mineral of the zeolite group, close to $Ca_2NaAl_5Si_{13}O_{36}$,-$14H_2O$. Monoclinic.

stillstand To remain stationary with respect to sea level or to the center of the earth.

still-water level The elevation of the surface of the water if all wave action were to cease.

stishovite A tetragonal mineral, SiO_2. A high-pressure, extremely dense polymorph of quartz, produced under static conditions at pressures above about 100 kb and found naturally associated with coesite and only in shock-metamorphosed quartz-bearing rocks. Its occurrence provides a criterion for meteorite impact.

stock A body of plutonic rock that covers less than 104 square kilometers has steep contacts (generally dipping outward), and although generally discordant may be concordant.

stock pile The ore accumulated at the surface when shipping is suspended, as on the iron ranges of Michigan and Minnesota during the winter months.

stockwork [*Ger.* Stockwerke] An ore deposit of such a form that it is worked in floors or stories. It may be a solid mass of ore, or a rock mass so interpenetrated by small veins of ore that the whole must be mined together.

Stokes' law A formula to express the rate of settling of spherical particles in a fluid. It is expressed as $V = Cr^2$ where V is the velocity in centimeters per second, r is the radius of the particle in centimeters, and C is a constant relating relative densities of fluid and particle, the acceleration due to gravity, and the viscosity of the fluid.

stomach stone Gastrolith, *q.v.*

stone *1.* Concreted earthy or mineral matter. A small piece of rock. Rock or rocklike material for building. Large natural masses of stone are generally called rocks; small or quarried masses are called stones; and the finer kinds, gravel or sand. *2.* A precious stone; a gem. *3.* In England, ironstone. *See* CLAY IRONSTONE

stone bubble Lithophysa, *q.v.*

stone circle *See* STONE RING

stone field *See* FELSENMEER

stone guano Breccia of agatelike fragments formed by leaching of guano and enrichment of deposit in insoluble phosphates.

stone iron *See* SIDEROLITE

stone line A line of angular or subangular rock fragments which parallels a sloping topographic surface at a depth of several feet.

stone net *See* STONE RING

stone polygon *See* STONE RING

stone ring A ring or polygon of stones surrounding a central area of fine debris in a bouldery soil region. *Syn:* STONE POLYGON; STONE NET; STONE CIRCLE; STONE WREATH; ROCK WREATH; FROST-HEAVED MOUND

stone stripe *See* STRIPED GROUND

stone wreath *See* STONE RING

stony meteorite Meteorite consisting mainly of rock-forming silicate minerals such as pyroxene, olivine, and feldspar.

stope *1.* An excavation from which the ore has been extracted, either above or below a level, in a series of steps. A variation of step. *2.* An underground excavation from which ore has been extracted, either above (overhand) or below (underhand) a level. Access to stopes is usually by way of adjacent raises.

stoping Overhand or magmatic stoping. *1.* A method of intrusion of light acidic magma into heavier basic rocks at Ascutney Mountain, Vermont. Blocks of the older rock are wedged loose overhead, settle in the magma, and are assimilated at depth. Thus the magma works its way upward. *See* STOPE. *2. Min:* The loosening and removal of ore in a mine either by working upward (overhead or overhand) or downward (underhand).

stoping, magmatic *See* MAGMATIC STOPING

stoping ground Part of an ore body opened by drifts and raises and ready for breaking down.

storm beach During exceptionally heavy storms coarse material is sometimes built by the waves into surprisingly strong ridges which stand some distance from the shore under normal conditions and are known as storm beaches.

storm berm A low beach ridge marking the upper limit of wave action during storms.

storm wave A rise of the sea over low coasts not ordinarily subject to overflow; it is caused primarily by wind and has no relation to the tide brought about by gravitational forces except that the two may combine.

stoss, *adj.* Facing the direction whence a glacier moves, as a rock or hill in its track; as, the stoss side of a crag; contrasted with lee. Applied to the struck side of a rounded ledge.

stoss-and-lee topography The persistently asymmetric arrangement of bosses and small hills in a strongly glaciated district, each hill having a comparatively gentle abraded slope on the stoss side and a somewhat steeper and rougher quarried slope on the lee side.

strain Deformation resulting from applied force; within elastic limits strain is proportional to stress.

strain ellipse In two-dimensional analysis of rock deformation, the imaginary ellipse whose half-axes are the greatest and least principal strains.

strain ellipsoid *1.* In elastic theory, a sphere under homogeneous strain is transformed into an ellipsoid with this property: the ratio of the length of a line, which has a given direction in the strained state, to the length of the corresponding line in the unstrained state, is proportional to the central radius vector of the surface drawn in the given direction. *2.* The ellipsoid whose half-axes are the principal strains.

strain gauges Mechanical, electrical, or optical devices which measure displacement or strain produced by force or stress.

strain-slip cleavage In slates and schists, a structure similar to fracture cleavage, *q.v.*, but with marked flexing of the earlier cleavage or foliation along the shear planes.

strait A relatively narrow waterway between two larger bodies of water.

strand The strand and beach are synonymous terms, applied to the portion of the shore between high and low water.

strand flat A low coastal platform that abuts inland against higher terrain. It may be partly submerged or slightly emerged and its local altitude depends partly upon recent vertical movements of the coast. Strand flats varying from a few hundred yards to many miles extend along hundreds of miles of arctic shore.

strandline The shoreline, where the sea meets the land; a beach.

strand plain A shore which has prograded by wave and current action leaving a smooth plain along the coast for some distance.

strata Plural of stratum.

strategic materials Those materials vital to the security of a nation which must be procured entirely or to a substantial degree from sources outside the continental limits of that nation because the available production will not be sufficient in quantity or quality to meet requirements in time of national emergency. *Cf.* CRITICAL MATERIALS

strath *1.* Generally used for a broad river valley. If it has been elevated and dissected, the erosion remnant is called a strath terrace. *2.* Valley deeply filled with alluvial deposits, particularly glacial outwash, not now occupied by stream.

stratification A structure produced by deposition of sediments in beds or layers (strata), laminae, lenses, wedges, and other essentially tabular units.

stratification index Number of beds in a stratigraphic unit times 100 divided by the unit's thickness.

stratified Formed or lying beds, layers, or strata.

stratified cone *See* COMPOSITE CONE

stratified drift Drift exhibiting both sorting and stratification, implying deposition from a fluid medium such as water or air.

stratified or **sedimentary rocks** Derivative or stratified rocks may be fragmental or crystalline; those which have been mechanically formed are all fragmental; those which have been chemically precipitated are generally crystalline, and those composed of organic remains are sometimes partially crystalline.

stratified water Standing water consisting of density layers differing in temperature or salinity.

stratiform *1.* Composed of layers. *2.* Resembling stratus clouds.

stratigrapher One who studies, or who has expert knowledge of, stratigraphy.

stratigraphic classification Classification of stratified rocks and geologic time into rock, timerock, time, and biostratigraphic units.

stratigraphic control The apparent localization of mineral deposition by stratigraphic features.

stratigraphic geology The study of stratified rocks.

stratigraphic leak Situation or process whereby microfossils, generally conodonts, are supposed to have descended through crevices or solution channels and lodged in a lower stratum where they may be associated with fossils of greater age.

stratigraphic map A map, involving a span of geologic time, that shows the distribution, configuration, or aspect of a stratigraphic unit or surface; e.g., a structurecontour map.

stratigraphic paleontology Study of fossils applied particularly to discrimination and correlation of fossiliferous strata.

stratigraphic range Geologic range, the spread or distribution of a given group of organisms through geologic time. Also, the persistence of a fossil genus or species through the stratigraphic sequence.

stratigraphic sequence A succession of sedimentary beds of interregional extent, chronologically arranged with the older strata below and the younger above.

stratigraphic trap A type of trap

which results from variation in lithology of the reservoir rock and a termination of the reservoir (usually on the updip extension) or other interruption of continuity.

stratigraphic unit Unit consisting of stratified mainly sedimentary rocks grouped for description, mapping, correlation, etc. *See* BIOSTRATIGRAPHIC; CHRONOSTRATIGRAPHIC

stratigraphy That branch of geology which treats of the formation, composition, sequence, and correlation of the stratified rocks as parts of the earth's crust.

stratosphere That part of the earth's atmosphere between the troposphere and the ionosphere.

stratovolcano A volcanic cone, generally of large dimension, built of alternating layers of lava and pyroclastic materials. Essentially synonymous with composite cone; stratified cone.

stratum *1.* A section of a formation that consists throughout of approximately the same kind of rock material; a stratum may consist of an indefinite number of beds, and a bed may consist of numberless layers; the distinction of layer and bed is not always obvious. *2.* A single sedimentary bed or layer, regardless of thickness

streak The color of the powder of a mineral as obtained by scratching the surface of the mineral with a knife or file or, if not too hard, by rubbing it on an unpolished porcelain surface.

streak plate A piece of unglazed porcelain for testing the streak of minerals.

stream Any body of flowing water or other fluid, great or small.

stream capacity The maximum amount of material that the stream is able to transport.

stream capture Piracy, *q.v.*

streamer A marine seismic recording cable containing grouped hydrophones designed for continuous towing; reflection surveys use streamers ranging from 5000 to 9000 feet in length.

streamflood An eroding agent made by transformation from a sheetflood that is depositional. The cause of the transformation may be differential uplift or a climatic change.

stream frequency *Geomorph:* Ratio of the number of stream segments of all orders to the area of a drainage basin; a measure of topographic texture. *Symbol:* F

stream gradient ratio Ratio of the gradient of a stream channel of one order to that of the next higher order in the same drainage basin; $R_s = g_n / g_{n+1}$. *Symbol:* R_s

streaming flow Glacier flow, *q.v.*, in which the ice flows without cracking or breaking into blocks.

stream length *Geomorph:* (Symbol L_u). Length of a stream segment of a particular order. Total stream length (symbol ΣL_u): the length of all stream segments of a particular order in a specified drainage basin. Mean stream length (symbol L_u): the mean of all stream lengths of a particular order within a specified drainage basin.

stream-length ratio *Geomorph:* Ratio of the mean length of stream segments of one order to the mean length of segments of the next lower order within a specified drainage basin. *Symbol:* R_L

streamline flow Laminar flow.

stream order *Geomorph:* First order streams are the smallest unbranched tributaries; second order streams are initiated by the confluence of two first order streams; third order streams are initiated by the confluence of two second order streams; etc. The symbol u refers to an order number.

stream segment *Geomorph:* The portion of a stream extending between the junctions of tributaries of different orders.

stream terrace (Shoshone type) Stream-cut rock terrace with thick cover of slope wash.

stream tin Tin ore (cassiterite) occurring in stream beds; distinguished from lode tin.

stream transportation The movement of weathered or eroded rock material by the action of stream flow including solution suspended load and bed load.

strength The limiting stress that a solid can withstand without failing by rupture or continuous plastic flow. Rupture strength or breaking strength refers to the stress at the time of rupture.

strength of magnetic field A vector quantity with a magnitude and direction defined as that of the force acting on a unit positive pole. The force between magnetic poles may be considered as the reaction of one pole on the magnetic field of the other. Thus a unit field strength exerts a force of 1 dyne on a unit pole, and such a field has a strength of 1 oersted. The magnetic field strength is represented by the density of the lines of force or the number of lines per square centimeter in a section perpendicular to their direction. These lines are maxwells, and the strength of the field in oersteds is the number of maxwells per square centimeter.

strength of magnetic pole The force between magnetic poles is proportional to the pole strength. Equal poles have unit strength when they exert unit force (1 dyne) when they are unit distance (1 cm.) apart.

stress *1.* Force per unit area, found by dividing the total force by the area to which the force is applied. *2.* The intensity at a point in a body of the internal forces or components of force which act on a given plane through the point. As used in product specifications, stress is calculated on the basis of the original dimensions of the cross section of the specimen.

stress difference The algebraic difference between the maximum and minimum principal stresses.

stress ellipsoid The ellipsoid whose half-axes are the principal stresses.

stress pressure *1.* Pressure resulting from nonhydrostatic stress. *2.* Mean of three principal components of nonhydrostatic stress.

stretched pebbles Pebbles in conglomerates that, more or less spherical originally, are now shaped like triaxial ellipsoids, oblate spheroids, or prolate spheroids because of rock deformation.

stretch thrust Thrust that forms when the inverted limb of an overturned or recumbent fold becomes so stretched that it ruptures.

stria *1.* A minute groove or channel. A threadlike line or narrow band. *See* GLACIAL STRIAE. *2.* A type of ornamentation or surface detail found on the shells of invertebrates: a very fine grooved line or thread. *3.* **striae,** *pl.* Parallel grooves or lines.

striation The markings with lines or striae, generally applied to the parallel scratches with which the bed of a glacier is scored by means of sharp angular fragments of rock embedded in the ice stream, which are forcibly dragged along with the ice.

striding level A spirit level, the frame of which carries at its two extremities inverted Ys below, so that it may be placed upon two concentric cylinders and straddle any small intervening obstacles.

strike The course or bearing of the outcrop of an inclined bed or structure on a level surface; the direction or bearing of a horizontal line in the plane of an inclined stratum, joint, fault, cleavage plane, or other structural plane. It is perpendicular to the direction of the dip.

strike fault A fault whose strike is parallel to the strike of the strata.

strike joint A joint that strikes parallel to the strike of the adjacent strata or schistosity if bedding is not present.

strike separation In a fault, the distance of apparent relative displacement of two formerly adjacent beds on either side of a fault, measured parallel to the strike of the fault.

strike shift The horizontal component of the shift measured parallel to the strike of the fault.

strike-shift fault A fault in which the component of the shift is parallel to the fault strike.

strike slip The component of the slip parallel with the fault strike, or the projection of the net slip on a horizontal line in the fault surface.

strike-slip or **transcurrent fault** A fault in which the net slip is practically in the direction of the fault strike.

string A driller's term for the drilling bit, jars, drill stem, rope socket, and other tools connected to the lower end of a drilling cable in standard or percussion drilling. Also used occasionally for the rig and complete drilling equipment.

stringer 1. A narrow vein or irregular filament of mineral traversing a rock mass of different material. 2. A thin layer of coal at the top of a bed, separating in places from the main coal by material similar to that comprising the roof.

stringer lead A small ore body, generally a vein leading to a more valuable one.

stringer lode A shattered zone containing a network of small nonpersistent veins. *Syn:* STRINGER ZONE

string galvanometer An instrument for measuring small electrical current, consisting of a very fine conducting fiber stretched loosely between the poles of a strong magnet. Current in the fiber causes it to move laterally, perpendicular to the magnetic lines of force; the motion is measured by projecting a magnified shadow of the fiber onto a photographic film or paper.

strip 1. To remove from a quarry, or other open working, the overlying earth and disintegrated or barren surface rock. 2. To mine coal, alongside a fault, or barrier.

striped ground Alternate stripes of fine and coarse debris on a slope. Other terms: stone stripes; stone-bordered strips; striate land; soil strips; soil stripes.

strip mining A process in which rock and top soil strata overlying ore or fuel deposits are scraped away by mechanical shovels.

stripped plain A plain composed of flat-lying or gently tilted sedimentary rocks from which sediments have been removed down to some resistant bed which seems to have controlled the depth of erosion.

stripped structural terrace Relatively level surface resulting from the removal of weak rock that overlay a resistant layer.

stripper well An oil well producing such small quantities of oil that it receives special attention. Apparently so called from the milk cow which must be "stripped" by special manual effort, in order to yield the small quantity of milk.

stromatolite Laminated but otherwise structureless calcareous objects; commonly called fossil calcareous algae. *Syn:* CRYPTO-ZOON; COLLENIA; GYMNOSOLEN

Stromatoporoidea Laminated, organic bodies made of calcium carbonate; probably hydrozoans.

strombolian Designating or pertaining to a type of volcanic eruption characterized by the appearance of fluid basaltic lava in a central crater, and in which the liquid lava is thrown up by explosions or fire fountains, some accumulating around the vent as spatter, scoria, and bombs. There is no eruption cloud.

strong *1.* Large; important; said of veins, dikes, etc. *2.* In Scotland: hard, not easily broken, e.g., strong coal, strong blaes.

strontianite A mineral, $SrCO_3$. Orthorhombic.

structural Pertaining to, part of, or consequent upon the geologic structure, as, a structural valley.

structural basin An elliptical or roughly circular structure in which the rock strata are inclined toward a central point.

structural control The apparent localization of mineral deposition by structural features.

structural crystallography Study of the internal arrangement and spacing of atoms and molecules composing solids.

structural fabric *See* FABRIC

structural feature Features produced in the rock by movements after deposition, and commonly after consolidation, of the rock.

structural geology Study of the structural (as opposed to the compositional) features of rocks, of the geographic distribution of the features and their causes.

structural petrology Study of structure within rocks, particu-

larly minute structure revealed by petrofabric investigation.

structural plain A gently sloping stratum plain.

structural terrace Where dipping strata locally assume a horizontal attitude.

structural trap One in which entrapment results from folding, faulting, or a combination of both.

structure *1.* The sum total of the structural features of an area. Not to be used as a synonym for structural feature, as, "this structure" meaning "this anticline." *2. Petrol:* One of the larger features of a rock mass, like bedding, flow banding, jointing, cleavage, and brecciation; also the sum total of such features. Contrasted with texture, *q.v.*

structure contour A contour line drawn through points of equal elevation on a stratum, key bed, or horizon, in order to depict the attitude of the rocks.

structure sections Diagrams to show the observed geologic structure on vertical faces or, more commonly, to show the inferred geologic structure as it would appear on the sides of a vertical trench cut into the earth.

stuffed minerals A mineral having large interstices in its structure may accommodate various foreign ions in these holes; such a mineral is then said to be "stuffed."

stylolite A term applied to parts of certain limestones which have a columnlike development; the "columns" being generally at right angles or highly inclined to the bedding planes, having grooved, sutured or striated sides, and irregular cross sections.

subaerial Formed, existing, or taking place on the land surface. Contrasted with subaqueous.

subage Geologic time unit smaller than an age, corresponding to the time-rock unit substage.

subalkalic *1.* Refers to igneous rocks lacking alkali minerals other than feldspars. *2.* Formerly used to describe Pacific series of igneous rocks.

suballuvial bench The outer extension of the pediment which is covered by alluvium with a thickness equivalent to the depth of stream scour during flood but which may be hundreds of meters thick in its distal portion.

subangular A roundness grade in which definite effects of wear are shown, the fragments retaining their original form and the faces virtually untouched, but the edges and corners rounded off to some extent. Secondary corners are numerous (10 to 20). Class limits 0.15 to 0.25.

subaqueous gliding Subaqueous solifluction.

subarctic *1.* The region immediately south of the Arctic Circle and those which have similar climate. *2. Oceanog:* That region in which arctic and nonarctic waters are found together at the surface. *3. Geog:* The regions in which mean temperature is not higher than 10° C. for more than 4 months of the year and the mean temperature of the coldest month not more than 0° C.

subarid Subhumid, *q.v.;* moderately arid.

subarkose Sandstone containing 10 to 25% feldspar.

subbituminous A coal Both weathering and nonagglomerating subbituminous coal having 11,000 or more, and less than 13,000 B.t.u. (moist, mineral-matter-free).

subbituminous B coal Both weathering and nonagglomerating subbituminous coal having 9500 or more, and less than

11,000 B.t.u. (moist, mineral-matter-free).

subbituminous C coal Both weathering and nonagglomerating subbituminous coal having 8300 or more, and less than 9500 B.t.u. (moist, mineral-matter-free).

subcapillary interstice An opening smaller than a capillary interstice, and theoretically so small that, at least in some parts, the attraction of the molecules of its walls extends through the entire space which it occupies.

subcrop *1.* Occurrence of strata on the undersurface of an inclusive stratigraphic unit that succeeds an important unconformity where overstepping is conspicuous. *2.* Area within which a formation occurs directly beneath an unconformity.

subcrop map *1.* Paleogeologic map. *2.* Geologic map showing distribution of formations immediately overlying an unconformity.

subduct In plate tectonics, *q.v.,* the depressing and passing of one plate margin of the earth under another plate.

subduction Descent of one tectonic unit under another. Most commonly used for descent of a slab of lithosphere, but appropriate at any scale. Refers to the process, not the site.

subduction zone An elongate region along which a crustal block descends relative to another crustal block, e.g., the descent of the Pacific plate beneath the Andean plate along the Andean trench. Deep oceanic trenches occur along subduction zones. *See also* SUBDUCTION

subfacies *1.* Subdivision of a broadly defined facies. *2.* Subdivision of a metamorphic facies based on compositional differences rather than pressure-temperature differences.

subgelisol Unfrozen ground below pergelisol.

subgenus; subgenera A group of species which is judged to have special characters in common and which is distinct from other such groups, or subgenera.

subgraywacke Similar to graywacke but has less feldspar and more and better rounded quartz grains.

subhedral Intermediate between anhedral and euhedral.

subjacent *1.* Lying under or below. *2. Petrol:* Applied to intrusive igneous bodies which enlarge downward and have no demonstrable base.

sublimation The transition of a substance directly from the solid state to the vapor state, or vice versa, without passing through the intermediate liquid stage.

sublittoral Refers to the benthonic zone extending from low tide level to a depth of about 100 meters.

submarginal resource The part of subeconomic resources that would require a substantially higher price (more than 1.5 times the current price) or a major cost-reducing innovation to become economic.

submarine bulge Fanlike sedimentary deposit, presumed to have been formed by turbidity currents, on the outer continental slope at the mouth of a submarine canyon.

submarine canyon Steep valley-like submarine depression crossing the continental margin region, except for isolated portions of outer ridges, less than 1.6 to more than 16 kilometers wide, less than 18 to more than 1830 meters deep. Commonest on the continental slope and shelf but some continue across the continental rise, also may cross marginal escarpments and landward slopes of trenches.

submarine delta Submarine sedimentary deposit formed at the mouth of a submarine canyon whose surface features resemble those of a subaerial delta.

submerged coast *See* SHORELINE OF SUBMERGENCE

submergence A term which implies that part of the land area has become inundated by the sea but does not imply whether the sea rose over the land or the land sank beneath the sea. *Syn:* EMERGENCE

submersible A small ship built to operate under water and designed for special functions.

submetallic Applied to minerals having an imperfect metallic luster, as, columbite, wolframite.

subrounded A roundness grade in which considerable wear is shown. Edges and corners are rounded to smooth curves, and the area of the original faces is considerably reduced, but the original shape of the grain is still distinct. Secondary corners are much rounded and reduced in number. Class limits 0.25 to 0.40.

subsequent Tributary to and subsequent in development to a primary consequent stream, but itself consequent upon structure brought out in the degradation of the region; subconsequent: said of some streams and their valleys, as, a subsequent valley.

subsequent streams Streams that have grown headward by retrogressive erosion along belts of weak structure, and also for streams which, having been thus developed in one cycle, persist in the same courses in a following cycle. *See* LONGITUDINAL VALLEY

subsequent valleys Valleys cut by those streams which have grown by headward erosion along belts of weak structure, without relation to the initial trough lines.

subsidence *1.* A sinking of a large part of the earth's crust. *2.* Movement in which there is no free side and surface material is displaced vertically downward with little or no horizontal component.

subsilicic A term to connote rocks having a silica content of less than 52%.

subsolidus Chemical system below its melting point; reactions may occur in the solid state.

subspecies Recognizable subdivision of a species that occupies a more or less definite geographic or ecologic range and grades into neighboring subspecies.

substage Time-stratigraphic unit next lower in rank than stage. Zone has been much used for such a subdivision.

substitution, ionic *See* IONIC SUBSTITUTION

substratum An underlayer or stratum; a stratum, as of earth or rock, lying immediately under another. The hypothetical vitreous basaltic substratum lying beneath the lithosphere or outer granitic shell of the earth.

subsurface Underground; zone below the surface whose geologic features, principally stratigraphic and structural, are interpreted on the basis of drill records and various kinds of geophysical evidence.

subsurface geology The study of structure, thickness, facies, correlation, etc., of rock formations beneath land or sea-floor surfaces by means of drilling for oil or water, core drilling, and geophysical prospecting.

subsurface water All the water that exists below the surface of the solid earth.

subsystem Time-stratigraphic unit proposed for Mississippian or Pennsylvanian rocks to harmonize American stratigraphic classification with European in which these divisions correspond to parts of the Carboniferous System.

subterranean Being or lying under the surface of the earth.

subterranean stream A body of flowing water that passes through a very large interstice, such as a cave, cavern, or a group of large communicating interstices.

subterranean water *See* GROUND WATER

sugarloaf A conical hill or mountain comparatively bare of timber.

suite *1.* Collection of rock specimens from a single area, generally representing related igneous rocks. *2.* Collection of rock specimens of a single kind, e.g., granites from all over the world. *3.* Succession of closely associated sedimentary strata, especially a repeated sequence.

sulcus *Palentol:* Sinus, *q.v.*

sulfate A salt of sulfuric acid; a compound containing the radical $(SO_4)^{-2}$.

sulfide A compound of sulfur with one other more positive element or radical.

sulfide enrichment The enrichment of a deposit by replacement of one sulfide by another of higher valuable metal content, as pyrite by chalcocite.

sulfide zone That part of a lode or vein not yet oxidized by the air or surface water and containing sulfide minerals.

sulfite A salt of sulfurous acid; a compound containing the radical $(SO_3)^{-2}$.

sulfur Sulphur. The native element, S. Orthorhombic.

sulfur bacteria Bacteria that obtain their metabolic energy by the oxidation of sulfur in hydrogen sulfide or various other compounds and the production of elemental sulfur or sulfate ions.

sulfur balls Round or irregularly angular masses of pyrite occurring as common impurities in many coals.

sulfur dioxide (SO_2) A heavy, pungent, colorless gas formed primarily by the combustion of fossil fuels. SO_2 damages the respiratory tract as well as vegetation and materials and is considered a major air pollutant.

summation method Method of correcting seismic reflection arrival times for time spent by the wave in the low-velocity zone. On a continuous, reversed, and interlocked reflection profile between two holes shot at reasonable depths below the low-velocity zone, the low-velocity time under each geophone is equal to one-half the sum of the first arrival times (first kicks) received at that geophone from both shot-holes minus the average high-velocity time, which is obtained by subtracting the uphole time from the first arrival time at the geophone at the shothole on the opposite end of the profile.

summit concordance The equal or almost equal elevation of ridge tops or mountain summits that is thought to indicate the existence of ancient erosion surfaces of which only scattered patches are preserved.

sump A hole or pit which serves for the collection of quarry or mine waters.

sun opal Fire opal.

sunspot A relatively dark area on the sun's surface representing lower temperature and consisting of a dark central umbra surrounded by a prenumbra which is intermediate in brightness between the umbra and the surrounding surface of photosphere.

sunspot cycle Cycle of increasing and decreasing intensity of magnetic storms on the sun's surface of about 11 years' duration.

sunstone A variety of oligoclase feldspar containing numerous small inclusions which cause a delicate play of colors. Used as a gem.

superanthracite Coal intermediate between anthracite and graphite.

supercapillary interstice An opening larger than a capillary interstice. It is so large that water will not be held in it far above the level at which it is held by hydrostatic pressure. Water moving in it may form crosscurrents and eddies.

supercooling Undercooling. The process of lowering the temperature of a phase or assemblage below the point or range at which a phase change should occur at equilibrium, i.e., making the system metastable by lowering the temperature.

supercritical At a temperature higher than the critical temperature.

superficial Surficial, q.v.

supergene Applied to ores or ore minerals that have been formed by generally descending water. Ores or minerals formed by downward enrichment. Cf. HYPOGENE

superheating 1. A process of adding more heat than is necessary to complete a given phase change (e.g., superheated steam). 2. In magmas, the accumulation of more heat than is necessary to cause essentially complete melting; in such cases the increase in temperature of the liquid above the liquidus temperature for any major mineral components is called the superheat.

superimposed drainage A natural drainage system that has been established on underlying rocks independently of their structure.

superimposed stream A stream that was established on a new surface and that maintained its course despite different preexisting lithologies and structures encountered as it eroded downward into the underlying rocks.

superindividual *Struct. petrol:* A fabric element composed of an aggregate of grains, commonly mineral grains produced by granulation of a single large crystal so that the smaller grains approximate the original orientation of the larger one.

superlattice A lattice some or all of whose translations are multiples of the translations of a particular lattice.

superposed stream Shortened version of superimposed stream.

superposition The order in which rocks are placed above one another.

supersaturated solution A solution which contains more of the solute than is normally present when equilibrium is established between the saturated solution and undissolved solute, in other words, more than could be dissolved by prolonged stirring.

superstructure The upper structural layer in an orogenic belt, subjected to relatively shallow or near-surface deformational processes.

supragelisol Material above pergelisol.

supralittoral zone Shore zone immediately above high tide level, commonly the zone kept more or less moist by waves and spray.

supratenuous fold *1.* A fold in which the beds thicken toward the syncline because the basin subsided during sedimentation. *2.* A fold which shows a thinning of the formations upward above the crest of the fold.

surf The wave activity in the area between the shoreline and the outermost limit of breakers.

surface *1.* The exterior part or

outside of a body. *2.* The top of the ground; the soil, clay, etc., on the top of strata.

surface anomalies Irregularities at the earth's surface, in the weathering zone, or in near-surface beds which interfere with geophysical measurements.

surface corrections Corrections of geophysical measurements for surface anomalies and ground elevations.

surface density Density of the surface material within the range of the elevation differences of the gravitational survey.

surface deposits Ore bodies that are exposed and can be mined from the surface.

surface geology The geology of the superficial deposits and of the surface of the fundamental rocks. *Cf.* AREAL GEOLOGY

surface interference Interference of geophysical measurements caused by surface anomalies or disturbances.

surface of no strain When a beam is bent the outer convex side is under tension, whereas the inner concave side is under compression. Somewhere near the middle is a surface that is neither lengthened nor shortened, hence is a surface of no strain.

surface of rupture The surface of rock from which the material of a landslide or slump was removed.

surface runoff That part of the precipitation that passes over the surface of the soil to the nearest surface stream without first passing beneath the surface.

surface tension That force that tends to reduce the total surface energy of a given phase; in general it results in a decrease in surface area.

surface thrust Erosion thrust.

surface velocity Initial velocity of the seismic wave in the earth's surface layer.

surface waves Waves which propagate along the earth's surface. Among these are the Love, Rayleigh, hydrodynamic, and coupled waves.

surficial; superficial Characteristic of, pertaining to, formed on, situated at, or occurring on the earth's surface; especially, consisting of unconsolidated residual, alluvial, or glacial deposits lying on the bedrock.

surf zone The area between the outermost breaker and the limit of wave uprush.

surge The name applied to wave motion with a period intermediate between that of the ordinary wind wave and that of the tide, say from 1/2 to 60 minutes. It is of low height; usually less than 10 centimeters.

surge channel Transverse channel cutting the outer edge of an organic reef in which the water level rises and falls as the result of wave and tidal action.

surge zone The region between the breaker zone and the 15–18 meters depth contour, where the effect of sea waves and swell produces oscillatory surges causing sediment transport and abrasive erosion.

survey To determine and delineate the form, extent, position, etc., of a tract of land, coast, harbor, or the like, by taking linear and angular measurements, and by applying the principles of geometry and trigonometry.

survey, geologic A survey or investigation of the character and structure of the earth, of the physical changes which the earth's crust has undergone or is undergoing, and of the causes producing those changes.

survey, geological A general term used to designate an organization making geologic surveys and investigations.

surveying That branch of applied mathematics which teaches the art of determining the area of any portion of the earth's surface, the lengths and direction of bounding lines, the contour of the surface, etc., and accurately delineating the whole on paper.

susceptibility, magnetic *See* MAGNETIC SUSCEPTIBILITY

suspended load In the process by which running water transports detritus, two factors are distinguished. The smaller particles are lifted far from the bottom, are sustained for long periods, and are distributed through the whole body of the current; they constitute the suspended load.

suspended sediment Sediment which remains in suspension in water for a considerable period of time without contact with the bottom.

suspended solids (SS) Small particles of solid pollutants in sewage that contribute to turbidity and that resist separation by conventional means. The examination of suspended solids and the BOD test constitute the two main determinations for water quality performed at waste water treatment facilities.

suspended turbidity current Turbidity current that overrides denser underlying water and is not in contact with the sea bottom.

suspended water Vadose water. Subsurface water occupying the zone of aeration.

suspension current Turbidity current, *q.v.*

suspension load That part of sediment moved in suspension by a stream rather than that transported by traction on the bottom.

suture A line or mark of splitting open; a groove marking of a natural division or union. The lengthwise groove of a plum or similar fruit.

suture joint Same as stylolite, q.v.

swale 1. A slight, marshy depression in generally level land. 2. A depression in glacial ground moraine.

swallow hole Swallet; swallet hole. Syn: SINK; SINKHOLE, q.v.

swamp A low, spongy land, generally saturated with moisture and unfit either for agricultural or pastoral purposes.

swamp theory The theory which holds that coal beds formed in the place where the plants grew. Syn: IN SITU THEORY

swash The rush of water up onto the beach following the breaking of a wave. Uprush; run up.

swash mark A thin wavy line of fine sand, mica flakes, bits of seaweed and other debris produced by the swash. Wave mark.

S-wave A transverse body wave which travels through the interior of an elastic medium. Originally applied in earthquake seismology, where it was the second (S) type of wave to arrive at a recording station. Propagates by local changes in shape, i.e., by shearing, in the medium. Syn: DISTORTIONAL WAVE; EQUIVOLUMNAR WAVE; SECONDARY WAVE; SHEAR WAVE; TRANSVERSE WAVE

sweet gas or oil Natural gas containing little or none of the sulfur compounds which, when present, cause it to be "sour." See SOUR GAS

swell 1. A low dome or quaquaversal anticline of considerable areal extent. 2. A large domed area within the nuclear part of the continent. 3. An essentially equidimensional uplift without connotation of size or origin.

swell-and-swale topography Topography of ground moraine having low relief and gentle slopes.

syenite A plutonic igneous rock consisting principally of alkalic feldspar usually with one or more mafic minerals such as hornblende or biotite. The feldspar may be orthoclase, microcline, or perthite.

syenodiorite Monzonite.

syenogabbro An intrusive rock which contains orthoclase in addition to the normal gabbroic minerals. An orthoclase gabbro. The plutonic equivalent of trachybasalt.

sylvanite A mineral, $(Au,Ag)Te_2$. Monoclinic. An ore of gold.

sylvinite Name given to mixtures of halite and sylvite mined as potassium ore.

sylvite A mineral, KCl. Isometric. The principal ore of potassium.

symbiosis The growth together of different species in a manner beneficial to the participants.

symbol 1: Crystallog: (1) Any letter or sign used to designate a group of similar faces. (2) Miller indices. 2. Geological maps are generally accompanied by special symbols to show the outcrops of formations and the attitude of bedding, foliation, faults, joints, etc.

symmetrical bedding Bedding characterized by lithologic types or facies that follow each other in a "retracing" arrangement illustrated by the sequence 1-2-3 2-1-2-3-2-1, etc.

symmetrical fold A fold the axial plane of which is essentially vertical, i.e., the two limbs dip at similar angles.

symmetry, plane of mirror; mirror plane of symmetry A plane through a crystal such that the pattern of atomic structure is precisely reproduced on each side of the plane as if it were the mirror image of that on the other side of the plane. In consequence, there is an equal tendency for pairs of crystal faces to form so that one of the pair is the mirror image of the other as reflected in this plane; likewise, vectorial properties, such as conductivity, optical constants, and elastic properties, are symmetrical with respect to such a mirror plane.

symmetry axis In a crystal, an imaginary line about which the crystal is rotated during which there may be 2, 3, 4, or 6 repetitions of its appearance (edges, corners, faces).

symmetry plane *See* PLANE OF SYMMETRY

symmetry 2 *Crystallog:* The repeat pattern of similar crystal faces that indicates the ordered internal arrangement of the substance.

synaeresis Separation of fluid from a gel.

synantetic A term applied to those primary minerals in igneous rocks which are formed by the reaction of two other minerals, as in kelyphite rims, reaction rims, etc.

synchronal Occurring at the same time.

synclinal *Geol:* Characteristic of, pertaining to, occurring or situated in or forming a syncline. *Ant:* ANTICLINAL

synclinal axis *Geol:* The central line of a syncline, toward which the beds dip from both sides.

synclinal mountain *See* ANTICLINAL MOUNTAIN

syncline A fold in rocks in which the strata dip inward from both sides toward the axis. *Ant:* ANTICLINE

synclinorium A broad regional syncline on which are superimposed minor folds.

synecology The study of the ecology of communities, associations or groups of organisms living together, as opposed to autecology.

syneresis A spontaneous throwing off of water by a gel during aging. In a hardened or set gel

the shrinkage resulting from loss of water causes cracking. *See* SEPTARIAN

synform A synclinal-type structure, the stratigraphic sequence of which is unknown. *Cf.* SYNCLINE

syngenesis The process by which mineral deposits were formed simultaneously and in a similar manner to the rock enclosing them. *See* EPIGENESIS

syngenetic A term now generally applied to mineral or ore deposits formed contemporaneously with the enclosing rocks, as contrasted with epigenetic deposits, which are of later origin than the enclosing rocks.

synonym *Tax:* Different names for one and the same thing.

synonymy (-ies) The list of all prior references to a genus or species including all names which have been used to refer to that particular form.

synoptic *Meteor:* Atmospheric conditions existing at a given time over an extended region, e.g., a synoptic weather map, which is drawn from observations taken simultaneously at a network of stations over a large area, thus giving a general view of weather conditions.

synorogenic; synkinematic; syntectonic; synchronous Adjectives to describe some process, usually the emplacement of plutons or the recrystallization of metamorphic rocks, that is contemporaneous with orogeny.

synplutonic dike Dike more or less contemporaneous with the plutonic rock in which it occurs.

synsedimentary Accompanying deposition.

syntaxial Applied to overgrowths retaining crystallographic continuity.

syntaxis Convergence of mountain ranges at a common center; *Ant:* VIRGATION

syntaxy Similar crystallographic

orientation in a mineral grain and its overgrowth.

syntectic A term applied to magmas produced by syntexis, and also used substantively to connote the magmas themselves.

syntectite Rock produced by melting of older rocks.

syntectonic Principal tectonic; synkinematic; synorogenic, *q.v.*

syntexis The formation of magma by melting of two or more rock types and assimilation of country rocks; anatexis of two or more rock types.

synthetic faults Subsidiary faults parallel to the master fault.

synthetic fuels Materials convertible into energy-producing fuels, including solid organic wastes, oil shale, and tar sand.

synthetic group Rock stratigraphic unit consisting of two or more formations which are associated because of similarities or close relationships between their fossils or lithologic characters.

syntype *1.* Any specimen of the author's original material when no holotype was designated. *2.* Any of a series of specimens described as "cotypes" of equal rank. *3.* One of two or more specimens to which a single name was equally attached in original publication. Term to be used in review and revision only.

system *1.* Designates rocks formed during a fundamental chronologic unit, a period, e.g., Devonian System. *2. Crystallog:* The division of first rank, in the classification of crystals according to form. The six systems ordinarily recognized are the isometric, tetragonal, hexagonal, orthorhombic (or rhombic), monoclinic, and triclinic; some divide the hexagonal system into hexagonal and trigonal.

systematics Study of similarities and differences in organisms and their relations; includes taxonomy and classification.

T

tabetisol Unfrozen ground above, within, or below pergelisol.

tableland *1.* Land elevated much above the level of the sea and generally offering no considerable irregularities of surface. *2.* A flat or undulating elevated area; a plateau or mesa.

tablemount A seamount (roughly circular or elliptical in plan) generally deeper than 100 fathoms, the top of which has a comparatively smooth platform. *Syn:* SEAMOUNT; GUYOT

tachygenesis The extreme crowding and eventually the loss of those primitive phyletic stages which are represented early in the life of the individual. *Syn:* ACCELERATION

tachylyte A volcanic glass of basaltic composition. *See* SIDEROMELANE

tachymeter A surveying instrument designed for use in rapid determination of distance, direction, and difference of elevation from a single observation, using a short base, which may be an integral part of the instrument. Range finders, *q.v.,* with self-contained bases belong to this class.

Taconic orogeny; Taconian orogeny An orogeny in the latter part of the Ordovician Period, named for the Taconic range of eastern New York State and well developed through most of the northern Appalachians in the United States and Canada.

taconite A ferruginous chert representing a complete replacement of greenalite rock by silica, iron ores, and ferruginous amphiboles. *See* ITABIRITE; JASPILITE

tactite A rock of complex mineralogical composition formed by contact metamorphism and metasomatism of carbonate rock. *See* CALC-FLINTA; CALC-SILICATE HORNFELS; LIMURITE; PNEUMATOLYTIC HORNFELS; SKARN

taiga [*Russ.*] The cold, swampy, forested region of the north which begins where the tundra leaves off.

tailings Those portions of washed ore that are regarded as too poor to be treated further.

talc A mineral, $Mg_3Si_4O_{10}(OH)_2$, commonly in foliated masses. Very soft (H=1), has a greasy or soapy feel, and is easily cut. Impure massive material is called steatite or soapstone. A common mineral of metamorphosed mafic rocks.

talcum Talc; soapstone.

talik [*Russ.*] *1.* A layer of unfrozen ground between the seasonal frozen ground (active layer) and the permafrost. *2.* An unfrozen layer within the permafrost. *Syn:* TABETISOL

talus A collection of fallen disintegrated material which has formed a slope at the foot of a steeper declivity. *Syn:* SCREE

talus-creep The slow downslope movement of a talus or scree, or of any of the material of a talus or scree.

talus slope *See* TALUS

Tamiskamian A sequence of oldest Precambrian rocks in the Lake Superior region and Canada.

tangential fault Fault with dominantly horizontal movement; contrasts with radial fault.

tangential section A section of a cylindrical organ, such as a stem, cut lengthwise and at right angles to the radius of the organ.

tangential stress *See* SHEARING STRESS

tangential wave Shear wave.

tank *1.* Tanks are natural depressions in an impervious stratum, in which rain or snow water collects and is preserved the greater portion of a year. *2.* A natural or artificial pool or water hole in a wash. Local in arid West.

tantalite A mineral, the Ta-rich member of the columbite-tantalite series, $(Fe,Mn)(Ta,Nb)_2O_6$. Principal ore of tantalum.

taphonomy The study of the death and decay of organisms, including the processes of preservation as fossils.

taphrogenesis Broad vertical movements with high-angle faulting.

taphrogeny Type of orogeny that forms rift valleys by tension.

taphrogeosyncline A sediment-filled, deeply depressed fault block bounded by one or more high-angle faults.

tapiolite A mineral, the tetragonal dimorph of tantalite, $(Fe,-Mn)(Ta,Nb)_2O_6$. An ore of tantalum.

tarn [<*Scand.*] A small mountain lake or pool, especially one that occupies an ice-gouged basin on the floor of a cirque.

tarnish *Mineral:* The thin film, of color different from that of a fracture, that forms on the exposed surface of a mineral, especially a metallic mineral, such as columbite.

tar sand Any sand body (including calcarenite) that is large enough to hold a commercial reserve of tar that represents, usually, the residue of a hydrocarbon deposit from which lighter volatiles have escaped; *e.g.*, Athabasca tar sands.

Tartarian The uppermost of five epochs of the Permian Period. Also the series of rocks deposited during that epoch.

tasmanite [coal] An impure coal intermediate between cannel coal and oil shale.

tautonomy Relations that exist if the same word is used for both the generic and specific name in the name of a species.

tautonym A name in which the specific name merely repeats the generic, as, Linaria Linaria Karst.

tautozonal facies Facies belonging to the same crystal zone.

taxion Taxon.

taxon A named group of organisms, i.e., a species, genus, family, etc.

taxonomic categories The principal grades of taxonomic units in descending order are kingdom, phylum, class, order, family, genus, and species. Intermediate grades are identified by these names modified by the prefixes super, sub, and infra. A few other grades, such as cohort, tribe, clan, and section also may be recognized.

taxonomy *1.* The science of the orderly arrangement of things. *2.* The systematic classification of plants and animals. *3.* The science of the classification and arrangement, according to relationships, of living and past organisms.

T-chert Tectonically controlled chert occurring in irregular masses related to fractures and ore bodies.

tchornozem; tscherosem *See* CHERNOZEM

T.D. Abbreviation for total depth. The greatest depth reached by a drill hole.

T.D. curve *See* TIME-DISTANCE CURVE

T △ T process A method of obtaining or measuring the vertical velocity of sound through subsurface sediments by use of the reflection seismograph technique.

t direction *Struct. petrol:* Direction of movement in gliding plane.

T-dolostone Tectonically controlled dolostone occurring in irregular masses related to fracture systems.

tear fault Strike-slip fault that trends transverse to the strike of the deformed rocks. *Syn:* TRANS-CURRENT; TRANSVERSE FAULT

tectofacies *1.* A group of strata of different tectonic aspect from laterally equivalent strata. *2.* Laterally varying tectonic aspects of a stratigraphic unit.

tectogene A deeply down-buckled belt of sediments within a eugeosyncline.

tectogenesis The processes by which rocks are deformed; more specifically, the formation of folds, faults, joints, and cleavage.

tectonic Of, pertaining to, or designating the rock structure and external forms resulting from the deformation of the earth's crust. As applied to earthquakes, it is used to describe shocks not due to volcanic action or to collapse of caverns or landslides.

tectonic axes *See* AXES, TECTONIC

tectonic basin A surface basin that owes its origin directly to deformation of the earth's crust, whether the result of warping or of fracture, or both.

tectonic cycle *1.* The cycle that relates the larger structural features of the earth's crust to gross crustal movements and to the kinds of rocks that form in the various stages of development of these features. *2.* Orogenic cycle, a geosynclinal cycle of three stages: peneplanation (widespread deposition on a relatively stable flat surface), geosynclinal (deposition

during subsidence), and orogenic (postgeosynclinal uplift commonly marked by faulting, after folding and magmatic intrusion in the geosyncline).

tectonic breccia An aggregation of angular coarse rocks formed as the result of tectonic movement.

tectonic conglomerate A coarse clastic rock produced by deformation of brittle, closely jointed rocks.

tectonic enclave Body of rock completely isolated by plastic structural disturbance from similar material with which it was once continuous.

tectonic fabric Particle or crystal arrangement determined by movement within a rock.

tectonic facies Rocks owing their present characters to tectonic movements, e.g., mylonites, some phyllites, etc.

tectonic framework The structural elements of a region including the rising, stable, and subsiding areas.

tectonic land Land raised by tectonic movements as contrasted with land formed by volcanism or sedimentary deposition.

tectonic map One on which are shown areas or lines of major structural features produced by uplift, downwarp, or faulting, together with the major lineation within such features.

tectonic rotation Movement in a tectonite that involves internal rotation in the direction of transport.

tectonics Study of the broader structural features of the earth and their causes.

tectonic transport Movement within a rock by flowage or slippage.

tectonism *1.* Crustal instability. *2.* The structural behavior of an element of the earth's crust during, or between, major cycles of sedimentation.

tectonite Rock whose minute structure has been produced by internal movement of its parts without these parts losing spacial continuity or the rock its individuality.

tectonosphere 1. Zone within the earth where crustal movements originate. 2. Crust of the earth in which tectonic adjustments occur. 3. Earth shell consisting of sial, salsima, and sima layers.

tectosilicates A structural class of silicate minerals in which the Si-O tetrahedra share all oxygens with adjacent tetrahedra to build up a three-dimensional network. Examples are quartz and feldspar.

tectotope A stratum or succession of strata with characteristics indicating accumulation in a common tectonic environment.

teilchron Geologic time unit corresponding to a teilzone.

teilzone 1. The local stratigraphic range of a given species or genus of plant or animal. 2. The subdivision of a zone.

tektite A small rounded, pitted, jet-black to olive-greenish or yellowish body of silicate glass of nonvolcanic origin, found usually in groups in several widely separated areas of the earth's surface. Most tektites have uniformly high silica and very low water content. Their composition is unlike that of obsidian and more like that of shale. They have various shapes, such as teardrop, dumbbell canoe, etc. Tektites are believed to be either of extraterrestrial origin or alternately the product of large hypervelocity meteorite impacts on terrestrial rocks.

telemagmatic Applied to deposits far from the intrusive center.

telescoped ore deposits Ore deposits that feature overlapping mineral zones within short distances, horizontally and/or vertically.

telescoping 1. Differential acceleration or retardation. 2. Overlap.

telethermal The ore deposits produced at or near the surface from ascending hydrothermal solutions and representing the terminal phase of its activity.

tellurate A salt of telluric acid; a compound containing the radical $(TeO_4)^{-2}$.

telluric Pertaining to the earth, particularly the depths of the earth.

telluric currents Natural electric currents that flow on or near the earth's surface in large sheets. *Syn:* EARTH CURRENTS

telluride A compound of tellurium with one other more positive element or radical.

tellurite A salt of tellurous acid; a compound containing the radical $(TeO_3)^{-2}$.

temblor A synonym of earthquake.

temperature coefficient A numerical value indicating the relation between the change in temperature and a simultaneous change in some other property, as, solubility, volume, electrical resistance, etc.

temperature correction Observations made with geophysical instruments, such as the magnetometer, pendulum, torsion balance, etc., are not independent of temperature as temperature changes result in expansion or contraction of parts of the delicate measuring instruments. The temperature correction is applied to the observed values to reduce them to a standard temperature.

temperature gradient Rate of change of temperature with distance in a specified direction. *Cf.* LAPSE RATE

temperature well logging A method of determining the temperature along the bore of a drill hole. A temperature well log is a graph showing the temperature as a function of depth in the drill hole. *Syn:* THERMAL LOGGING

template; templet   A gauge, pattern, or mold; commonly a thin plate or board or a light frame used as a guide to the form of work to be executed. In gravity and magnetic interpretation a chart with holes in a certain array to select values from a map to be used in calculation of derivatives or other functions of the field.

temporary base level   The lowest level to which the stream can bring its valley under the conditions which exist when the flat is developed.

temporary hardness   Hardness of water resulting from the presence of dissolved calcium bicarbonate; it can be removed by boiling.

tennantite   A mineral, the As-rich member of the series, tetrahedrite-tennantite, $(Cu,Ag,Fe)_{12}(Sb,As)_4$-S_{13}. Isometric. An ore of copper and silver. *Syn:* GRAY COPPER ORE; FAHLORE

tenor   The percentage or average metallic content of an ore, matte, or impure metal.

tensile strength   The ability of a material to resist a stress tending to stretch it or to pull it apart. *Cf.* YOUNG'S MODULUS; HOOKE'S LAW

tensile stress   A normal stress that tends to pull apart the material on the opposite sides of a real or imaginary plane.

tension   A system of forces tending to draw asunder the parts of a body, especially of a line, cord, or sheet, combined with an equal and opposite system of resisting forces of cohesion holding the parts of the body together; stress caused by pulling. Opposed to compression, and distinguished from torsion.

tension fault   A fault produced by tension; sometimes used incorrectly as synonymous with gravity fault or normal fault.

tension fracture   A fracture that is the result of stresses that tend to pull material apart.

tension joint   A joint that is a tension fracture, *q.v.*

tepee butte   *1.* A conical erosion hill, so named from its resemblance to the Indian wigwam or tepee. *2.* A hill formed by a columnar bioherm found in the Cretaceous Pierre shale of Colorado, containing enormous numbers of the small pelecypod, *Lucina.*

tepetate   *See* CALICHE

tephra   A collective term for all clastic volcanic materials which during an eruption are ejected from a crater or from some other type of vent and transported through the air; includes volcanic dust, ash, cinders, lapilli, scoria, pumice, bombs, and blocks. *Syn:* VOLCANIC EJECTA

tephrochronology   A chronology based on the dating of volcanic ash layers.

tephroite   A mineral, Mn_2SiO_4, a member of the olivine group. Orthorhombic.

terminal moraine   A moraine formed across the course of a glacier at its farthest advance, at or near a relatively stationary edge, or at places marking the termination of important glacial advances.

terminus   The outer or distal margin of the ablation area of a glacier.

ternary, diagram   A triangle diagram which is a graphic depiction of a three-component mixture.

ternary system   A system of three components, e.g., $CaO-Al_2O_3$-SiO_2.

terra   A bright upland or mountainous region on the surface of the moon, characterized by a lighter color than that of a mare, the moon, characterized by a rough texture formed by large intersecting or overlapping craters. It may represent a remnant of an ancient lunar surface, sculptured largely by impact of meteorites; it may also be attributed to igneous and volcanic activity from within the moon. *Syn:* CONTINENT

terrace Benches and terraces are relatively flat, horizontal, or gently inclined surfaces, sometimes long and narrow, which are bounded by a steeper ascending slope on one side and by a steeper descending slope on the opposite side.

terraced pools Shallow circular pools with rims arranged terracelike fashion built by calcareous secreting algae on coral reef surfaces.

terracettes Ledges of earth on steep hillsides, varying from a few centimeters to one or two meters in height and averaging one meter in width, formed as a result of the development of slippage planes in, and subsequent slumping of, the soil or mantle. *Syn:* TERRACETTE SLOPES

terra cotta The "baked earth" of the Italians. Kiln-burnt clay assuming a peculiar reddish-brown color fashioned into vases, statuettes, and other moldings.

terra-cotta clay A loose term that might include any clay used in the manufacture of terra cotta.

terrain *1.* A complex group of strata accumulated within a definite geologic epoch. *2.* Area of ground considered as to its extent and natural features in relation to its use for a particular operation. *3.* The tract or region of ground immediately under observation.

terrain analysis The process of interpreting a geographical area to determine the effect of the natural and man-made features on military operations.

terrain or **topographic correction** A correction applied to observed values obtained in geophysical surveys in order to remove the effect of variations to the observations due to the topography in the vicinity of the sites of observation.

terrain factors Consist of land forms, drainage features, ground, vegetation, and cultural features on man-made changes in the earth's surface.

terrane *1.* A formation or group of formations. *2.* The area or surface over which a particular rock or group of rocks is prevalent. *3.* An area or region considered in relation to its fitness or suitability for some specific purpose.

terra rossa Residual red clay mantling limestone bedrock.

terrestrial Consisting of or pertaining to the land.

terrestrial magnetism *1.* The magnetic field of the earth as a whole. *2.* The science which treats the laws of magnetism and the application of these laws in surveying, navigation, etc.

terrigenous Produced from or of the earth. *Geol:* Deposited in or on the earth's crust.

terrometer The name of one of the devices designed for the detection of buried metallic objects at shallow depths.

Tertiary The older of the two geologic periods comprising the Cenozoic Era; also the system of strata deposited during that period.

test *1.* Hard covering or supporting structure of some invertebrate animals; may be enclosed within an outer layer of living tissue; a shell. *2.* An oil well, particularly a wildcat.

test pit A shallow shaft or excavation made to determine the existence, extent, or grade of a mineral deposit or to determine the fitness of an area for engineering works such as buildings or bridges.

tetartohedral Having or requiring one-fourth the number of planes or faces required by the symmetry of the holohedral class of the same system.

Tethys Elongated east-west seaway that separated Eurasia from Gondwanaland, except perhaps for a western junction, from at least the Early Paleozoic to Late Cretaceous.

tetracoral Coral with fourfold symmetry.

tetragonal system That system of crystals in which the forms are referred to three mutually perpendicular axes, two of which are of equal length and the third longer or shorter.

tetrahedral radius The radius of an atom which has four covalent bonds with other atoms.

tetrahedrite A mineral, the Sb-rich member of the series, tetrahedrite-tennantite, $(Cu,Ag,Fe)_{12}(Sb,As)_4S_{13}$. Isometric. An ore of copper and silver. *Syn:* GRAY COPPER ORE; FAHLORE

tetrahedron A crystal form in the isometric system, having four faces each with equal intercepts on all three axes.

tetrahedron hypothesis The hypothesis that the earth, because of shrinking, tends to assume the form of a tetrahedron.

tetrahexahedron Isometric crystal form with twenty-four faces, each parallel to one crystallographic axis and cutting the others at unequal distances.

Tetrapoda Subphylum of vertebrates; animals equipped with four limbs, mostly terrestrial.

texture *1.* Geometrical aspects of the component particles of a rock, including size, shape, and arrangement. *2.* The relative spacing of drainage lines in country which has undergone fluvial and pluvial dissection.

texture ratio *Geomorph:* Ratio of the greatest number of channels crossed by a contour line to the length of the upper basin perimeter intercept; a measure of topographic scale or texture, closely related to drainage density. *Symbol:* T

thalassic Of or pertaining to the sea.

thalassophile elements Elements whose amount in sea water is greater than, or a large fraction of, the total amount supplied to the sea by weathering and erosion. Carbon (C), bromine (Br), iodine (I), boron (B), sulfur (S), and sodium (Na).

Thallophyta Division of nonvascular plants, those without differentiated roots, stems, or leaves; includes algae and fungi.

thalweg [*Ger.*] *1. Hydraul:* The line joining the deepest points of a stream channel. *2.* By many geomorphologists the term is used as a synonym for valley profile.

thanatocoenose; thanatocoenosis A group of organisms brought together after death.

thaw depression Hollow formed by the melting of ice in perennially frozen ground.

thaw lake *1.* Lake or pond in permafrost area whose basin is formed by thawing of ground ice. *See* THERMOKARST. *2.* A pool of water on the surface of sea ice or large glaciers formed by accumulation of melt water.

theca External skeleton of a coelenterate.

thenardite A mineral, Na_2SO_4. Orthorhombic.

theodolite An instrument for measuring horizontal and vertical angles.

therm A quantity of heat equivalent to 100,000 British thermal units.

thermal Hot; warm. Applied to

springs which discharge water heated by natural agencies.

thermal analysis Determination of the temperatures at which phase changes occur during the heating of clay and other substances by measuring the heat evolved or absorbed. Differential thermal analysis.

thermal aureole Aureole.

thermal or **heat conductivity** A quantity for classifying materials according to their ability to conduct heat. The conductivity is expressed in calories per centimeter per second per degree centigrade.

thermal-detection methods Geophysical thermal-detection methods involve the location of objects by their heat radiation.

thermal diffusivity See DIFFUSIVITY, THERMAL

thermal logging See TEMPERATURE WELL LOGGING

thermal metamorphism Metamorphism in which heat is the principal agent causing reconstitution.

thermal pollution Degradation of water quality by the introduction of a heated effluent. Primarily a result of the discharge of cooling waters from industrial processes, particularly from electrical power generation.

thermal prospecting A system of geophysical prospecting based on measuring underground temperatures or temperature gradients and relating their irregularities to geological deformation.

thermal stratification The stratification of a lake produced by changes in temperature at different depths and resulting in horizontal layers of differing densities.

thermal unit A unit chosen for the comparison or calculation of quantities of heat, as, the calorie, or the British thermal unit.

thermionic emission The emission of electrons from a hot cathode, as in a vacuum tube. Syn: RICHARDSON EFFECT

thermistor A heat-sensitive device used in bolometric measurements of power in high-frequency electric circuits.

thermocline 1. A temperature gradient. 2. The horizontal plane in a thermal stratified lake located at the depth where temperature decreases most rapidly with depth. 3. A vertical, negative gradient of temperature that is characteristic of a layer of ocean water.

thermocouple A union of two conductors, as, bars or wires of dissimilar metals joined at their extremities for producing a thermoelectric current. Syn: THERMOJUNCTION; THERMOELECTRIC COUPLE

thermodiffusion A process whereby certain dissolved "molecules" (or their constituent ions) diffuse toward the chilled margins of a magma chamber; the driving force is the concentration gradient established in the liquid owing to the precipitation of material at the margins.

thermodynamic process If on comparing the state of a thermodynamic system at two different times there is a difference in any macroscopic property of the system, then a process has taken place.

thermodynamics The mathematical treatment of the relation of heat to mechanical and other forms of energy.

thermograph A self-registering thermometer.

thermohaline Deep-water circulation, largely controlled by temperature and salinity of the water.

thermokarst Settling or caving of the ground due to melting of ground ice.

thermokarst pit Steep-walled depression formed by thermokarst processes.

thermokarst topography Irregular land surface containing depressions formed by thermokarst processes; it resembles the karst topography resulting from solution of limestone.

thermoluminescence The property possessed by many minerals of emitting visible light when heated. It results from release of energy stored as electron displacements in the crystal lattice.

thermometry The measurement of temperature.

thermopile An apparatus consisting of a number of thermo-electric couples, as of antimony and bismuth or copper sulfide and German silver, combined so as to multiply the effect.

thick bands A field term that, in accordance with an arbitrary scale established for use in describing banded coal, denotes vitrain bands with a range of thickness from 5.0 to 50.0 mm.

thick-bedded Relative term applied to strata occurring in uniformly constituted beds variously defined as exceeding 6.5 cm. to 1.2 meters in thickness.

thin bands A field term that, in accordance with an arbitrary scale established for use in describing banded coal, denotes vitrain bands with a range of thickness from 0.5 to 2.0 mm. thick.

thin-bedded *1.* Applied to shale to indicate it is easy to split. *2.* Occurring in relatively thin layers or laminae.

thinolite A tufa deposit of calcium carbonate occurring on an enormous scale in northwestern Nevada; also occurs about Mono Lake, California. It forms layers of interlaced crystals of a pale yellow or light-brown color and often skeleton structure except when covered by a subsequent deposit of calcium carbonate.

thin out A stratum is said to thin out when it becomes thinner and thinner as it is traced in any direction, till it finally disappears and its place is taken by some other stratum. *Syn:* PINCH OUT; WEDGE OUT; LENSING

thin section A fragment of rock or mineral ground to paper thinness (usually 0.03 mm.), polished, and mounted between glasses as a microscopical slide.

thixotropy The property exhibited by some gels of becoming fluid when shaken. The change is reversible.

tholeiite A group of basalts primarily composed of plagioclase (approximately an. 50), pyroxene (especially augite or subcalcic augite), and iron oxide minerals as phenocrysts in a glassy groundmass or intergrowth of quartz and alkali feldspar; also, any rock in that group; little or no olivine is present.

tholeiitic magma A type of basaltic magma containing little or no olivine and yielding oversaturated late differentiates.

thorax Central part of the arthropod body consisting of several segments that generally are movable.

thorianite A mineral, ThO_2, commonly containing some uranium. Isometric.

thorite A mineral, essentially $ThSiO_4$, commonly altered and metamict. Tetragonal.

three-dimension dip *Seis. prospecting:* The true dip of a reflection or refraction horizon found by exploration and calculation. *Syn:* TRUE DIP

three-faceted stone Dreikanter, *q.v.*

three-layer structure Structure in minerals composed of repeated layered units each consisting of an aluminum octahedral layer between two silicon tetrahedral layers as in muscovite.

three-point method *1.* Geometric determination of dip and strike in any regular plane whose elevation is known at three accurately located points. *2.* Determination of geographic position inside or outside of the triangle formed by the intersection of bearing lines from three triangulation stations.

threshold Low necks of sand which frequently divide lakes into two basins and in the desert separate the bajirs one from the other.

threshold pressure Yield point. The stress at which plastic deformation begins.

threshold velocity (wind erosion) The minimum velocity at which wind will begin moving particles of sand or other soil material.

throw 1. The amount of vertical displacement occasioned by a fault. 2. More generally, used for the vertical component of the net slip.

throwing clay Clay plastic enough to be shaped on a potter's wheel.

thrust Fault occurring in place of the overturned limb of a fold.

thrust fault A reverse fault that is characterized by a low angle of inclination with reference to a horizontal plane. *See* FAULT

thrust plane The plane of a thrust or reversed fault.

thrust scarps Sinuous scarps marking the front of a low-angle thrust sheet or block.

thrust sheet The block above a thrust fault.

thunder egg Geodelike body commonly containing opal, agate, or chalcedony weathered out of welded tuff or lava.

tidal bore *See* BORE

tidal channel *See* TIDAL INLET

tidal compartment The tidal compartment of a river may be defined as that portion of the stream which intervenes between the area of unimpeded tidal action and that in which there is a complete cessation or absence of tidal action.

tidal constant Either of two parameters, which combined completely specify a simple tide, the first being the amplitude of a tide (the elevation above mean sea level) and the second its epoch or the time between the moon's meridian passage and the ensuing high tide.

tidal correction A correction applied to gravitational observations to remove the effect of earth tides on gravimetric observations.

tidal datum A plane defined by reference to a certain phase of tide.

tidal day The time of the rotation of the earth with respect to the moon, or the interval between two successive upper transits of the moon over the meridian of a place, about 24.84 solar hours (24 hours and 50 minutes) in length or 1.035 times as great as the mean solar day.

tidal delta Deltas formed in the seaward and lagoonal mouths of a tidal inlet through a barrier island or baymouth bar by tidal currents that sweep sand in and out of the inlet.

tidal flat A marshy or muddy land area which is covered and uncovered by the rise and fall of the tide.

tidal friction The frictional effect of the tides, especially in shallow waters, lengthening the tidal epoch and tending to retard the rotational speed of the earth and so increase very slowly the length of the day.

tidal inlet 1. A natural inlet maintained by tidal flow. 2. Loosely, any inlet in which the tide ebbs and flows. Also tidal outlet.

tidal marsh *See* TIDAL FLAT

tidal pool A pool of water remaining on a beach or reef after recession of the tide.

tidal prism The total amount of water that flows into the harbor or out again with movement of the tide, excluding any freshwater flow.

tidal range The difference between the level of water at high tide and low tide.

tidal theory A theory of origin of the solar system involving tidal forces set upon the sun by the near approach of another star; the J. H. Jeans and Harold Jeffreys theory.

tidal wave 1. In astronomical usage, restricted to the periodic variations of sea level produced by the gravitational attractions of the sun and the moon. 2. Commonly and incorrectly used for a large sea wave caused by a submarine earthquake or volcanic eruption, properly called a tsunami, *q.v.* 3. Sometimes used for a large sea wave caused by a hurricane wind or a severe gale, properly called a storm wave.

tide The periodic rise and fall of oceans and bodies of water connecting them, caused chiefly by the attraction of the sun and moon.

tied island An island connected to the mainland or to another island by a tombolo.

tie line A line at constant temperature connecting the compositions of any two phases that are in equilibrium at the temperature of the tie line.

tierra blanca *See* CALICHE

tie-time The reflection times obtained by shooting in opposite directions over an interval resulting in a common reflection point. When this time is corrected for uphole time, or corrected to datum, the resulting comparison of corrected reflection times is the tie-time.

tiff A sparry mineral; calcite in southwest Missouri; barite in southeast Missouri.

tiger's-eye A chatoyant stone, usually yellow-brown, much used for ornament. It is silicified crocidolite.

tight fold Closed fold.

till 1. Nonsorted, nonstratified sediment carried or deposited by a glacier. 2. *V:* To plow, harrow, seed, or otherwise work soil.

tillite A sedimentary rock composed of cemented till.

tilt 1. *Photogram:* The distortion in a photograph caused by tilting of the photographic plane due to variable winds and air currents. 2. The rotation of the photograph about the axis parallel to the line of flight.

tilt blocks Blocks that have received a marked tilt in regions of block faulting.

tiltmeter 1. *Earthquake seismol:* A device for observing surface disturbances on a bowl of mercury, employed in an attempt to predict earthquakes. 2. An instrument used to measure displacement of the ground surface from the horizontal. *Volcanol:* Used to indicate the degree and intensity of tumescence or doming-up of a volcano by magmatic pressure.

tilt slide Gravity slide of rocks or sediments down the slope of an uptilted surface.

time Duration; a period in which something occurs, or endures. *Geol:* Any division of geologic chronology.

time at shot point *Seis. explor:* The time required for the seismic impulse to travel from the charge in the shothole to the surface of the earth. *Syn:* UPHOLE TIME

time break An indication on a seismic record showing the instant of detonation of a shot or charge. *Syn:* SHOT MOMENT; SHOT INSTANT. *Cf.* TIME SIGNAL

time constant The time taken for a current in a circuit having a steady e.m.f. to reach a definite fraction of its final value after the circuit is closed.

time-depth chart A graphical expression of the functional relation between the velocity function and the times observed in the seismic method of geophysical exploration.

time-depth curve See TIME-DEPTH CHART

time-distance curve In refraction seismic computations, a graph, usually with arrival times of seismic events plotted as ordinates and distances along the surface of the earth plotted as abscissas. In earthquake studies, the times of arrival of seismic waves at recording stations may be known but the time of initiation of the waves may be unknown.

time-distance graph In refraction seismic computations, a plot of the arrival times of refracted events against the shot point to detector distance. The reciprocal slopes of the segments plotted are the refraction velocities for the refracting bed. *Syn:* ODO-GRAPH

time domain Measurements as a function of time or operations in which time is the variable, in contrast to the frequency domain.

time gradient In the reflection seismic methods applied to dipping reflectors, the travel-time curves may not be straight lines, i.e., the apparent velocity observed varies with the spread from shot point to detectors. The time gradient is the reciprocal of the apparent velocity. *Seis. prospecting:* The rate of change of travel time with depth.

time integral In regard to any variable f which is a function of the time, the definite integral of the product of the variable by the element of time between specified limits, viz.: f dt. *Cf.* IMPULSE

time lag In refraction seismic interpretation, where arrival times are plotted against shot-detector distances, if some of the paths from shot point to detector include a low-speed bed, the corresponding arrival times will be abnormally long, and the departure from normal travel time is called a time lag. *Seis. prospecting:* Time delays in arrivals due to phase shifts in filtering, to shothole fatigue, etc.

time leads In a method of interpretation of refraction seismic records where the arrival times are plotted against shot-detector distances, if some of the paths from shot point to detector include a high-speed segment, the corresponding travel times will not fall on a smooth curve. The departure in this case from the curve is called a time lead, and it is proportional to the horizontal extent of the high-speed segment.

time line Line indicating equal time in a geologic cross section or correlation diagram.

time plane Stratigraphic horizon identifying an instant in geologic time.

time-rock unit Time-stratigraphic unit.

time section A representation of reflection seismology records in which the vertical axis is time and the horizontal axis is surface distance.

time-stratigraphic Term applied to rock units with boundaries based on geologic time, i.e., with synchronous boundaries.

time tie In seismograph continuous profiling, a coincident travel path for seismic energy initiated at opposite ends of the path. The use of such coincident travel paths on adjacent reflection layouts facilitates correlation from one layout to the next as the shot point or recording position is changed.

time transgressive formation Formation whose sediment accumulated in an environment that shifted geographically with time;

consequently its age varies from place to place.

timing lines Marks or lines placed on seismic records at precisely determined intervals of time (usually at intervals of 0.01 or 0.005 seconds) for the purpose of measuring the time of events recorded. The timing mechanism commonly includes an accurate tuning fork for the determination of small time intervals.

tinguaite A dike rock composed of alkalic feldspars, nepheline, and alkalic pyroxene and amphibole. The rock is commonly porphyritic and the mafic constituents have a characteristic crisscross orientation in the groundmass. A textural variety of phonolite.

tinstone Cassiterite.

tip *1.* The pile of snow formed by an avalanche which has come to rest. *2. Photogram:* The rotation of a photograph about the axis perpendicular to the line of flight. *Cf.* TILT, 2.

titaniferous Carrying titanium, as titaniferous iron ore. *See* ILMENITE

titanite *Syn:* SPHENE

tjäle [*Sw.*] *See* FROZEN GROUND

toe The most distal part of the landslide; the downslope edge of a landslide or slump.

toluene A liquid hydrocarbon of the aromatic series, formula C_7H_8.

tombolo [*It.*] A bar connecting an island with the mainland or with another island.

tonalite Quartz diorite.

tongue *1.* A long narrow strip of land, projecting into a body of water. *2.* Part of a formation that is known to wedge out laterally, between sediments of a different lithologic constitution, and in the other direction, thickens and becomes part of a larger body of like sediments.

top The contact between the uppermost bedding-plane surface of a stratum and separating two geologic formations used in correlation, especially for the purpose of compiling structure maps, e.g., calling "tops" in well logs.

topaz A mineral, $Al_2SiO_4(F,OH)_2$. Orthorhombic. Used as a gem.

topaz quartz The yellow variety of quartz, citrine, used as a gem.

topocline A cline whose members vary regularly and progressively from place to place.

topographic adjustment A tributary (stream) is in topographic adjustment when its gradient is harmonious with that of its main.

topographic adolescence A stage in the stream erosional cycle when lakes have mostly disappeared and river drainage is well established, stream channels being comparatively narrow and well marked and falls occurring characteristically.

topographic correction *See* TERRAIN CORRECTION

topographic high Frequently used in the oil fields to indicate the higher elevations, regardless of age; opposed to topographic low which indicates a lower elevation. *Cf.* GEOLOGIC HIGH

topographic infancy A stage in the stream erosional cycle characterized by a smooth nearly level surface of deposit, lakes abounding in slight depressions, shallow streams, and drainage systems not well established.

topographic low *See* TOPOGRAPHIC HIGH

topographic map Map showing the topographic features of a land surface generally by means of contour lines.

topographic maturity A stage in the stream erosional cycle of maximum diversity of form when valleys have greatly increased and the river channels are widely opened.

topographic old age A stage in the stream erosional cycle in which there is a featureless surface, differing from the earliest stage (topographic infancy) in having a system of drainage streams, separated by faintly swelling hills.

topographic texture The disposition, grouping, or manner of assembly of the topographic units in a stream-dissected district.

topographic unconformity Adjacent topographic differences. Preferable term: topographic discontinuity.

topographic youth *See* YOUNG VALLEY

topography [*Gr.* topos, place; graphein, to write] The physical features of a district or region, such as are represented on maps, taken collectively; especially, the relief and contour of the land.

topology Topographical study of a particular place; specifically, the history of a region as indicated by its topography.

topotype A specimen from the original locality from which a species was described.

top-set beds The material laid down in horizontal layers on top of a delta. *See* FORE-SET BEDS; BOTTOM-SET BEDS

topsoil The fertile, dark-colored surface soil, or A horizon.

tor Roche moutonnée.

torbernite A mineral, $Cu(UO_2)_2(PO_4)_2.8H_2O$. Tetragonal, in green tabular crystals.

toreva-block slide A landslide consisting essentially of a single large mass of unjostled material which, during descent, has undergone a backward rotation toward the parent cliff about a horizontal axis which roughly parallels it.

torose load casts Elongate ridges on undersurfaces of sandstone layers, which pinch and swell along their trends and may ter-

minate in bulbs, teardrops, or spiral forms.

torque Product of a force and the perpendicular distance between its line of application and the axis of rotation. *See* COUPLE

torrent A stream of water flowing with great velocity or turbulence, as during a freshet or down a steep incline; cascade; freshet; hence, any similar stream, as of lava.

torrential cross-bedding Fine, horizontally laminated strata alternating with uniformly cross-bedded strata composed of coarser materials. Believed to originate under desert conditions of concentrated rainfall, abundant wind action, and playa lake deposition.

torrent tract Mountain tract.

torsion A body is under torsion when subjected to force couples acting in parallel planes about the same axis of rotation but in opposite senses.

torsion balance An instrument for measuring force fields in which the field being measured is opposed by a known force. The torsion balance measures small forces, such as gravitational or electrical, by determining the amount of torsion or twisting they cause in a slender wire or filament.

torsion coefficient A measure of the resistance offered by a material to a torsional stress.

torsion fault Wrench fault, *q.v.*

torsion head *1.* That part of a torsion balance from which the filament or wire is suspended. *2.* A rotary cap, often graduated in degrees, atop the vertical tube supporting a torsion suspension.

torsion period The natural period of oscillation of the suspended system in a torsion balance.

torsion wire The filament or wire supporting the beam in a torsion balance or gravity meter.

total displacement Slip, *q.v.*

total field The vector sum or combination of all the components of the field under consideration, such as the magnetic or gravitational fields.

total intensity *See* TOTAL FIELD

total porosity Porosity, *q.v.*

total reflection Internal reflection in which the angle of incidence exceeds a value known as the critical angle, whose sine is the relative refractive index from the more to the less refractive medium, so called because all the energy is reflected and none is transmitted in the steady state.

total time correction *Seis. prospecting:* The sum of all corrections applied to a reflection travel time to express the time in reference to a selected datum plane. The main corrections are those for the low-velocity layer and the so-called elevation correction to datum.

tough Having the quality of flexibility without brittleness; yielding to force without breaking.

tourmaline Schorl. A mineral, a complex borosilicate of Na, Li, Mg, Fe, and Al, occurring commonly in granitic pegmatites. Hexagonal rhombohedral. Used as a gem.

tower A peak rising with precipitous slopes from an elevated tableland; a towerlike formation. Local in Northwest.

township The unit of survey of the public lands of the United States and of Canada. Normally a quadrangle approximately 6 miles on a side with boundaries conforming to meridians and parallels. It is further subdivided into 36 sections, each approximately one mile square.

T-phase *Earthquake seismol:* A phase designation applied to a short-period (1 sec. or less) wave which travels through the ocean with the speed of sound in water,

and is occasionally identified on the records of earthquakes in which a large part of the path from epicenter to station is across the deep ocean.

T plane Plane of movement in crystal gliding.

trace *1.* The record made by a recording device on paper or film, as, as, one of the traces of a seismograph record. *2.* A line on one plane representing the intersection of another plane with the first one, e.g., a fault trace. *3.* A very small quantity of a constituent, especially when not quantitatively determined, owing to its minuteness.

trace elements Elements present in minor amount in the earth's crust. All elements except the eight abundant rock-forming elements, Oxygen (O), silicon (Si), aluminum (Al), iron (Fe), calcium (Ca), sodium (Na), potassium (K), and magnesium (Mg). *Syn:* MINOR ELEMENTS; ACCESSORY ELEMENTS

trace slip Component of net slip parallel to the trace of an index plane (vein, bedding, etc.) on plane of the fault.

trace-slip fault A fault on which the net slip is parallel to the trace of a bed (or some other index plane) on the fault.

trachyandesite An extrusive rock containing sodic plagioclase and a considerable amount of alkali feldspar, with one or more mafic constituents (biotite, amphibole, or pyroxene). Its composition is intermediate between andesite and trachyte.

trachybasalt An extrusive rock intermediate between trachyte and basalt and consisting primarily of calcic plagioclase, sanidine, augite, and olivine. Analcite or leucite may be minor constituents.

trachyte An extrusive rock composed essentially of alkali feldspar and minor biotite, hornblende, or pyroxene. Small amounts of sodic plagioclase may be present. The extrusive equivalent of syenite.

trachytic A textural term applied to the groundmasses of volcanic rocks in which neighboring microlites of feldspar are arranged in parallel or subparallel fashion, bending around phenocrysts, and corresponding to the flow lines of the lava from which they were formed. The texture is common in trachytes.

trachytoid A textural term applied to phaneritic igneous rocks in which the feldspars have a parallel or subparallel disposition, as in many varieties of nepheline syenite. Corresponds to the trachytic texture of some lava flows.

traction The entire complex process of carrying material along the bottom of a stream.

tractional load Bottom load; bed load, *q.v.*

tractive current Current in standing water that transports sediment along the bottom, as in a river, contrasted with turbidity current or current not in contact with the bottom.

trafficability A rather loose term that usually means the capacity of the soil to support moving vehicles, but is also used to refer to the suitability of the terrain as a whole for cross-country movement of military forces. The more precise terms soil trafficability, *q.v.*, and cross-country movement, *q.v.*, are preferred.

trail Track; trace or sign of the passing of one or many animals, generally used for markings in sedimentary rocks made by moving invertebrates.

trail of the fault Crushed material of a bed or vein that indicates the direction of the fault movement; valuable as a guide to the miner in search of the main vein.

train *1.* Rows of large stones, some perched, some dropped and broken, which probably fell from the drifting ice. If so, the lines point out the course of the moving rafts and the run of the stream which moved them, but this test is uncertain. *2.* A series of successive repetitive events, as, a train of waves. *3. Seismol:* A series of reflections on a seismograph record.

transcurrent fault Strike-slip fault, *q.v.*

transection glacier A glacier that entirely fills a valley system overflowing the divides between the valleys.

transfer Process occurring in the frame of space-time consisting of erosion, sediment transportation, and deposition.

transfer percentage For any element, the ratio of the amount present in sea water to the amount supplied to sea water during geologic time by weathering and erosion, multiplied by 100.

transformation In phase studies, used interchangeably with inversion, *q.v.*

trans-formational breccia Breccia in more or less vertical bodies cutting across strata, produced by collapse, as above a dissolved salt bed.

transform fault A strike-slip fault characteristic of mid-oceanic ridges and along which the ridges are offset. Analysis of transform faults is based on the concept of sea-floor spreading.

transformist One who believes that all granites had a metasomatic or palingenetic origin. *Cf.* MAGMATIST

transgression Gradual expansion of a shallow sea resulting in the progressive submergence of land, as when sea level rises or land subsides. *Syn:* OVERLAP; PROGRESSIVE OVERLAP

transgressive reef A series of

bioherms or reefs that transgress time in a shoreward direction, becoming younger.

transient methods Electrical methods of geophysical exploration that depend on either the introduction into the ground of a sharp current pulse, such as may be produced by suddenly closing or opening an electrical circuit connected to grounded electrodes, or upon impressing an electric current of a certain wave form on the ground. Measurements are made either of the form of the current or more commonly of the form of the resulting potential.

transit *1.* A surveying instrument with the telescope mounted so that it can be transited; called also a transit theodolite. *2.* The passage of one heavenly body over the disk of another, or the apparent passage of a heavenly body over the meridian of a place.

transition metals Elements in the middle of the long periods of the periodic table. Usage varies, but most commonly the transition elements are taken to include those from scandium (Sc) to zinc (Zn) in the first long period, from yttrium (Y) to cadmium (Cd) in the second, and from lanthanum (La) to mercury (Hg)—excluding the 14 rare earth metals from cerium (Ce) to lutecium (Lu)—in the third.

transit theodolite *See* TRANSIT, *1*

translation Homogeneous sideward motion.

translational movement Refers to movement along faults. All straight lines on opposite sides of the fault and outside the dislocated zone that were parallel before faulting are parallel after faulting.

translation gliding That type of single-crystal slip, produced either by compression or tension, by which displacement on preferred lattice planes takes place, in a given direction or directions, without reorientation or rupture of the deformed parts.

translation plane Gliding plane.

translatory fault Rotary fault, *q.v.*

translucent Admitting the passage of light, but not capable of being seen through. Transmitting light diffusely.

transmutation The transformation of one element into another.

transparent May be seen through. Transmitting light without diffusing or scattering its rays.

transpiration The process by which water vapor escapes from a living plant and enters the atmosphere.

transport, tectonic A general term for differential movement in tectonites.

transportation *Geol:* The shifting of material from one place to another on the earth's surface by moving water, ice, or air.

transported soils Alluvial soils and dune field soils, immature.

transverse crevasse A crack in a glacier at right angles (approximately) to the direction of ice flow.

transverse dune A strongly asymmetrical dune ridge extending transverse to the direction of dominant sand-moving winds; the leeward slope stands at or near the angle of repose of sand if the dune is active, while the windward slope is comparatively gentle.

transverse fault A fault whose strike is transverse to the general structure.

transverse fold Cross fold.

transverse joint A joint that is transverse to the strike of the strata or schistosity.

transverse valley A valley having

a direction at right angles to the strike of the rocks.

transverse wave *1. Seismol:* A wave motion in which the motion of the particles, or the entity that vibrates, is perpendicular to the direction of progression of the wave train. *2. Geophys:* A body seismic wave advancing by shearing displacements. *Syn:* DISTORTIONAL WAVE; EQUIVOLUMNAR WAVE; SECONDARY WAVE; S-WAVE; SHEAR WAVE

trap *1.* Trap in the Dutch language signifies stairs, a staircase [*Sw.* trappa; *Ger.* Trappe]. In basaltic lava fields a remarkable steplike or terracelike appearance is observable. This configuration is due to the abrupt terminations of the successive flows. *2.* A body of reservoir rock completely surrounded by impervious rock; a closed reservoir.

trap-door fault A circular fault, hinged at one edge: it is an intrusion displacement structure that is common in the Little Rockies of Montana.

trapezohedron *1.* In the isometric system, the same as tetragonal trisoctahedron. *See* TRISOCTAHEDRON (b). *2.* In the tetragonal and hexagonal systems, any of several forms having principal and lateral axes of symmetry, but no planes of symmetry, and enclosed by six, eight, or twelve quadrilateral faces each having unequal intercepts on all the axes.

trap rock A term applied to dark-colored dike and flow rocks, chiefly basalt and diabase. Also spelled trapp.

traverse *1.* A line surveyed across a plot of ground. *2. V:* To make a traverse survey. *3.* Line across a thin section or other sample in which elements are counted or measured.

travertine Calcium carbonate, $CaCO_3$, of light color and usually concretionary and compact, deposited from solution in ground and surface waters. Extremely porous or cellular varieties are known as calcareous tufa, calcareous sinter, or spring deposit. Compact, banded varieties, capable of taking a polish, are called onyx marble. Travertine forms the stalactites and stalagmites of limestone caves, and the filling of some veins and hot spring conduits.

tread The flat part of a step-like natural land form; can be applied to a glacial stairway, RISER stream or marine terraces. *Cf.* **treasure finder** Terrometer, *q.v.*

trellised or **grapevine drainage** A stream pattern in which master and tributary streams are arranged nearly at right angles with respect to one another.

Tremadocian The lowest of six epochs of the Ordovician Period; also the series of strata deposited during that epoch.

tremolite A mineral of the amphibole group, $Ca_2Mg_5Si_8O_{22}(OH)_2$.

tremor An earthquake having small intensity. A quick vibratory movement. Any quivering or trembling.

trench (marine) *1.* A long but narrow depression of the deep-sea floor having relatively steep sides. *2.* A long narrow intermontane depression occupied by two or more streams (whether expanded into lakes or not) alternately draining the depression in opposite directions. *Ger.* Graben; *Fr.* ravine.

trend The direction or bearing of the outcrop of a bed, dike, sill, or the like, or of the intersection of the plane of a bed, dike, joint, fault, or other structural feature with the surface of the ground.

treppen concept The idea that on a surface reduced to old age by streams and then uplifted, the rejuvenated streams develop second-cycle valleys first near their mouths and that these young valleys are extended headward to form "stair steps."

triangular diagram A method of plotting compositions in terms of the relative amounts of three materials or components, involving a triangle wherein each apex represents a pure component. The perpendicular distances of a point from each of the three sides (in an equilateral triangle) will then represent the relative amounts of each of the three materials represented by the apexes opposite those sides.

triangular facets Truncated spur ends with broad base and apex pointing upward. Usually associated with gravity faults, but may also characterize fault-line scarps. Triangular facets may also be formed by other processes such as wave erosion of a mountain front and truncation of spurs by a valley glacier.

triangulate To divide into triangles; to survey by triangulation; having triangular markings.

triangulation The laying out and accurate measurement of a network of triangles, especially on the surface of the earth, as in surveying.

Triassic The earliest of the three periods of the Mesozoic; also the system of strata deposited during that period.

triboluminescence The property displayed by some specimens of zinc sulfide of emitting sparks when scratched.

tributary Any stream which contributes water to another stream.

triclinic symmetry *Struct. petrol:* Refers to either symmetry of fabric or symmetry of movement in which there are no planes or axes of symmetry.

triclinic system That system of crystals in which the forms are referred to three unequal, mutually oblique axes. The characteristic of all triclinic crystals is the total lack of symmetry other than a possible center.

tridymite A mineral, SiO_2, trimorphous with quartz and cristobalite. Hexagonal at high temperature, orthorhombic at low.

trigonal Having, in the ideal or symmetrically developed form, triangular faces, as, the trigonal trisoctahedron.

trigonal system *See* RHOMBOHEDRAL SYSTEM

trigonometrical survey A survey accomplished by the trigonometrical calculation of lines after careful measurement of a base line and of the angles made with this line by the lines toward points of observation; generally preliminary to a topographical survey. *See* TRIANGULATION

trilling A compound crystal consisting of three twinned individuals.

Trilobita Class of arthropods with a dorsal skeleton consisting of cephalon, thorax, and pygidium, and divided longitudinally into a central axis and two pleural regions.

trimline A line marking the former extent of the margins of a glacier.

trimorphism The presence of three forms in every species (of foraminifer), two of them representing the megalospheric forms and the last the microspheric form.

trioctahedral Refers to the structure of phyllosilicate minerals in which all possible octahedral positions are occupied by divalent ions, e.g., biotite.

tripartite method A method of

determining the apparent surface velocity and direction of propagation of microseisms or earthquake waves by determining the times at which a given wave passes three separated points.

triphylite A mineral, the part with Fe>Mn of the lithiophilite-triphylite series, $Li(Fe,Mn)PO_4$. Orthorhombic.

triple junction A point or small region where three plates meet.

tripoli; tripolite An incoherent, highly siliceous sedimentary rock composed of the shells of diatoms or of radiolaria, or of finely disintegrated chert. Used as a polishing powder and for filters.

trisoctahedron In the isometric system, either of two forms of normal symmetry, enclosed by 24 faces: (a) the trigonal or ordinary trisoctahedron, having triangular faces, each with equal intercepts on two axes and a greater intercept on the third axis; (b) the tetragonal trisoctahedron (also called trapezohedron and icositetrahedron), having trapezial faces, each with equal intercepts on two axes and a less intercept on the third axis.

tritium An isotope of hydrogen with three neutrons; it is radioactive with a half-life of 12.26 and an atmospheric residence time of 1.5 years. See DEUTERIUM

trivariant, *adj.* Referring to a system having three degrees of freedom, i.e., having a variance of three.

trivial name Second or specific name in the name of a species.

troilite A mineral, FeS, a variety of pyrrhotite occurring in nodular masses and thin veins in meteorites. Hexagonal.

trona An impure form of hydrous sodium carbonate. $Na_2CO_3.NaHCO_3.2H_2O$.

troposphere That portion of the atmosphere next to the earth's surface in which temperature generally rapidly decreases with altitude, clouds form, and convection is active. In middle latitudes the troposphere generally includes the first 10 to 12 km. above the earth's surface.

trough *1.* An elongate and wide depression, with gently sloping borders. *Ger.* Mulde; *Fr.* vallée. *2.* A long narrow channel or depression between ridges on land or between crests of waves at sea. *3.* In brachiopods, the furrow on posterior part of pedicle valve which provides space for the pedicle beneath the apex.

trough of wave The lowest part of a wave form between successive crests. Also that part of a wave below still-water level.

true dip A synonym of dip, used in comparison with apparent dip.

true folding Same as flexure folding, *q.v.*

truncate, *v.* To cut the top or end from; to terminate abruptly as if cut or broken off.

truncated spur The widening of a stream valley by a glacier results in the truncation of the spurs which extend into it from the two sides.

trunk glacier The main ice stream in a system of tributary valley glaciers.

tschernosem; tchornozem See CHERNOZEM

tsunami A great sea wave produced by a submarine earthquake or volcanic eruption. Commonly misnamed tidal wave, *q.v.*

tufa A chemical sedimentary rock composed of calcium carbonate or of silica, deposited from solution in the water of a spring or of a lake or from percolating ground water; sinter.

tuff A rock formed of compacted volcanic fragments, generally smaller than 4 mm. in diameter.

tuff ball Accretionary lapilla, *q.v.*

tuffite Indurated rocks composed of a mixture of pyroclastic and sedimentary detritus, especially ash and fine sediment.

tuff lava A term applied to consolidated, lavalike tuffa consisting primarily of lenses of black and gray obsidian lying in a tuffaceous matrix that displays a streaky, varicolored banding or eutaxitic structure. Essentially synonymous with welded tuff.

tuff palagonite A bedded aggregate of dust and fragments of basaltic lava, among which are conspicuous angular pieces and minute granules of pale yellow, green, red, or brown altered basic glass called palagonite.

tumescence *Volcanol:* The swelling or uparching of a volcano during periods of rising magma preceding an eruption.

tumulose Full of small hills and mounds.

tundra One of the level or undulating treeless plains characteristic of arctic regions, having a black muck soil and a permanently frozen subsoil.

tundra soil Soil of the tundra biome; usually extremely thin and rocky.

tungstate A salt of tungstic acid; a compound containing the radical WO_4^{-2}.

tunnel *1.* Strictly speaking, a passage in a mine open at both ends. Often used loosely as a synonym for adit, drift, gallery. *2.* A spiral opening in the plane of coiling bounded by chromata, in the Fusulinidae (a family of Foraminifera).

tunnel valley Sizable streams flowing beneath the ice and not loaded with coarse sediment which cut shallow trenches in the till and other loose material at the surface.

turbidimeter A device used to measure the amount of suspended solids in a liquid.

turbidite Turbidity current deposit.

turbidity current Density current, *q.v.*

turbidity size analysis Size analysis based upon the amount of turbidity in a suspension, the turbidity decreasing as the grains settle.

turbulent flow That type of flow in which the stream lines are thoroughly confused through heterogeneous mixing of flow. The head loss varies approximately with the second power of the velocity. *See* LAMINAR FLOW; REYNOLDS NUMBER

turf Peat that has been dried for use as fuel.

turkey-fat ore In Missouri, a name for a variety of smithsonite, colored yellow; so called from its appearance.

turnover *1.* In production, the rate at which something is used up and replaced. *2.* In a limnological sense, the mixing of layers of lentic waters and the redistribution of oxygen and nutrients.

turquoise A mineral, $CuAl_6(PO_4)_4(OH)_8.4H_2O$. Triclinic. Used as a gem.

turrelite A Texas asphaltic shale.

turtleback Large smoothly curved topographic surface underlain by folded metamorphic rocks in the Death Valley region; resembles a structural nose with amplitude up to several thousand feet.

turtleback fault Low-angle fault which has brought Cenozoic sedimentary and volcanic rocks into contact with metamorphics at a turtleback surface; interpreted as a folded overthrust or as a plane along which normal faulting or extensive landsliding has occurred.

turtle stone Large nodular concretion found in certain clays and marls. In form it has a rough resemblance to turtles, and this appearance is increased by its being divided into angular compartments by cracks filled with spar, reminding one of the plates on the shell of a turtle. *See* SEPTARIUM

twin; twinned crystal; twin crystal A nonparallel, rational, symmetrical intergrowth of two or more grains of the same crystalline species. Contact twin. *See* TWIN LAW; TWINNING

twin-gliding That type of single-crystal slip, produced either by compression or tension, by which displacement on preferred lattice planes takes place, with a fixed direction sense, producing reorientation of part of the crystal so that it is in a twin-position with respect to the stationary part. Rupture is not involved.

twin law The statement of the relation between the parts of a twin, including (1) the orientation of the twinning axis or of the twinning plane, one of which is necessarily rational, (2) the nature of the twinning relationship or operation, such as reflection across the twinning plane or rotation about the twinning axis, (3) the nature of the surface of contact between the parts of the twin, including the orientation if it is a plane surface.

twinning The formation of twins.

twinning, cyclic *See* CYCLE TWINNING

twinning, polysynthetic *See* POLYSYNTHETIC TWINNING

twinning axis Any direction in a twin that has the same relation to the lattices of both parts of the twin. It is always normal to a twinning plane, and at least one of these is always rational with respect to the lattices.

twinning displacement Movement along a crystallographic plane that results in twinning.

twinning plane Any plane that bears the same relation to the lattices of both parts of a twin. It is always normal to a twinning axis. Either the twinning plane or the twinning axis or both are always rational with respect to the lattices.

two-cycle valley A valley that has been subjected to two cycles of erosion. This is evidenced by narrow inner valley bordered by high-level terraces.

two-layer structure Structure of minerals composed of repeated layered units each consisting of an aluminum octahedral layer and a silicon tetrahedral layer as in kaolinite. Both Al and Si can be replaced by other elements.

two-phase flow Flow of two associated liquids such as oil and water.

Tyler standard scale A grade scale for the determination of size grades of sediment particles which is based on $\sqrt{2}$, and in which the midpoint values of each class turn out to be simple whole numbers or fractions.

type The term type, used alone and unqualified, generally refers to the holotype. *See* HOLOTYPE; TYPE SPECIMEN; TYPE SPECIES

type concept To associate each specific name and description with a definite preserved specimen, each generic name and description with a named species, etc.

type fossil Occasionally used as equivalent to index fossil, *q.v. Syn:* INDEX FOSSIL

type genus The generic name from which a family or subfamily name is formed.

type locality 1. The place at which a formation is typically displayed. 2. The locality from which the holotype of a species was collected.

type material All of the fossil specimens whose study provided the basis for the description of a new species.

type section Stratigraphic section recognized as the standard, generally the one from which a stratigraphic unit received its name.

type species The species which must always be included in any concept of a genus.

type specimen That specimen, generally the holotype, that provides the basis for description and recognition of a species of organisms; used loosely for other less important types.

typology Taxonomy governed by the concept that species are defined by the morphology of individual type specimens.

typonym A later generic name which has the same genotype as an earlier, valid name.

tyuyamunite A mineral, $Ca(UO_2)_2$-$(VO_4)_2.7-10H_2O$. Orthorhombic. usually massive, powdery, in yellow incrustations. An ore of uranium and vanadium.

U

Udden scale A logarithmic scale for size classification of sediments starting from 1 mm. and progressing by the ratio ½ in one direction and 2 in the other. This was the scale adopted by C. K. Wentworth, and by the Committee on Sedimentation.

uintahite *See* GILSONITE

ulexite Cotton ball. A mineral, $NaCaB_5O_9 \cdot 8H_2O$. Triclinic, fibrous.

ultimate base level Ultimate base level is a base level at sea level or below, to which lands may be reduced by the processes acting upon them.

ultimate form *See* CYCLE OF EROSION

ultimate recovery The quantity of oil or gas that a well, a pool, a field, or a property will produce. It is the total obtained or to be obtained from the beginning to final abandonment.

ultimate strength The maximum differential stress that a material can sustain under the conditions of deformation. Beyond this point, rock failure occurs.

ultisol Deeply weathered yellow and red soils associated with old land surfaces in humid tropical or humid temperate climates.

ultrabasic, *adj.* Some igneous rocks and most varieties of meteorites containing less than 45% silica; containing virtually no quartz or feldspar and composed essentially of ferromagnesian silicates, metallic oxides and sulfides, and native metals, or of all three. *Syn:* ULTRAMAFIC—the preferred term.

ultrahaline Hypersaline.

ultramafic Ultrabasic.

ultrametamorphism Melting of rock and creation of magma *in situ*.

ultrasima Layer within the earth underlying and heavier than the sima and presumably consisting of more basic material.

ultraviolet Applied to radiation outside of the visible spectrum of light at its violet end; said of rays more refrangible than the extreme violet rays.

umbilical plug *See* PLUG, UMBILICAL

umbo *1.* A projection arising from the surface. *2.* In pelecypods the elongate raised or highest part of the valve when laid with the interior down.

unaka Monadnock. The term monadnock is applied to a single isolated residual, such as Mount Monadnock in New Hampshire, which stands alone. More massive residuals of greater size and height would seem to require a different name, and the term unaka has been proposed for such large residuals as the Unaka Mountains.

unary or **unicomponent system** A system of one component.

unavailable moisture Moisture, held in soil by adsorption or other forces, that cannot be utilized by plants.

unbalanced force A force or system of forces that results in a change of motion, i.e., causes acceleration, deceleration, or a change in direction of a moving body.

unconformable Having the relation of unconformity to the underlying rocks; not succeeding the underlying strata in immediate order of age and in parallel position.

unconformity A surface of erosion that separates younger strata from older rocks. *See* ANGULAR UNCONFORMITY, DISCONFORMITY, NONCONFORMITY

unconsolidated material From the standpoint of workability, this constitutes the "earth," *q.v.*, of the A.S.C.E. classification system.

unda That part of the floor of the ocean or other water body that lies in the zone of wave action.

undaform The subaqueous land surface produced in the unda environment.

undaform zone That part of the ocean floor which lies in the zone of wave action and in which, therefore, the bottom is repeatedly stirred and reworked by storm waves.

undathem The rock unit deposited in the unda environment.

undation *1.* Large wavelike fold in the earth's crust. *2.* Theoretical rhythmic oscillation of the earth's surface in broad waves.

undation theory A theory of mountain building that assumes that long broad anticlines of basement rock rose like huge waves in the crust. The sedimentary cover and sometimes the basement itself slid off to form the folds and faults observed in orogenic belts.

underclay A stratum of clay beneath a coal bed often containing roots of coal plants, especially *Stigmaria;* it may contain a large amount of quartz silt and is thought by some to be a paleosol. *See* FIRE CLAY

undercooling Reduction in temperature of a liquid to the point where viscosity increases to such a degree that the liquid behaves like a solid. Glass is an example

of an undercooled liquid. *Syn:* SUPERCOOLING

undercut slope *See* SLIP-OFF SLOPE

underfit stream A stream that appears too small to have eroded the valley in which it flows.

underflow *1.* The movement of ground water in an underflow conduit. *2.* Movement of water through a pervious subsurface stratum; the flow of percolating water; the flow of water under ice, or under a structure.

underlie *Geol:* To occupy a lower position than, or to pass beneath; said of stratified rocks over which other rocks are spread out.

undersaturated *Petrol:* Applied to igneous rocks consisting wholly or in part of unsaturated minerals. The class of rocks is subdivided into nonfeldspathoidal and feldspathoidal divisions.

underthrust A thrust fault in which the footwall was the active element.

underthrust fold A fold in which the axial planes dip away from the force producing them.

undertow A supposed undersurface flow return of surface wave water taking place after the wave has broken on the beach. These returning waters are now thought to be concentrated into definite surface currents and called rip currents *q.v.*

undiscovered resources Unspecified bodies of mineral-bearing material surmised to exist on the basis of broad geologic knowledge and theory.

undisturbed Rocks that lie in the positions in which they were originally formed. *Cf.* DISTURBED

undulatory extinction Irregular darkening of crystals in thin section when rotated between crossed Nicols; results from distortion of the crystal lattice.

unfolding Deformation process that reduces or obliterates previous folds.

uniaxial Having but one direction in which light passing through the crystal is not doubly refracted. Typical of tetragonal and hexagonal minerals. *Cf.* BIAXIAL

unicellular One-celled; refers to an organism the entire body of which consists of a single cell.

unicomponent system *See* UNARY SYSTEM

uniformitarianism The concept that the present is a key to the past, and that past geologic events are to be explained by those same physical principles that govern the present.

uniformity coefficient An expression of variety in sizes of grains that constitute a granular material.

unilocular Containing a single chamber or cell.

uniserial Consisting of a single series as, the plates of primitive crinoid arms.

unit cell The smallest volume or parallelopiped within the three-dimensional repetitive pattern of a crystal that contains a complete sample of the atomic or molecular groups that compose this pattern; crystal structure can be described in terms of the translatory repetition of this unit in space in accordance with one of the space lattices.

unit form A crystal form in a system other than the cubic, having intercepts on the chosen crystal axis that define the axial ratio. Unit forms have Miller indices, {111}, {110}, and so forth.

unitization Combination of adjacent oil leases for efficient operation in which the value of oil, regardless of where the producing wells are located, is allocated among the properties according to some reasonable formula.

univalve 1. Animal protected by a shell consisting of one piece, particularly a gastropod. 2. Shell of such an animal.

universal stage Microscope stage that can be rotated through both horizontal and vertical planes. The ordinary universal stage has four axes of rotation in addition to that of the common petrographic microscope.

unmixing 1. Natural separation of unlike things from any mixture. 2. Separation of chemical compounds from mixcrystals when temperature falls and these become unstable; exsolution. 3. Segregation and concentration, as in the diagenesis of some sediments.

unpaired terraces Remnant of a former continuous alluvial terrace, which (because of differential erosion) is no longer duplicated on opposite sides of a valley.

unrestricted movement Movement accompanying rock flowage in which elongation of particles is in the direction of movement.

unsaturated 1. Applied to minerals that are incapable of crystallizing from rock magmas in the presence of an excess of silica. Such minerals are said to be unsaturated with regard to silica and include the feldspathoids, analcime, magnesian olivine, melanite, pyrope, perovskite corundum, calcite, and perhaps spinel. 2. State of a chemical compound, particularly an organic one, that is capable of holding additional atoms, especially hydrogen, without change in its basic molecular structure.

unstable equilibrium Not in true or in metastable equilibrium; in the process of change, as, a piece of ice in hot water.

unstratified Not formed or deposited in beds or strata.

updip block Block on the updip side of a strike fault.

uphole shooting *Seis. explor:* The setting off of successive shots in a hole at varying depths in order to determine velocities and velocity variation of the materials forming the hole walls.

uphole time Term used to denote the observed travel time of a seismic wave from the point of generation at a given depth in a shothole to a detector at the surface, and the observed time equivalent of the corresponding shot depth. *Syn:* T-AT-SHOT POINT

uplift Elevation of any extensive part of the earth's surface relative to some other parts. *Ant:* SUBSIDENCE

upper *Geol:* Designates a later period or formation; so called because the strata are normally above those of the earlier formations.

upright fold Fold with vertical axial plane.

uprush Swash. The rush of water up onto the beach following the breaking of a wave.

upthrow The block or mass of rock on that side of a fault which has been displaced relatively upward. The term should be used with the definite understanding that it refers merely to a relative and not an absolute displacement.

upwarp An area that has been uplifted; generally used for broad anticlines.

upwelling *Oceanog:* Light surface water transported away from a coast (by action of winds parallel to it) and replaced near the coast by heavier subsurface water.

uralite A fibrous or acicular variety of hornblende occurring in altered rocks and pseudomorphous after pyroxene. *See* AMPHIBOLE

uralitization The conversion of pyroxene into hornblende; usually as a finely fibrous aggregate. It is usually considered to be a late magmatic process.

uraninite Pitchblende. A mineral, essentially UO_2, but usually containing Th and rare earths as substituents, and UO_3, Ra, and Pb formed by radioactive decay. Isometric. Pitchblende is the massive variety found in sulfide-bearing veins. The chief ore of U.

uranium-235 A rare isotope of uranium that will fission when bombarded by neutrons.

uranophane; uranotil A mineral, $Ca(UO_2)_2Si_2O_7.6H_2O$. Orthorhombic.

urstromtal [<*Ger.*] Trenchlike valley cut by a temporary stream flowing along the margin of a former ice sheet.

urtite A plutonic rock composed largely of nepheline with minor aegirine, and apatite. Urtite is transitional into ijolite with increasing aegirine and decreasing nepheline.

U-shaped valley A valley carved by glacial erosion and having a characteristic parabolic cross section.

U-tube *See* PITOT TUBE

uvala *1.* A large, broad sinking in the karst with uneven floor, formed by the breaking down of the wall between a series of dolinas. These uvalas possess the chief characteristics of polyes, for their major axes agree with the strike, but they differ from them in their irregular floor, and they lack the special hydrographical condition of polyes. *2.* Large sinkhole formed by the coalescence of several doline sinks.

uvarovite A mineral of the garnet group, $Ca_3Cr_2(SiO_4)_3$.

vadose water Suspended water. A term used to designate subsurface water above the zone of saturation in the zone of aeration.

vadose-water discharge Discharge of soil water not derived from the zone of saturation.

vagrant animal An animal that customarily moves about by its own volition, either continuously or intermittently.

vagrant benthos Bottom-dwelling organisms which are capable of movement on, in, or above the substratum.

valid name An available name whose "title" to a species is clear, i.e., which is neither a synonym nor a homonym of an earlier name.

valley [*Lat.* vallis] Any hollow or low-lying land bounded by hill or mountain ranges, and usually traversed by a stream or river which receives the drainage of the surrounding heights.

valley braid A term proposed for an individual runway of a valley which is in anastomosis, its valley parts passing about features in bas-relief or about upland tracts.

valley fill A valley underlain by unconsolidated rock waste derived from the erosion of the bordering mountains.

valley flat The low flat land between valley walls bordering a stream channel.

valley glacier A glacier occupying a valley. Mountain glacier; Alpine glacier.

valley plug A local constriction in a stream channel formed by any of several types of channel obstructions and which may cause rapid deposition.

valley profile The longitudinal profile of a valley.

valley sink An elongated sink or series of interconnecting sinks forming a valleylike depression.

valley system A valley and its tributaries.

valley tract Middle part of a stream course characterized by moderate gradient and a fairly wide valley.

valley train A long narrow body of outwash confined within a valley.

valve *1.* Any device or arrangement that is used to open or close a passage to permit or stop the flow of a substance; a device for rectifying an alternating current. *2.* A single part of the two-piece shell of the clams, ostracods, and brachiopods.

vanadate A salt or ester of vanadic acid; a compound containing the radical VO_4^{-3} or VO_2^-.

vanadinite A mineral, $Pb_5(VO_4)_3$Cl, commonly containing As and P replacing V. Hexagonal. An ore of vanadium.

Van Allen radiation zone Powerful doughnut-shaped zone of radiation 1600 to 4800 kilometers above the earth's surface and parallel with the equator.

van der Waals forces The weak attraction exerted by a-1 atoms on one another, resulting from the mutual interaction of the electrons and nuclei of the atom. Sometimes called the van der Waals bond.

van't Hoff's law The law that when a system is in equilibrium, of the two opposed interactions the endothermic is promoted by raising the temperature, the exothermic by lowering it.

vapor pressure The pressure at which a liquid and its vapor are in equilibrium at a given temperature. *Syn:* VAPOR TENSION

variation *1.* The angle by which the compass needle deviates from the true north. 2. One of the laws of organic nature; organisms vary in time, from place to place, and also in one locality and time; they vary also in their morphology.

variation diagram A name given a method of plotting the chemical compositions of rocks in an igneous rock series, designed to reveal genetic relationships and the nature of the processes that have affected the series. Usually the weight per cent silica is plotted as the abscissa and the weight per cent of individual other oxides as the ordinates. Also called a Harker diagram.

variety *Morphol:* A distinctive group of individuals within a population (a species), differentiated from other parts of the population by possession of some character or combination of characters lacking in the others; an artificial group, not a sort of subspecies.

variometer *Geophys:* A device for measuring or recording variations in terrestrial magnetism; a variable inductance provided with a scale.

Variscan orogeny Late Paleozoic orogenic era of Europe, extending through the Carboniferous and Permian. By current usage, it is synonymous with the Hercynian orogeny.

variscite A mineral, $AlPO_4 \cdot 2H_2O$, used as a gemstone. Yellow green used occasionally as a substitute for turquoise. Orthorhombic.

varve *1.* Any sedimentary bed or lamination that is deposited within one year's time. 2. A pair of contrasting laminae representing seasonal sedimentation, as, summer (light) and winter (dark) within a single year.

vascular plant Plant with well-developed circulatory system and structural differentiation into roots, stem, and leaves; includes majority of terrestrial plants.

vectorial *Struct. petrol:* Applied to the physical features of a fabric which are directional in character, e.g., lattice and dimensional orientation are vectorial features.

vein A tubular body, long in two dimensions and short in the third. An occurrence of ore minerals, usually disseminated throughout gangue, or veinstone. A vein and a lode are generally the same thing; the former being more scientific, the latter, a miner's name for it. *See* LODE; FISSURE; FISSURE VEIN

vein dike A pegmatitic intrusion that has the characteristics of both a vein and a dike.

veined gneiss Metamorphic rock formed by intrusion of magma into nonfissile country rock in numerous veins and dikelets extending in all directions.

velocity analysis Calculation of subsurface P-wave velocity distributions using observed reflection moveout times at large geophone-shot point distances.

velocity discontinuity An abrupt change of the rate of propagation of seismic waves within the earth, as at an interface.

velocity of waves The speed with which an individual wave advances.

velocity profile A seismic reflection spread designed to record data which may be used to compute average velocities in the earth to reflecting horizons by observation of time variations compared with geometrical ray paths traveled.

vent agglomerate Agglomerate that is localized within a volcanic vent. *See* AGGLOMERATE

vent breccia Volcanic breccia that is localized within a volcanic vent. A conduit filling or neck of volcanic breccia.

ventifact A general term for any stone or pebble shaped, worn, faceted, cut, or polished by the abrasive or sandblast action of windblown sand, generally under desert conditions; e.g., a dreikanter. *See also* WINDKANTER

ventral, *adj.* Front; relating to the inner face or part of an organ; opposite the back or dorsal part.

Venus hair Fine rutile crystals occurring as inclusions in quartz. *See* SAGENITE

verde antique A dark-green rock composed essentially of serpentine (hydrous magnesium silicate). Usually crisscrossed with white veinlets of magnesium and calcium carbonates. Used as an ornamental stone. In commerce often classed as a marble.

vermiculite A group of micaceous clay minerals closely related to chlorite and having the general formula, $(Mg,Fe,Al)_9(Al,Si)_4O_{10}$- $(OH)_2 \cdot 4H_2O$. Derived generally from the alteration of biotite and phlogopite. Characterized by marked exfoliation when heated above 150°.

vernier An auxiliary scale used in conjunction with the main scale of a measuring device to obtain one more significant figure of a particular measurement.

vertical exaggeration In a stereoscopic image, the increase in relief seen by the eye.

vertical intensity The magnitude of the vertical component of any vector; the strength of intensity

of the vertical component of the earth's magnetic or gravitational field at any point.

vertical photograph An aerial photograph made with the camera axis vertical or as nearly vertical as practicable.

vertical range The local sequence of strata through which a certain species or genus is found. *Syn:* TEILZONE

vertical separation In faulting, the separation between the two parts of the displaced index plane (bed, vein, dike, etc.) measured in a vertical direction.

vertical shift The vertical component of the shift. *See* SHIFT

vertical slip In faulting, the vertical component of the net slip; this is the same as the vertical component of the dip slip.

vertical-variability map Map showing areal relations of vertical variability in a stratigraphic unit; may show (a) degree of differentiation of the unit in subunits of different lithologic types, or (b) vertical distribution or concentration in the unit of one lithologic type.

vertisol A soil order characterized by at least 30 per cent clay, and found in regions having pronounced dry seasons, causing deep, wide surface cracks.

vesicle *1.* A small, circular, enclosed space. *2.* A small cavity in an aphanitic or glassy igneous rock, formed by the expansion of a bubble of gas or steam during the solidification of the rock.

vesicular *1.* Characteristic of, or characterized by, pertaining to, or containing vesicles. *2.* Containing many small cavities.

vestigial Pertaining to organic structures whose embryonic start is ordinary but whose later development is retarded and often so much reduced that these structures are functionless.

vestigial structure Nonfunctional structure of an organism of little or no use to the individual that was inherited from ancestors to whom it was useful; vestigial organs commonly become smaller and eventually may be lost entirely by distant descendants.

vesuvianite Idocrase.

vibration gravimeter A device which affords a measurement of gravity by observation of the period of transverse vibration of a thin wire tensioned by the weight of a known mass, useful for observations at sea.

vibroseis A seismic method using a vibrator to generate seismic waves, where the frequency of the vibrator is varied continuously. A Continental Oil Co. trademark.

vicinal forms *Crystallog:* Forms taking the place of the simple fundamental forms to which they approximate very closely in angular position.

virgation Divergence of mountain ranges from a common center. *Ant:* SYNTAXIS

vis-à-vis Mirror image.

viscosity Internal friction due to molecular cohesion in fluids. The internal properties of a fluid which offer resistance to flow. *See* POISE

viscosity, absolute The force which will move 1 sq. cm. of plane surface with a speed of 1 cm. per second relative to another parallel plane surface from which it is separated by a layer of the liquid 1 cm. thick. This viscosity is expressed in dynes per square centimeter, its unit being the poise, which is equal to 1 dyne-second per square centimeter. *See* CENTIPOISE; POISE

viscosity coefficient *Hydrol:* A quantitative expression of the friction between the molecules of water when in motion. It is the amount of force necessary to maintain a unit difference in velocity between two layers of water at a unit distance apart. It decreases rapidly with increase in temperature.

viscous *1.* Adhesive or sticky, and having a ropy or glutinous consistency. *2.* Imperfectly fluid; designating a substance that, like tar or wax, will change its form under the influence of a deforming force, but not instantly, as more perfect fluids appear to do.

viscous flow Flowage that occurs upon application of any unbalanced force however small providing that the stress difference continues to operate for a sufficient length of time. *See* PLASTIC FLOW

vitrain [*Fr.* vitre, glass] Thin horizontal bands in coal, visible to the naked eye, up to 20 mm. thick, but may also be in thicker lenticels. It has brilliant gloss, strong rectangular fracture perpendicular to bedding, conchoidal fracture in other directions, and clean specular reflection.

vitreous *1.* Having the luster of broken glass, quartz, calcite. *2.* Having no crystalline structure; amorphous.

vitric tuff An indurated deposit of volcanic ash dominantly composed of glassy fragments blown out during a volcanic eruption. The term should properly be restricted to tuffs containing more than 75 per cent by volume of glass particles. *See* TUFF

vitrifaction *See* VITRIFICATION

vitrification *1.* Act, art, or process of vitrifying; state of being vitrified; also, a vitrified body. *2.* Any process tending to make a body more vitreous, or glassy. *Syn:* VITRIFACTION

vitrify To convert into, or cause to resemble, glass or a glassy substance, by heat and fusion.

vitro- [<*Lat.* vitrum, glass] A combining form meaning glassy.

vitroclastic Pertaining to a structure typical of fragmental glassy rocks, in which the particles usually have crescentic, rudely triangular outline, or somewhat concave borders.

vitrophyre Porphyritic volcanic glass.

vitrophyric Of, pertaining to, formed of, or characterized by vitrophyre.

vivianite Blue iron earth. A mineral, $Fe_3(PO_4)_2.8H_2O$. Monoclinic.

vly; vlei; vley 1. A small swamp, usually open and containing a pond. 2. A valley where water collects. Local in Middle Atlantic States.

vogesite A lamprophyre consisting primarily of hornblende or augite or both and either orthoclase or sanidine. Plagioclase is common in many varieties, and biotite and olivine are occasionally present.

void Interstice. A general term for pore space or other openings in rock. In addition to pore space, the term includes vesicles, solution cavities, or any openings either primary or secondary. *See* PORE, 1

voidal concretion Hollow limonitic concretion resulting from the weathering of clay ironstone.

void ratio Ratio of intergranular voids to volume of solid material in a sediment or sedimentary rock.

volatile matter Those products, exclusive of moisture, given off by a material as gas and vapor, determined by definite prescribed methods which may vary according to the nature of the material.

volatiles The volatile constituents (or "rest magma") remaining after the less volatile ores have crystallized as igneous rocks.

volcanic Of, pertaining to, like, or characteristic of a volcano; characterized by or composed of volcanoes, as, a volcanic region, volcanic belt; produced, influenced, or changed by a volcano or by volcanic agencies; made of materials derived from volcanoes, as, a volcanic cone.

volcanic action [<*Lat.* Vulcanus, god of fire] Igneous action at the surface of the earth, in contradistinction to plutonic action which takes place beneath the surface.

volcanic agglomerate Coarse volcanic material produced by explosions. Occurs in necks or pipes of old volcanoes. Not stratified.

volcanic belt A linear or arcuate arrangement of volcanoes, generally of great extent and confined to orogens along the margins of the continents or within the ocean basins; e.g., the volcanoes of the Aleutian Island chain comprise a volcanic belt.

volcanic block A subangular, angular, round, or irregularly shaped mass of lava, varying in size up to several feet or yards in diameter. *Cf.* VOLCANIC BOMB; BLOCK

volcanic bombs Detached masses of lava shot out by volcanoes, which, as they fall, assume rounded forms like bombshells.

volcanic breccia A more or less indurated pyroclastic rock consisting chiefly of accessory and accidental angular ejecta 32 mm. or more in diameter lying in a fine tuff matrix.

volcanic chain Volcanic belt.

volcanic cinders *See* CINDERS, VOLCANIC

volcanic cloud *See* ERUPTION CLOUD

volcanic cluster A group of volcanoes, volcanic cones, or volcanic vents without any apparent systematic arrangement.

volcanic conduit *See* CONDUIT, 2

volcanic cone A cone-shaped eminence formed by volcanic discharges.

volcanic conglomerate A rock composed mainly or entirely of rounded or subangular fragments, chiefly or wholly of volcanic rocks, in a paste of the same material.

volcanic crater See CRATER, VOLCANIC

volcanic or cumulo dome A steep-sided protrusion of viscous lava forming a more or less dome-shaped or bulbous mass over and around a volcanic vent.

volcanic dust See DUST, VOLCANIC

volcanic earthquake Seismic disturbances which are due to direct action of volcanic force or one whose origin lies under or near a volcano, whether active, dormant, or extinct.

volcanic ejecta Tephra, q.v.

volcanic eruption See ERUPTION, VOLCANIC

volcanic focus The supposed seat or center of activity in a volcanic region or beneath a volcano.

volcanic gases The primary magmatic gases emitted from lavas, either quietly or with explosive violence, at the earth's surface, chief among which are water vapor, hydrogen, oxygen, nitrogen, hydrogen sulfide, sulfur dioxide, sulfur trioxide, carbon dioxide, carbon monoxide, gaseous hydrochloric acid, chlorine, methane, gaseous hydrogen fluoride, argon, and helium.

volcanic glass Natural glass produced by the cooling of molten lava, or some liquid fraction of molten lava, too rapidly to permit crystallization, and forming such material as obsidian, pitchstone, sideromelane, and the glassy mesostasis in the groundmass of many effusive rocks.

volcanicity; vulcanicity The quality or state of being volcanic; volcanism.

volcanic mud Mud formed by the mixture of water with volcanic dust, ash, or other fragmental products of volcanic eruptions, often initially hot and flowing down the flanks of a volcanic cone as a hot lahar or mudflow.

volcanic mudflow Lahar.

volcanic neck The solidified material filling a vent or pipe of a dead volcano.

volcanic pipe Sometimes the streams of lava are very fluid, and they cool at the bottom and upper surfaces much more rapidly than in the interior. The rocks thus formed remain, while the interior molten lava flows on and caves are formed in this manner which are known as volcanic pipes.

volcanic plug The term is restricted by some to necks consisting of a monolithic mass of solidified igneous rock. See VOLCANIC NECK

volcanic prediction The more or less accurate timing of volcanic eruptions from scientific data. Although not now possible, many scientists believe that volcanic prediction is an achievable goal within a few decades.

volcanic rent or fissure trough A great volcanic depression, usually concentric in plan, caused by the tearing apart of volcanic cones by movements that are mainly horizontal. The sliding may be induced by the injection of dike swarms or by the overloading of cones on a weak substratum, either sedimentary or volcanic.

volcanic rift zone A narrow zone of fissures extending down the flanks of a volcano, ordinarily reaching from the summit crater to the foot of the mountain and beyond.

volcanic rocks *1.* The class of igneous rocks that have been poured out or ejected at or near the earth's surface. *Syn:* EXTRUSIVE ROCKS; EFFUSIVE ROCKS. *2.* One of the three great subdivisions of rocks under a classification proposed by H. H. Read. It includes the effusive rocks and associated intrusive rocks. Dominantly basic, magmatic, igneous. Nonorogenic.

volcanics General collective term for extrusive igneous and pyroclastic material and rocks.

volcanic sand Sand-sized volcanic debris of either pyroclastic or detrital origin.

volcanic slag *See* SLAG, VOLCANIC

vocanic spine A slender, pointed monolithic protrusion of lava squeezed up in viscous condition on the surface of a thick lava flow or on the surface of a viscous volcanic dome through an opening in the solidified upper crust or carapace. They range in height from a few inches to many hundreds of feet. The classic example of a large spine is that of Mt. Pelée, in Martinique.

volcanic tuff *See* TUFF

volcanic vent An opening or channel in the earth's crust through which magmatic materials are transported and out of which volcanic materials (lava, pyroclastic detritus) are erupted at the surface.

volcanic water Water in or derived from magma at the earth's surface or at a relatively shallow depth.

volcanism; vulcanism Volcanic power or activity; volcanicity. The term ordinarily includes all natural processes resulting in the formation of volcanoes, volcanic rocks, lava flows, etc.

volcano *1.* A vent in the earth's crust from which molten lava, pyroclastic materials, volcanic gases, etc., issue. *2.* A mountain which has been built up by the materials ejected from the interior of the earth through a vent.

volcanologist; vulcanologist One versed in the study of volcanic phenomena. A volcanist.

volcanology; vulcanology The branch of science treating of volcanic phenomena. Volcanological.

volchonskoite A clay mineral. A chromium-bearing montmorillonite.

volt-second The unit of magnetic flux in the M.K.S. and practical systems.

volume elasticity Bulk modulus, *q.v.*

volume law, Lindgren's *See* LINDGREN'S VOLUME LAW

volume susceptibility (magnetic) The ratio of the magnetization of the material to the strength of the magnetizing field. Thus defined, the magnetic susceptibility is a dimensionless ratio.

volumetric measurements The determination of the quantity of oil or gas contained in a reservoir. Porosity, thickness, saturation (for oil), area, temperature, and reservoir pressure (for gas) are elements in the calculation.

von Wolff's classification A chemicomineralogical classification of igneous rocks.

V's, rule of In regions where there is relief, the outcrop of a horizontal bed extends up the valleys to form a V or U that points upstream. This is also true if the bed dips upstream or downstream, except where the bed dips downstream at a steeper angle than the slope of the stream, in which case the V points downstream.

V-shaped A gorge with evenly sloping sides is often called V-shaped and the V is narrow or broad according to the amount of wasting which has taken place.

vug A cavity, often with a mineral lining of different composition from that of the surrounding rock. *See* GEODE

vuggy porosity Porosity due to vugs in calcareous rock.

vugular Vuggy.

vulcanian *1.* Designating or pertaining to a type of volcanic eruption in which the phenomena are explosive, with the emission of much fine ash and ash-laden gases which ascend to form a voluminous, cauliflowerlike eruption cloud. *2.* Of or pertaining to plutonism; plutonic.

wad Bog manganese. An impure mixture of manganese and other oxides. It contains 10 to 20 per cent water, and is generally soft, soiling the hand.

wady; wadi; ouady A ravine or watercourse, dry except in the rainy season.

wall *1.* The side of a level or drift. *2.* The country rock bounding a vein laterally. *3.* The side of a lode.

walled lake A lake with an accumulation of boulders resembling walls about its shores.

wall rock The rock forming the walls of a vein or lode; the country rock.

wandering The compound movement of sweeping meanders in a swinging meander belt.

waning slope The concave lower slope beneath the scarp which may include the pediment.

warping The gentle bending of the earth's crust without forming pronounced folds or dislocations.

wash *1.* A Western miner's term for any loose, surface deposits of sand, gravel, boulders, etc. *2.* Auriferous gravel. *3.* Coarse alluvium; an alluvial cone.

wash load Wash load is that part of the total sediment load composed of all particles finer than limiting size, which is normally washed into and through the reach under consideration.

washover Small delta built on the landward side of a bar separating a lagoon from the open sea.

wastage The process or processes by which glaciers lose substance. Wastage is usually considered as including melting, wind erosion, evaporation, and calving, *q.v.*, but is sometimes used as a synonym for ablation, *q.v.*

waste *Also see* SOLID WASTE. *1.* Bulky waste—items whose large size precludes or complicates their handling by normal collection, processing, or disposal methods. *2.* Construction and demolition waste—building materials and rubble resulting from construction, remodeling, repair, and demolition operations. *3.* Hazardous waste—wastes that require special handling to avoid illness or injury to persons or damage to property. *4.* Special waste—those wastes that require extraordinary management. *5.* Wood pulp waste—wood or paper fiber residue resulting from a manufacturing process. *6.* Yard waste—plant clippings, prunings, and other discarded material from yards and gardens. Also known as yard rubbish.

waste plain The debris cones along the foot of a mountain range usually so completely coalesce that they form a true plain, called often a waste plain or waste slope. *See* PIEDMONT ALLUVIAL PLAIN

water, film Water held tenaciously by the soil particles, not free to move in the interstices.

water, interstitial Water that exists in the interstices or voids in a rock, or other porous medium.

water, juvenile Water from the interior of the earth which is new or has never been a part of the general system of ground-water circulation, e.g., magmatic water.

water, meteoric Water that previously existed as atmospheric moisture, or surface water, and that entered from the surface into the voids of the rock.

water-bearing bed See AQUIFER

water-bearing stratum See AQUIFER

water bed A bed of coarse gravel or pebbles occurring in the lower part of the upper till in the Upper Mississippi Valley.

water content Water contained in porous sediment or sedimentary rock, generally expressed as a ratio of water weight to dry sediment weight.

watercourse 1. A stream of water; a river or brook. 2. A natural channel for water; also a canal for the conveyance of water, especially in draining lands.

water cycle Hydrologic cycle, q.v.

waterfall A point in the course of a stream or river where the water descends perpendicularly or nearly so.

waterflooding The secondary-recovery operation in which water is injected into a petroleum reservoir for the purpose of effecting a water drive.

water gap A pass in a mountain ridge through which a stream flows.

water humus Organic matter deposited in water.

water lime Impure limestone which can be burned without addition of other material to produce cement that sets when mixed with water. Cf. HYDRAULIC CEMENT

water of dehydration Water that was once in chemical combination with certain minerals and has been by later chemical changes set free as water.

water of imbibition 1. The proportionate amount of water that a rock can contain above the line of water level or saturation. Quarry water. 2. Water of saturation.

water of retention That part of the interstitial water in a sedimentary rock which remains in the pores under a definite capillary pressure differential and conditions of unhindered flow. Usually called connate water.

water parting The high land which forms the divisional line between two contiguous river basins is called the water parting. Term suggested as a substitute for watershed.

water pollution The addition of sewage, industrial wastes, or other harmful or objectionable material to water in concentrations or in sufficient quantities to result in measurable degradation of water quality.

water quality criteria The levels of pollutants that affect the suitability of water for a given use. Generally, water use classification includes: public water supply; recreation; propagation of fish and other aquatic life; agricultural use and industrial use.

watershed The area contained within a drainage divide above a specified point on a stream.

water source A body of surface water, a spring, or a well from which raw water can be obtained for a water supply, q.v.

water supply A volume of water that has been treated or is safe, and is ready for distribution. Cf. WATER SOURCE

water table 1. The upper surface of a zone of saturation except where that surface is formed by an impermeable body. 2. Locus of points in soil water at which the pressure is equal to atmospheric pressure.

water witch One who purportedly locates underground water with a divining instrument.

Waucoban The lowest of three Cambrian epochs in North America; also the series of strata deposited during that epoch.

wave *1.* An oscillatory movement in a body of water manifested by an alternate rise and fall of the surface. *2. Geophys:* A disturbance of the equilibrium of a body or of a medium in which the disturbance is propagated from point to point through the medium with a continuous recurring motion.

wave amplitude *1. Hydrodyn:* One-half the wave height. *2. Eng:* Loosely, the wave height from crest to trough.

wave base *1.* The plane to which waves may degrade the bottom in shallow water. *2.* The depth at which wave action ceases to stir the sediments.

wave-built terrace An embankment extending seaward or lakeward from the shoreline produced by wave deposition.

wave-current ripple mark Ripple mark supposed to have been produced by wave modification of previously existing transverse ripples.

wave-cut bench A bench extending seaward from the base of a sea cliff produced by wave erosion.

wave-cut notch *See* NOTCH

wave-cut scarp *See* SEA CLIFF

wave-cut terrace Marine-cut terrace; plain of marine abrasion; shore platform; wave-cut plain, wave platform.

wave delta *See* WASHOVER

wave forecasting The theoretical determination of future wave characteristics, usually from observed or predicted meteorological phenomena.

wavefront *Seismol:* The surface of equal time lapse from the point of detonation to the position of the resulting outgoing signal at any given time after the charge has been detonated. In a more restricted sense, the surface along which phase is constant at a given instant.

wavefront chart A plot of horizontal distance versus depth on which have been plotted wavefronts and raypaths emanating from a point. Their position is determined by a particular velocity distribution and is identified by two-way travel and moveout times.

wavellite A mineral, $Al_3(PO_4)_2(OH,F)_3.5H_2O$. Orthorhombic, radiating.

wavemark *See* SWASH MARK

wave normal optics Line rising perpendicularly from a point in a plane that is tangent to the wave surface at that point.

wave of translation Waves in which there is a pronounced forward movement.

wave platform Marine-cut terrace; plain of marine abrasion; shore platform; wave-cut plain; wave-cut terrace.

wave refraction The process by which the direction of a train of waves moving in shallow water at an angle to the contours is changed.

wave ripple marks Ripple marks with symmetrical slopes, sharp crests, and rounded troughs produced by oscillatory waves. Oscillation ripple mark; symmetrical oscillation ripple mark; oscillatory ripple mark; symmetrical ripple mark; aqueous oscillation ripple mark.

wave steepness The ratio of a wave's height to its length.

wave surface *Optics:* The surface that includes all loci of light in the same phase that originated at a given point.

wave velocity The speed with which an individual wave advances.

wavy extinction Irregular extinction of a mineral under the polarizing microscope, due to bending or distortion of the crystal, or due to a subparallel aggregate of crystals. *Syn:* UNDULATING EXTINCTION

Wealdian In northern Europe the Upper Jurassic and Lower Cretaceous are continental. Wealdian is used for the Cretaceous part of that continental sequence.

weather *Meteor: 1.* The state of the atmosphere, defined by measurement of the six meteorological elements, viz., air temperature, barometric pressure, wind velocity, humidity, clouds, and precipitation. *2. Geol:* To undergo or endure the action of the atmosphere; to suffer meteorological influences.

weathered layer In seismic work, a zone extending from the surface to a limited depth, usually characterized by a low velocity of transmission which abruptly changes to a higher velocity in the underlying rock. More properly called the low-velocity layer.

weathering The group of processes, such as the chemical action of air and rain water and of plants and bacteria and the mechanical action of changes of temperature, whereby rocks on exposure to the weather change in character, decay, and finally crumble into soil.

weathering correction In seismic work, a time correction applied to reflection and refraction data to correct for the travel time of the observed signals in the low-velocity layer, or weathered layer.

weathering index A measure of the weathering characteristics of coal, according to a standard laboratory procedure.

weathering-potential index A measure of the degree of susceptibility to weathering of a rock or a mineral, computed from a chemical analysis.

weathering profile Succession of layers in unconsolidated surface material produced by prolonged weathering; where well developed it consists of surface soil, chemically decomposed layer, leached and oxidized layer, oxidized but unleached layer, and unaltered material. *Cf.* SOIL PROFILE

weathering velocity The velocity with which a seismic compressional wave passes through the low-velocity layer, or weathered layer. It ranges from 53 to 1373 meters per second in different parts of the earth.

wedge out To thin out.

wedgework of ice If abundant moisture is present in the pores and cracks of the rock a change of temperature from 7.22° to 1.67° C. might be far less effective in breaking the rock than a change from 1.67° to −3.89° C. in the same time, for in the latter case the sudden and very considerable expansion (about one-tenth) which water undergoes on freezing is brought into play. This may be called the wedgework of ice.

Wegener, Alfred (1880–1930) A German meteorologist who became intrigued with the similarity of coastal shapes across the Atlantic. He later widely promoted the Theory of Continental Drift.

Wegener theory *See* CONTINENTAL DRIFT

Weissenberg camera A single crystal goniometer involving a coupled motion of the crystal and film in conjunction with a screen by means of which layers of the reciprocal lattice are X-rayed individually.

welded tuff, welded pumice A tuff which has been indurated by the combined action of the heat retained by the particles and the enveloping hot gases. Ignimbrite.

welding Welding is consolidation by pressure due either to the weight of superincumbent material or to earth movement.

well core Sample of rock penetrated in a well or other borehole obtained by use of a hollow bit that cuts a circular channel around a central column or core.

well cuttings Rock chips cut by a bit in the process of well drilling and removed from the hole by pumping or bailing. Well cuttings collected at closely spaced intervals provide a record of the strata penetrated.

well-graded soil A coarse-grained unconsolidated material with a continuous distribution of grain sizes from the coarsest to the finest components in such proportions that the successively smaller grains just fill the spaces between the larger grains.

well log Record of a well, generally a lithologic record of the strata penetrated. See CALIPER, ELECTRIC, and GAMMA-RAY LOG

well-rounded A roundness grade in which no original faces, edges, or corners are left and the entire surface consists of broad curves.

well sample Cuttings produced in well drilling that are collected and saved as a record of the kinds of rock penetrated in the hole.

well shooting *Seismol:* A method or methods of logging wells so that average velocities, continuous velocities, or interval velocities are obtained by lowering geophones into the hole. Shots are usually fired from surface shotholes, but may be fired in the well itself, or perforating-gun detonations may be used. In continuous logging, a sound source is lowered in the hole together with recording geophones.

well spacing The geographic distribution of well locations on the surface above a reservoir; the number of acres per well, as, "20-acre spacing."

well ties The comparison of seismic datum points with geologic datum points at well locations, being the measure of the reliability of the seismic map.

welt A relatively narrow but sharp uplift. Part of the "welts and furrows" or geanticline-geosynclinal couple.

Wenlockian The middle of three epochs of the Silurian; also the series of strata deposited during that epoch.

Wenner array A direct current resistivity method using a linear arrangement of 4 equally spaced electrodes. Current is supplied by the outside electrodes, while potential differences are observed between the inside ones.

Wentworth scale A logarithmic grade scale for size classification of sediment particles, starting at 1 mm. and using the ratio 1/2 in 1 direction (and 2 in the other), providing diameter limits to the size classes of 1, 1/2, 1/4, etc., and 1, 2, 4, etc.

Wernerian Of or pertaining to A. G. Werner (1750-1817), a German mineralogist and geologist who classified minerals according to their external characters and advocated the theory that the strata of the earth's crust were formed by deposition from water; neptunian.

wet gas Natural gas that contains more or less oil vapors. It occurs with or immediately above the oil. Also sometimes called casing-head gas.

wetted perimeter *Hydraul.* and *Geomorph:* The part of a channel surface that is below the water level of a stream.

wetting The property of a fluid and a solid whereby the fluid adsorbs on the surface of the solid in a relatively unbroken film.

wetting agent A substance, usually fluid, introduced into another liquid in order to reduce surface tension of the latter.

whaleback dune A general, self-explanatory descriptive term for elongate dunes with a rounded crest.

wheelerite A yellowish resin filling fissures and interstratified with coal in the Cretaceous lignite beds of New Mexico.

whetstone Natural rock shaped into a sharpening stone.

whitecap On the crest of a wave, the white froth caused by wind.

white coal 1. Water power; first so called by the French (houille blanche). 2. Tasmanite.

wildcat Applied to a mining or oil company organized, or to a mine or well dug, to develop unproven ground far from previous production. Any risky venture in mining or the petroleum industry.

wildflysch A type of flysch, which displays small-scale folding and much twisting and confusion in the beds. The constituents are siliceous shales, clays, and sandstones, with included exotic blocks.

willemite A mineral, Zn_2SiO_4, commonly containing manganese. Hexagonal rhombohedral. A minor ore of zinc.

Williston's law Evolution tends to reduce the number of similar parts in organisms and render them more different from each other.

win 1. To extract ore or coal. To mine, to develop, to prepare for mining. 2. To recover metal from an ore.

wind In general, air in natural motion relative to the surface of the earth, in any direction whatever and with any velocity.

wind corrasion Wind abrasion.

wind gap The low depressions or notches in the ridges where streams formerly flowed are now called wind gaps and are utilized for highways in crossing the ridges.

windkanter A ventifact, usually highly polished, bounded by one or more, curved or nearly flat, smooth faces or facets ending or intersecting in one or more sharp edges or angles. The faces may be cut at different times, as when the wind changes seasonally or the pebble is undermined and turned over on its flattened face thereby permitting another face to be cut. *Syn:* FACETED PEBBLE; WIND-FACETED STONE

window Circular or ellipsoidal erosional break in an overthrust sheet whereby the rocks beneath the overthrust are exposed. Fenster, *q.v.*

wind polish The high gloss or luster developed on a rock in desert areas as a result of abrasion by sand blown against it.

wind ripple Ripples created by wind.

wind scale A numerical scale for expressing the different degrees of wind speed, in a fashion suitable for easy communication and rapid plotting on a weather map. The form in almost universal use is the Beaufort wind scale, *q.v.*

wind set-up The vertical rise in the still-water level on the leeward side of a body of water caused by wind stresses on the surface of the water.

wind shadow That portion of a scarp or slope which is protected from the direct action of the wind blowing over it.

wind tide *See* WIND SET-UP

wineglass valley Hourglass valley.

winnowing Separation of fine particles from coarser ones by action of the wind.

winze A steeply inclined shaft driven to connect one mine level with a lower level.

wire gold or **silver** (or other metals) Native metal in the form of wires or threads.

Wisconsin The last of four classical glacial stages in the Pleistocene of North America.

witch *1. V:* To search for underground water, ore, etc., with a divining instrument. *2. N:* One who searches in this way.

witherite A mineral, $BaCO_3$. Orthorhombic.

witness corner A post set near a corner of a mining claim with the distance and direction of the true corner indicated thereon. Used when the true corner is inaccessible.

wold *See* CUESTA

Wolfcampian The oldest of four provincial epochs of the Permian in the United States and northern Mexico; also the series of strata deposited during that epoch.

wolframite A mineral series, $(Fe,Mn)WO_4$, ranging from $FeWO_4$ (ferberite) to $MnWO_4$ (huebnerite). Monoclinic. The principal ore of tungsten.

wollastonite A mineral, $CaSiO_3$, commonly found in contact-metamorphosed limestones. Triclinic.

wood coal *1.* Woody lignite. *See* BOARD COAL. *2.* Charcoal.

wood opal A variety of opal consisting of wood in which the organic matter has been replaced by silica; silicified wood. *Syn:* XYLOPAL

wood tin A botryoidal, or nodular variety of finely crystalline cassiterite, cream to dark brown. Colloform texture with thin growth layers of radial cassiterite crystals, commonly intergrown with hematite and silica.

Worden gravimeter A compact, small, temperature-compensated gravity meter in which a system is held in unstable equilibrium about an axis, so that an increase in the gravitational pull on a mass at the end of a weight arm causes a rotation opposed by a sensitive spring.

worm's-eye map *1.* A map showing overlap of sediments. *2.* A map showing progressive transgressions of a sea over a given surface. *3.* A map representing the beds immediately overlying and in contact with the formations forming the geological surface at the time in question. *4.* The pattern of formations visible to an observer who would look upward at the bottom of the rocks overlying the surface in question.

wrench fault A nearly vertical strike-slip fault.

wrinkle ridge A sinuous, irregular, segmented, apparently smooth elevation occurring within the borders of a mare region of the moon's surface and characterized by dikelike outcrops, crest-top craters, and longitudinal rifts. Wrinkle ridges are up to 35 km. wide and 100 m. high, and may extend for hundreds of kilometers. They probably originated in fissure eruptions or from volcanic activity along fractures. *Syn:* MARE RIDGE

wulfenite A mineral, $PbMoO_4$. Tetragonal.

Wulff net A coordinate system used in crystallography to plot a stereographic projection.

Würm The fourth of the four classical glacial stages of the Pleistocene of Europe; it is now known that there was glaciation prior to the first (that is, the Günz).

wurtzilite A black, massive infusible, asphaltic pyrobitumen, insoluble in turpentine and derived from the metamorphosis of petroleum.

wurtzite A mineral, ZnS, dimorphous with sphalerite. Hexagonal.

wye level A leveling instrument having the telescope with attached spirit level supported in wyes (Ys), in which it may be rotated about its longitudinal axis (collimation axis), and from which it may be lifted and reversed, end for end.

X

X *Geophys:* The letter used in equations to designate a positive direction from the origin to the north. *Seis. prospecting:* The distance from the shot point to the center of the spread, or to any particular geophone.

X^2 - T^2 analysis A method of interpreting seismic reflection data which yields subsurface velocity information.

xeno- [*Gr.*] A combining form meaning guest, stranger, strange, foreign.

xenoblastic A term applied to a texture of metamorphic rocks in which the constituent mineral grains lack proper crystal facies. Corresponds to the allotriomorphic- or xenomorphic-granular texture of igneous rocks.

xenocryst A term applied to allothogenic crystals in igneous rocks that are foreign to the body of rock in which they occur.

xenolith A term applied to allothogenic rock fragments that are foreign to the body of igneous rock in which they occur. An inclusion.

xenomorphic The fabric or texture of an igneous rock in which the crystals do not have their own characteristic form, but an irregular shape imposed by the interference of surrounding minerals; an igneous rock fabric or texture characterized by anhedral crystals.

xenothermal Deposit formed at high temperature, but at shallow to moderate depth.

xenotime A mineral YPO_4, usually containing rare earths, thorium and uranium. Tetragonal.

xerophyte A plant with very low water requirements as opposed to hydrophyte.

xerothermic period A historical warm dry period.

X rays Non-nuclear electromagnetic radiation with a very short wavelength, i.e., shorter wavelength than ultraviolet and longer than gamma rays.

xylopal *See* WOOD OPAL

Y

Y The letter used in geophysical equations to designate a positive direction to the east from the origin.

yardang; yarding; jardang Irregular ridges, commonly alternating with round-bottomed troughs, formed by eolian erosion.

yarding Var. of yardang, *q.v.*

Yarmouth Post Kansan interglacial stage in North America.

yazoo stream A tributary that flows parallel to the main stream for a considerable distance before joining it at a deferred junction; esp. such a stream forced to flow along the base of a natural levee formed by the main stream. Type example: Yazoo River in western Mississippi, joining the Mississippi River at Vicksburg. Also spelled: Yazoo stream. *Syn:* YAZOO; YAZOO-TYPE TRIBUTARY; DEFERRED TRIBUTARY

yellow ground Oxidized kimberlite of yellowish color found at the surface of diamond pipes (e.g., South Africa), above the zone of blue ground.

yellow ochre *1.* A mixture of limonite usually with clay and silica, used as a pigment. *See also* YELLOW EARTH [MINERAL]. *Syn:* SIL. *2.* A soft, earthy, yellow variety of limonite or of goethite.

yield [lake] *N: 1.* The amount of water that can be taken continuously from a lake for any economic purpose. *2.* The amount of organic matter (plant and animal) produced by a lake, either naturally or under management.

yield point *See* ELASTIC LIMIT. *Syn:* YIELD STRESS

Y-level *See* WYE LEVEL

yoked basin Zeugogeosyncline.

young; youthful Being in the stage of increasing vigor and efficiency of action: said of some streams; also, being in the stage of accentuation of and a tendency toward complexity of form: said of some topography resulting from land sculpture. Contrasted with mature and old.

young plain A plain that has a level surface poorly defined, and perhaps swampy divides, and shallow lakes.

young river *Geol:* A river which has begun to form a drainage system in newly raised or newly deformed land.

Young's modulus Stretch modulus. Expressed in dynes/cm.^{2} or lbs./ft.^{2}. *Syn:* STRETCH MODULUS

young valley A valley in its early stages, when it is relatively straight, has steep slopes, a high gradient, and a V-shaped cross section, while its tributaries are short.

youth [streams] The first stage in the development of a stream, at which it has just entered upon its work of erosion and is increasing in vigor and efficiency, being able everywhere to erode its channel and having not reached a graded condition. It is characterized by: an ability to carry a load greater than the load it is actually carrying; active and rapid downcutting, forming a deep, narrow, steep-walled, V-shaped valley (gorge or canyon) with a steep and irregular gradient and rocky outcrops; numerous waterfalls, rapids, and lakes; a swift current and clear water; a few, short, straight tributaries; an absence of flood plains as the stream occupies all or nearly all of the valley floor; and an ungraded bed.

youthful Pertaining to the stage of youth of the cycle of erosion; esp. said of a topography or region, and of its landforms (such as a plain or plateau), having undergone little erosion or being in an early stage of development. *Syn:* YOUNG; JUVENILE

youthful stage of erosion cycle *See* YOUTH

youthful topography One in which the rivers flow in gorges and canyons, rather than in ordinary valleys, the drainage system is but partially developed, and the divides between the streams are broad, flat-topped areas, while the streams are obstructed by many falls and rapids.

Z

Z *Geophys:* The letter used in equations to designate the depth below the origin or datum of a point under consideration.

zebra dolomite Hydrothermally altered dolomite in the Leadville district of Colorado consisting of bands, generally parallel to bedding, that are light gray and coarsely crystalline, alternating with darker fine-grained bands.

zenith *Geod:* The point on the celestial sphere that is directly above the observer and directly opposite to the nadir. In a more general sense, the term denotes the stretch of sky overhead.

zenithal map projection *See* AZIMUTHAL MAP PROJECTION

zeolite A group of hydrous aluminosilicate minerals containing sodium, calcium, barium, strontium, and potassium, and characterized by their ease of exchange of these ions. Members of the group are natrolite, heulandite, analcime, chabazite, stilbite, phillipsite, mordenite, clinoptilolite, and others.

zero length spring A spring so coiled that if adjacent coils were cut, they would collapse precisely into a plane. Such springs exert a restoring force that is proportional to their physical length, rather than to their extension. Gravimeters using such springs achieve extreme sensitivity.

zeuge Earth pillar. It consists of soft rock with a layer of hard rock at the summit, and beneath the hard cap the soft rock is carved into a slope which resembles the typical denudation curve.

zeugogeosyncline A parageosyncline that receives its sediment from eroded complementing highlands within the craton.

zigzag fold An accordion fold, the limbs of which are of unequal length. *Cf.* CHEVRON FOLD

zinc blende Sphalerite.

zincite A mineral, ZnO, usually containing some Mn. Hexagonal. A minor ore of zinc.

zinc oxide; zinc white A white pulverulent oxide ZnO, made by burning zinc in air. It is used as a pigment, chiefly as a substitute for white lead. Called also flowers of zinc, nihil album, philosopher's wool, and zinc bloom.

zircon A mineral, $ZrSiO_4$. Tetragonal. Used as a refractory and as the gem, hyacinth. The chief ore of zirconium.

zoarium Skeleton of a bryozoan colony.

zoic *Geol:* Containing fossils, or yielding evidence of contemporaneous plant or animal life; said of rocks.

zoisite A mineral, $Ca_2Al_3(SiO_4)_3$(OH). Orthorhombic. Found in metamorphic rocks.

zonal axis Straight line parallel to all faces of a crystal zone.

zonal guide fossil Species of known limited vertical range in the local section. *See* INDEX FOSSIL

zonal soil An old soil classification term, denoting widespread soils that have well-developed characteristics that reflect the influence of climate and living organisms more than of parent rock.

zonal theory A theory of ore

deposition which holds that the ores originate in a zone of differentiation in the lower part of the zone of crystallization, where the siliceous-aqueous-metalliferous residues are formed, which in passing upward through faults and fractures deposit the ores in successive zones, each marked by its distinctive mineral associations.

zonation *Stratig:* The condition of being arranged or distributed in bands or zones, generally more or less parallel to the bedding.

zone *1. Stratig:* A group of beds, of an inferior status, characterized by one or several special fossils, which serve as indices. *2.* An area or region more or less set off or characterized as distinct from surrounding parts, as in a metalliferous region, the mineral zone. *3. Crystallog:* A series of faces whose intersection lines with each other are all parallel.

zone axis That line or crystallographic direction through the center of a crystal which is parallel to the intersection edges of the crystal faces defining the crystal zone. *Syn:* ZONAL AXIS

zone chart A template for making terrain corrections in gravity surveys.

zone of ablation Ablation area.

zone of accumulation *Snow: 1.* Accumulation area. *2.* In respect to an avalanche, a syn. of accumulation zone. *Soil:* B horizon.

zone of aeration A subsurface zone containing water under pressure less than that of the atmosphere, including water held by capillarity; and containing air or gases generally under atmospheric pressure. This zone is limited above by the land surface and below by the surface of the zone of saturation, i.e., the water table.

zone of capillarity An area that overlies the zone of saturation and contains capillary voids, some, or all, of which are filled with water that is held above the zone of saturation by molecular attraction acting against gravity.

zone of discharge The zone embracing that part of the belt of saturation which has a means of horizontal escape.

zone of eluviation The A horizon of soils, *q.v.*

zone of equilibrium *See* PROFILE OF EQUILIBRIUM

zone of flow *1. Interior Earth:* Zone of plastic flow. *2. Glaciol:* The inner, mobile main body or mass of a glacier, in which most of the ice flows without fracture. *Cf.* ZONE OF FRACTURE

zone of fracture *1. Interior Earth:* The upper part of the earth's crust which is brittle, i.e., in which deformation is by fracture rather than by plastic flow; that region of the crust in which fissures can exist. *Cf.* ZONE OF PLASTIC FLOW; ZONE OF FRACTURE AND PLASTIC FLOW. *Syn:* ZONE OF ROCK FRACTURE. *2. Glaciol:* The outer, rigid part of a glacier, in which the ice is much fractured. *Cf.* ZONE OF FLOW

zone of fracture and plastic flow That region of the earth's crust which is intermediate in depth and pressure between the zone of fracture and zone of plastic flow, and in which deformation of the weaker rocks is by plastic flow, and of the stronger rocks, by fracture.

zone of leaching A horizon of soils.

zone of mobility Asthenosphere.

zone of plastic flow That part of the earth's crust which is under sufficient pressure to prevent fracturing, i.e., is ductile, so that deformation is by flow. *Cf.* ZONE OF FRACTURE; ZONE OF FRACTURE AND PLASTIC FLOW. *Syn:* ZONE OF FLOW; ZONE OF ROCK FLOWAGE; ZONE OF FLOWAGE

zone of rock flowage Zone of plastic flow.

zone of rock fracture Zone of fracture.

zone of saturation A subsurface zone in which all the interstices are filled with water under pressure greater than that of the atmosphere. Although the zone may contain gas-filled interstices or interstices filled with fluids other than water, it is still considered saturated. This zone is separated from the zone of aeration (above) by the water table. *Syn:* SATURATED ZONE; PHREATIC ZONE

zone of weathering The superficial layer of the earth's crust above the water table that is subjected to the destructive agents of the atmosphere, and in which soils develop.

zoning of crystals Refers to those solid solution crystals which do not have a uniform composition throughout, but possess irregularities in composition from one point to another, occurring in the form of more or less concentric zones. Also called zonal structure.

zooecology The branch of ecology concerned with the relationships between animals and their environment. *Cf.* PHYTOECOLOGY

zoogene *Geol:* Of, pertaining to, consisting of, resulting from, or indicative of animal life or structure.

zoogenic rock A biogenic rock produced by animals or directly attributable to the presence or activities of animals; e.g., shell limestone, coral reefs, guano, and lithified calcareous ooze. *Syn:* ZOOGENOUS ROCK

zoolite; zoolith A fossil animal.

zooplankton The animal forms of plankton, e.g., jellyfish. They consume the phytoplankton.

Z phenomenon The possible time lag (a few seconds or less) between the issuance of P and S waves from an earthquake locus.

zweikanter A windkanter or stone having two faces intersecting in two sharp edges. *Etymol: Ger.* Zweikanter, "one having two edges." *Pl:* zweikanters; zweikanter.